Approximation Theory X

Wavelets, Splines, and Applications

Approximation Theory X

Wavelets, Splines, and Applications

Edited by

Charles K. Chui
Department of Mathematics
University of Missouri–St. Louis
St. Louis, Missouri

Larry L. Schumaker
Department of Mathematics
Vanderbilt University
Nashville, Tennessee

Joachim Stöckler
Department of Mathematics
University of Dortmund
Dortmund, Germany

VANDERBILT UNIVERSITY PRESS
Nashville

ISBN 0-8265-1416-2 (alk. paper)

INNOVATIONS IN APPLIED MATHEMATICS
An international series devoted to the latest research in
modern areas of mathematics, with significant applications
in engineering, medicine, and the sciences.

SERIES EDITOR
Larry L. Schumaker
Stevenson Professor of Mathematics
Vanderbilt University

CONTENTS

v

PREFACE

The *Tenth International Symposium on Approximation Theory* was held March 26 – 29, 2001 at the Sheraton Conference Center in Westport Plaza, St. Louis. Previous conferences in this series were held in 1973, 1976, 1980, and 1992 in Austin; 1983, 1986, 1989, and 1995 in College Station; and 1998 in Nashville. The conference was attended by 124 active participants in addition to over twenty professional mathematicians and students from the local St. Louis area. Twenty-two different countries were represented.

The program included seven invited one-hour survey talks and the one-hour presentation of the 2001 Popov Prize winner, Emmanuel Candès. In addition to a large number of contributed talks, there was a special poster session on interdisciplinary topics of current research. Due to the large number of submitted papers, these proceedings appear in two separate volumes. The first, subtitled *Abstract and Classical Analysis*, covers several areas of classical and modern approximation theory, while the second, subtitled *Wavelets, Splines, and Applications*, is devoted to recent applied and computational developments.

The conference began with opening remarks by George Lorentz which included an overview of the history of the previous nine meetings, a review of the impact of the meetings, especially on international cooperation, and a call for everyone in the field to continue this cooperation by attending each other's meetings and publishing in each other's journals.

This conference was sponsored by the National Science Foundation and the U.S. Army Research Office. We are also indebted to the Department of Mathematics and Computer Science of the University of Missouri – St. Louis for their support. In particular, we would like to thank Kai Bittner for assistance in the preparation of the conference program, Michael Schulte and Jan Stöckler for creating and managing the web site, and Deloris Licklider for her assistance in registration and other logistics.

We would also like to acknowledge the support of the Departments of Mathematics at Vanderbilt University and the University of Dortmund in connection with the preparation of these proceedings volumes. Thanks are also due to the invited speakers and other presenters, as well as everyone who attended for making the conference a success. Finally, we would like to thank our reviewers who helped select articles for the proceedings, and also Gerda Schumaker who assisted in the preparation of these volumes.

Charles K. Chui   March 1, 2002
Larry L. Schumaker
Joachim Stöckler

CONTRIBUTORS

Numbers in parentheses indicate pages on which the author's contribution(s) begin. Page numbers marked with ** refer to this volume, while those marked with a * refer to the companion volume *Approximation Theory X: Abstract and Classical Analysis*.

GEORGE A. ANASTASSIOU (1*), *Department of Mathematical Sciences, The University of Memphis, Memphis, TN 38152* [ganastss@memphis.edu]

GERARD AWANOU (1**), *Department of Mathematics, University of Georgia, Athens, Ga 30602* [gawanou@math.uga.edu]

ROBERT BAIER (9*), *Chair of Applied Mathematics, University of Bayreuth, D-95440 Bayreuth, Germany* [Robert.Baier@uni-bayreuth.de]

M. W. BARTELT (23*), *Department of Mathematics, Christopher Newport University, Newport News, VA 23606* [mbartelt@pcs.cnu.edu]

R. K. BEATSON (17**), *Department of Mathematics and Statistics, University of Canterbury, Christchurch, New Zealand* [R.Beatson@math.canterbury.ac.nz]

AURELIAN BEJANCU (27**), *Department of Applied Mathematics, University of Leeds, Leeds LS2 9JT, UK* [A.Bejancu@amsta.leeds.ac.uk]

KAI BITTNER (41**), *Department of Mathematics and Computer Science, University of Missouri – St. Louis, St. Louis, MO 63121* [bittner@math.umsl.edu]

BORISLAV BOJANOV (31*), *Dept. of Mathematics, University of Sofia, Blvd. James Boucher 5, 1164 Sofia, Bulgaria* [boris@fmi.uni-sofia.bg]

LASSE BORUP (51**), *Dept. of Mathematical Sciences, Aalborg University, Aalborg, DK-9220, Denmark* [lasse@math.auc.dk]

MARCIN BOWNIK (63**), *Department of Mathematics, University of Michigan, 525 East University Ave., Ann Arbor, MI 48109–1109* [marbow@umich.edu]

YU. BRUDNYI (91*), *Department of Mathematics, Technion – Israel Institute of Technology, 32000 Haifa, Israel* [ybrudnyi@math.technion.ac.il]

EMMANUEL J. CANDÈS (87**), *Applied and Computational Mathematics, California Institute of Technology, Pasadena, California 91125* [emmanuel@acm.caltech.edu]

BRUCE CHALMERS (113*), *Department of Mathematics, University of California, Riverside, Riverside, California 92521* [blc@math.ucr.edu]

B. CHANANE (155**), *Department of Mathematical Sciences, K. F. U. P. M., Dhahran 31261, Saudi Arabia* [chanane@kfupm.edu.sa]

MARGARET CHENEY (167**), *Department of Mathematical Sciences, Rensselaer Polytechnic Institute, Troy, NY 12180 USA* [cheney@rpi.edu]

HOI LING CHEUNG (179**), *Department of Mathematics, City University of Hong Kong, Kowloon, Hong Kong* [96473695@plink.cityu.edu.hk]

CHARLES K. CHUI (41**,187**,207**), *Department of Mathematics and Computer Science, University of Missouri – St. Louis, St. Louis, MO 63121* [cchui@stat.stanford.edu]

STEVEN B. DAMELIN (117*), *Department of Mathematics and Computer Science, Georgia Southern University, Post Box 8093, Statesboro, GA 30460-8093, USA* [damelin@gasou.edu]

OLEG DAVYDOV (231**), *Mathematisches Institut, Justus-Liebig Universität, D-35392 Giessen, Germany* [oleg.davydov@math.uni-giessen.de]

UWE DEPCZYNSKI (241**), *FACT Unternehmensberatung GmbH, Goetheplatz 5, D-60313 Franfurt/Main, Germany* [u.depczynski@fact.de]

ASEN L. DONTCHEV (261**), *Mathematical Review, Ann Arbor, MI 48107, USA* [ald@ams.org]

P. D. DRAGNEV (137*), *Department of Mathematics, Indiana-Purdue University, Fort Wayne, IN 46825, USA* [dragnevp@ipfw.edu]

DIMITER DRYANOV (145*), *Département des Mathématiques et de Statistique, Université de Montréal, C. P. 6128, Montréal, Québec H3C 3J7, Canada* [dryanovd@dms.umontreal.ca]

NIRA DYN (9*), *School of Mathematical Sciences, Sackler Faculty of Exact Sciences, Tel Aviv University, Tel Aviv 69978, Israel* [niradyn@post.tau.ac.il]

TAMÁS ERDÉLYI (153*), *Department of Mathematics, Texas A & M University, College Station, Texas 77843, USA* [terdelyi@math.tamu.edu]

ELZA FARKHI (9*), *School of Mathematical Sciences, Sackler Faculty of Exact Sciences, Tel Aviv University, Tel Aviv 69978, Israel* [elza@post.tau.ac.il]

GREG FASSHAUER (271**), *Department of Applied Mathematics, Illinois Institute of Technology, Chicago, IL 60616* [fass@amadeus.math.iit.edu]

MICHAEL S. FLOATER (197*), *SINTEF, Post Box 124, Blindern, N-0314 Oslo, Norway* [mif@math.sintef.no]

MICHAEL I. GANZBURG (211*), *Dept. of Mathematics, Hampton University, Hampton, VA 23668, USA* [michael.ganzburg@hamptonu.edu]

DAVID E. GILSINN (283**), *National Institute of Standards and Technology, 100 Bureau Drive, Stop 8910, Gaithersburg, MD 20899-8910* [dgilsinn@nist.gov]

MANFRED V. GOLITSCHEK (295**), *Institut für Angewandte Mathematik und Statistik, Universität Würzburg, 97074 Würzburg, Germany* [goli@mathematik.uni-wuerzburg.de]

WERNER HAUSSMANN (145*), *Department of Mathematics, Gerhard–Mercator–University, 47048 Duisburg, Germany* [haussmann@math.uni-duisburg.de]

TIAN-XIAO HE (305**), *Department of Mathematics and Computer Science, Illinois Wesleyan University, Bloomington, IL 61702-2900* [the@sun.iwu.edu]

WENJIE HE (187**), *Department of Mathematics and Computer Science, University of Missouri–St. Louis, St. Louis, MO 63121-4499* [he@neptune.cs.umsl.edu]

WALTER HENGARTNER (223*), *Université de Laval, Département de Mathématique, Québec G1K 7P4, Canada* [walheng@mat.ulaval.ca]

KAI HORMANN (197*), *Computer Graphics Group, University of Erlangen-Nürnberg, Am Weichselgarten 9, D-91058 Erlangen, Germany* [hormann@cs.fau.de]

ROBERT HUOTARI (239*), *Mathematics Department, Glendale Community College, Glendale AZ 85302* [r_huotari@yahoo.com]

FRANKLIN KEMP (247*), *3004 Pataula Lane, Plano, TX 75074-8765* [lfkemp@attbi.com]

BRUCE KESSLER (323**), *1 Big Red Way, Department of Mathematics, Western Kentucky University, Bowling Green, KY 42101* [bruce.kessler@wku.edu]

C. K. KOBINDARAJAH (267*), *Department of Mathematics, Eastern University, Chenkalady, Sri Lanka* [kshant@yahoo.com]

OGNYAN KOUNCHEV (333**), *Institute of Mathematics, Bulgarian Akademy of Sciences, Bonchev Str. 9, 1143 Sofia, Bulgaria* [kounchev@bas.bg]

BORIS I. KVASOV (355**), *School of Mathematics, Suranaree University of Technology, 111 University Avenue, Nakhon Ratchasima 30000, Thailand* [boris@math.sut.ac.th]

MING-JUN LAI (1**,369**,385**), *Mathematics Department, University of Georgia, Athens, GA 30602* [mjlai@math.uga.edu]

JOHN E. LAVERY (283**), *Computing and Information Sciences Division, Army Research Office, Army Research Laboratory, P. O. Box 12211, Research Triangle Park, NC 27709-2211* [lavery@arl.aro.army.mil]

J. LEVESLEY (17**,277*), *Department of Mathematics and Computer Science, University of Leicester, Leicester LE1 7RH, UK* [jl1@mcs.le.ac.uk]

JIAN-AO LIAN (207**), *Dept. of Mathematics, Prairie View A&M University, Prairie View, TX 77446* [unurbs@yahoo.com]

A.-J. LÓPEZ-MORENO (287*), *Departamento de Matemáticas, Universidad de Jaén, Spain* [ajlopez@ujaen.es]

JOSÉ L. MARTÍNEZ-MORALES (309*), *Instituto de Matemáticas, Universidad Nacional Autónoma de México, Ap. Post. 6-60, Cuernavaca, Mor. 62131, Mexico* [martinez@matcuer.unam.mx]

J. MARTÍNEZ-MORENO (287*), *Departamento de Matemáticas, Universidad de Jaén, Spain* [jmmoreno@ujaen.es]

F. J. MUÑOZ-DELGADO (287*), *Departamento de Matemáticas, Universidad de Jaén, Spain* [fdelgado@ujaen.es]

MORTEN NIELSEN (51**), *Dept. of Mathematics, University of South Carolina, Columbia, SC 29208* [nielsen@math.sc.edu]

YURI L. NOSENKO (325*), *Dept. of Mathematics, Donetsk National Technical University, 58, Artema str., Donetsk, 83000, Ukraine* [yuri@nosenko.dgtu.donetsk.ua]

GÜNTHER NÜRNBERGER (405**), *Institut für Mathematik, Universität Mannheim, D-618131 Mannheim, Germany* [nuern@euklid.math.uni-mannheim.de]

GERHARD OPFER (223*), *University of Hamburg, Department of Mathematics, Bundesstr. 55, 20146 Hamburg, Germany* [opfer@math.uni-hamburg.de]

PETAR PETROV (145*), *Department of Mathematics, University of Sofia, 1164 Sofia, Bulgaria* [peynov@fmi.uni-sofia.bg]

ALEXANDER PETUKHOV (425**), *Department of Mathematics, University of South Carolina, Columbia, SC 29208* [petukhov@math.sc.edu]

ALLAN PINKUS (333*), *Dept. of Mathematics, Technion – Israel Institute of Technology, Haifa, 32000, Israel* [pinkus@tx.technion.ac.il]

MICHAEL P. PROPHET (241*), *Department of Mathematics, University of Northern Iowa, Cedar Falls, IA 50614* [prophet@math.uni.edu]

HOUDUO QI (261**), *Department of Applied Mathematics, The Hong Kong Polytechnic University, Hung Hom, Kowloon, Hong Kong* [mahdqi@polyu.edu.hk]

LIQUN QI (261**), *Department of Applied Mathematics, The Hong Kong Polytechnic University, Hung Hom, Kowloon, Hong Kong* [maqilq@polyu.edu.hk]

DAVID L. RAGOZIN (277*), *University of Washington, Department of Mathematics, Seattle, WA 98195-4350, USA* [rag@math.washington.edu]

MARTIN REIMERS (197*), *Institutt for informatikk, Universitetet i Oslo, Post Box 1080, Blindern, N-0316 Oslo, Norway* [martinre@ifi.uio.no]

HERMANN RENDER (333**), *Fachbereich Mathematik, Gerhard-Mercator Universität Duisburg, D- 47048 Duisburg, Germany* [render@math.uni-duisburg.de]

JASON RIBANDO (239*), *Department of Mathematics, University of Northern Iowa, Cedar Falls, IA 50614* [jribando@math.uni.edu]

DAVID W. ROACH (369**), *Mathematics and Statistics Department, Murray State University, Murray, KY 42071* [david.roach@murraystate.edu]

Contributors

THOMAS SAUER (353*), Lehrstuhl für Numerische Mathematik, Justus–
Liebig–Universität Gießen, Heinrich–Buff–Ring 44, D–35092
Gießen, Germany [Tomas.Sauer@math.uni-giessen.de]

R. SCHABACK (433**), Institut für Numerische und Angewandte Mathematik, Zentrum für Informatik, Lotzestraße 16–18, D-37083
Göttingen, Germany [schaback@math.uni-goettingen.de]

LARRY L. SCHUMAKER (405**), Center for Constructive Approximation,
Department of Mathematics, Vanderbilt University, Nashville,
TN 37205, USA [s@mars.cas.vanderbilt.edu]

BORIS SHEKHTMAN (113*,367*), Department of Mathematics,
University of South Florida, Tampa, FL 33620-5700, USA
[boris@math.usf.edu]

DARRIN SPEEGLE (63**), Department of Mathematics, Saint Louis
University, 221 N. Grand Boulevard, Saint Louis, MO 63103
[speegled@slu.edu]

JOACHIM STÖCKLER (187**), Universität Dortmund, Fachbereich
Mathematik, 44221 Dortmund, Germany
[joachim.stoeckler@math.uni-dortmund.de]

HANS STRAUSS (441**), Institut für Angewandte Mathematik,
Universität Erlangen–Nürnberg, Martensstr. 3, D–91058
Erlangen, Germany [strauss@am.uni-erlangen.de]

JOHN J. SWETITS (23*), Department of Mathematics and Statistics, Old
Dominion University, Norfolk, VA 23529 [jswetits@odu.edu]

CANQIN TANG (179**), Department of Mathematics, Changde Teacher's
College, Hunan, P. R. China [tangcq2000@yahoo.com.cn]

V. N. TEMLYAKOV (373*), Department of Mathematics, University of
South Carolina, Columbia, SC 29208, USA
[temlyak@math.sc.edu]

JIANZHONG WANG (453**), Sam Houston State University, Huntsville,
TX 77341-2206, USA [mth_jxw@shsu.edu]

HOLGER WENDLAND (473**), Institut für Numerische und Angewandte
Mathematik, Universität Göttingen, Lotzestr. 16-18, 37083
Göttingen, Germany [wendland@math.uni-goettingen.de]

PAUL WENSTON (385**), Department of Mathematics, The University
of Georgia, Athens, GA 30602 [paul@math.uga.edu]

YUESHENG XU (353*), Department of Mathematics, West Virginia
University, Morgantown, WV 26506, USA [yxu@math.wvu.edu]

LUNG-AN YING (385**), *School of Mathematical Sciences, Peking University, Beijing, China* [yingla@pku.edu.cn]

FRANK ZEILFELDER (405**), *Institut für Mathematik, Universität Mannheim, D-618131 Mannheim, Germany* [zeilfeld@euklid.math.uni-mannheim.de]

C^1 Quintic Spline Interpolation Over Tetrahedral Partitions

Gerard Awanou and Ming-Jun Lai

Abstract. We discuss the implementation of a C^1 quintic super-spline method for interpolating scattered data in $\mathbb{R}^3$ based on a modification of Alfeld's generalization of the Clough-Tocher scheme described by Lai and LeMéhauté [4]. The method has been implemented in MATLAB, and we test for the accuracy of reproduction on a basis of quintic polynomials. We present numerical evidences that when the partition is refined, the spline interpolant converges to the function to be approximated.

§1. Introduction

There are a few trivariate spline spaces available for interpolation over a tetrahedral partition $\triangle$ of a polygonal domain in $\mathbb{R}^3$. We would like to mention a direct polynomial interpolation by Zenisek in [9]. This scheme requires piecewise polynomials of degree 9 and is globally C^1 over Ω while C^4 around the vertices and C^2 around the edges of $\triangle$. Another scheme is the Alfeld scheme (cf. [1]) which uses polynomials of degree 5 to construct spline functions over a 3D Clough-Tocher refinement of a tetrahedral partition $\triangle$. The scheme produces spline interpolants which are globally C^1 over Ω while C^2 around the vertices and C^1 around the edges of $\triangle$. A further generalization of the Clough-Tocher refinement enables Worsey and Farin in [7] to construct interpolation by C^1 cubic splines. Worsey and Piper constructed C^1 quadratic spline functions based on special tetrahedral partitions in [8].

The present paper is concerned with the implementation of the modification introduced in [4] of Alfeld's Clough-Tocher scheme. It uses polynomials of degree 5 over the Alfeld version of Clough-Tocher refinement

Approximation Theory X: Wavelets, Splines, and Applications   1
Charles K. Chui, Larry L. Schumaker, and Joachim Stöckler (eds.), pp. 1–16.
Copyright ⊖ 2002 by Vanderbilt University Press, Nashville, TN.
ISBN 0-8265-1416-2.

of $\triangle$ to construct spline interpolants globally C^1 over Ω and locally C^2 around the vertices and C^1 around the edges of $\triangle$. The main difference is that the Alfeld scheme reproduces only polynomials of degree 3, while the new scheme reproduces polynomials of degree 5. Let us mention that the Alfeld scheme was implemented in [2]. Our program like others of the kind involves a great computational complexity.

The paper is organized as follows: First we begin by a review of the B-form of polynomials on tetrahedra. Then we review the construction of [4] in §3. The details on how to compute C^1 quintic spline interpolants are given in §4. We then give the properties of the interpolant. In §6, we present numerical evidence that the scheme reproduces all polynomials of degree 5 and that the interpolation error reduces when the partition is refined. Finally, we will point out our future research topics.

§2. B-form of Polynomials on Tetrahedra

We assume the reader is familiar with the Bernstein representation of polynomials on tetrahedra. An introduction to this topic can be found in [3]. Here, we give only a brief account.

We first recall how to represent trivariate polynomials

$$p(x, y, z) = \sum_{0 \leq i+j+k \leq d} \alpha_{ijk} x^i y^j z^k, \qquad \alpha_{ijk} \in \mathbb{R},$$

of degree d in terms of the barycentric coordinates of the evaluation point (x, y, z) with respect to a given tetrahedron $T = \langle v_1, v_2, v_3, v_4 \rangle$. Recall that any $v \in \mathbb{R}^3$ can be written uniquely in the form

$$v = b_1 v_1 + b_2 v_2 + b_3 v_3 + b_4 v_4$$

with

$$b_1 + b_2 + b_3 + b_4 = 1,$$

where b_1, b_2, b_3 and b_4 are the barycentric coordinates of v with respect to T. Let

$$B_{ijkl}^d(v) = \frac{d!}{i!j!k!l!} b_1^i b_2^j b_3^k b_4^l, \qquad i + j + k + l = d$$

be the Bernstein polynomials of degree d. They form a basis of the space of polynomials of degree less than or equal d. As a consequence, any such polynomial can be written uniquely on T in the so-called B-form

$$p = \sum_{i+j+k+l=d} c_{ijkl} B_{ijkl}^d, \qquad c_{ijkl} \in \mathbb{R}.$$

As usual, the c_{ijkl} are associated with the domain points

$$\left\{ \frac{iv_1 + jv_2 + kv_3 + lv_4}{d} \right\}_{i+j+k+l=d}.$$

A polynomial of total degree d is uniquely determined by its values on the domain points, i.e. the c_{ijkl}, $i + j + k + l = d$ are completely determined by interpolation on the domain points.

Let $\mathbf{u} = y - x$ be a vector with x and y having barycentric coordinates $(\alpha_1, \alpha_2, \alpha_3, \alpha_4)$ and $(\beta_1, \beta_2, \beta_3, \beta_4)$, respectively. One refers to $a = (\beta_1 - \alpha_1, \beta_2 - \alpha_2, \beta_3 - \alpha_3, \beta_4 - \alpha_4) = (a_1, a_2, a_3, a_4)$ as the T-coordinates of $\mathbf{u}$. In terms of a and the $c_{ijkl}, i + j + k + l = d$, the derivative of p in the direction $\mathbf{u}$ can be written in B-form on T as

$$D_{\mathbf{u}}p = d \sum_{i+j+k+l=d-1} \Delta_a(c_{ijkl}) B_{ijkl}^{d-1}$$

with

$$\Delta_a(c_{ijkl}) = a_1 c_{i+1,j,k,l} + a_2 c_{i,j+1,k,l} + a_3 c_{i,j,k+1,l} + a_4 c_{i,j,k,l+1}.$$

Given a spline, *i.e.*,, a piecewise polynomial defined on a collection of tetrahedra, we work with the B-form of each polynomial piece. So it makes sense to look for conditions on the coefficients that will ensure that the spline has global smoothness properties. We explicitly derive the smoothness conditions for a model case.

Let $v_5 = (v_1 + v_2 + v_3 + v_4)/4$, and let

$$p_4 = \sum_{i+j+k+l=d} c_{ijkl}^4 B_{ijkl}^d \qquad \text{on } T_4 = \langle v_1, v_2, v_3, v_5 \rangle$$

and

$$p_1 = \sum_{i+j+k+l=d} c_{ijkl}^1 B_{ijkl}^d \qquad \text{on } T_1 = \langle v_2, v_3, v_4, v_5 \rangle.$$

For p_1 and p_4 to be joined continuously across the common face $\langle v_2, v_3, v_5 \rangle$ they must agree on that face. Since p_4 and p_1 are uniquely determined by their coefficients on that face, the condition of C^0 continuity reads

$$c_{0jkl}^4 = c_{jk0l}^1.$$

To ensure continuity of the first order derivatives, we need only to check continuity of the directional derivatives $D_{v_3-v_2}$, $D_{v_5-v_2}$ and $D_{v_1-v_2}$. We already have

$$D_{v_3-v_2}p_4 = D_{v_3-v_2}p_1; \quad D_{v_5-v_2}p_4 = D_{v_5-v_2}p_1 \text{ on } \langle v_2, v_3, v_5 \rangle$$

since those derivatives depend only on the values of the polynomials on the common face.

Notice that $v_1 - v_2$ has T-coordinates $(1, -1, 0, 0)$ with respect to T_4 and $(-2, -1, -1, 4)$ with respect to T_1. So to ensure C^1 continuity, we need

$$\Delta_{(1,-1,0,0)}(c_{0jkl}) = \Delta_{(-2,-1,-1,4)}(c_{jk0l}), \qquad i + j + k + l = d - 1$$

or equivalently

$$c^4_{1jkl} - c^4_{0,j+1,k,l} = -2c^1_{j+1,k,0,l} - c^1_{j,k+1,0,l} - c^1_{jk1l} + 4c^1_{j,k,0,l+1}.$$

Using the C^0 continuity conditions, this gives

$$c^4_{1jkl} = -c^1_{j+1,k,0,l} - c^1_{j,k+1,0,l} - c^1_{jk1l} + 4c^1_{j,k,0,l+1}.$$

Finally we give the following subdivision formulas that give the B-form of p on $\langle v_1, v_2, v_3, w \rangle$ for any point w in $\mathbf{R}^3$. The B-form of p on $\langle v_1, v_2, v_3, w \rangle$ is

$$p = \sum_{i+j+k+l=d} d_{ijkl} B^d_{ijkl}$$

with

$$d_{ijkl} = \sum_{\mu+\nu+\kappa+\delta=l} c_{i+\mu,j+\nu,k+\kappa,\delta} B^l_{\mu,\nu,\kappa,\delta}(w).$$

For example with $w = v_5$,

$$d_{ijkl} = \sum_{\mu+\nu+\kappa+\delta=l} c_{i+\mu,j+\nu,k+\kappa,\delta} \frac{l!}{\mu!\nu!\kappa!\delta!} \left(\frac{1}{4}\right)^l.$$

§3. Description of the Scheme

We describe the new scheme for 3D scattered data interpolation we have implemented.

Let us introduce more notation. For each edge e, let m_e be the midpoint of e and let e_1 and e_2 be two directions which are perpendicular to e and are linearly independent. For each face $f = \langle v_1, v_2, v_3 \rangle$, let f_1, f_2, f_3 be the three domain points $\{(iv_1 + jv_2 + kv_3)/5, (i, j, k) = (2, 2, 1), (1, 2, 2), (2, 1, 2)\}$ on f. Let n_f be a unit normal vector to f. For each tetrahedron t, let u_t be the center point of t.

For a tetrahedral partition $\triangle$, for each tetrahedron t, we split t into four subtetrahedra at the center u_t by connecting u_t to any two of four vertices of t. This generates a 3D Clough-Tocher refinement of $\triangle$. For simplicity, we call it the Alfeld refinement and denote it by $A(\triangle)$.

The C^1 quintic spline interpolant on $A(\triangle)$ can then be described as follows. Given a function $g \in C^2(\Omega)$, the interpolant S_g satisfies the following conditions:

1) For each vertex $v \in \triangle$

$$D_x^{\alpha_1} D_y^{\alpha_2} D_z^{\alpha_3} S_g(v) = D_x^{\alpha_1} D_y^{\alpha_2} D_z^{\alpha_3} g(v), \qquad \forall \, |\alpha| \le 2.$$

2) For each edge e of $\triangle$,

$$D_{e_i} S_g(m_e) = D_{e_i} g(m_e), \qquad i = 1, 2,$$

where D_{e_i} denotes the derivative along the direction e_i and m_e the midpoint of e;

3) For each face f of $\triangle$,

$$D_{n_f} S_g(f_j) = D_{n_f} g(n_f), \qquad j = 1, 2, 3,$$

where D_{n_f} denotes the derivative along the direction n_f;

4) For each tetrahedron t,

$$D_x^{\alpha_1} D_y^{\alpha_2} D_z^{\alpha_3} S_g(u_t) = D_x^{\alpha_1} D_y^{\alpha_2} D_z^{\alpha_3} g(u_t), \qquad \forall \, |\alpha| \le 1.$$

This is the scheme which was introduced in [4].

§4. Details of Computation

This is best presented by looking at the case of a single tetrahedron $T = \langle v_1, v_2, v_3, v_4 \rangle$. The faces, edges and vertices of T will be referred to as boundary faces, edges and vertices, respectively, in this case. Let $u_t = v_5 = (v_1 + v_2 + v_3 + v_4)/4$ be the center of T. It subdivides T into 4 subtetrahedra:

$$T_1 = \langle v_2, v_3, v_4, v_5 \rangle, \quad T_2 = \langle v_1, v_3, v_4, v_5 \rangle,$$

$$T_3 = \langle v_1, v_2, v_4, v_5 \rangle, \quad T_4 = \langle v_1, v_2, v_3, v_5 \rangle.$$

The problem then is to determine the approximating polynomial p on each subtetrahedron. Given a polynomial p of degree 5, for each $s = 1, \ldots, 4$, let

$$p(v) = \sum_{i+j+k+l=5} c_{ijkl}^s B_{ijkl}^5(v)$$

be the B-form of p on T_s. For simplicity we show how to compute the c_{ijkl}^4's. The other coefficients c_{ijkl}^s, $s = 1, 2, 3$, can be computed in a similar fashion.

Clearly $c_{5000}^4 = p(v_1)$, and similar relations hold at the other vertices of T. Next we compute coefficients in the first ring around v_1. Because

$$D_{v_2-v_1}p(v_1) = 5(c_{4100}^4 - c_{5000}^4),$$

we have

$$c_{4100}^4 = \frac{1}{5}D_{v_2-v_1}p(v_1) + c_{5000}^4,$$

where $D_{v_2-v_1}p(v_1)$ is computed using the first order derivatives at v_1. Using other directional derivatives, we get

$$c_{4010}^4 = \frac{1}{5}D_{v_3-v_1}p(v_1) + c_{5000}^4,$$

$$c_{4001}^4 = \frac{1}{5}D_{v_4-v_1}p(v_1) + c_{5000}^4.$$

Proceeding with the coefficients in the ring of radius 2, we need to use second order directional derivatives:

$$c_{3200}^4 = \frac{1}{20}D_{v_2-v_1}^2p(v_1) + 2c_{4100}^4 - c_{5000}^4,$$

$$c_{3020}^4 = \frac{1}{20}D_{v_3-v_1}^2p(v_1) + 2c_{4010}^4 - c_{5000}^4,$$

$$c_{3002}^4 = \frac{1}{20}D_{v_5-v_1}^2p(v_1) + 2c_{4001}^4 - c_{5000}^4,$$

$$c_{3110}^4 = \frac{1}{20}D_{v_3-v_1}D_{v_2-v_1}p(v_1) + c_{4010}^4 + c_{4100}^4 - c_{5000}^4,$$

$$c_{3101}^4 = \frac{1}{20}D_{v_5-v_1}D_{v_2-v_1}p(v_1) + c_{4001}^4 + c_{4100}^4 - c_{5000}^4,$$

$$c_{3011}^4 = \frac{1}{20}D_{v_5-v_1}D_{v_3-v_1}p(v_1) + c_{4001}^4 + c_{4010}^4 - c_{5000}^4.$$

It is then clear how to obtain similar formulas in the ring of radius 2 around the other vertices v_3 and v_4.

It is convenient to view the B-net of p over T_1, T_2, T_3 and T_4 as composed of layers. Thus the coefficients on the boundary faces form the first layer. The face $\langle v_1, v_2, v_3 \rangle$ of T_4 is also a boundary face of T. On that face only 3 coefficients are to be determined: c_{2120}^4, c_{1220}^4 and c_{2210}^4. To compute those coefficients, we use the given data at the midpoints of boundary edges to find directional derivatives along the edges at the midpoints of those edges. The coefficients to be found are then simply expressed in terms of the later derivatives. We show how to compute c_{1220}^4 for example.

We consider the edge $\langle v_2, v_3 \rangle$, with midpoint $v_{23} := m_{\langle v_2, v_3 \rangle}$. At this point a directional derivative along this edge can be computed as

$$D_{v_3-v_2}p(v_{23}) = 5\sum_{j=0}^4 (c_{0,j,5-j,0}^4 - c_{0,j+1,4-j,0}^4)B_{0,j,4-j,0}^4(v_{23}).$$

In the equation

$$D_{v_2-v_1}p(v_{23}) = 5 \sum_{j=0}^{4} (c^4_{0,j+1,4-j,0} - c^4_{1,j,4-j,0})B^4_{0,j,4-j,0}(v_{23}),$$

there are two unknowns, $D_{v_2-v_1}p(v_{23})$ and c^4_{1220}. We have

$$D_{v_2-v_1}p(v_{23}) = \alpha D_{v_3-v_2}p(v_{23}) + \beta D_{e_1}p(v_{23}) + \gamma D_{e_2}p(v_{23})$$

or equivalently

$$D_{v_2-v_1}p(v_{23}) = \alpha D_{v_3-v_2}p(v_{23}) + \beta D_{e_1}g(v_{23}) + \gamma D_{e_2}g(v_{23})$$

for some constants α, β, and γ. e_1 and e_2 are two directions perpendicular to the edge e= $\langle v_2, v_3 \rangle$. Now c^4_{1220} can be computed. Proceeding the same way with the edges $\langle v_1, v_2 \rangle$ and $\langle v_1, v_3 \rangle$, we get c^4_{2120} and c^4_{2210}. We compute the coefficients on the second layer and get three of them by smoothness conditions

$$c^4_{0221} = \frac{1}{4}(c^1_{2210} + c^4_{1220} + c^4_{0320} + c^4_{0230}),$$

$$c^4_{2021} = \frac{1}{4}(c^2_{2210} + c^4_{2120} + c^4_{3020} + c^4_{2030}),$$

$$c^4_{2201} = \frac{1}{4}(c^3_{2210} + c^4_{2210} + c^4_{3200} + c^4_{2300}).$$

To get the other coefficients in T_4 on the second layer, namely c^4_{1121}, c^4_{2111} and c^4_{1211}, we use values of the normal derivative to $\langle v_1, v_2, v_3 \rangle$ at 3 points on that face: f_1, f_2 and f_3. More precisely, since all coefficients on $f = \langle v_1, v_2, v_3 \rangle$ are determined, we can compute the values of $D_{v_1-v_2}p$ and $D_{v_3-v_2}p$ at the points f_1, f_2 and f_3. Since the $D_{\eta_f}p(f_i) := D_{\eta_f}g(f_i)$, $i = 1, \ldots, 3$ are given by interpolation conditions, by expressing $v_5 - v_2$ in terms of η_f, $v_1 - v_2$ and $v_3 - v_2$, we can compute the $D_{v_5-v_2}p(f_i)$, $i = 1, 2, 3$. Notice that

$$D_{v_5-v_2}p(f_1) = 5 \sum_{i+j+k=4} (c^4_{ijk1} - c^4_{i,j+1,k,0})\frac{4!}{5^4 i! j! k!}2^{i+j},$$

$$D_{v_5-v_2}p(f_2) = 5 \sum_{i+j+k=4} (c^4_{ijk1} - c^4_{i,j+1,k,0})\frac{4!}{5^4 i! j! k!}2^{j+k},$$

$$D_{v_5-v_2}p(f_3) = 5 \sum_{i+j+k=4} (c^4_{ijk1} - c^4_{i,j+1,k,0})\frac{4!}{5^4 i! j! k!}2^{i+k}.$$

These form a system of equations with unknowns c^4_{1121}, c^4_{2111} and c^4_{1211} which can be solved easily. This finishes the computations of all coefficients on the second layer.

On the third layer there are six coefficients in T_4 which are simply computed by using smoothness conditions across interior faces. Those coefficients are $c_{2102}^4, c_{1202}^4, c_{0212}^4, c_{0122}^4, c_{2012}^4, c_{1022}^4$.

At this stage, the only coefficients on this layer which remain to be computed are c_{1112}^s, $s = 1, \ldots, 4$. This layer is viewed as the B-net of a polynomial of degree 3. There are 4 data to be computed and we get them by using the 4 data at the center v_5 of the tetrahedron. More precisely, using the given information at the center, one can compute $p(v_5)$, $D_{v_2-v_1}p(v_5)$, $D_{v_3-v_1}p(v_5)$ and $D_{v_5-v_1}p(v_5)$. It is not difficult to see that they can be expressed in terms of the c_{1112}^s, $s = 1, \ldots, 4$. We therefore have a system of 4 equations in 4 unknowns which had to be solved.

The coefficients in the remaining two layers are obtained by using the subdivision method. In this way, the interpolant will be C^3 at v_5.

§5. Properties of the Interpolant

We prove in this section that the scheme reproduces all quintic polynomials and that the interpolant thus constructed is C^2 around the vertices, C^1 around the edges, C^3 at the center of each tetrahedron, and globally C^1.

Property 1: The scheme reproduces all quintic polynomials. This follows from

Lemma 1. *A polynomial p of degree 5 on $T = \langle v_1, v_2, v_3, v_4 \rangle$ with center v_5 is uniquely determined by the following data:*

$$D_x^{\alpha_1} D_y^{\alpha_2} D_z^{\alpha_3} p(v_i), \quad |\alpha| \le 2, \quad i = 1, \ldots, 4,$$

the values of derivatives in two independent directions perpendicular to each edge of T at the midpoint of the edge and

$$D_x^{\alpha_1} D_y^{\alpha_2} D_z^{\alpha_3} p(v_5), \quad |\alpha| \le 1.$$

The proof of the lemma is given in the appendix.

Property 2: The interpolant is C^2 around the vertices. If two tetrahedra share the same vertex v, by construction the polynomial pieces share the same values $D_x^{\alpha_1} D_y^{\alpha_2} D_z^{\alpha_3} g(v)$. So S_g is C^2 at v.

Property 3: The interpolant is C^1 around the edges. Assume for example that two tetrahedra share the common edge $e = \langle v_1, v_2 \rangle$. The coefficients of each polynomial piece on e are the same since they are computed using data at the vertices v_1 and v_2. This gives continuity across the edge and also continuity of $D_{v_2-v_1}$ across e. To prove C^1 continuity, we need to show continuity of derivatives in three independent directions. Notice that the interpolation conditions at the vertices determine all derivatives up to order 2 at vertices. Any of the derivatives $D_{e_i} S_g$, $i = 1, 2$ reduce to the same univariate quartic polynomial on e. $D_{e_i} S_g$ is uniquely determined by the 5 pieces of data

$$D_{e_i} S_g(m_e), D_{e_i} S_g(v_1), D_{v_2 - v_1} D_{e_i} S_g(v_1), D_{e_i} S_g(v_2), D_{v_2 - v_1} D_{e_i} S_g(v_2).$$

This assures continuity of $D_{e_i} S_g, i = 1, 2$ across e.

We show below continuity across interior faces of a tetrahedron which implies continuity across the interior edges.

Property 4: The interpolant is globally C^1. Recall that each tetrahedron is subdivided into 4 subtetrahedra so we study differentiability across tetrahedra in the original partition and differentiability across the subtetrahedra obtained after refinement.

Intertetrahedral continuity: Assume two tetrahedra share a common face f. By construction, coefficients on such a face are determined either by using data at the vertices or data at the midpoints of edges of that face. Hence the continuity of the interpolant S_g across f follows.

Intertetrahedral continuity of derivatives: To prove that the interpolant is C^1 across a face f, it is enough to check continuity of the normal derivative. We show that the restriction of such a derivative to f does not depend on the polynomial pieces. $D_{\eta_f} S_g$ is a polynomial of degree 4 on f which is uniquely determined by the following 15 data:

$$D_{\eta_f} S_g(v_i), \quad i = 1, 2, 3; \qquad D_{v_2 - v_1} D_{\eta_f} S_g(v_i), \quad i = 1, 2, 3;$$

$$D_{v_3 - v_1} D_{\eta_f} S_g(v_i), \quad i = 1, 2, 3; \qquad D_{\eta_f} g(f_i), \quad i = 1, 2, 3$$

and the values of $D_{\eta_f} S_g$ at the midpoints of the edges of f.

Internal continuity: This is obtained by construction since coefficients on internal faces are computed by using data independent of the faces. So they do not depend on the polynomial piece.

Internal continuity of derivatives: We explicitly show how C^1 smoothness is built across the interior face $\langle v_2, v_3, v_5 \rangle$ which is common to $T_1 = \langle v_2, v_3, v_4, v_5 \rangle$ and $T_4 = \langle v_1, v_2, v_3, v_5 \rangle$. We have

$$c_{0jkl}^4 = c_{jk0l}^1$$

and

$$c_{1jkl}^4 = -c_{j+1,k,0,l}^1 - c_{j,k+1,0,l}^1 - c_{jk1l}^1 + 4c_{j,k,0,l+1}^1$$

for $j + k + l = d$. We group the c_{1jkl}^4 into 6 categories:

(1) c_{1400}^4, c_{1310}^4 and c_{1301}^4 are determined by data at v_2. They satisfy the conditions since they are entirely computed using these data.

(2) c_{1130}^4, c_{1040}^4 and c_{1031}^4 are determined by data at v_3. They satisfy the conditions as explained above.

(3) c_{1022}^4 and c_{1202}^4 are computed by requiring smoothness conditions across $\langle v_2, v_3, v_5 \rangle$.

(4) c_{1220}^4 is computed by using data on the edge $\langle v_2, v_3 \rangle$. It enters the smoothness condition used to set c_{0221}^4 above.

(5) c_{1121}^4 and c_{1211}^4 are determined by using the values of the normal derivative. They enter the smoothness conditions used to set c_{0122}^4 and c_{0212}^4 respectively. Explicitly

$$c_{0122}^4 = \frac{c_{1211}^1 + c_{1121}^4 + c_{0221}^4 + c_{0131}^4}{4}$$

and

$$c_{0212}^4 = \frac{c_{2111}^1 + c_{1211}^4 + c_{0221}^4 + c_{0311}^4}{4}.$$

(6) $c_{1112}^4, c_{1103}^4, c_{1013}^4$ and c_{1004}^4 are computed by considering a layer as the B-net of a polynomial of degree 3. Such a polynomial is already smooth.

Property 5: The interpolant is C^3 at the center of each tetrahedron. This follows from the construction process.

§6. Numerical Experiments

We have implemented the interpolation scheme in MATLAB. To make sure that our implementation is correct, we have checked that our programs reproduce all polynomials of degree ≤ 5 by testing all 56 basis functions. Starting with a cube subdivided into 12 subtetrahedra by connecting the midpoint of the cube to a diagonal of each face of the cube, the maximum errors of spline interpolants of the 56 basis functions are about $.6661 \times 10^{-15}$. When each of the 12 subtetrahedra is subdivided into 8 subtetrahedra, the maximum errors are around $.7772 \times 10^{-15}$. The slight increase in the maximum errors is probably due to round-off errors.

Next we demonstrate how well this scheme approximates given functions and how the interpolation error evolves when the partition is refined. Starting with a single tetrahedron, we refine this tetrahedral partition 3 times and in a few cases, 4 times. Each time, we subdivide each tetrahedron t into 8 subtetrahedra by using the midpoints of six edges of t and dividing the central octahedron into four subtetrahedra. The central octahedron has three diagonals. The choice of a diagonal determines the kind of refinement one has. A common measure of degeneracy used for a tetrahedron T is

$$\sigma = \frac{h}{\rho},$$

Number of tetrahedra	1	8	64	512
Dimension	68	254	1346	8762

Tab. 1. Numbers of tetrahedra and dimension of spline spaces.

where h is the diameter of T and ρ the diameter of the largest sphere inscribed in T. From the three possible tetrahedral partitions that could arise from the choice of the diagonal of the central octahedron, we choose the diagonal that yields the smallest σ. With the tetrahedron with vertices of coordinates $(0,0,0), (1,0,0), (0,1,0)$ and $(1,0,1)$ this leads to a uniform refinement in the sense that all tetrahedra have the same measure of degeneracy. The choice of this model tetrahedron was suggested by Ong [6]. We first display the dimension of the spline spaces that were used for interpolation. The formula to compute the dimension was given in [4]. If V denotes the number of vertices, E the number of edges, F the number of faces and T the number of tetrahedra in a tetrahedral partition, the dimension of the corresponding spline space is given by

$$10V + 2E + 3F + 4T.$$

The dimensions of the first three refinement levels are given in Table 1. The fourth level of refinement involve 4096 tetrahedra and the dimension of the corresponding spline space is 63338. The limitation of computational power at hand prevents us from displaying additional levels of refinement. We have tested the code on the following functions:

$$f_1(x,y,z) = \exp(x + y + z), \quad f_2(x,y,z) = \sin(x^3 + y^3 + z^3),$$

$$f_3(x,y,z) = x^6 + y^6 + z^6, \quad f_4(x,y,z) = 10\exp(-x^2 - y^2 - z^2).$$

The results are presented in Table 2. We also checked the results of interpolating the homogeneous polynomials of degree 6 and the polynomials $x^7 + y^7 + z^7$ and $x^8 + y^8 + z^8$ of degree 7 and 8, respectively, see Table 3.

The maximum errors of the spline interpolants computed by evaluation on each tetrahedron at the domain points

$$\{\psi_{ijkl}\}_{i+j+k+l=10}$$

are displayed as well as the numerical rate of convergence.

These results show that the errors decrease like $O(h^6)$ when the partition is refined. The convergence rate is specially good for homogeneous polynomials of degree 6.

Number of tetrahedra	$f_1(x, y, z)$	Rate	$f_2(x, y, z)$	Rate
1	3.9822×10^{-3}	0	4.9375×10^{-2}	0
8	9.8569×10^{-5}	40.40	1.5402×10^{-2}	32.05
64	1.9578×10^{-6}	50.34	6.4640×10^{-4}	23.82
512	3.4577×10^{-8}	56.62	1.3401×10^{-5}	48.23
4096	5.7476×10^{-10}	60.16	2.1887×10^{-7}	61.23

Number of tetrahedra	$f_3(x, y, z)$	Rate	$f_4(x, y, z)$	Rate
1	4.6875×10^{-2}	0	1.2769×10^{-1}	0
8	7.3242×10^{-4}	64	6.9864×10^{-3}	18.27
64	1.1444×10^{-5}	64	1.5328×10^{-4}	45.57
512	1.7881×10^{-7}	64	2.6073×10^{-6}	58.78
4096	2.7940×10^{-9}	64	4.1614×10^{-8}	62.65

Tab. 2. Numerical maximum errors of the interpolation scheme.

x^6	Rate	x^5y	Rate	x^5z	Rate
1.5625×10^{-2}	0	1.5625×10^{-2}	0	1.5625×10^{-2}	0
2.4966×10^{-4}	62.59	2.4659×10^{-4}	63.36	2.4414×10^{-4}	64
3.9349×10^{-6}	63.45	3.8618×10^{-6}	63.85	3.8525×10^{-6}	63.37
6.1482×10^{-8}	64	6.0341×10^{-8}	64	6.0341×10^{-8}	63.84

x^4y^2	Rate	x^4yz	Rate	x^4z^2	Rate
1.5625×10^{-2}	0	1.5625×10^{-2}	0	1.5625×10^{-2}	0
2.4529×10^{-4}	63.70	2.4414×10^{-4}	64	2.4414×10^{-4}	64
3.8334×10^{-6}	63.99	3.8147×10^{-6}	64	3.8334×10^{-6}	63.69
5.9897×10^{-8}	64	5.9605×10^{-8}	64	5.9897×10^{-8}	64

x^3y^3	Rate	x^3y^2z	Rate	x^3yz^2	Rate
1.5625×10^{-2}	0	1.5625×10^{-2}	0	1.5625×10^{-2}	0
2.4414×10^{-4}	64	2.4414×10^{-4}	64	2.4414×10^{-4}	64
3.8147×10^{-6}	64	3.8147×10^{-6}	64	3.8147×10^{-6}	64
5.9605×10^{-8}	64	5.9605×10^{-8}	64	5.9605×10^{-8}	64

x^3z^3	Rate	x^2y^4	Rate	x^2y^3z	Rate
1.5625×10^{-2}	0	1.5625×10^{-2}	0	1.5625×10^{-2}	0
2.4414×10^{-4}	64	2.4414×10^{-4}	64	2.4414×10^{-4}	64
3.8147×10^{-6}	64	3.8303×10^{-6}	63.74	3.8147×10^{-6}	64
5.9605×10^{-8}	64	5.9884×10^{-8}	63.97	5.9605×10^{-8}	64

Tab. 3a. Numerical maximum errors of the interpolation scheme.

$x^2y^2z^2$	Rate	x^2yz^3	Rate	x^2z^4	Rate
1.5625×10^{-2}	0	1.5625×10^{-2}	0	1.5625×10^{-2}	0
2.4414×10^{-4}	64	2.4414×10^{-4}	64	2.4414×10^{-4}	64
3.8147×10^{-6}	64	3.8147×10^{-6}	64	3.8147×10^{-6}	64
5.9605×10^{-8}	64	5.9605×10^{-8}	64	5.9897×10^{-8}	63.69

xy^5	Rate	xy^4z	Rate	xy^3z^2	Rate
1.5714×10^{-2}	0	1.5625×10^{-2}	0	1.5625×10^{-2}	0
2.4554×10^{-4}	64	2.4414×10^{-4}	64	2.4414×10^{-4}	64
3.8618×10^{-6}	63.58	3.8147×10^{-6}	64	3.8147×10^{-6}	64
6.0341×10^{-8}	64	5.9605×10^{-8}	64	5.9605×10^{-8}	64

xy^2z^3	Rate	xyz^4	Rate	xz^5	Rate
1.5625×10^{-2}	0	1.5625×10^{-2}	0	1.5625×10^{-2}	0
2.4414×10^{-4}	64	2.4414×10^{-4}	64	2.4414×10^{-4}	64
3.8147×10^{-6}	64	3.8147×10^{-6}	64	3.8365×10^{-6}	63.64
5.9605×10^{-8}	64	5.9605×10^{-8}	64	6.0341×10^{-8}	63.59

y^6	Rate	y^5z	Rate	y^4z^2	Rate
1.6048×10^{-2}	0	1.5625×10^{-2}	0	1.5625×10^{-2}	0
2.5183×10^{-4}	63.72	2.4414×10^{-4}	64	2.4414×10^{-4}	64
3.9349×10^{-6}	64	3.8525×10^{-6}	63.37	3.8334×10^{-6}	63.69
6.1482×10^{-8}	64	6.0341×10^{-8}	63.84	5.9897×10^{-8}	64

y^3z^3	Rate	y^2z^4	Rate	yz^5	Rate
1.5625×10^{-2}	0	1.5625×10^{-2}	0	1.5625×10^{-2}	0
2.4414×10^{-4}	64	2.4414×10^{-4}	64	2.4554×10^{-4}	63.64
3.8147×10^{-6}	64	3.8237×10^{-6}	63.85	3.8484×10^{-6}	63.80
5.9605×10^{-8}	64	5.9897×10^{-8}	63.83	6.0341×10^{-8}	63.77

z^6	Rate	$x^7 + y^7 + z^7$	Rate	$x^8 + y^8 + z^8$	Rate
1.5625×10^{-2}	0	1.6406×10^{-1}	0	3.6328×10^{-1}	0
2.5075×10^{-4}	62.31	2.9907×10^{-3}	54.86	8.2966×10^{-3}	43.79
3.9179×10^{-6}	64	5.0068×10^{-5}	59.73	1.6653×10^{-4}	49.82
6.1482×10^{-8}	63.72	8.0839×10^{-7}	61.93	2.9523×10^{-6}	56.41

Tab. 3b. Numerical maximum errors of the interpolation scheme.

§7. Future Research Problems

The authors plan to use the energy minimization method to construct C^1 quintic spline interpolants for given scattered data. This method will not require higher order derivatives information at vertices. Also, the authors plan to apply this interpolation scheme to some real life data sets from oceanography and/or meteorology.

§8. Appendix

We now give the proof of Lemma 1 of Section 5. Let

$$p = \sum_{i+j+k+l=5} c_{ijkl} B^5_{ijkl}$$

be the B-form of a polynomial of degree 5 with respect to T. For a face of T, say $f = \langle v_1, v_2, v_3 \rangle$, from the given data we can determine the following 21 degrees of freedom on f.

1) $p(v_1), p(v_2)$ and $p(v_3)$
2) $D_{v_1-v_2}p(v_i), D_{v_3-v_2}p(v_i), i = 1, 2, 3.$
3) $D^2_{v_1-v_2}p(v_i), D_{v_1-v_2}D_{v_3-v_2}p(v_i)$ and $D^2_{v_3-v_2}p(v_i), i = 1, 2, 3.$
4) Values of the outward normal derivative at the midpoints of the three edges of f.

These data (fifth-degree Argyris element) determine completely p on the given face f. Similarly, p is determined on other faces of T. It remains to determine the coefficients of p which are not associated with domain points on any face of T. These coefficients are $c_{2111}, c_{1211}, c_{1121}$ and c_{1112}. We therefore write

$$p = c_{2111}B^5_{2111} + c_{1211}B^5_{1211} + c_{1121}B^5_{1121} + c_{1112}B^5_{1112} + q$$

where all coefficients in q are determined, i.e. q is known. Now for any point v with barycentric coordinates (b_1, b_2, b_3, b_4),

$$B^5_{ijkl}(v) = \frac{5!}{i!j!k!l!} b_1^i b_2^j b_3^k b_4^l,$$

and so

$$p(v) = \frac{5!}{2} b_1 b_2 b_3 b_4 (c_{2111}b_1 + c_{1211}b_2 + c_{1121}b_3 + c_{1112}b_4) + q(v).$$

The data

$$p(v_5), \quad \frac{\partial}{\partial x}p(v_5), \quad \frac{\partial}{\partial y}p(v_5), \quad \frac{\partial}{\partial z}p(v_5),$$

at the center v_5 determine $L(v_5), D_{v_1-v_2}L(v_5), D_{v_1-v_3}L(v_5), D_{v_2-v_3}L(v_5)$, where

$$L(v) = c_{2111}b_1 + c_{1211}b_2 + c_{1121}b_3 + c_{1112}b_4.$$

This is because v_5 is not on any face of T. For example,

$$L(v_5) = \left(\frac{5!}{2}b_1b_2b_3b_4|_{v_5}\right)^{-1}(p(v_5) - q(v_5)) = \frac{4^4 \times 2}{5!}(p(v_5) - q(v_5))$$

and

$$D_{v_1-v_2}p(v_5) = D_{v_1-v_2}(\frac{5!}{2}b_1b_2b_3b_4)|_{v_5}L(v_5) + (\frac{5!}{2}b_1b_2b_3b_4)|_{v_5}D_{v_1-v_2}L(v_5)$$

from which we can compute $D_{v_1-v_2}L(v_5)$ since its coefficient in this last equation is $\dfrac{5!}{2!}\left(\dfrac{1}{4}\right)^4$. Using the formula for directional derivatives of Section 2 with $v_5 = (v_1 + v_2 + v_3 + v_4)/4$, we get

$$c_{2111} + c_{1211} + c_{1121} + c_{1112} = 4L(v_5),$$
$$c_{2111} - c_{1211} = \frac{1}{5}D_{v_1-v_2}L(v_5),$$
$$c_{2111} - c_{1121} = \frac{1}{5}D_{v_1-v_3}L(v_5),$$
$$c_{1211} - c_{1121} = \frac{1}{5}D_{v_2-v_3}L(v_5).$$

These equations form a system of equations which can be readily solved. This determines all the coefficients of p and concludes the proof of the lemma. $\square$

References

1. Alfeld, P., A trivariate Clough-Tocher scheme for tetrahedral data, Comp. Aided Geom. Design **1** (1984), 169–181.

2. Alfeld, P. and B. Harris, Microscope: A software system for multivariate analysis, manuscript, 1984.

3. de Boor, C., *B*-form Basics, *Geometric Modeling*, edited by G. Farin, SIAM Publication, Philadelphia, (1987), 131–148.

4. Lai, M. J. and A. Le Méhauté, A new kind of trivariate C^1 spline space, manuscript, 1999.

5. Lai, M. J. and L. L. Schumaker, *Splines on Triangulations*, in preparation, 2002.

6. Ong, M. E. G., Uniform refinement of a tetrahedron, SIAM J. Sci. Comput., **15** (1994), 1134–1144.

7. Worsey, A. J. and G. Farin, An n-dimensional Clough-Tocher interpolant, Constr. Approx., **3** (1987), 99–110.

8. Worsey, A. J. and B. Piper, A trivariate Powell-Sabin interpolant, Comp. Aided Geom. Design, **5** (1988), 177–186.

9. Ženíček, A., Polynomial approximation on tetrahedrons in the finite element method, J. Approx. Theory, **7** (1973), 334–351.

Gerard Awanou
University of Georgia
Math Department
Athens, Ga 30602
gawanou@math.uga.edu

Ming-Jun Lai
University of Georgia
Math Department
Athens, Ga 30602
mjlai@math.uga.edu

Good Point/Bad Point Iterations for Solving the Thin–Plate Spline Interpolation Equations

R. K. Beatson and J. Levesley

Abstract. Preconditioned methods for solving the thin–plate spline interpolation equations are disadvantaged by data points at which it is difficult to construct almost Lagrange functions. Here we give an algorithm which separates such bad points from the remaining good points, and iterates to convergence, solving on the good points using some iterative method, and solving on the small set of bad points directly. Numerical results are given demonstrating the effectiveness of this method.

§1. Introduction

Interpolation using radial basis functions has become popular because data need not have any particular structure, and the method is not dimension-dependent. One of the most (if not the most) popular radial basis function is the thin–plate spline, which is the minimal energy interpolant in two dimensions.

Suppose we have a finite point data set $\Lambda \subset \mathbb{R}^2$ of cardinality N, and we wish to interpolate a function f at the points in Λ. We construct an interpolant s using the univariate function $\phi(r) = r^2 \log r$ in the following way:

$$s(x, y) = \sum_{(w,z)\in\Lambda} c_{(w,z)}\Phi(x - w, y - z) + p_1(x, y), \qquad (1)$$

where p_1 is a linear polynomial and $\Phi(x, y) := \phi(\sqrt{x^2 + y^2})$. The coefficients $\{c_{(w,z)} : (w, z) \in \Lambda\}$ also satisfy the side conditions

$$\sum_{(w,z)\in\Lambda} c_{(w,z)}q(w, z) = 0, \qquad (2)$$

Approximation Theory X: Wavelets, Splines, and Applications   17
Charles K. Chui, Larry L. Schumaker, and Joachim Stöckler (eds.), pp. 17–25.
Copyright © 2002 by Vanderbilt University Press, Nashville, TN.
ISBN 0-8265-1416-2.

for every linear polynomial q. These conditions serve two purposes: the first is to produce an interpolant which behaves like a linear polynomial near to infinity, and the second is to use up the extra degrees of freedom introduced by the introduction of a linear polynomial to the interpolant.

If we apply the interpolation conditions $s(x, y) = f(x, y)$, $(x, y) \in \Lambda$, we arrive, in conjunction with (2), at a full square system of size $N + 3$. Using a direct method of solution is too costly for any but moderate (a few thousand) numbers of interpolation points. To be competitive as a method, iterative solution techniques need to be used.

In recent years a number of iterative methods have been explored [1,2,4,5,7,8]. Several of these methods are based on the idea of finding a good basis for expanding the solution. Using the good localisation properties of thin-plate splines one hopes to construct good approximations $\overline{\chi}_\eta$ to Lagrange functions χ_η with

$$\overline{\chi}_\eta(\nu) \approx \chi_\eta(\nu) = \delta_{\eta\nu}, \quad \eta, \nu \in \Lambda,$$

which are also thin–plate splines. Unfortunately, for some of the points $\eta \in \Lambda$ it is difficult to construct a good almost Lagrange function. Such points we will call **bad points**. The rest of the points we call **good points**.

The method we describe in Section 3 iterates between the good and bad points, using the almost Lagrange functions to expand the solution at good points, and interpolating at the bad points using a thin–plate spline as in (1). We use some iterative method to solve on the good points, which we expect to be rapid since the associated interpolation matrix is almost diagonal due to the use of almost Lagrange functions. We expect there to be few bad points, so that direct solution at these points is relatively inexpensive. One advantage of this method is that it is relatively easy to implement.

In Section 2 we describe the case when data is gridded, as this is where the idea for the good point/bad point iteration arose. In Section 4 we give some numerical examples.

§2. Motivation via the Gridded Case

In this section we consider the problem of interpolating on a uniform grid in $([0, 1])^2$. For $h = 1/n$, with $n \in \mathbb{N}$, the interpolation point set is $\Lambda_h := \{(x, y) : x = ih \text{ and } y = jh, \ 0 \leq i, j \leq n\}$. In this case it is straightforward to construct good basis functions for the expansion of the solution as long as we are on the subgrid $G_h := \{(x, y) : x = ih \text{ and } y = jh, \ 2 \leq i, j \leq n - 2\}$. For $(x, y) \in G_h$ let $\overline{\Delta}_h \phi(x, y) = h^{-2}((\phi(x + h, y) + \phi(x - h, y) + \phi(x, y + h) + \phi(x, y - h) - 4\phi(x, y))$ be a discretisation of the Laplacian $\Delta\phi$. Then, since $\Delta^2\phi = 8\pi\delta$, $\overline{\Delta}_h^2\phi$ should

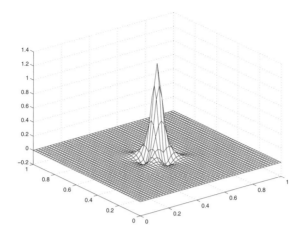

Fig. 1. $\psi_{1/16}(x - 1/2, y - 1/2)$.

be a good approximation to $8\pi\delta$. Let $\psi_h = h^2/(8\pi)\overline{\Delta}_h^2\phi$. In Figure 1 we graph $\psi_h(x - 1/2, y - 1/2)$, for $h = 1/16$, on $[0,1]^2$.

As we can see, this function decays rapidly away from its centre, and, in fact, using the results of Rabut [8], it is straightforward to show that $|\psi_h(x,y)| \le K(1 + h^{-2}(x^2 + y^2))^{-2}$, for some positive constant K. These functions were first used as a basis for solving the thin–plate spline equations by Dyn et. al. [6].

If we interpolate $f \in \mathbf{C}([0,1]^2)$ on G_h with an interpolant of the form

$$s_{G_h}(x, y) = \sum_{(w,z) \in G_h} c_{(w,z)} \psi(x - w, y - z), \tag{3}$$

then we get good approximation for $(x,y) \in [2h, 1 - 2h]^2$ but poor approximation at the edges. In fact, because of the rapid decay of ψ, s_{G_h} is small on $B_h = \Lambda_h \setminus G_h$; see Figure 2, where we have interpolated $\exp((x - 0.5)^2 + (y - 0.5)^2)$ on G_h.

Thus the following algorithm for solving the thin–plate spline equations suggests itself:

Step 1: Compute the thin plate spline interpolant $s_{B_h}^1$ to f on B_h, and evaluate the residual $r_{B_h}^1 = f - s_{B_h}^1$ on G_h.

Step 2: Compute an approximate interpolant $s_{G_h}^1$ to $r_{B_h}^1$ on G_h using an approximation of the form (3). Calculate the residual $r_{G_h}^1 = r_{B_h}^1 - s_{G_h}^1$ on B_h.

Step 3: Set $k = 2$. Iterate the following steps until the error at the grid points is sufficiently small:

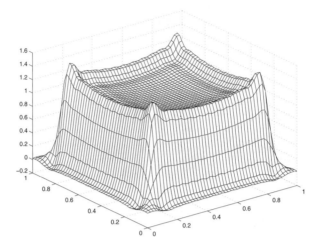

Fig. 2. Approximation to $\exp((x - 1/2)^2 + (y - 1/2)^2)$.

Step 3a: Compute the thin plate spline interpolant $s_{B_h}^k$ to $r_{G_h}^{k-1}$ on B_h, and evaluate the residual $r_{B_h}^k = r_{G_h}^{k-1} - s_{B_h}^k$ on G_h.

Step 3b: Compute an approximate interpolant $s_{G_h}^k$ to $r_{B_h}^k$ on G_h using an approximation of the form (3). Calculate the residual $r_{G_h}^k = r_{B_h}^k - s_{G_h}^k$ on B_h.

Step 3c: Increment k by 1 if a satisfactory error is not achieved.

Step 4: Set the final (approximate) interpolant to

$$s_h = \sum_{i=1}^{k} s_{B_h}^i + s_{G_h}^i.$$

We should note here that the number of points at which we compute the thin-plate spline interpolant in Steps 1 and 3a is of the order of $N^{1/2}$ if N is the total number of interpolation points. If we solve this interpolation problem using a direct method, we have order $N^{3/2}$ flops. The solution of the system arising in Steps 2 and 3b will usually be by some iterative method, requiring, with the use of some fast evaluation algorithm (see [3]), order N or $N \log N$ flops. Thus we expect this method to be of order $N^{3/2}$. We are investigating the reduction of the cost of the solution of the systems in Steps 1 and 3a in order to get an order N or $N \log N$ algorithm, which is the best we can expect.

In the next section we generalise the idea of *good points*, i.e. those at which we can construct a good basis function, and *bad points*, which are the remaining points. An algorithm based on iterating between the good and bad points will be described.

§3. Scattered Data, Good and Bad Points, The Algorithm

In this section we consider the problem of interpolation on a set Λ of scattered data in $\mathbb{R}^2$. The main aim of this section is to describe how good points and bad points are distinguished, and to describe the algorithm for interpolation which arises.

We wish to generate a good basis with which to expand the solution of the interpolation problem. For $\eta \in \Lambda$ let us call the set of q nearest neighbours to $\eta = (\eta_x, \eta_y)$, including η itself, $N_q(\eta)$. Using the good localisation properties of thin-plate spline interpolation, we construct approximate Lagrange functions

$$\overline{\chi}_\eta(x, y) = \sum_{\nu \in N_q(\eta)} c_\nu \phi(x - \nu_x, y - \nu_y), \tag{4}$$

which satisfy, for $\nu \in \Lambda$,

$$\overline{\chi}_\eta(\nu) \approx \delta_{\eta\nu} = \begin{cases} 1, & \text{if } \eta = \nu, \\ 0, & \text{otherwise.} \end{cases}$$

Several different types of approximate Lagrange functions have been studied in [1,5]. For our current application the **decay element approximate Lagrange functions** of [1] have proven most suitable. For $\eta \in \Lambda$ we choose χ_η to satisfy the following constrained optimisation problem:

$$\overline{\chi}_\eta(\nu) \approx \delta_{\eta\nu}, \qquad \nu \in N_q(\eta),$$
$$|\overline{\chi}_\eta(\xi)| = \mathcal{O}(\|\xi\|^{-3}), \text{ as } \xi \to \infty,$$

for which we obtain a least squares solution. In order to satisfy these decay conditions, we need to satisfy 14 polynomial conditions on the coefficients c_ν in (4), the details of which are given in [1]. Suffice it to say that the coefficients act like a differential operator much as $\overline{\Delta}^2$ did in the previous section.

We now use the computed decay element to test whether the point is a good point or a bad point. We calculate

$$E_\eta = |1 - \overline{\chi}_\eta(\eta)| + \sum_{\nu \in N_\eta \setminus \{\eta\}} |\overline{\chi}_\eta(\nu)|,$$

and if $E_\eta <$ tolerance then η is a good point, otherwise η is a bad point. Here we are measuring how close to a local Lagrange function $\overline{\chi}_\eta$ is. Let G be the set of good points and B be the set of bad points. In Figure 3 we see that the majority of the bad points (marked with circles) are at the edge of the domain exactly as in the gridded case. Here $N = 1000$,

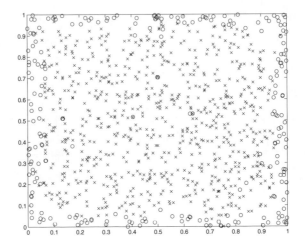

Fig. 3. Good points (x) and bad points (o): $N = 1000$, $q = 50$.

Tab. 1. Number of bad points for $q = 50$.

# Λ	# B
200	64
400	108
800	156
1600	206
3200	312
6400	418

and $q = 50$. In Table 1 we see that, for $q = 50$, the number of bad points grows like $\sqrt{N}$, again as in the gridded case.

We now follow the same idea as in the previous section. We interpolate on B using the thin-plate spline interpolant (see (1)) and approximate on the good points using a function of the form

$$s_G = \sum_{\nu \in G} c_\nu \overline{\chi}_\nu. \tag{5}$$

We compare two different strategies for computing a function s_G. The first strategy (LAG) is to let $c_{(w,z)} = f(w, z)$ i.e. treat the $\chi_{(w,z)}$ as Lagrange functions. The second (GMG) is to use a few iterations of GMRES to compute an approximate interpolant. The first method is cheaper per iteration but we might expect it to take more iterations to converge.

The analogue of the algorithm of the previous section is:

Step 1: Compute the thin plate spline interpolant s_B^1 to f on B, and evaluate the residual $r_B^1 = f - s_B^1$ on G.

Step 2: Compute an approximate interpolant s_G^1 to r_B^1 on G using an approximation of the form (5). Calculate the residual $r_G^1 = r_B^1 - s_G^1$ on B.

Step 3: Set $k = 2$. Iterate the following steps until satisfied:

> **Step 3a:** Compute the thin plate spline interpolant s_B^k to r_G^{k-1} on B, and evaluate the residual $r_B^k = r_G^{k-1} - s_B^k$ on G.

> **Step 3b:** Compute an approximate interpolant s_G^k to r_B^k on G using an approximation of the form (5). Calculate the residual $r_G^k = r_B^k - s_G^k$ on B.

> **Step 3c:** Increment k by 1 if an acceptable error is not achieved.

Step 4: Set the final (approximate) interpolant to

$$s = \sum_{i=1}^{k} s_B^i + s_G^i.$$

§4. Numerical Results

In this section we demonstrate the effectiveness of the algorithms LAG and GMG of Section 3 when interpolating the PEAKS function from MATLAB:

$$f(x, y) = 3(1 - 3x)^2 \exp(-9x^2 - (3y + 1)^2) - 1/3 \exp(-(3x + 1)^2 - 9y^2)$$
$$- 10(3x/5 - 27x^3 - 243y^5) \exp(-9(x^2 + y^2)).$$

In Table 2 we give iteration counts for the two algorithms LAG and GMG. If we use a fast evaluation method for matrix–vector products (see [3]) then the cost per iteration of the two methods is comparable in terms of order, though LAG is still cheaper per iteration. In all cases $q = 50$. The final maximum norm residual achieved is 10^{-7}. The algorithm is implemented in MATLAB which has restricted the size of the test problems.

Tab. 2. Numerical results for LAG and GMG.

# Λ	LAG	GMG
100	13	6
200	14	8
400	8	8
800	8	7
1600	37	9
3200	16	8

What we notice immediately is that GMG is much more stable in terms of number of iterations than LAG, and that the flop count of GMG is never significantly more than that of LAG. GMG appears to converge with a bounded number of iterations regardless of the number of interpolation points. Thus, if we can find a faster method to solve the interpolation problem on the bad points we have the opportunity to produce an algorithm which is either optimal or near optimal in terms of order, in other words, either order N or $N \log N$.

§5. Discussion

In this paper we have described an algorithm for generating an approximate thin–plate spline interpolant which involves the partition of the data set into good points and bad points. One advantage of this good/bad approach to solving radial basis function interpolation problems is that it is relatively easy to implement. Even without a fast multiply for the radial basis function involved, it can be used to solve systems with tens of thousands of points, since the matrix corresponding to the good points or good approximate cardinal functions need not be calculated or stored. Only its action needs to be calculated, and doing this is just evaluation of a radial basis function. Further it is clear that the direct solution procedure on the bad points can itself be replaced by some iterative method, thus removing the size restrictions imposed by storing a factorization of the bad point matrix.

References

1. Beatson, R. K., J. B. Cherrie, and C. T. Mouat, Fast fitting of radial basis functions: Methods based on preconditioned GMRES iteration, Advances in Comp. Math. **11** (1999), 253–270.

2. Beatson, R. K., G. Goodsell, and M. J. D. Powell, On multigrid techniques for thin plate spline interpolation in two dimensions, in *The Mathematics of Numerical Analysis (Park City, UT, 1995)*, Lectures in Appl. Math., 32, Amer. Math. Soc., Providence, RI, 1996, 77–97.

3. Beatson, R. K., and W. A. Light, Fast evaluation of radial basis functions: methods for two-dimensional polyharmonic splines, IMA J. Numer. Anal. **17** (1997), 343–372.

4. Beatson, R. K., W. A. Light, and W. A. S. Billings, Fast solution of the radial basis function interpolation equations: domain decomposition methods, SIAM J. Numer. Anal. **22** (2000), 1717–1740.

5. Beatson, R. K., and M. J. D. Powell, An iterative method for thin plate spline interpolation that employs approximations to Lagrange

functions, in *Numerical Analysis 1993 (Dundee, 1993)*, Chapman & Hall, 1994, 17–39.

6. Dyn, N., D. Levin, and S. Rippa, Numerical procedures for surface fitting of scattered data by radial functions, SIAM J. Sci. Statist. Comput. **7** (1986), 639–659.

7. Faul, A. C., and M. J. D. Powell, Proof of convergence of an iterative technique for thin plate spline interpolation in two dimensions, *Radial Basis Functions and their Applications*, Advances in Comp. Math. **11** (1999), 183–192.

8. Faul, A. C., and M. J. D. Powell, Krylov subspace methods for radial basis function interpolation, in *Numerical Analysis 1999 (Dundee, 1999)*, Chapman & Hall, 2000, 115–141.

9. Rabut, C., Elementary m-harmonic cardinal B-splines. Numer. Algorithms **2** (1992), 39–61.

J. Levesley
Department of Mathematics and Computer Science
University of Leicester
Leicester LE1 7RH, UK
jl1@mcs.le.ac.uk

R. K. Beatson
Department of Mathematics and Statistics
University of Canterbury
Christchurch, New Zealand
R.Beatson@math.canterbury.ac.nz

On Semi-Cardinal Interpolation
for Biharmonic Splines

Aurelian Bejancu

Abstract. The study of semi-cardinal interpolation for polyharmonic splines of several variables was initiated by the author in a previous article [2]. The present paper continues this study in the two-dimensional case, showing that the cardinal interpolation scheme for biharmonic splines of two variables can be approximated by a sequence of corresponding semi-cardinal interpolation schemes. The two-dimensional situation also allows for a simpler exposition of the ideas originally introduced in the general multi-variable case.

§1. Introduction

As is well-known, cardinal interpolation means interpolation to data on the multi-integer mesh $\mathbb{Z}^d$, for a given dimension d. The study of cardinal interpolation for univariate splines was begun by Schoenberg in his paper [8], in which he proposed a Fourier transform approach for the construction and analysis of the associated cardinal schemes. The subject of the present paper is related to the generalization of Schoenberg's approach to radial basis functions in the multi-dimensional case $d \geq 2$, a topic developed by Madych and Nelson [7] for polyharmonic splines, and by Buhmann [3] for a wide class of basis functions.

In particular, the problem of cardinal interpolation for biharmonic splines of two variables was formulated by Madych and Nelson as follows: Given a sequence $\{f_j\}_{j \in \mathbb{Z}^2} \subset \mathbb{R}$, find a biharmonic cardinal spline s, *i.e.*, a tempered distribution $s \in C^1(\mathbb{R}^2)$ for which $\Delta^2 s = 0$ on $\mathbb{R}^2 \setminus \mathbb{Z}^2$ in the distributional sense ($\Delta = \frac{\partial^2}{\partial x_1^2} + \frac{\partial^2}{\partial x_2^2}$ being the Laplace operator), such that

$$s(j) = f_j, \qquad j \in \mathbb{Z}^2. \tag{1}$$

Approximation Theory X: Wavelets, Splines, and Applications
Charles K. Chui, Larry L. Schumaker, and Joachim Stöckler (eds.), pp. 27–40.

Assuming that the data sequence $\{f_j\}_{j\in\mathbb{Z}^2}$ is of polynomial growth, Madych and Nelson's solution to the above problem is given by the absolutely and locally convergent Lagrange series representation

$$s(x) = \sum_{j\in\mathbb{Z}^2} f_j L(x-j), \qquad x \in \mathbb{R}^2, \tag{2}$$

where $L : \mathbb{R}^2 \to \mathbb{R}$ is the unique biharmonic cardinal spline satisfying

$$L(k) = \delta_{k0} = \begin{cases} 1, & k = 0, \\ 0, & k \in \mathbb{Z}^2 \setminus \{0\}. \end{cases} \tag{3}$$

In turn, L is constructed as a classical inverse Fourier transform

$$L(x) := \frac{1}{(2\pi)^2} \int_{\mathbb{R}^2} e^{ixt} \frac{\omega(t)}{\|t\|^4} \, dt, \quad x \in \mathbb{R}^2, \tag{4}$$

where

$$\omega(t) := \frac{1}{\sum_{k\in\mathbb{Z}^2} \|t + 2\pi k\|^{-4}}, \qquad t \in \mathbb{R}^2. \tag{5}$$

Here and in the sequel, the symbol $\|\cdot\|$ denotes the Euclidean norm in $\mathbb{R}^2$ and i is the usual root of -1. Note that ω is even and 2π-periodic in each variable and has a zero of order 4 at the origin. Also, ω admits an absolutely convergent Fourier series expansion

$$\omega(t) = \sum_{k\in\mathbb{Z}^2} a_k e^{-ikt}, \qquad t \in Q^2, \tag{6}$$

where, as throughout the paper, $Q = [-\pi, \pi]$. An important observation is that the distributional Fourier transform of $\Delta^2 L$ is ω. This fact and (6) imply $\Delta^2 L = 0$ on $\mathbb{R}^2 \setminus \mathbb{Z}^2$, so L is indeed a biharmonic cardinal spline. On the other hand, the cardinal conditions (3) follow easily from the form (5) of ω by periodization of the integrand in (4) with respect to t (the one-dimensional version of this argument dates back to Schoenberg's paper [8]).

Let $\mathbb{Z}_+ = \{0, 1, \ldots\}$. For $d \geq 2$, in [2] we considered interpolation by polyharmonic splines to data on the mesh $\mathbb{Z}^{d-1} \times \mathbb{Z}_+$, referred to as semi-cardinal interpolation. We mention that similar problems had been treated previously in the spline and radial basis function literature only in the case $d = 1$, corresponding to the mesh $\mathbb{Z}_+$. Specifically, semi-cardinal interpolation for natural splines was investigated by Schoenberg [9], whereas Buhmann [4] considered a class of univariate decaying radial basis functions, including the inverse multiquadric. The approach of both these authors is based on the Wiener-Hopf factorization technique for solving semi-infinite systems of difference equations, but, unlike the cardinal case, it does not involve Fourier transform constructions.

The absolute novelty of the work in [2] consists in combining Wiener-Hopf factorizations with the Fourier transform approach for radial basis functions. We first illustrated this idea in [1], in which a Fourier transform construction for semi-cardinal interpolation with one-dimensional natural splines was obtained. In the simplest multivariable case, we introduced the problem of semi-cardinal interpolation for biharmonic splines of two variables ($d = 2$) as follows: Given a sequence $\{f_j\}_{j \in \mathbb{Z} \times \mathbb{Z}_+} \subset \mathbb{R}$, find a biharmonic semi-cardinal spline s, *i.e.*, a tempered distribution $s \in C^1(\mathbb{R}^2)$ for which $\Delta^2 s = 0$ on $\mathbb{R}^2 \setminus (\mathbb{Z} \times \mathbb{Z}_+)$ in the distributional sense, such that

$$s(j) = f_j, \qquad j \in \mathbb{Z} \times \mathbb{Z}_+. \tag{7}$$

Under the assumption that the data sequence $\{f_j\}_{j \in \mathbb{Z} \times \mathbb{Z}_+}$ is absolutely summable, a solution to this problem is given in [2] by means of the absolutely and locally convergent Lagrange series representation

$$s(x) = \sum_{j \in \mathbb{Z} \times \mathbb{Z}_+} f_j L_j(x), \qquad x \in \mathbb{R}^2, \tag{8}$$

where each L_j is a biharmonic semi-cardinal spline satisfying

$$L_j(k) = \delta_{kj} = \begin{cases} 1, & k = j, \\ 0, & k \in (\mathbb{Z} \times \mathbb{Z}_+) \setminus \{j\}. \end{cases} \tag{9}$$

As it turns out, the construction of the Lagrange functions L_j, $j \in \mathbb{Z} \times \mathbb{Z}_+$, is substantially more involved than that of the cardinal Lagrange function L. The next two sections are devoted to the derivation of some of the main properties of this construction. Although the exposition follows the approach adopted in [2] for the general polyharmonic case, we include several new features, such as the definition (31), Lemma 1 and Lemma 4. In the present two-dimensional biharmonic case it is also possible to avoid the finite difference operators which were used in [2] for proving the semi-cardinal Lagrange conditions (9). The construction is applied in the last section to establish approximation relationships between cardinal and semi-cardinal interpolation schemes. For a discussion of other potential applications and some open problems generated by this construction, we refer the reader to the article [2].

§2. Wiener-Hopf Factorizations on Semi-Plane Lattices

In the following, by a "+" type function we shall mean a continuous function which is 2π-periodic in each of its two variables and whose sequence of Fourier coefficients is absolutely summable and supported only on $\mathbb{Z} \times \mathbb{Z}_+$.

The natural approach for constructing a biharmonic semi-cardinal Lagrange function L_j, for each $j \in \mathbb{Z} \times \mathbb{Z}_+$, is to replace $\omega(t)$ by $\omega_j(-t)$, and L by L_j in the inverse Fourier transform formula (4), where ω_j is a "+" type function. The idea is to ensure that the distributional Fourier transform of $\Delta^2 L_j$ is a periodic function all whose Fourier coefficients indexed by $\mathbb{Z} \times \{1, 2, \ldots\}$ are zero, which would provide $\Delta^2 L_j = 0$ on $\mathbb{R}^2 \setminus (\mathbb{Z} \times \mathbb{Z}_+)$ by distribution theory rules (see Friedlander [6]). Thus it would only remain to select the "+" type function ω_j in such a way so that the semi-cardinal Lagrange conditions (9) are satisfied.

Guided by this approach, in [2] we obtained a suitable "+" type function ω_j, for each $j \in \mathbb{Z} \times \mathbb{Z}_+$, by means of the Wiener-Hopf technique. This is based on a factorization

$$\omega(t) = \omega_+(t)\omega_-(t), \qquad t \in Q^2, \tag{10}$$

where $\omega_\pm$ admit absolutely convergent Fourier series expansions of the form

$$\omega_\pm(t) = \sum_{k \in \mathbb{Z} \times \mathbb{Z}_+} \beta_k^\pm e^{\pm ikt}, \qquad t \in Q^2, \tag{11}$$

with real coefficients $\beta_k^\pm$, $k \in \mathbb{Z} \times \mathbb{Z}_+$. In other words, ω_+ and $\overline{\omega_-}$ are "+" type functions (here and in the sequel, the overline denotes complex conjugation). Once such a factorization is constructed, the Wiener-Hopf technique defines ω_j, for each $j \in \mathbb{Z} \times \mathbb{Z}_+$, by the formula

$$\omega_j(t) := \omega_+(t)P_+\{e^{ijt}\omega_-(t)\}, \qquad t \in Q^2, \tag{12}$$

where P_+ is the orthogonal projection operator associating to each absolutely convergent Fourier series its "+" type truncation.

We point out that the standard method for obtaining (10)–(11) requires a well-defined $\log \omega$, but in our case definition (5) implies $\omega(0) = 0$. This obstruction in applying the Wiener-Hopf technique was overcome in [2] by incorporating the zero of ω in a periodic factor. More precisely, for each $t = (t_1, t_2) \in Q^2$, formula (5) is rewritten as

$$\omega(t) = \{4[\sin(t_1/2)]^2 + 4[\sin(t_2/2)]^2\}^2 \widetilde{\omega}(t)$$
$$- \{4 - e^{-it_1} - e^{it_1} - e^{-it_2} - e^{it_2}\}^2 \widetilde{\omega}(t), \tag{13}$$

where

$$\widetilde{\omega}(t) = \left\{ \frac{t_1^2 + t_2^2}{4[\sin(t_1/2)]^2 + 4[\sin(t_2/2)]^2} \right\}^2 \frac{1}{1 + \|t\|^4 \sum' \|t + 2\pi k\|^{-4}}. \tag{14}$$

The notation $\sum'$ means that the summation index k ranges through $\mathbb{Z}^2 \setminus \{0\}$. The required factorization of ω is performed by first finding factorizations of the type (10)–(11) separately for $\widetilde{\omega}$ and for the "zero"-factor $\{4 - e^{-it_1} - e^{it_1} - e^{-it_2} - e^{it_2}\}^2$, and then grouping together their corresponding subfactors.

Since $\widetilde{\omega}$ takes only strictly positive values, its logarithm can be used to construct a standard Wiener-Hopf factorization

$$\widetilde{\omega}(t) = \Omega_+(t)\Omega_-(t), \qquad t \in Q^2, \tag{15}$$

where Ω_+, $\overline{\Omega_-}$ and their inverses are "+" type functions. Specifically, the procedure uses the Fourier series representation

$$\log \widetilde{\omega}(t) = \sum_{k \in \mathbb{Z}^2} \lambda_k e^{ikt}, \qquad t \in Q^2, \tag{16}$$

whose absolute convergence is guaranteed by the fact that the Fourier coefficients λ_k, $k \in \mathbb{Z}^2$, of $\log \widetilde{\omega}$ satisfy ([2, Lemma 2.2])

$$|\lambda_k| = O(\|k\|^{-4}), \quad \text{as } \|k\| \to \infty. \tag{17}$$

The factorization (15) is obtained by splitting the series (16) in two parts,

$$\log \widetilde{\omega}(t) = \Lambda_+(t) + \Lambda_-(t), \tag{18}$$

where

$$\Lambda_\pm(t) := \frac{1}{2} \sum_{k_1 \in \mathbb{Z}} \lambda_{(k_1,0)} e^{ik_1 t_1} + \sum_{k \in \mathbb{Z} \times \{1,2,\ldots\}} \lambda_k e^{\pm ikt}, \quad t \in Q^2, \tag{19}$$

and by setting

$$\Omega_\pm(t) := \exp[\Lambda_\pm(t)], \qquad t \in Q^2. \tag{20}$$

It is now straightforward to check the stated properties for $\Omega_\pm$ (see [2] for details and references).

To obtain a factorization of the type (10)–(11) for the remaining factor in (13) containing the zero of ω, we treat this factor as a function of e^{it_2} which depends on the parameter t_1. Specifically, we write

$$\begin{aligned}
\{4 - e^{-it_1} &- e^{it_1} - e^{-it_2} - e^{it_2}\}^2 \\
&= e^{-2it_2}\{1 - 2[1 + 2\sin^2(t_1/2)]e^{it_2} + e^{2it_2}\}^2 \tag{21} \\
&= e^{-2it_2}\{[e^{it_2} - \rho(t_1)][e^{it_2} - 1/\rho(t_1)]\}^2,
\end{aligned}$$

where

$$\rho(t_1) := 1 + 2[\sin(t_1/2)]^2 - 2|\sin(t_1/2)|\sqrt{1 + [\sin(t_1/2)]^2}, \qquad t_1 \in Q. \tag{22}$$

Note that

$$3 - 2\sqrt{2} \le \rho(t_1) \le 1, \qquad \forall\, t_1 \in Q. \tag{23}$$

Since the Fourier coefficients θ_κ, $\kappa \in \mathbb{Z}$, of the periodic and even function ρ are given by the formula

$$\theta_\kappa = \frac{1}{\pi} \int_0^\pi \rho(t_1) \cos(\kappa t_1)\, dt_1, \qquad \kappa \in \mathbb{Z}, \tag{24}$$

using integration by parts twice and the values of ρ' at 0 and π, we find

$$|\theta_\kappa| = O(|\kappa|^{-2}), \quad \text{as } |\kappa| \to \infty, \tag{25}$$

which guarantees an absolutely convergent Fourier series representation for ρ. The same property is also seen to hold for $1/\rho$.

Our aim is now to write (21) as a product of two factors, the first being a polynomial in e^{it_2} and the second a polynomial in e^{-it_2}, but note that there are several different such factorizations available, modulo factors which depend only on t_1. Some of these factorizations are easily discarded. For example, if the first factor was $[e^{it_2} - \rho(t_1)][e^{it_2} - 1/\rho(t_1)]^2$ and the second $e^{-it_2}[1 - \rho(t_1)e^{-it_2}]$, then ω_- would contain the factor e^{-it_2}. Hence $P_+\{e^{ijt}\omega_-(t)\} \equiv 0$ for each $j \in \mathbb{Z} \times \{0\}$, and (12) would imply $\omega_j(t) \equiv 0$ in this case, which is unacceptable. On the other hand, any factorization that assigns the factor $1 - [\rho(t_1)]^{-1}e^{-it_2}$ to ω_- is ruled out by our method of proof of the semi-cardinal Lagrange conditions (9) in the next section (see Remark 3). Thus we are led to the following factorization of the zero of ω:

$$\{4 - e^{-it_1} - e^{it_1} - e^{-it_2} - e^{it_2}\}^2 = \{[\rho(t_1)]^{-1} - e^{it_2}\}^2\{1 - \rho(t_1)e^{-it_2}\}^2. \tag{26}$$

Now we can group together the corresponding factors in (15) and (26) by setting

$$\begin{aligned}
\omega_+(t) &:= \{[\rho(t_1)]^{-1} - e^{it_2}\}^2 \Omega_+(t), \\
\omega_-(t) &:= \{1 - \rho(t_1)e^{-it_2}\}^2 \Omega_-(t),
\end{aligned} \tag{27}$$

which provides the required factorization (10) of ω. Accordingly, for each $j \in \mathbb{Z} \times \mathbb{Z}_+$, by (12) there exists an absolutely summable sequence of real numbers $\{b_{jk}\}_{k \in \mathbb{Z} \times \mathbb{Z}_+}$, such that ω_j has the "+" type representation

$$\omega_j(t) = \sum_{k \in \mathbb{Z} \times \mathbb{Z}_+} b_{jk} e^{ikt}, \qquad t \in Q^2. \tag{28}$$

In addition, noting that $\omega_+(t)$ has a zero of order 2 in $t = 0$, it also follows from (12) that there are two positive constants M_1, M_2 such that, for all $j \in \mathbb{Z} \times \mathbb{Z}_+$ and $t \in Q^2$, we have

$$|\omega_j(t)| \le M_1 \quad \text{and} \quad |\omega_j(t)| \le M_2\|t\|^2. \tag{29}$$

For more details and other properties of the functions ω_j, $j \in \mathbb{Z} \times \mathbb{Z}_+$, we refer the reader to [2, Lemma 2.4].

§3. Semi-Cardinal Lagrange Functions

Having obtained the set $\{\omega_j\}_{j \in \mathbb{Z} \times \mathbb{Z}_+}$ of "+" type functions, the next stage in the semi-cardinal construction is to define, for each $j \in \mathbb{Z} \times \mathbb{Z}_+$, a C^1 tempered distribution L_j satisfying (9), such that the distributional Fourier transform of $\Delta^2 L_j$ is $\overline{\omega_j}$. As already mentioned at the begining of the previous section, the reason for the last condition is that it guarantees that L_j is a biharmonic semi-cardinal spline (note that $\overline{\omega_j}(t) = \omega_j(-t)$, since the Fourier coefficients of ω_j are real).

The difficulty at this point is that mere replacement of ω by $\overline{\omega_j}$ in (4) does not work, since the second inequality of (29) is not sufficient to ensure the absolute integrability of $\omega_j(-t)\|t\|^{-4}$ in a neighbourhood of $t = 0$. Nevertheless, we observe that, for $q \in \{1, 2\}$, the expression $it_q\omega_j(-t)\|t\|^{-4}$, which is formally the Fourier transform of a gradient component of the required biharmonic semi-cardinal Lagrange function L_j, is absolutely integrable over $\mathbb{R}^2$ due to (29). Therefore we let

$$G_{j,q}(x) := \frac{1}{(2\pi)^2} \int_{\mathbb{R}^2} e^{ixt} \frac{it_q\omega_j(-t)}{\|t\|^4} \, dt, \qquad x \in \mathbb{R}^2, \qquad (30)$$

and, for each $x \in \mathbb{R}^2$, we integrate along the line segment joining x and 0 in order to define the following candidate for the Lagrange function L_j:

$$L_j(x) := \delta_{j0} + \int_0^1 [x_1 G_{j,1}(\tau x) + x_2 G_{j,2}(\tau x)] \, d\tau, \qquad x \in \mathbb{R}^2. \qquad (31)$$

This ensures $L_j \in C^1(\mathbb{R}^2)$ and automatically provides $L_j(0) = \delta_{j0}$, but to verify all of the semi-cardinal Lagrange conditions (9) we need a simpler expression for L_j.

Lemma 1. For any $j \in \mathbb{Z} \times \mathbb{Z}_+$ and $x \in \mathbb{R}^2$, we have

$$L_j(x) = \delta_{j0} + \frac{1}{(2\pi)^2} \int_{\mathbb{R}^2} \left(e^{ixt} - 1\right) \frac{\omega_j(-t)}{\|t\|^4} \, dt. \qquad (32)$$

Proof: Applying Fubini's theorem for Lebesgue integrals,

$$\begin{aligned}
\int_0^1 G_{j,q}(\tau x) \, d\tau &= \frac{1}{(2\pi)^2} \int_0^1 \left(\int_{\mathbb{R}^2} e^{i\tau xt} \frac{it_q\omega_j(-t)}{\|t\|^4} \, dt\right) d\tau \\
&= \frac{1}{(2\pi)^2} \int_{\mathbb{R}^2} \left(\int_0^1 e^{i\tau xt} \, d\tau\right) \frac{it_q\omega_j(-t)}{\|t\|^4} \, dt \qquad (33) \\
&= \frac{1}{(2\pi)^2} \int_{\mathbb{R}^2} \frac{e^{ixt} - 1}{x_1 t_1 + x_2 t_2} \frac{t_q\omega_j(-t)}{\|t\|^4} \, dt,
\end{aligned}$$

where $q \in \{1, 2\}$. Thus (32) follows from (31) and (33). $\square$

Note that formula (32) has been generalized and taken as definition for L_j in the wider case of d-dimensional polyharmonic semi-cardinal splines treated in [2].

Proposition 2. For each $j \in \mathbb{Z} \times \mathbb{Z}_+$, the function L_j satisfies the "semi-cardinal Lagrange conditions" (9).

Proof: Letting $x = k \in \mathbb{Z} \times \mathbb{Z}_+$ in the integral of (32) and using Lebesgue's dominated convergence theorem, we have

$$
\int_{\mathbf{R}^2} (e^{ikt} - 1) \frac{\omega_j(-t)}{\|t\|^4} \, dt = \sum_{l \in \mathbb{Z}^2} \int_{Q^2} (e^{-ikt} - 1) \frac{\omega_j(t)}{\|t - 2\pi l\|^4} \, dt
$$
$$
= \int_{Q^2} (e^{-ikt} - 1) \, \omega_j(t) \sum_{l \in \mathbb{Z}^2} \|t - 2\pi l\|^{-4} \, dt. \tag{34}
$$

At this stage, unlike the cardinal case, further work is needed beyond the above periodization argument in order to obtain the semi-cardinal Lagrange conditions (9). Note that (5), (10) and (12) imply

$$
\omega_j(t) \sum_{l \in \mathbb{Z}^2} \|t - 2\pi l\|^{-4} = \frac{\omega_j(t)}{\omega(t)} = \frac{P_+\{e^{ijt}\omega_-(t)\}}{\omega_-(t)}, \quad t \in Q^2 \setminus \{0\}. \tag{35}
$$

Further, by expansion (11),

$$
P_+\{e^{ijt}\omega_-(t)\} = e^{ijt}\omega_-(t) - \sum_{n=1}^{\infty} \sum_{l=-\infty}^{\infty} \beta^-_{(l,n+j_2)} e^{-ilt_1} e^{-int_2}. \tag{36}
$$

Thus (32), (34)–(36) and the definition (27) for ω_- imply

$$
L_j(k) = \delta_{jk} - \frac{1}{(2\pi)^2} \int_{Q^2} (e^{-ikt} - 1) \frac{\sum_{n=1}^{\infty} \sum_{l=-\infty}^{\infty} \beta^-_{(l,n+j_2)} e^{-ilt_1} e^{-int_2}}{[1 - \rho(t_1)e^{-it_2}]^2 \Omega_-(t)} \, dt. \tag{37}
$$

Treating the last integrand as a function of t_2 that depends on the parameter t_1, we note that, for each $t_1 \neq 0$, we have $0 < \rho(t_1) < 1$ (cf. (22),(23)), which ensures the expansion

$$
[1 - \rho(t_1)e^{-it_2}]^{-2} = \sum_{n=0}^{\infty} (n+1)[\rho(t_1)]^n e^{-int_2}. \tag{38}
$$

Since the Fourier expansion of $[\Omega_-(t)]^{-1}$ can also be rearranged as an absolutely convergent series of non-negative powers of e^{-it_2}, we deduce that, for each $t_1 \neq 0$, the fractional expression in the integrand of (37) admits an absolutely convergent expansion in *strictly* positive powers of e^{-it_2},

$$
\frac{\sum_{n=1}^{\infty} \sum_{l=-\infty}^{\infty} \beta^-_{(l,n+j_2)} e^{-ilt_1} e^{-int_2}}{[1 - \rho(t_1)e^{-it_2}]^2 \Omega_-(t)} =: \sum_{n=1}^{\infty} \Upsilon_n(t_1) e^{-int_2}, \tag{39}
$$

where each Υ_n also depends on j. Consequently, for $t_1 \neq 0$ we have

$$
\int_Q (e^{-ikt} - 1) \sum_{n=1}^{\infty} \Upsilon_n(t_1) e^{-int_2} \, dt_2
$$
$$
= \sum_{n=1}^{\infty} \Upsilon_n(t_1) \int_Q (e^{-ik_2 t_2} e^{-ik_1 t_1} - 1) e^{-int_2} \, dt_2 = 0,
$$
(40)

since $k_2 \geq 0$ and $n \geq 1$. Applying Fubini's theorem for Lebesgue integrals, it follows from the last two displays that the integral of (37) is zero, which yields the desired semi-cardinal Lagrange conditions. $\square$

Remark 3. Note that an absolutely convergent expansion of the form (39) could no longer hold if ω_- contained the factor $1 - [\rho(t_1)]^{-1} e^{-it_2}$. This observation shows that (26) is the natural choice for the Wiener-Hopf factorization of the zero of ω.

To verify that each function L_j is a biharmonic semi-cardinal spline, we need to prove first that L_j is a tempered distribution. It is sufficient to show that L_j is of polynomial growth. This can be established directly from (29)–(31), but it is also a consequence of the next lemma.

Lemma 4. *There exists a constant $M > 0$ such that*

$$
|L_j(x)| \leq M(1 + |x_2|), \quad \forall \, x \in \mathbb{R}^2, \; \forall \, j \in \mathbb{Z} \times \mathbb{Z}_+,
$$
(41)

where x_2 is the second component of x.

Proof: For any $x \in \mathbb{R}^2$ and any $j \in \mathbb{Z} \times \mathbb{Z}_+$, we can choose a multi-integer $m_{j,x} \in \mathbb{Z} \times \mathbb{Z}_+$, $m_{j,x} \neq j$, such that

$$
\|x - m_{j,x}\| \leq \max\{1, \sqrt{1 + x_2^2}\} \leq 1 + |x_2|.
$$
(42)

Then formula (32) and the property $L_j(m_{j,x}) = 0$ imply

$$
L_j(x) = L_j(x) - L_j(m_{j,x})
$$
$$
= \frac{1}{(2\pi)^2} \int_{\mathbb{R}^2} \left(e^{ixt} - e^{im_{j,x}t} \right) \frac{\omega_j(-t)}{\|t\|^4} \, dt.
$$
(43)

Since, by (42),

$$
\left| e^{ixt} - e^{im_{j,x}t} \right| = \left| e^{i(x - m_{j,x})t} - 1 \right| \leq (1 + |x_2|)\|t\|, \quad \forall \, t \in \mathbb{R}^2,
$$
(44)

using the inequalities (29) we obtain the estimate

$$
|L_j(x)| \leq \frac{(1 + |x_2|)M_2}{(2\pi)^2} \int_{\|t\| \leq 1} \|t\|^{-1} \, dt + \frac{2M_1}{(2\pi)^2} \int_{\|t\| > 1} \|t\|^{-4} \, dt.
$$
(45)

Thus (41) is proved. □

Proposition 5. *For each $j \in \mathbb{Z} \times \mathbb{Z}_+$, the function L_j is a biharmonic semi-cardinal spline with the additional property that all of its distributional partial derivatives of order two are square integrable over $\mathbb{R}^2$.*

Proof: From (30), it follows that, for $q \in \{1, 2\}$, the Fourier transform of the tempered distribution $\frac{\partial L_j}{\partial x_q}$ coincides with the L^1-function $it_q \omega_j(-t) \|t\|^{-4}$ (defined everywhere on $\mathbb{R}^2$ except 0). Therefore, using the hat to denote the Fourier transform, we have the following identities in the sense of distributions:

$$\widehat{(\Delta^2 L_j)} = -\|t\|^2 \widehat{(\Delta L_j)} = -\|t\|^2 \left[it_1 \widehat{\left(\frac{\partial L_j}{\partial x_1}\right)} + it_2 \widehat{\left(\frac{\partial L_j}{\partial x_2}\right)} \right] = \omega_j(-t).$$
(46)

Since the Fourier series expansion of the periodic distribution ω_j is given by (24), taking inverse Fourier transforms in (46) we obtain

$$\Delta^2 L_j = \sum_{k \in \mathbb{Z} \times \mathbb{Z}_+} b_{jk} \delta_k,$$
(47)

where δ_k is the Dirac distribution centred in k and the series converges in the sense of distributions. Thus $\Delta^2 L_j = 0$ on $\mathbb{R}^2 \setminus (\mathbb{Z} \times \mathbb{Z}_+)$, so each L_j is indeed a biharmonic semi-cardinal spline.

On the other hand, for any $q, p \in \{1, 2\}$, we obtain the distributional identities

$$\widehat{\left(\frac{\partial^2 L_j}{\partial x_q \partial x_p}\right)} = it_q \widehat{\left(\frac{\partial L_j}{\partial x_p}\right)} = -\frac{t_q t_p \omega_j(-t)}{\|t\|^4}.$$
(48)

Since the above right-hand side is square integrable over $\mathbb{R}^2$ due to (29), we deduce (Plancherel's theorem) that $\frac{\partial^2 L_j}{\partial x_q \partial x_p}$ is also square integrable over $\mathbb{R}^2$. This completes the proof of Proposition 5. □

We conclude this section with an open problem already formulated in [2] for the general polyharmonic case.

Conjecture 6. *Each function L_j constructed above is the unique biharmonic semi-cardinal spline satisfying both the interpolation conditions (9) and the additional property stated in Proposition 5.*

§4. Approximation Results

Let $\mathcal{L}_2^2(\mathbb{R}^2)$ be the space of tempered distributions all whose partial derivatives of order 2 are square integrable functions over $\mathbb{R}^2$. This space is endowed with the natural seminorm

$$|g|_{2,2} := \left\{ \int_{\mathbf{R}^2} \left(\left| \frac{\partial^2 g}{\partial x_1^2}(x) \right|^2 + 2 \left| \frac{\partial^2 g}{\partial x_1 \partial x_2}(x) \right|^2 + \left| \frac{\partial^2 g}{\partial x_2^2}(x) \right|^2 \right) dx \right\}^{1/2},$$

(49)

for each $g \in \mathcal{L}_2^2(\mathbf{R}^2)$. It is well-known that any element of $\mathcal{L}_2^2(\mathbf{R}^2)$ is in fact a continuous function and that convergence in the seminorm $|\cdot|_{2,2}$ implies, modulo linear polynomials, uniform convergence on compact sets of $\mathbf{R}^2$ (see Duchon [5]).

It follows by Proposition 5 that $L_j \in \mathcal{L}_2^2(\mathbf{R}^2)$, for each $j \in \mathbb{Z} \times \mathbb{Z}_+$. Further, it is proved in [2, Theorem 3.4] that, if the data sequence $\{f_j\}_{j \in \mathbb{Z} \times \mathbb{Z}_+}$ is absolutely summable, then the series s defined by (8), which converges absolutely and uniformly over compact sets of $\mathbf{R}^2$, is a biharmonic semi-cardinal spline satisfying the interpolation conditions (7), and $s \in \mathcal{L}_2^2(\mathbf{R}^2)$.

Our main goal in this section is to show that the cardinal scheme (2) can be approximated in the seminorm of $\mathcal{L}_2^2(\mathbf{R}^2)$ by a sequence of suitable semi-cardinal schemes of the type (8). We start with a result concerning the building blocks of these schemes.

Theorem 7. *If $j = (j_1, j_2) \in \mathbb{Z} \times \mathbb{Z}_+$, then*

$$\lim_{j_2 \to \infty} |L(\cdot - j) - L_j|_{2,2} = 0,$$

(50)

independently of the parameter j_1.

Proof: Firstly we notice that definition (4) of L implies that the distributional Fourier transform of $\frac{\partial^2 L}{\partial x_q \partial x_p}$ coincides with the L^2-function $-t_q t_p \omega(t) \|t\|^{-4}$, for any $q, p \in \{1, 2\}$. Thus $L \in \mathcal{L}_2^2(\mathbf{R}^2)$. By Plancherel's theorem (see [6, Theorem 9.2.2]), it follows from (48) that the square of the Duchon seminorm of $L(\cdot - j) - L_j$ is given by the formula

$$|L(\cdot - j) - L_j|_{2,2}^2 = \frac{1}{(2\pi)^2} \int_{\mathbf{R}^2} \frac{|e^{ijt} \omega(t) - \omega_j(t)|^2}{\|t\|^4} dt.$$

(51)

Next, using formulas (10), (12) and (27), we have

$$e^{ijt} \omega(t) - \omega_j(t) = [\rho^{-1}(t_1) - e^{it_2}]^2 \Omega_+(t) \sum_{n=1}^{\infty} \sum_{l=-\infty}^{\infty} \beta_{(l,n+j_2)}^- e^{-ilt_1} e^{-int_2},$$

(52)

which implies

$$|e^{ijt} \omega(t) - \omega_j(t)| \le |\rho^{-1}(t_1) - e^{it_2}|^2 |\Omega_+(t)| \sum_{n=1}^{\infty} \sum_{l=-\infty}^{\infty} |\beta_{(l,n+j_2)}^-|.$$

(53)

Consequently, we obtain the estimate

$$|L(\cdot - j) - L_j|_{2,2} \le C \sum_{n=1}^{\infty} \sum_{l=-\infty}^{\infty} |\beta^-_{(l,n+j_2)}|, \qquad (54)$$

where

$$C := \frac{1}{2\pi} \left(\int_{\mathbf{R}^2} \frac{|\rho^{-1}(t_1) - e^{it_2}|^4}{\|t\|^4} \, dt \right)^{1/2} \max_{t \in [-\pi,\pi]^2} |\Omega_+(t)|. \qquad (55)$$

Since the absolute convergence of (11) implies

$$\lim_{j_2 \to \infty} \sum_{n=1}^{\infty} \sum_{l=-\infty}^{\infty} |\beta^-_{(l,n+j_2)}| = 0, \qquad (56)$$

the proof is complete. □

The next result can be deduced from Theorem 7 using the relationship between convergence in the seminorm $|\cdot|_{2,2}$ and uniform convergence on compact sets of $\mathbf{R}^2$, but for more clarity we give a short and direct proof below.

Theorem 8. *In the notation of the previous theorem, given any bounded set $\mathcal{B} \subset \mathbf{R}^2$, we have*

$$\lim_{j_2 \to \infty} \sup_{x \in \mathcal{B}} |L(x - j) - L_j(x)| = 0, \qquad (57)$$

independently of the parameter j_1.

Proof: From (3), (4) and (32) we obtain (after substituting $t \to -t$)

$$L(x - j) - L_j(x) = \frac{1}{(2\pi)^2} \int_{\mathbf{R}^2} \left(e^{-ixt} - 1 \right) \frac{e^{ijt}\omega(t) - \omega_j(t)}{\|t\|^4} \, dt. \qquad (58)$$

Therefore the inequality $|e^{-ixt} - 1| \le \|x\| \|t\|$ and (53) provide the estimate

$$|L(x - j) - L_j(x)| \le C' \|x\| \sum_{n=1}^{\infty} \sum_{l=-\infty}^{\infty} |\beta^-_{(l,n+j_2)}|, \qquad (59)$$

where

$$C' := \frac{1}{(2\pi)^2} \int_{\mathbf{R}^2} \frac{|\rho^{-1}(t_1) - e^{it_2}|^2 |\Omega_+(t)|}{\|t\|^3} \, dt. \qquad (60)$$

The required conclusion (57) now follows from (59) and (56). □

Before stating the main result, we observe that one consequence of our analysis so far is that the series (2) and (8) converge in the seminorm $|\cdot|_{2,2}$ if their corresponding data sequences are absolutely summable (the straightforward justification based on the uniform estimates (29) is left to the reader).

Theorem 9. *Assume that the sequence* $\{f_j\}_{j\in\mathbb{Z}^2} \subset \mathbb{R}$ *satisfies*

$$\sum_{j\in\mathbb{Z}^2} |f_j| =: A < \infty, \tag{61}$$

and let the biharmonic cardinal spline s *be defined by (2). For each* $n \in \mathbb{Z}_+$, *define the biharmonic semi-cardinal spline* s_n *by the formula*

$$s_n(x) := \sum_{j\in\mathbb{Z}\times\mathbb{Z}_{\geq -n}} f_j L_{j+(0,n)} \left(x + (0,n)\right), \qquad x \in \mathbb{R}^2, \tag{62}$$

where $\mathbb{Z}_{\geq -n} = \{-n, -n+1, \dots\}$. *Then*

$$\lim_{n\to\infty} |s - s_n|_{2,2} = 0. \tag{63}$$

Proof: By the convergence of the series (2) and (62) in the seminorm $|\cdot|_{2,2}$, we have

$$|s - s_n|_{2,2} \leq \sum_{j\in\mathbb{Z}\times\mathbb{Z}_{< -n}} |f_j||L(\cdot - j)|_{2,2} + \sum_{j\in\mathbb{Z}\times\mathbb{Z}_+} |f_{j-(0,n)}||L(\cdot - j) - L_j|_{2,2},$$
$$\tag{64}$$

where we have used the relation

$$|L(\cdot - j) - L_{j+(0,n)}\left(\cdot + (0,n)\right)|_{2,2} = |L(\cdot - j - (0,n)) - L_{j+(0,n)}|_{2,2}, \tag{65}$$

together with a re-indexing of the second sum.

The first sum of the right-hand side of (64) tends to zero when n tends to infinity, due to (61) and the fact that

$$|L(\cdot - j)|_{2,2} = |L|_{2,2}, \qquad j \in \mathbb{Z}^2. \tag{66}$$

We split the second sum in two parts. Firstly, we note that (54) implies

$$\sum_{j_1\in\mathbb{Z}} \sum_{0\leq j_2\leq n/2} |f_{j-(0,n)}||L(\cdot - j) - L_j|_{2,2} \leq C\Gamma \sum_{j\in\mathbb{Z}\times\mathbb{Z}_{\leq -n/2}} |f_j|, \tag{67}$$

where $\Gamma := \sum_{k\in\mathbb{Z}\times\mathbb{Z}_+} |\beta_k^-|$ and the right-hand side of (67) tends to zero for $n \to \infty$ due to (61). Next, from (54) and (61) we obtain the estimate

$$\sum_{j\in\mathbb{Z}\times\mathbb{Z}_{>n/2}} |f_{j-(0,n)}||L(\cdot - j) - L_j|_{2,2} \leq CA \sum_{k>n/2} \sum_{l=-\infty}^{\infty} |\beta_{(l,k)}^-|. \tag{68}$$

By the absolute convergence of (11), the last double sum tends to zero as n increases, which completes the proof. $\square$

We remark that the convergence of the sequence of biharmonic semi-cardinal splines $\{s_n\}_{n=0}^\infty$ to the cardinal spline (2) can also be established in the uniform norm on compact sets of $\mathbb{R}^2$. All the results of this section can be extended to the more general setting of [2].

References

1. Bejancu, A., A new approach to semi-cardinal spline interpolation, East J. Approx. **6**, No. 4 (2000), 447–463.

2. Bejancu, A., Polyharmonic spline interpolation on a semi-space lattice, East J. Approx. **6**, No. 4 (2000), 465–491.

3. Buhmann, M. D., Multivariate cardinal interpolation with radial-basis functions, Constr. Approx. **6** (1990), 225–255.

4. Buhmann, M. D., On quasi-interpolation with radial basis functions, J. Approx. Theory **72** (1993), 103–130.

5. Duchon, J., Interpolation des fonctions de deux variables suivant le principe de la flexion des plaques minces, Rev. Française Automat. Informat. Rech. Opér., Anal. Numer. **10**, No. 12 (1976), 5–12.

6. Friedlander, G., *Introduction to the Theory of Distributions*, Cambridge University Press, Cambridge, 1998, 2nd edition, with additional material by M. Joshi.

7. Madych, W. R., and S. A. Nelson, Polyharmonic cardinal splines, J. Approx. Theory **60** (1990), 141–156.

8. Schoenberg, I. J., Contributions to the problem of approximation of equidistant data by analytic functions, Quart. Appl. Math. **4** (1946), 45–99.

9. Schoenberg, I. J., Cardinal interpolation and spline functions VI. Semi-cardinal interpolation and quadrature formulae, J. Analyse Math. **27** (1974), 159–204.

Aurelian Bejancu
Department of Applied Mathematics
University of Leeds
Leeds LS2 9JT, UK
A.Bejancu@amsta.leeds.ac.uk

Gabor Frames with Arbitrary Windows

Kai Bittner and Charles K. Chui

Abstract. A sufficient condition under which a window function is guaranteed to generate a Gabor frame for sufficiently high sampling rates is given. Furthermore, an estimate for the frame bounds that depends only on decay and smoothness conditions of the window function is derived.

§1. Introduction

The investigation of frame bounds of Gabor frames is an important problem in time-frequency analysis. Let $d \in \mathbb{N}$. A Gabor system

$$g_{jk}(x) := g(x - aj)e^{2\pi i bkx}, \quad j, k \in \mathbb{Z}^d,$$

generated by a single window function $g \in L^2(\mathbb{R}^d)$ consists of time shifts and frequency shifts (or phase modulations) of g with sampling parameters $a, b > 0$, respectively. If $\{g_{jk}\}$ is a frame of $L^2(\mathbb{R}^d)$, i.e., if there exist constants A, B with $0 < A \leq B < \infty$ so that

$$A\|f\|_{L^2}^2 \leq \sum_{j,k \in \mathbb{Z}^d} \left|\langle f, g_{jk}\rangle\right|^2 \leq B\|f\|_{L^2}^2, \quad f \in L^2(\mathbb{R}^d),$$

then $\{g_{jk}\}$ is called a Gabor frame. To emphasize the dependence on the parameters a and b, we say that (g, a, b) generates a Gabor frame if $\{g_{jk}\}$ is a frame of $L^2(\mathbb{R}^d)$.

A fundamental result is that no (g, a, b) generates a frame if $ab > 1$ [2,11]. Moreover, $\{g_{jk}\}$ cannot be complete in this case. For $ab = 1$ there exist window functions $g \in L^2(\mathbb{R}^d)$ which generate Gabor frames. For these (and only these) parameters the frame is exact, i.e., $\{g_{jk}\}$ is a Riesz basis. But from the Balian-Low Theorem [1,2,9] it follows that if $(g, a, \frac{1}{a})$

Approximation Theory X: Wavelets, Splines, and Applications
Charles K. Chui, Larry L. Schumaker, and Joachim Stöckler (eds.), pp. 41–50.
Copyright ⓒ 2002 by Vanderbilt University Press, Nashville, TN.
ISBN 0-8265-1416-2.

generates a Gabor frame, then g has poor time-frequency localization. The most interesting case is therefore $ab < 1$. Then, for any choice of a, b there exist Gabor frames with excellent time-frequency localization as evidenced from the example of the Gaussian $g(x) = e^{-x^2}$ in [10,13]. In the literature, there is a wide range of results on Gabor analysis. For an introduction the reader is refered to [7], and for a more advanced overview, see [6].

Here we are interested in the existence of Gabor frames and on estimates of their frame bounds. Previous results for compactly supported or band limited windows can be found in [4]. In [3, Theorems 2.5 and 2.6], Daubechies has given more general conditions for the existence of Gabor frames and estimates for the corresponding frame bounds. Though these results play an important role in the investigation of Gabor frames, the conditions are rather technical and in general hard to verify for a particular g. Based on Daubechies' result, Heil and Walnut [8,14] found more practicable conditions under which (g, a, b) generates a Gabor frame. In particular, they showed that $\{g_{jk}\}$ is a Bessel family (meaning the existence of a finite upper frame bound B), if the window or its Fourier transform is contained in the Wiener space $W(\mathbb{R}^d)$. However, to secure a positive lower frame bound A as well (meaning a true frame), an additional condition which depends on a is needed. If $\frac{1}{ab}$ is an integer, then an explicit formula for the computation of frame bounds is given in [3, p. 981]. This result can be generalized to rational $\frac{1}{ab}$ (see [15]). Further generalizations can be found in [12].

In this paper we give a sufficient condition (see Theorem 3) on g, which does not depend on a or b, so that (g, a, b) generates a Gabor frame if a and b are sufficiently small. Our second result (to be stated as Theorem 6) enables us to choose proper values for a and b to obtain a Gabor frame with desired frame bounds A and B, under slightly stronger conditions on g. These results constitute an exact formulation of the following somewhat loose but commonly believed statement in Gabor analysis (see e.g. [7, Sec. 6.5]):

If g satisfies suitable smoothness and decay conditions and if $a, b > 0$ are sufficiently small, then (g, a, b) generates a Gabor frame.

This rather vague formulation is actually inaccurate, as is demonstrated by examples in [5]. Here, we will not only give sufficient conditions on the smoothness and rate of decay of g to obtain a Gabor frame, but will also show that such a frame can be arbitrarily close to a tight frame if a and b are sufficiently small. This property enables us to control the stability of the frame by increasing the sampling rate $\frac{1}{ab}$.

From general frame theory, it is known that (g, a, b) generates a Gabor frame with frame bounds A and B if and only if the frame operator defined by

$$Sf = \sum_{j,k \in \mathbf{Z}^d} \langle f, g_{jk} \rangle g_{jk} \tag{1}$$

satisfies $A\mathrm{I} \leq S \leq B\mathrm{I}$ (see e.g. [8, Theorem 2.1.3]). The frame operator can be regarded as a discretization of the reconstruction formula for the short-time Fourier transform $V_g f(t, \xi) = \int_{\mathbf{R}^d} f(x) \overline{g(x - t)} e^{-2\pi i x \xi} dx$, which is

$$f(x) = \frac{1}{\|g\|_{L^2}} \iint_{\mathbf{R}^{2d}} V_g f(t, \xi) g(x - t) e^{2\pi i x \xi} d\xi \, dt.$$

Indeed, using the midpoint rule in numerical integration, we obtain

$$\iint_{\mathbf{R}^{2d}} V_g f(t, \xi) g(x - t) e^{2\pi i x \xi} d\xi \, dt \sim (ab)^d \sum_{j,k \in \mathbf{Z}^d} \langle f, g_{jk} \rangle g_{jk}.$$

Hence, under certain smoothness and decay conditions on the integrand, it is natural to expect the quadrature formula to converge if $a, b \to 0$, i.e,

$$Sf(x) \to (ab)^{-1} \iint_{\mathbf{R}^{2d}} V_g f(t, \xi) g(x - t) e^{2\pi i x \xi} d\xi \, dt = \frac{\|g\|_{L^2}}{(ab)^d} f(x).$$

In what follows we will show that we can assure convergence for all $f \in L^2(\mathbf{R}^d)$ under certain mild conditions on g. In particular, the rate of convergence can be controlled by the smoothness and the decay conditions of the window $g \in L^2(\mathbf{R}^d)$. To achieve this goal we need to investigate the frame operator in more detail.

§2. A Sufficient Condition on Window Functions

In [8], Heil and Walnut showed that the frame operator is bounded if the window function g is in the Wiener space

$$W(\mathbf{R}^d) := \left\{ g \in L^\infty(\mathbf{R}^d) : \|g\|_W := \sum_{n \in \mathbf{Z}^d} \operatorname{ess\,sup}_{x \in [0,1]^d} |g(x + n)| < \infty \right\}.$$

Furthermore, for $g \in W(\mathbf{R}^d)$ the frame operator has the representation

$$Sf(x) = b^{-d} \sum_{m \in \mathbf{Z}^d} f(x + \tfrac{m}{b}) G_m(x), \tag{2}$$

where

$$G_m(x) := \sum_{n \in \mathbf{Z}^d} g(x + an) \overline{g(x + an + \tfrac{m}{b})}.$$

(See [8, Theorem 4.2.1] for the univariate case or [7, Theorem 6.3.2] for the multivariate formulation.) The following estimates for the frame bounds can be derived by using the representation of S in (2) (cf. [2,3,7,8]).

Theorem 1. (Daubechies [3], Heil-Walnut [8]). *Let $g \in W(\mathbb{R}^d)$. Suppose that*

$$\sum_{\substack{m \in \mathbb{Z}^d \\ m \neq 0}} \|G_m\|_\infty < \operatorname*{ess\,inf}_{x \in \mathbb{R}^d} |G_0(x)| \qquad (3)$$

and $\|G_0\|_\infty < \infty$. Then (g, a, b) generates a Gabor frame with frame bounds

$$A_1 := b^{-d} \left(\operatorname*{ess\,inf}_{x \in \mathbb{R}^d} |G_0(x)| - \sum_{\substack{m \in \mathbb{Z}^d \\ m \neq 0}} \|G_m\|_\infty \right) \qquad (4)$$

$$B_1 := b^{-d} \sum_{m \in \mathbb{Z}^d} \|G_m\|_\infty. \qquad (5)$$

In particular, if $g \in W(\mathbb{R}^d)$ satisfies

$$\operatorname*{ess\,inf}_{x \in \mathbb{R}^d} |G_0(x)| = \operatorname*{ess\,inf}_{x \in \mathbb{R}^d} \sum_{n \in \mathbb{Z}^d} |g(x + an)|^2 > 0 \qquad (6)$$

for some $a > 0$, then there always exists some $b_0 > 0$ so that (g, a, b) generates a Gabor frame for every $0 < b < b_0$.

For a complete proof of this result, we refer the reader to [7, Chap. 6]. Let us only mention that (4) and (5) follow from the representation (2) of the frame operator. The sufficiency of (6) is due to the following lemma which we will need for our investigations.

Lemma 2. (Walnut [14]). *If $g \in W(\mathbb{R}^d)$, then for any given $\eta > 0$, there is a $b_0 > 0$, depending only on g and η, such that for every $0 < b \leq b_0$ and $0 < a \leq 1$,*

$$\sum_{\substack{m \in \mathbb{Z}^d \\ m \neq 0}} \|G_m\|_\infty < a^{-d} \eta.$$

Although the original lemma [14, Lemma 2.1] is only stated for the univariate setting, its proof can be easily generalized to the multivariate case, as is done in [7, Lemma 6.5.2].

Theorem 1 implies that there are window functions g in $W(\mathbb{R}^d)$ that generate Gabor frames for suitable parameters $a, b > 0$, since it is easy to see that (6) is fulfilled for a large class of functions g for some a. Once g is fixed, one can determine a suitable translation parameter a, and then b_0 has to be found in dependence of a. One problem is that there is no known result for describing the function $b_0(a)$. From Theorem 1, it is not even clear if there exists a rectangle $R = (0, a_0) \times (0, b_0)$ so that (g, a, b) generates a frame for every $(a, b) \in R$. On the other hand, a desired result would be to describe a class of window functions which generate a Gabor frame for all sufficiently small $a, b > 0$ (cf. [7, Sec. 6.5]).

To motivate this approach, let us first make several observations. It is well known that if (g, a, b) generates a Gabor frame, then so does $(\hat{g}, b, a)$. This fact suggests that all conditions on g should be invariant under the Fourier transform. However, the conditions in Theorem 1 are not. Furthermore, $a^d \sum_{n \in \mathbb{Z}^d} |g(x + an)|^2$ can be interpreted as a quadrature formula for the integral $\|g\|_{L^2}^2 = \int_{\mathbb{R}^d} |g(x)|^2 \, dx$. Under suitable smoothness and decay conditions, we would expect these quadrature formulae to converge. It turns out that $\hat{g} \in W(\mathbb{R}^d)$ is already a sufficient condition for the convergence of the sum above. The first result in this paper is the following

Theorem 3. *If g and $\hat{g}$ are contained in $W(\mathbb{R}^d)$, then there exist values $a_0, b_0 > 0$ so that (g, a, b) generates a Gabor frame for all $0 < a < a_0$ and $0 < b < b_0$.*

Proof: Let $\varepsilon_1 > 0$ and $\varepsilon_2 > 0$ be given such that $\varepsilon_1 + \varepsilon_2 \leq \|g\|_{L^2}^2$. The Poisson summation formula implies that

$$G_0 = \sum_{n \in \mathbb{Z}^d} |g(x + an)|^2 = a^{-d} \sum_{k \in \mathbb{Z}^d} \hat{\gamma}(\tfrac{k}{a}) \, e^{2\pi i \frac{k}{a} x}$$

with $\gamma(x) = |g(x)|^2$. Note that $g, \hat{g} \in W(\mathbb{R}^d)$ guarantee pointwise and absolute convergence of the above sums. By Parseval's identity, we have

$$\hat{\gamma}(\eta) = \int_{\mathbb{R}^d} g(x) \overline{g(x)} \, e^{-2\pi i x \eta} \, dx = \int_{\mathbb{R}^d} \hat{g}(\xi) \overline{\hat{g}(\xi - \eta)} \, d\xi,$$

and hence,

$$\|g\|_{L^2}^2 - a^d G_0(x) = \sum_{\substack{k \in \mathbb{Z}^d \\ k \neq 0}} \int_{\mathbb{R}^d} \hat{g}(\xi) \overline{\hat{g}(\xi + \tfrac{k}{a})} \, d\xi \, e^{2\pi i \frac{k}{a} x}.$$

Taking the absolute value, we obtain

$$\left| \|g\|_{L^2}^2 - a^d G_0(x) \right| \leq \sum_{\substack{k \in \mathbb{Z}^d \\ k \neq 0}} \left| \int_0^1 \sum_{n \in \mathbb{Z}^d} \hat{g}(\xi + n) \overline{\hat{g}(\xi + n + \tfrac{k}{a})} \, d\xi \right|$$

$$\leq \sum_{\substack{k \in \mathbb{Z}^d \\ k \neq 0}} \operatorname*{ess\,sup}_{\xi \in [0,1]} \left| \sum_{n \in \mathbb{Z}^d} \hat{g}(\xi + n) \overline{\hat{g}(\xi + n + \tfrac{k}{a})} \right|. \tag{7}$$

Since $\hat{g} \in W(\mathbb{R}^d)$, Lemma 2 (with g replaced by $\hat{g}$) implies that there is an a_0, depending only on $\hat{g}$ and $\varepsilon_1 > 0$, so that for every $0 < a \leq a_0$

$$\sum_{\substack{k \in \mathbb{Z}^d \\ k \neq 0}} \operatorname*{ess\,sup}_{\xi \in \mathbb{R}^d} \left| \sum_{n \in \mathbb{Z}^d} \hat{g}(\xi + n) \overline{\hat{g}(\xi + n + \tfrac{k}{a})} \right| \leq \varepsilon_1.$$

Thus, we obtain

$$\frac{\|g\|_{L^2}^2 - \varepsilon_1}{a^d} \leq G_0(x) \leq \frac{\|g\|_{L^2}^2 + \varepsilon_1}{a^d}.$$

Again from Lemma 2, we see that for $g \in W(\mathbb{R}^d)$ and for $\varepsilon_2 > 0$, there is some b_0 depending only on g and $\varepsilon_2 > 0$, such that for every $0 < b \leq b_0$ and $0 < a \leq 1$, we have

$$\sum_{\substack{m \in \mathbb{Z}^d \\ m \neq 0}} \|G_m\|_\infty < a^{-d}\varepsilon_2.$$

From the assumption $\varepsilon_1 + \varepsilon_2 \leq \|g\|_{L^2}^2$, it follows that

$$\operatorname*{ess\,inf}_{x \in \mathbb{R}^d} |G_0(x)| - \sum_{\substack{m \in \mathbb{Z}^d \\ m \neq 0}} \|G_m\|_\infty > \frac{\|g\|_{L^2}^2 - \varepsilon_1 - \varepsilon_2}{a^d} \geq 0,$$

i.e., (3) is satisfied for all $0 < a < a_0$ and $0 < b < b_0$. Now, by Theorem 1 we know that (g, a, b) generates a Gabor frame and the proof of Theorem 3 is complete. $\square$

From the proof one sees immediately that the frame bounds in Theorem 1 can be estimated by

$$\frac{\|g\|_{L^2}^2 - \varepsilon_1 - \varepsilon_2}{(ab)^d} \leq A_1 \leq B_1 \leq \frac{\|g\|_{L^2}^2 + \varepsilon_1 + \varepsilon_2}{(ab)^d} \tag{8}$$

for any $\varepsilon_1, \varepsilon_2 > 0$, if a and b are sufficiently small. That is, we can bring the ratio $\frac{B_1}{A_1}$ as close to 1 as we want by increasing the sampling rate $\frac{1}{ab}$. However, it is not obvious how one should choose a and b to obtain a Gabor frame with desired frame bounds corresponding to a given function g.

§3. Frame Bound Estimates

In the previous section we have shown that for $g, \hat{g} \in W(\mathbb{R}^d)$ the ratios $\frac{B_1}{A_1} \to 1$ if $a, b \to 0$. In order to describe the rate of convergence, we now consider window functions with faster decay and stronger smoothness condition. In other words, we introduce the spaces

$$W^s(\mathbb{R}^d) := \left\{ f : w_s f \in W(\mathbb{R}^d) \right\}, \quad s \geq 0,$$

with the weight function $w_s(x) := (1 + |x|)^s$, and norm given by $\|f\|_{W^s} := \|w_s f\|_W$.

Lemma 4. Let $g \in W^\beta(\mathbb{R}^d)$, $\beta > 0$. Then

$$\sum_{\substack{m \in \mathbb{Z}^d \\ m \neq 0}} \|G_m\|_\infty \leq \lceil \tfrac{1}{a} \rceil^d \lceil b \rceil^d b^\beta \|g\|^2_{W^\beta}.$$

Proof: With $1 + |x - y| \leq (1 + |x|)(1 + |y|)$, we have

$$\sum_{\substack{m \in \mathbb{Z}^d \\ m \neq 0}} \|G_m\|_\infty \leq b^\beta \sum_{m \in \mathbb{Z}^d} (1 + |\tfrac{m}{b}|)^\beta \|G_m\|_\infty$$

$$\leq b^\beta \sum_{m \in \mathbb{Z}^d} \operatorname*{ess\,sup}_{x \in \mathbb{R}^d} (1 + |\tfrac{m}{b}|)^\beta \sum_{n \in \mathbb{Z}^d} |g(x + an)| \, |g(x + an + \tfrac{m}{b})|$$

$$\leq b^\beta \sum_{m \in \mathbb{Z}^d} \operatorname*{ess\,sup}_{x \in \mathbb{R}^d} \sum_{n \in \mathbb{Z}^d} (1 + |x + an|)^\beta |g(x + an)|$$

$$\times \; (1 + |x + an + \tfrac{m}{b}|)^\beta |g(x + an + \tfrac{m}{b})|$$

$$\leq b^\beta \operatorname*{ess\,sup}_{x \in \mathbb{R}^d} \sum_{n \in \mathbb{Z}^d} |w_\beta g(x + an)| \sum_{m \in \mathbb{Z}^d} |w_\beta g(x + an + \tfrac{m}{b})|.$$

For $\alpha > 0$, any function $f \in W(\mathbb{R}^d)$ satisfies the inequality

$$\operatorname*{ess\,sup}_{x \in \mathbb{R}^d} \sum_{n \in \mathbb{Z}^d} |f(x + \tfrac{n}{\alpha})| \leq \lceil \alpha \rceil^d \|f\|_W.$$

Since $w_\beta g \in W(\mathbb{R}^d)$, it follows that

$$\sum_{\substack{m \in \mathbb{Z}^d \\ m \neq 0}} \|G_m\|_\infty \leq \lceil \tfrac{1}{a} \rceil^d \lceil b \rceil^d b^\beta \|g\|^2_{W^\beta}. \quad \square$$

Lemma 5. Let $\hat{g} \in W^\alpha(\mathbb{R}^d)$, $\alpha > 0$. Then

$$\left| \|g\|^2_{L^2} - a^d G_0(x) \right| \leq \lceil a \rceil^d a^\alpha \|\hat{g}\|^2_{W^\alpha}.$$

Proof: From the proof of Theorem 3 (particularly inequality (7)), we already know that

$$\left| \|g\|^2_{L^2} - a^d G_0(x) \right| \leq \sum_{\substack{k \in \mathbb{Z}^d \\ k \neq 0}} \operatorname*{ess\,sup}_{\xi \in [0,1]} \left| \sum_{n \in \mathbb{Z}^d} \hat{g}(\xi + n) \, \overline{\hat{g}(\xi + n + \tfrac{k}{a})} \right|.$$

Now Lemma 5 follows immediately from Lemma 4 with g replaced by $\hat{g}$.
$\square$

Theorem 6. Let $g \in W^\beta(\mathbb{R}^d)$, $\beta > 0$, and $\hat{g} \in W^\alpha(\mathbb{R}^d)$, $\alpha > 0$. Suppose that the lattice parameters $0 < a, b \le 1$ satisfy

$$a^\alpha \|\hat{g}\|_{W^\alpha}^2 + (1 + a)^d b^\beta \|g\|_{W^\beta}^2 < \|g\|_{L^2}^2. \tag{9}$$

Then (g, a, b) generates a Gabor frame with frame bounds

$$A_2 := (ab)^{-d} \left(\|g\|_{L^2}^2 - a^\alpha \|\hat{g}\|_{W^\alpha}^2 - (1 + a)^d b^\beta \|g\|_{W^\beta}^2 \right), \tag{10}$$

$$B_2 := (ab)^{-d} \left(\|g\|_{L^2}^2 + a^\alpha \|\hat{g}\|_{W^\alpha}^2 + (1 + a)^d b^\beta \|g\|_{W^\beta}^2 \right). \tag{11}$$

Proof: From Lemma 5, it follows that

$$\|g\|_{L^2}^2 - \lceil a \rceil^d a^\alpha \|\hat{g}\|_{W^\alpha}^2 \le a^d G_0(x) \le \|g\|_{L^2}^2 + \lceil a \rceil^d a^\alpha \|\hat{g}\|_{W^\alpha}^2.$$

Since $\lceil a \rceil = 1$, we have

$$a^{-d} \left(\|g\|_{L^2}^2 - a^\alpha \|\hat{g}\|_{W^\alpha}^2 \right) \le G_0(x) \le a^{-d} \left(\|g\|_{L^2}^2 + a^\alpha \|\hat{g}\|_{W^\alpha}^2 \right), \tag{12}$$

Also, since $\lceil b \rceil = 1$ and $\lceil \frac{1}{a} \rceil \le 1 + \frac{1}{a}$, we obtain from Lemma 4 that

$$\sum_{\substack{m \in \mathbb{Z}^d \\ m \ne 0}} \|G_m\|_\infty \le a^{-d}(1 + a)^d b^\beta \|g\|_{W^\beta}^2. \tag{13}$$

Thus, if condition (9) holds, then (3) is satisfied and (g, a, b) generates a Gabor frame. Applying (12) and (13) to (4) and (5), we may conclude that $0 < A_2 \le A_1 \le B_1 \le B_2 < \infty$, i.e., A_2 and B_2 are frame bounds for $\{g_{jk}\}$. $\square$

Obviously, the assumption $g \in W^s(\mathbb{R}^d)$, $s > 0$, is a decay condition on g. Fast decay of the Fourier transform (e.g. if $\hat{g} \in W^s(\mathbb{R}^d)$, $s > 0$) implies a certain smoothness condition on g. That is, for sufficiently smooth functions g which decay sufficiently fast we can always find parameters a and b, so that (g, a, b) generates a Gabor frame.

Acknowledgments. The research of K. Bittner was supported by the Deutsche Forschungsgemeinschaft under the Emmy Noether program. The research of C. K. Chui was partially supported by NSF Grants CCR-9988289 and CCR-0098331, and ARO Grant DAAD 19-01-1-0512.

References

1. Balian, R., Un principe d'incertitude fort en théorie du signal ou en mécanique quantique, C. R. Acad. Sci. Paris **292** (1981), 1357–1362.

2. Daubechies, I., *Ten Lectures on Wavelets*, SIAM, Philadelphia, PA, 1992.

3. Daubechies, I., The wavelet transform, time-frequency localization and signal analysis, IEEE Trans. Inform. Theory **36** (1990), 961–1005.

4. Daubechies, I., A. Grossmann, and Y. Meyer, Painless nonorthogonal expansions, J. Math. Phys. **27** (1986), 1271–1283.

5. Feichtinger, H. G., and A. J. E. M. Janssen, Validity of WH-frame bound conditions depends on lattice parameters, Appl. Comput. Harmonic Anal. **8** (2000), 104–112.

6. Feichtinger, H. G., and T. Strohmer (eds.), *Gabor Analysis and Algorithms: Theory and Applications*, Birkhäuser, Boston, 1998.

7. Gröchenig, K., *Foundations of Time-Frequency Analysis*, Birkhäuser, Boston, 2001.

8. Heil, C. E., and D. F. Walnut, Continuous and discrete wavelet transforms, SIAM Review **31** (1989), 628–666.

9. Low, F., Complete sets of wave packets, in *A Passion for Physics – Essays in Honor of Geoffrey Chew*, C. DeTar. (ed.), World Scientific, Singapore, 1985.

10. Lyubarskii, Y. I., Frames in the Bargmann space of entire functions, in *Advances in Soviet Mathematics* Vol. 11, B. Y. Levin (ed.), Springer-Verlag, Berlin, 1992.

11. Rieffel, M. A., Von Neumann algebras associated with pairs of lattices in Lie groups, Math. Ann. **257** (1981), 403–418.

12. Ron, A., and Z. Shen, Weyl-Heisenberg frames and Riesz bases in $L_2(\mathbb{R}^d)$, Duke Math. J. **89** (1997), 237–282.

13. Seip, K., and R. Wallstén, Density theorems for sampling and interpolation in the Bargmann-Fock space II, J. Reine Angew. Math. **429** (1992), 107–113.

14. Walnut, D. F., Continuity properties of the Gabor frame operator, J. Math. Anal. Appl. **165** (1992), 479–504.

15. Zibulski, M., and Y. Zeevi, Oversampling in the Gabor scheme, IEEE Trans. Signal Processing **41** (1993), 2679–2687.

Kai Bittner
Department of Mathematics and Computer Science
University of Missouri – St. Louis
St. Louis, MO 63121
bittner@math.umsl.edu
http://www.math.umsl.edu/~bittner

Charles K. Chui
Department of Mathematics and Computer Science
University of Missouri – St. Louis
St. Louis, MO 63121
and
Department of Statistics
Stanford University
Stanford, CA 94305
cchui@stat.stanford.edu

Nonseparable Wavelet Packets

Lasse Borup and Morten Nielsen

Abstract. We introduce wavelet packets in the setting of a multiresolution analysis of $L^2(\mathbb{R}^d)$ generated by an arbitrary dilation matrix A satisfying $|\det A| = 2$, and note that the best-basis algorithm of wavelet packets in the standard one dimensional case can be generalized to this multidimensional setting. We then consider the application of the best basis algorithm in the two-dimensional case for a specific dilation matrix to image compression. The performance of the algorithm is compared to the separable wavelet algorithm for twelve test images.

§1. Introduction

Wavelet analysis was originally introduced in order to improve seismic signal processing by switching from short-time Fourier analysis to new algorithms better suited to detect and analyze abrupt changes in signals. It corresponds to a decomposition of phase space in which the trade-off between time and frequency localization has been chosen to provide better and better time localization at high frequencies in return for poor frequency localization. This makes the analysis well adapted to the study of transient phenomena and has proven a very successful approach to many problems in signal processing, numerical analysis, and quantum mechanics. Nevertheless, for stationary signals, wavelet analysis is outperformed by short-time Fourier analysis. Wavelet packets were introduced by R. Coifman, Y. Meyer, and M. V. Wickerhauser [2] to improve the poor frequency localization of wavelet bases at high frequencies and thereby provide a more efficient decomposition of signals containing both transient and stationary components.

So far most work on wavelet packets has been done in one dimension or using separable wavelet packets in higher dimensions (i.e. tensor products of one dimensional wavelet packets). However, separable wavelet and

Approximation Theory X: Wavelets, Splines, and Applications   51
Charles K. Chui, Larry L. Schumaker, and Joachim Stöckler (eds.), pp. 51–61.

wavelet packet bases both have several drawbacks for the application to fields like image analysis since they impose an unavoidable line structure on the plane. For example, the zero set of a separable wavelet packet at high frequencies will contain a large number (same order of magnitude as the frequency) of horizontal and vertical lines that may create artifacts in the reconstructed image. Another potential problem is in the Fourier domain where separable two-dimensional wavelet packets have four characteristic peaks making it hard to selectively localize a unique frequency. R. Coifman and F. Meyer introduced the so-called Brushlets in [3] to remove the "uncertainty" in frequency localization, however the Brushlets are essentially Fourier transforms of smooth local trigonometric bases and are therefore no longer functions associated with a multiresolution structure.

In the present paper we will introduce wavelet packets on $\mathbb{R}^d$ associated with a multiresolution analysis generated by a general dilation matrix A with determinant ± 2. The (nonseparable) wavelets associated with such structures have been investigated in several recent papers [1] and they have also been applied to image analysis. As in the one dimensional case, wavelet packet analysis extends the wavelet expansion of a signal to a whole library of orthonormal expansions providing a more adaptable scheme to effectively represent the signal.

The structure of the paper is as follows. In Section 2 we introduce wavelet packets associated with the class of multiresolution analyses of $\mathbb{R}^d$ for which there are associated wavelet bases generated by only one wavelet. Such functions provide the same large number of orthonormal bases as wavelet packets in one-dimension do, and should provide a good platform for doing image analysis using the well known best basis algorithm of Coifman and Wickerhauser. Section 3 contains the numerical results we obtained from applying the algorithm, in the two-dimensional case with a specific dilation matrix, to image compression. We use twelve "standard" test images to compare the effectiveness of the algorithm to the associated nonseparable wavelet decomposition at different compression rates.

§2. Non-stationary Wavelet Packets for a General Dilation Matrix

We begin by recalling some facts about multiresolution analyses associated with a general dilation matrix that we will use later in this section to define the wavelet packets we have in mind. The reader can find a more extensive discussion of the topic in [4].

Let A be a $(d \times d)$-matrix such that $A : \mathbb{Z}^d \to \mathbb{Z}^d$. If the eigenvalues of A all have absolute value strictly greater than 1 then we call A a dilation matrix. We can define a multiresolution analysis associated with such A:

Definition 1. *A* multiresolution associated with a dilation matrix A *is a sequence of closed subspaces* $(V_j)_{j \in \mathbb{Z}}$ *of* $L^2(\mathbb{R}^d)$ *satisfying*

(i) $V_j \subset V_{j+1}, \quad \forall j \in \mathbb{Z}$,

(ii) $\overline{\bigcup_{j \in \mathbb{Z}} V_j} = L^2(\mathbb{R}^d)$ *and* $\bigcap_{j \in \mathbb{Z}} V_j = \{0\}$,

(iii) $f \in V_j \Leftrightarrow f(A \cdot) \in V_{j+1}, \quad \forall j \in \mathbb{Z}$,

(iv) *there exists a function* $\phi \in V_0$, *called a* scaling function, *such that the system* $\{\phi(\cdot - \gamma)\}_{\gamma \in \mathbb{Z}^d}$ *is an orthonormal basis for* V_0.

The wavelet spaces W_j associated with such a multiresolution analysis are given by $W_j = V_{j+1} \cap V_j^{\perp}$, and one can easily check that $f \in W_j \Leftrightarrow f(A \cdot) \in W_{j+1}$ and $L^2(\mathbb{R}^d) = \bigoplus_{j \in \mathbb{Z}} W_j$. A family of wavelets associated with the multiresolution analysis is a collection of s functions $\{\Psi^r\}_{r=1}^s$ for which $\{\Psi^r(\cdot - \gamma) : r = 1, \ldots, s, \gamma \in \mathbb{Z}^d\}$ is an orthonormal basis for W_0. Suppose $|\det A| = q$. It turns out that the number of wavelets needed to generate such a basis for W_0 is exactly $q - 1$. This makes the case $|\det A| = 2$ especially interesting since the wavelet basis is generated by only one function just as in the one-dimensional case.

Let $\{V_j\}_{j \in \mathbb{Z}}$ be a multiresolution analysis of $L^2(\mathbb{R}^d)$ associated with a dilation matrix A satisfying $|\det A| = 2$. Suppose (Φ, Ψ) is an associated scaling function/wavelet pair. Then there exist $2\pi \mathbb{Z}^d$-periodic functions m_0 and m_1 such that

$$\hat{\Phi}(\xi) = m_0(D\xi)\hat{\Phi}(D\xi)$$

$$\hat{\Psi}(\xi) = m_1(D\xi)\hat{\Phi}(D\xi),$$

with $D = (A^*)^{-1}$. Since $|\det A| = 2$ we can find $\Gamma \in \mathbb{Z}^d$ such that $\Gamma + \mathbb{Z}^d / A^* \mathbb{Z}^d$ satisfies $\mathbb{Z}^d = A^* \mathbb{Z}^d \cup (\Gamma + A^* \mathbb{Z}^d)$. Then it is easy to check that the matrix

$$\begin{bmatrix} m_0(\xi) & m_0(\xi + 2\pi D\Gamma) \\ m_1(\xi) & m_1(\xi + 2\pi D\Gamma) \end{bmatrix}$$

is unitary for a.a. $\xi \in \mathbb{R}^d$. This observation leads to the following definition. We let A and Γ be related as above.

Definition 2. *Let* m_0 *and* m_1 *be* $2\pi \mathbb{Z}^d$ *periodic functions for which*

$$\begin{bmatrix} m_0(\xi) & m_0(\xi + 2\pi D\Gamma) \\ m_1(\xi) & m_1(\xi + 2\pi D\Gamma) \end{bmatrix}$$

is unitary a.e. Then we call (m_0, m_1) *a pair of* orthogonal quadrature filters *associated with* (A, Γ).

We can now define the natural generalization of wavelet packets to the setting of a multiresolution analysis associated with a dilation matrix A with $|\det A| = 2$.

Definition 3. Let $\{(m_0^{(p)}, m_1^{(p)})\}_{p=1}^{\infty}$ be a sequence of orthogonal quadrature filters associated with (A, Γ). We define the basic nonstationary wavelet packets $\{w_n\}_{n=0}^{\infty}$ by $w_0 = \Phi$, $w_1 = \Psi$, and

$$\hat{w}_n(\xi) = \left[\prod_{j=1}^{k+1} m_{\epsilon_j}^{(k-j+2)} (D^j \xi) \right] \hat{\Phi}(D^{k+1}\xi)$$

for $2^k \leq n < 2^{k+1}$ with binary expansion $n = \sum_{j=1}^{k+1} \epsilon_j 2^{j-1}$.

Let us state two most important facts about the wavelet packets from the above definition. The two theorems below show how to extract orthonormal bases from the wavelet packet construction above, and thus gives us some new tools for signal and image processing. We have included the proofs for convenience. However, the reader should notice that everything works exactly as in the one-dimensional case; only the multiresolution structure matters.

Theorem 4. The basic wavelet packets

$$\{w_n(\cdot - k): \ 0 \leq n < 2^j, k \in \mathbb{Z}^d\}$$

form a basis for V_j. Furthermore,

$$\{w_n(\cdot - k): \ n \in \mathbb{N}_0, k \in \mathbb{Z}^d\}$$

form an orthonormal basis for $L^2(\mathbb{R}^d)$.

Proof: Let $\Omega_n = \overline{Span}\{w_n(\cdot - k)\}_{k \in \mathbb{Z}^d}$, and define $\delta f(x) = \sqrt{2}f(Ax)$. Using the QMF-condition it is not hard to verify that $\delta \Omega_n = \Omega_{2n} \oplus \Omega_{2n+1}$ (see eg. [4] p. 112). Thus,

$$\delta \Omega_0 \ominus \Omega_0 = \Omega_1$$
$$\delta^2 \Omega_0 \ominus \delta \Omega_0 = \delta \Omega_1 = \Omega_2 \oplus \Omega_3$$
$$\delta^3 \Omega_0 \ominus \delta^2 \Omega_0 = \delta \Omega_2 \oplus \delta \Omega_3 = \Omega_4 \oplus \Omega_5 \oplus \Omega_6 \oplus \Omega_7$$

$$\vdots$$

$$\delta^k \Omega_0 \ominus \delta^{k-1} \Omega_0 = \Omega_{2^{k-1}} \oplus \Omega_{2^{k-1}+1} \oplus \cdots \oplus \Omega_{2^k-1}.$$

By telescoping the above equalities we finally get the desired result

$$\delta^k \Omega_0 \equiv \delta^k V_0 = V_k = \Omega_0 \oplus \Omega_1 \oplus \cdots \oplus \Omega_{2^k-1},$$

and $\cup_{k \geq 0} V_k$ is dense in $L^2(\mathbb{R}^d)$ by the definition of a multiresolution analysis. $\square$

The above theorem can be generalized considerably. The following construction gives us a whole library of orthonormal bases each with different time-frequency properties. The collection of bases will be used for the best basis algorithm.

Theorem 5. *Let $\{w_n\}$ be a family of non-stationary wavelet packets associated with the dilation matrix A. For every partition P of $\mathbb{N}_0$ into sets of the form $I_{nj} = \{n2^j, \ldots, (n+1)2^j - 1\}$ with $n, j \in \mathbb{N}_0$, the family*

$$\{2^{j/2} w_n(A^j \cdot -k)\}_{k \in \mathbb{Z}^d, I_{nj} \in P}$$

is an orthonormal basis for $L^2(\mathbb{R}^d)$.

Proof: An argument similar to the one in Theorem 4 shows that

$$\delta^k \Omega_n = \Omega_{2^k n} \oplus \Omega_{2^k n+1} \oplus \cdots \oplus \Omega_{2^k(n+1)-1}.$$

Moreover, the functions $\{2^{j/2} w_n(A^j \cdot -q)\}_{q \in \mathbb{Z}^d}$ span the space $\delta^j \Omega_n$ and

$$\sum_{I_{nj} \in P} \delta^j \Omega_n = \bigoplus_{q \geq 0} \Omega_q = L^2(\mathbb{R}^d),$$

which proves the theorem. $\square$

2.1 The best basis algorithm

One of the main advantages of wavelet packets in the nonseparable setting as described above is that one can use the implementation of the best-basis algorithm of Coifman and Wickerhauser for the one dimensional wavelet packet case without any modification whatsoever. We will not describe the algorithm here but instead refer the reader to [2].

2.2 A family of nonseparable wavelets

The nonseparable wavelets we are going to use in the following section are examples of the family of wavelets constructed by E. Belogay and Y. Wang in [1]. Using the dilation matrix

$$A = \begin{bmatrix} 0 & 2 \\ 1 & 0 \end{bmatrix},$$

they constructed a family of nonseparable wavelets with arbitrary smoothness. We denote the corresponding family of coefficient masks of size $4(r-1) \times 2$, $r > 0$ by $\mathcal{C}_r$.

§3. Implementation

In this section we are going to test the performance of the nonseparable wavelet packets constructed using Belogay and Wang's family of wavelets. We will compare the performance with a wavelet transform using the same wavelets and with a standard separable wavelet packets construction using Daubechies univariate wavelets. The tests are made on the following 12 standard images

image	pixel size	image	pixel size
Lena	512×512	Plane	512×512
Barbara	512×512	Milkdrop	512×512
Goldhill	512×512	Kilauea	416×512
Baboon	512×512	Boats	512×512
Bridge	512×512	France	480×672
Peppers	512×512	Library	352×448

All the images are given with 8-bit gray-scale values. In Section 3.1 we describe in details the test on the image "Lena", followed by the results from the tests on the other images. In the following "Daub#" means Daubechies 1D filter of length #.

We have implemented the nonseparable best basis algorithm in MAT-LAB, using the same construction as in the wavelet toolbox "Uvi_wave" (see http://www.tsc.uvigo.es/~wavelets/uvi_wave.html). Since we have just one scaling function and mother wavelet in our nonseparable construction, only minor modification of the one dimensional best basis algorithm is needed. To get perfect reconstruction of a transformed image, we have chosen to use circular transformations, i.e. we represent an image as a periodic function. When we make a best basis search we only allow transformations down to a certain level k_{max}. If, say a particular image is of size $M \times N$ and $m \times n$ is the size of a given filter, then k_{max} must satisfy $M2^{-k_{max}} > m$ and $N2^{-k_{max}} > n$. This restriction only ensures that it is enough to copy the image once at the boundary. As cost function in the best basis algorithm we use Shannon's entropy function (other cost functions such as the ℓ_p-norm, $p \leq 1$, have been used as well, but the overall picture was the same).

3.1 Test results

To introduce the method used to test the nonseparable wavelet packets, we begin by describing a comparison of $\mathcal{C}_5$ and Daub20 when used on the image "Lena". On this image we do a best basis search, using $\mathcal{C}_5$ and Daub20, followed by hard thresholding keeping only a certain percentage of the coefficients. Then we perform the inverse transformation using only the remaining coefficients. Figure 1 (left) displays the signal to noise ratio

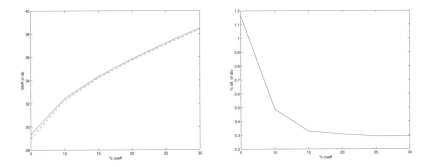

Fig. 1. Left: Error due to threshold using Best-WPT with $\mathcal{C}_5$ (solid) and Daub20 (dashed). Right: Relative improvement using $\mathcal{C}_5$ instead of Daub20.

(SNR) for different values of the threshold using the best basis wavelet packets transformation (best-WPT) with $\mathcal{C}_5$ and Daub20, respectively. The SNR is given by

$$\text{SNR} := 10 \log_{10} \left(\frac{\|x^{exact} - x^{thres}\|_{L^2}^2}{\|x^{exact}\|_{L^2}^2} \right),$$

where x^{exact} is the original image and x^{thres} is the reconstructed image after thresholding. Figure 1 (right) shows the relative improvement using $\mathcal{C}_5$ instead of Daub20, measured in SNR frame i.e.

$$100 \cdot \frac{\text{SNR}^{\mathcal{C}_5} - \text{SNR}^{\text{Daub20}}}{\text{SNR}^{\mathcal{C}_5}} \%.$$

As can be seen from the plot, the largest improvement is given at the highest compression rates.

To see if there is any visual difference between the result of using $\mathcal{C}_5$ and Daub20, we plot the reconstructed image at compression rate 1:50 using the two methods (see Figure 2). Notice that the left image seems less blurred than the right.

3.2 Remaining results

We now display the test results corresponding to the other test images. Shannon's entropy function is still used as cost function. We first test the improvement of using best wavelet packets transform instead of using an ordinary wavelet transform (WT).

The following table gives the relative improvement using best-WPT with $\mathcal{C}_5$ instead of a WT using $\mathcal{C}_5$. The top row shows the percentage of kept coefficients before reconstruction.

best–WPT, compression rate 1:50, SNR = 25.2649 db.   best–WPT, compression rate 1:50, SNR = 24.7839 db.

Fig. 2. Reconstruction of "Lena" with a compression rate of 1:50 using $\mathcal{C}_5$ (left) and Daub20 (right).

% coeff	1	2	5	10	15	20
Lena	2.146	2.209	2.058	1.400	0.990	0.770
Barbara	8.091	10.598	10.956	9.301	7.550	6.071
Goldhill	3.439	4.153	4.309	3.984	3.704	3.509
Baboon	2.093	3.029	4.418	4.931	4.823	4.502
Bridge	3.173	2.939	2.586	2.432	2.290	2.176
Peppers	5.099	4.167	2.034	1.114	0.896	0.732
Plane	3.502	3.978	2.642	1.968	1.548	1.269
Milkdrop	6.306	4.763	2.209	1.516	1.288	1.187
Kilauea	3.162	2.161	0.733	-0.156	-0.741	-1.208
Boats	3.484	3.872	4.326	4.073	3.904	3.637
France	4.163	4.874	3.320	0.795	0.618	0.233
Library	4.078	3.744	3.118	2.729	2.038	1.431

Notice that best-WPT performs better on all twelve images at the highest compression rates, compared to the WT. Especially there is a big difference on "Barbara" possibly due to the textures. In Figure 3 a 1:20 compressed version of "Barbara" is shown using best-WPT (left) and WT (right). Notice the different performances in reconstructing the pattern, say on the trousers.

As an apparent paradox, WT performs better on "Kilauea" at compression rate 1:10. This is because the algorithm chooses the best basis relative to a specific cost function different from the SNR-measure.

We now return to the test on "Lena" described above. The next table gives the relative improvements using best-WPT with $\mathcal{C}_5$ instead of Daub20.

Fig. 3. The image "Barbara" after a 1:20 compression using Best-WPT (left) and WT (right), both using $\mathcal{C}_5$.

% coeff	1	2	5	10	15	20
Lena	1.996	1.904	1.163	0.487	0.329	0.307
Barbara	1.004	0.774	0.332	-0.043	-0.307	-0.512
Goldhill	0.889	0.920	0.756	0.494	0.422	0.356
Baboon	0.596	0.743	0.619	0.382	0.266	0.153
Bridge	0.492	0.431	0.375	0.317	0.333	0.336
Peppers	4.311	2.777	1.173	0.690	0.545	0.477
Plane	2.009	2.174	1.725	1.209	0.900	0.737
Milkdrop	4.025	3.083	1.457	0.798	0.584	0.484
Kilauea	1.594	5.418	10.196	7.025	2.766	2.965
Boats	1.347	1.785	1.916	1.480	1.091	0.802
France	1.614	3.094	5.691	8.341	10.081	11.237
Library	1.596	2.049	3.204	3.853	3.948	3.987

Notice again that the overall best improvements are given at the highest compression rates. A particularly good performance is on the images "Kilauea" and "France". This might be because both of these images contain text. Figure 4 displays the result of a 3:50 compression of "Kilauea". Observe that the text seems sharper on the left image than on the right.

On the images "Peppers" and "Milkdrop" we also experience a relatively large improvement when keeping only about 1% of the transformed coefficients. This might be due to the relatively simple structure of both images. The difference is hardly visible on the compressed images though (see Figure 5).

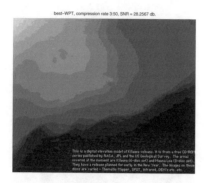 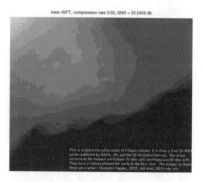

Fig. 4. The image "Kilauea" after a 3:50 compression using Best-WPT, $\mathcal{C}_5$ (left) and daub20 (right).

Fig. 5. The image "Peppers" after a 1:100 compression using Best-WPT, $\mathcal{C}_5$ (left) and daub20 (right).

§4. Conclusion

We have introduced wavelet packets in the setting of a multiresolution analysis associated with a dilation matrix with determinant 2. This setting allows us to use the well known wavelet packets analysis from the one dimensional case. The wavelet packets analysis associated with this setting is in general nonseparable, and thus different from what can be obtained by taking tensor products of one dimensional wavelet packets. The construction is especially interesting in two dimensions, since it makes it possible to perform image analysis using methods very similar to one dimensional methods. Due to this equivalence, it is easier to index function expansions in this setting than in a separable one, thus simplifying implementation and image analysis.

We have tested the new algorithm on 12 standard images. The nonseparable best wavelet packets transform (best-WPT) has been compared to a separable best-WPT, to give an impression of the performance of the new algorithm. The choice of separable wavelets has solely been made such that the two algorithms have comparable complexity. We notice that the new algorithm shows some advantages when compressing images with a simple structure or images containing text. We have also tested the best-WPT against a wavelet transform (both in the nonseparable case). In almost all of the tests, the best-WPT gave very positive results. On the test image "Barbara" the relative improvement exceeded 10% at certain compression rates, although in a few tests, the WT performed better. This is due to an inherent problem with the best basis algorithm; the algorithm finds the optimal choice with respect to the cost function and not the L^2-norm. Finally we should notice that any dilation matrix with determinant 2 and corresponding MRA would fit into this scheme.

References

1. Belogay, Eugene, and Yang Wang, Arbitrarily smooth orthogonal nonseparable wavelets in $\mathbb{R}^2$, SIAM J. Math. Anal. **30** (3) (1999), 678–697.

2. Coifman, Ronald R., Yves Meyer, Steven Quake, and M. Victor Wickerhauser, Signal processing and compression with wavelet packets, in *Progress In Wavelet Analysis And Applications (Toulouse, 1992)*, Frontières, Gif, 1993, 77–93.

3. Meyer, François G., and Ronald R. Coifman, Brushlets: a tool for directional image analysis and image compression, Appl. Comput. Harmonic Anal. **4**(2) (1997), 147–187.

4. Wojtaszczyk, P., *A Mathematical Introduction To Wavelets*, Cambridge University Press, Cambridge, 1997.

Lasse Borup
Dept. of Mathematical Sciences
Aalborg University
Aalborg, DK-9220
Denmark
lasse@math.auc.dk

Morten Nielsen
Dept. of Mathematics
University of South Carolina
Columbia, SC 29208
nielsen@math.sc.edu

The Wavelet Dimension Function for Real Dilations and Dilations Admitting non-MSF Wavelets

Marcin Bownik and Darrin Speegle

Abstract. The wavelet dimension function for arbitrary real dilations is defined and used to address several questions involving the existence of MRA wavelets and well-localized wavelets for irrational dilations. The theory of quasi-affine frames for rational dilations and the existence of non-MSF wavelets for certain irrational dilations play an important role in this development. Expansive dilations admitting non-MSF wavelets are characterized, and an example of a wavelet with respect to a non-expansive matrix is given.

§1. Introduction

The wavelet dimension function, which is sometimes referred to as the multiplicity function, is an important subject in the theory of wavelets that has been extensively studied by a number of authors, e.g., [2,4,8,20]. However, up to the present time, the wavelet dimension function has been considered only for integer expansive dilations. In this work we plan to initiate the study of the wavelet dimension function for non-integer dilations by investigating the situation in one dimension for all real dilation factors.

This subject appears to be quite intricate with many unexpected twists that combine seemingly unrelated matters including, among other things, the development of the notion of quasi-affine systems for rational dilations and the characterization of dilations admitting non-MSF wavelets. One of the biggest surprises is the fact that the usual formula for

Approximation Theory X: Wavelets, Splines, and Applications
Charles K. Chui, Larry L. Schumaker, and Joachim Stöckler (eds.), pp. 63–85.
Copyright ⊖ 2002 by Vanderbilt University Press, Nashville, TN.
ISBN 0-8265-1416-2.

the wavelet dimension function is often valid also for real dilation factors, but not always. In fact, it is valid precisely in two distinct situations: when a dilation factor is rational or a particular wavelet is MSF. Otherwise, the core space of a GMRA generated by a wavelet is not shift invariant, and the usual wavelet dimension function may not be even integer valued.

The consequences of the existence of the wavelet dimension function for real dilation factors are far reaching, as was demonstrated by Auscher [2] for expansive integer dilations. In particular, it is used in this paper to answer Daubechies' question [13] by showing the non-existence of well-localized wavelets for irrational dilations by different methods than the original argument used by the first author in [5]. Another consequence is the non-existence of MRA wavelets (not necessarily well-localized) for irrational dilations, which was expected much earlier, see [1], but lacked a proof.

We close our considerations by giving a complete characterization of expansive dilations that admit non-MSF wavelets. This turns out to be intimately connected to the topics mentioned above for several reasons. First, this characterization gives a converse to the theorem by Chui and Shi [6], which says that for dilations a such that $a^j \notin \mathbf{Q}$ for all integers $j \geq 1$, the only wavelets that exist are MSF wavelets. Secondly, it shows that the results obtained in this paper and concerning the wavelet dimension function are optimal, e.g., it provides examples of wavelets for which the usual formula for the wavelet dimension function is not valid. Thirdly, it shows that an approach used in this paper and in [5] is necessary for the complete solution to Daubechies' question, as well as providing limits on what types of negative results can be proved in higher dimensions. The idea of the proof of this characterization is to use the techniques of Meyer as explained in [15] and the procedure for constructing interpolation families of wavelet sets [14,21] to construct the Fourier transform of a wavelet. The key element that makes it possible is the recent characterization of wavelets via their Fourier transforms [9,10].

The paper is organized as follows. In Section 2 we introduce quasi-affine systems for rational dilations and we show the equivalence of affine and quasi-affine tight frames. In the next section we build the wavelet dimension function for arbitrary real dilation factors and we derive its basic properties. In Section 4 we apply the wavelet dimension function to show several results about well-localized wavelets. In Section 5 we give a characterization of all expansive dilations that admit non-MSF wavelets and we use these results to show the sharpness of the results obtained in Section 3. Finally, we give some explicit examples of wavelets illustrating the ideas underlying this paper.

We begin by recalling several definitions and theorems that will be useful in the sequel.

Definition 1.1. *Suppose* $\Psi = \{\psi^1, \ldots, \psi^L\} \subset L^2(\mathbb{R})$ *and* $a \in \mathbb{R}$, $|a| > 1$. *The* affine system $X(\Psi)$ *associated with the dilation* a *is defined as*

$$X(\Psi) = \{\psi^l_{j,k} : j, k \in \mathbb{Z}, \, l = 1, \ldots, L\}.$$

Here for $\psi \in L^2(\mathbb{R})$ *we set* $\psi_{j,k}(x) = |a|^{j/2}\psi(a^j x - k)$ *for* $j, k \in \mathbb{Z}$.

Definition 1.2. *Suppose* $a \in \mathbb{R}$ *and* $|a| > 1$. *We say that a measurable subset* E *of* $\mathbb{R}$ *is* a-multiplicatively invariant *if* $aE = E$ *modulo sets of measure zero. Given such* E *we introduce the closed subspace* $\check{L}^2(E) \subset L^2(\mathbb{R})$ *by*

$$\check{L}^2(E) = \{f \in L^2(\mathbb{R}) : \operatorname{supp} \hat{f} = \{\xi : \hat{f}(\xi) \neq 0\} \subset E\}.$$

We say that $\Psi = \{\psi^1, \ldots, \psi^L\} \subset \check{L}^2(E)$ *is a* multiwavelet for $\check{L}^2(E)$ *associated with* a, *or shortly* a-multiwavelet for $\check{L}^2(E)$, *if* $X(\Psi)$ *is an orthonormal basis of* $\check{L}^2(E)$.

The following theorem shown by Chui and Shi when $E = \mathbb{R}$ characterizes affine systems $X(\Psi)$ that are tight frames for arbitrary real dilation factors, see [10]. The general case of Theorem 1.3 is an easy generalization of the arguments given in [10].

Theorem 1.3. *Suppose that* $\Psi = \{\psi^1, \ldots, \psi^L\} \subset \check{L}^2(E)$. *Then* $X(\Psi)$ *is a tight frame with constant 1 for* $\check{L}^2(E)$ *if and only if*

$$\sum_{l=1}^{L} \sum_{\substack{(j,m) \in \mathbb{Z} \times \mathbb{Z}, \\ \alpha = a^{-j}m}} \hat{\psi}^l(a^j \xi)\overline{\hat{\psi}^l(a^j(\xi + \alpha))} = \delta_{\alpha,0}\mathbf{1}_E(\xi), \qquad \text{for a.e. } \xi \in \mathbb{R},$$

(1.1)

and for all α *belonging to the set of all* a-adic numbers, i.e.,

$$\{\alpha \in \mathbb{R} : \alpha = a^{-j}m \quad \text{for some } (j, m) \in \mathbb{Z} \times \mathbb{Z}\}.$$

The above result can be also generalized to higher dimensions, see [9]. Indeed, let A be an expansive matrix, i.e., A is $n \times n$ matrix such that all eigenvalues of A have modulus bigger than one. An A-wavelet is a function $\psi \in L^2(\mathbb{R}^n)$ such that $\{|\det A|^{j/2}\psi(A^j x - k) : j \in \mathbb{Z}, k \in \mathbb{Z}^n\}$ is an orthonormal basis for $L^2(\mathbb{R}^n)$.

Theorem 1.4. *Let* A *be an expansive matrix. A function* $\psi \in L^2(\mathbb{R}^n)$ *is an* A-wavelet *if and only if* $\|\psi\| = 1$ *and*

$$\sum_{\substack{(j,m) \in \mathbb{Z} \times \mathbb{Z}^n, \\ \alpha = (A^*)^{-j}m}} \hat{\psi}(A^{*j}\xi)\overline{\hat{\psi}(A^{*j}(\xi + A^{*-j}m))} = \delta_{\alpha,0}, \qquad \text{for a.e. } \xi \in \mathbb{R}^n$$

(1.2)

and for all α *belonging to the set of* A-adic vectors, i.e.,

$$\Lambda = \{\alpha \in \mathbb{R}^n : \alpha = (A^*)^{-j}m \quad \text{for some } (j, m) \in \mathbb{Z} \times \mathbb{Z}^n\}. \qquad (1.3)$$

§2. Quasi-affine Systems for Rational Dilations

In this section we introduce a notion of a quasi-affine system for rational dilations. Quasi-affine systems have been introduced and studied for integer dilations by Ron and Shen [19]. Their importance stems from the fact that the frame property is preserved when moving from an affine system to its corresponding quasi-affine system, and vice versa. Therefore, even though affine systems are dilation invariant (and thus not easy to study directly), instead one can work with quasi-affine systems which are shift invariant (and thus much easier to study).

The main idea behind quasi-affine systems is to oversample negative scales of the affine system at a rate adapted to the scale in order for the resulting system to be shift invariant. Even though the orthogonality of the affine system is not preserved by the corresponding quasi-affine system, however, it turns out that the frame property is preserved.

In order to define quasi-affine systems for rational dilation factors, we must oversample not only negative scales of the affine system (again at a rate proportional to the scale), but also, the positive scales. Since the resulting system coincides with the usual affine system only at the scale zero (for rational non-integral dilations), it is less clear (than in the case of integer dilations where both systems coincide at all non-negative scales) whether the quasi-affine system will share any common properties with its affine counterpart. Nevertheless, as was the case for integer dilations, the affine and quasi-affine systems share again the frame property. This turns out to be of critical importance in this work. The major consequence of this fact is the existence of the wavelet dimension function for the class of rational factors instead of the standard class of integer factors.

We start by defining the quasi-affine system for rational factors, and by showing that the property of being a tight frame is preserved when moving between these two systems. Our proof will be based on the characterization of tight affine systems by Chui and Shi [10], see Theorem 1.3.

Definition 2.1. *Suppose* $\Psi = \{\psi^1, \ldots, \psi^L\} \subset L^2(\mathbb{R})$ *and* $a = p/q \in \mathbf{Q}$, *where* $\gcd(p, q) = 1$. *The* quasi-affine system $X^q(\Psi)$ associated with the dilation a is defined as

$$X^q(\Psi) = \{\tilde{\psi}^l_{j,k} : j, k \in \mathbb{Z}, \ l = 1, \ldots, L\}.$$

Here for $\psi \in L^2(\mathbb{R})$ *and* $j, k \in \mathbb{Z}$ *we set*

$$\tilde{\psi}_{j,k}(x) = \begin{cases} \frac{p^{j/2}}{q^j} \psi(a^j x - q^{-j} k) & \text{if } j \geq 0, \\ \frac{p^j}{q^{j/2}} \psi(a^j x - p^j k) & \text{if } j < 0. \end{cases}$$

We will use the following standard notation. The translation by $k \in \mathbb{R}$ is $T_k f(x) = f(x - k)$. Given a family $\Phi \subset L^2(\mathbb{R})$ define the shift invariant (SI) system $E(\Phi)$ and SI space $S(\Phi)$ by

$$E(\Phi) = \{T_k \varphi : k \in \mathbb{Z}, \varphi \in \Phi\}, \qquad S(\Phi) = \overline{\operatorname{span}} E(\Phi). \qquad (2.1)$$

Theorem 2.2. *Suppose that* $a \in \mathbb{Q}$ *and* $\Psi = \{\psi^1, \ldots, \psi^L\} \subset \check{L}^2(E)$, *where* E *is an* a-*multiplicatively invariant subset of* $\mathbb{R}$. *Then the affine system* $X(\Psi)$ *is a tight frame with constant 1 for* $\check{L}^2(E)$ *if and only if its quasi-affine counterpart* $X^q(\Psi)$ *is a tight frame with constant 1 for* $\check{L}^2(E)$.

Proof: Theorem 1.3 gives a characterization of $X(\Psi)$ being a tight frame with constant 1 in terms of equation (1.1). In the case when a is rational, Chui and Shi [10] have shown that (1.1) can be written as

$$\sum_{l=1}^{L} \sum_{j \in \mathbb{Z}} |\hat{\psi}^l(a^j \xi)|^2 = 1_E(\xi), \qquad \text{for a.e. } \xi \in \mathbb{R}, \qquad (2.2)$$

$$\sum_{l=1}^{L} \sum_{j=0}^{s} \hat{\psi}^l(a^j \xi) \overline{\hat{\psi}^l(a^j(\xi + q^s t))} = 0, \qquad \text{for a.e. } \xi \in \mathbb{R} \qquad (2.3)$$

for all $s = 0, 1, 2, \ldots$ and all $t \in \mathbb{Z}$ not divisible by p nor q $(p, q \nmid t)$.

On the other hand, since the system $X^q(\Psi)$ is shift invariant, we can phrase a necessary and sufficient condition for $X^q(\Psi)$ to be a tight frame in terms of dual Gramians following the work of Ron and Shen [18]. Our goal is to show that this condition is equivalent to (2.2) and (2.3). Indeed, for simplicity, assume that a quasi affine system $X^q(\Psi)$ is generated by a single function $\Psi = \{\psi\}$. Let Φ be a set whose integer translates generate $X^q(\Psi)$, i.e.,

$$X^q(\Psi) = E(\Phi) := \{T_k \varphi : k \in \mathbb{Z}, \varphi \in \Phi\}.$$

Let D_j denote the set of representatives of different cosets of $\mathbb{Z}/(p^j \mathbb{Z})$ if $j \geq 0$ and of $\mathbb{Z}/(q^{-j} \mathbb{Z})$ if $j < 0$. Clearly, we can take the generating set Φ to be

$$\Phi = \{\tilde{\psi}_{j,d} : j \in \mathbb{Z}, d \in D_j\}.$$

Recall from [7, 18] that the dual Gramian of the system $E(\Phi)$ at a point ξ is defined as an infinite matrix $\tilde{G}(\xi) = (\tilde{G}(\xi)_{k,l})_{k,l \in \mathbb{Z}}$ given as

$$\tilde{G}(\xi)_{k,l} = \sum_{\varphi \in \Phi} \hat{\varphi}(\xi + k) \overline{\hat{\varphi}(\xi + l)}.$$

Therefore,

$$\tilde{G}(\xi)_{k,l} = \sum_{j\in\mathbb{Z}}\sum_{d\in D_j} \widehat{\tilde{\psi}_{j,d}}(\xi+k)\overline{\widehat{\tilde{\psi}_{j,d}}(\xi+l)}$$

$$= \sum_{j\geq 0}\sum_{d\in D_j} p^{-j}\hat{\psi}(a^{-j}(\xi+k))\overline{\hat{\psi}(a^{-j}(\xi+l))}e^{-2\pi i q^{-j}da^{-j}(k-l)}$$

$$+ \sum_{j<0}\sum_{d\in D_j} q^{j}\hat{\psi}(a^{-j}(\xi+k))\overline{\hat{\psi}(a^{-j}(\xi+l))}e^{-2\pi i p^{j}da^{-j}(k-l)}$$

$$= \sum_{j\in\mathbb{Z}} \hat{\psi}(a^{-j}(\xi+k))\overline{\hat{\psi}(a^{-j}(\xi+l))} \times \begin{cases} \sum_{d\in D_j} p^{-j}e^{-2\pi i p^{-j}d(k-l)}, & j\geq 0 \\ \sum_{d\in D_j} q^{j}e^{-2\pi i q^{j}d(k-l)}, & j<0 \end{cases}$$

$$= \sum_{j=m_-}^{m_+} \hat{\psi}(a^{-j}(\xi+k))\overline{\hat{\psi}(a^{-j}(\xi+l))},$$

where

$$m_+ = \max\{j\in\mathbb{Z}: k-l\in p^{j}\mathbb{Z}\}, \quad m_- = \min\{j\in\mathbb{Z}: k-l\in q^{-j}\mathbb{Z}\}.$$

By a result of [18], see also [7, Theorem 2.5], $X^q(\Psi) = E(\Phi)$ is a tight frame with constant 1 for $\check{L}^2(E)$ if and only if the dual Gramian $\tilde{G}(\xi)$ is an orthogonal projection onto the range function $J(\xi)$ corresponding to $\check{L}^2(E)$ for a.e. ξ. Recall that $J(\xi)$ is given by

$$J(\xi) = \{v = (v(k))_{k\in\mathbb{Z}} \in l^2(\mathbb{Z}) : v(k)\neq 0 \Rightarrow \xi+k\in E\}. \qquad (2.4)$$

This in turn is equivalent to the fact that

$$\tilde{G}(\xi)_{k,l} = \begin{cases} 1, & \text{if } k=l \text{ and } \xi+k\in E \\ 0, & \text{otherwise} \end{cases} \qquad \text{for a.e. } \xi. \qquad (2.5)$$

Now if $k = l \in \mathbb{Z}$, then $m_+ = \infty$, $m_- = -\infty$, and by the calculation above and (2.5)

$$\tilde{G}(\xi)_{k,k} = \sum_{j=-\infty}^{\infty} \hat{\psi}(a^{-j}(\xi+k))\overline{\hat{\psi}(a^{-j}(\xi+k))} = \mathbf{1}_E(\xi+k), \qquad \text{for a.e. } \xi,$$

which is just (2.2). If $k \neq l \in \mathbb{Z}$, then $l - k = p^{m_+}q^{-m_-}t$, where $t \in \mathbb{Z}$ is not divisible by p nor q. Hence by the calculation above and (2.5),

$$\tilde{G}(\xi)_{k,l} = \sum_{j=m_-}^{m_+} \hat{\psi}(a^{-j}(\xi+k))\overline{\hat{\psi}(a^{-j}(\xi+l))}$$

$$= \sum_{j=0}^{m_+-m_-} \hat{\psi}(a^{j-m_+}(\xi+k))\overline{\hat{\psi}(a^{j-m_+}(\xi+l))}$$

$$= \sum_{j=0}^{m_+-m_-} \hat{\psi}(a^j a^{-m_+}(\xi+k))\overline{\hat{\psi}(a^j(a^{-m_+}(\xi+k)+a^{-m_+}(l-k)))}$$

$$= \sum_{j=0}^{s} \hat{\psi}(a^j\check{\xi})\overline{\hat{\psi}(a^j(\check{\xi}+q^s t))} = 0,$$

where $s = m_+ - m_-$, $\check{\xi} = a^{-m_+}(\xi+k)$. Hence given $s = 0, 1, \ldots$, and t not divisible by p or q, we can let $k = 0$ and $l = q^s t$ to obtain (2.3).

This shows that, if $X^q(\Psi)$ is a tight frame (with constant 1 for $\check{L}^2(E)$), then (2.2) and (2.3) hold, and hence $X(\Psi)$ is also a tight frame. Vice versa, if $X(\Psi)$ is a tight frame, then by (2.2), (2.3) and by the above calculation, the dual Gramian $\tilde{G}(\xi)$ of $X^q(\Psi)$ satisfies (2.5), i.e., $X^q(\Psi)$ is a tight frame. This completes the proof of Theorem 2.2. $\square$

Theorem 2.2 can be generalized to the case of general (dual) frames, as is the case for integer dilations [11,19]. However, these results will not be needed here.

§3. The Wavelet Dimension Function for Real Dilation Factors

The goal of this section is to show the existence of the dimension function associated with arbitrary real dilation factors. Originally, the dimension function of wavelets, sometimes referred to as a multiplicity function, was studied only for integer dilation factors [2,4,8,20].

Suppose $W \subset L^2(\mathbb{R})$ is a shift invariant (SI) subspace of $L^2(\mathbb{R})$, i.e., W is a closed subspace of $L^2(\mathbb{R})$ such that $f \in W$ implies that $T_k f \in W$ for $k \in \mathbb{Z}$. The dimension function of W is a 1-periodic function $\dim_W : \mathbb{R} \to \mathbb{N} \cup \{0, \infty\}$ which measures the size of W over the fibers of $\mathbb{R}/\mathbb{Z}$. The precise definition in terms of a range function can be found in [3,7]. However, the reader who is not familiar with the dimension function of a general SI space can take Proposition 3.1 as a definition for the purposes of this work.

Proposition 3.1. *Suppose* $\Phi \subset L^2(\mathbb{R})$. *If the system* $E(\Phi)$ *is a tight frame with constant 1 for the space* $W = S(\Phi)$, *then*

$$\dim_W(\xi) := \dim \overline{\text{span}}\{(\hat{\varphi}(\xi+k))_{k\in\mathbb{Z}} : \varphi \in \Phi\} = \sum_{\varphi\in\Phi}\sum_{k\in\mathbb{Z}} |\hat{\varphi}(\xi+k)|^2.$$

$$(3.1)$$

The above proposition is folklore, and its proof follows easily from [3,18], see also [7, Theorem 2.5].

Suppose next that $\Psi \subset L^2(\mathbb{R})$ is a multiwavelet associated with a real dilation factor a, $|a| > 1$. Let $(V_j)_{j \in \mathbb{Z}}$ be a generalized multiresolution analysis (GMRA), see [4], associated with the multiwavelet Ψ which is given by

$$V_j = \overline{\text{span}}\{\psi^l_{i,k} : i < j, k \in \mathbb{Z}, l = 1, \dots, L\}. \tag{3.2}$$

Clearly, $(V_j)_{j \in \mathbb{Z}}$ is a sequence of closed subspaces of $L^2(\mathbb{R})$ that satisfies all the usual properties of a GMRA, e.g.,

$$V_j \subset V_{j+1}, \quad \overline{\bigcup_{j \in \mathbb{Z}} V_j} = L^2(\mathbb{R}), \quad \bigcap_{j \in \mathbb{Z}} V_j = \{0\}, \tag{3.3}$$

except one. It is not evident at all that the core space V_0 must be SI.

In the case when a is an integer (or more generally A is an expansive dilation matrix with integer entries), it is well known that V_0 must be SI. In that case, the dimension function of V_0 coincides with the wavelet dimension function $D_\Psi(\xi)$ of a multiwavelet Ψ,

$$\dim_{V_0}(\xi) = D_\Psi(\xi) \qquad \text{for a.e. } \xi \in \mathbb{R},$$

where

$$D_\Psi(\xi) := \sum_{l=1}^{L} \sum_{j=1}^{\infty} \sum_{k \in \mathbb{Z}} |\hat{\psi}^l(a^j(\xi + k))|^2. \tag{3.4}$$

A natural question is what happens for non-integer dilation factors a. Can we still expect V_0 to be SI? If yes, is the dimension function of V_0 given by $D_\Psi(\xi)$ as above? We will see that the answer to the first question is positive if either a is rational or Ψ is a combined MSF multiwavelet. In that case, the dimension function of V_0 is indeed given by (3.4). What is more interesting, these are the only cases when we can expect the space V_0 to be SI, see Theorem 3.4. Furthermore, in Section 5 we exhibit examples of wavelets for which V_0 is not SI and the wavelet dimension function $D_\Psi(\xi)$ given by (3.4) is not integer valued. Despite these setbacks, we can still define a meaningful wavelet dimension function which measures the dimensions of certain fibers of V_0 and is integer valued, see Definition 3.6.

Theorem 3.2. *Suppose* $\Psi = \{\psi^1, \dots, \psi^L\} \subset L^2(\mathbb{R})$ *is a multiwavelet associated with the real dilation factor* a, $|a| > 1$. *Then the space*

$$V = \overline{\text{span}}\{\psi^l_{jj_0,k} : j < 0, k \in \mathbb{Z}, l = 1, \dots, L\} \tag{3.5}$$

is shift invariant. Furthermore, its dimension function is given by

$$\dim_V(\xi) = \sum_{l=1}^{L} \sum_{j=1}^{\infty} \sum_{k \in \mathbb{Z}} |\hat{\psi}^l(a^{jj_0}(\xi + k))|^2, \tag{3.6}$$

where

$$j_0 = j_0(a, \Psi) = \begin{cases} 1, & \text{if } \Psi \text{ is combined MSF,} \\ \inf\{j \geq 1 : a^j \in \mathbf{Q}\}, & \text{otherwise.} \end{cases} \quad (3.7)$$

Proof: Note that $j_0 = j_0(a, \Psi)$ is well defined for any multiwavelet Ψ and any dilation factor a, $|a| > 1$. Indeed, if a is such that $a^j \notin \mathbf{Q}$ for all $j \geq 1$, then by the result of Chui and Shi [10] for $L = 1$ and [6] for general L, Ψ must be a combined MSF multiwavelet. Recall that a multiwavelet Ψ is said to be **combined minimally supported frequency** if $\bigcup_{l=1}^{L} \operatorname{supp} \hat{\psi}^l$ has minimal Lebesgue measure.

Suppose first that Ψ is combined MSF, i.e., the set $K = \bigcup_{l=1}^{L} \operatorname{supp} \hat{\psi}^l$ has minimal Lebesgue measure. Therefore, see [6],

$$\overline{\operatorname{span}}\{T_k \psi^l : k \in \mathbb{Z}, \ l = 1, \ldots, L\} = \check{L}^2(K)$$

and

$$\sum_{l=1}^{L} |\hat{\psi}^l(\xi)|^2 = \mathbf{1}_K(\xi).$$

Since, $\{a^j K\}_{j \in \mathbb{Z}}$ is a partition of $\mathbb{R}$, we have $V = \check{L}^2(\bigcup_{j=-\infty}^{-1} a^j K)$ is shift invariant under all translations. Therefore, by Proposition 3.1,

$$\dim_V(\xi) = \sum_{k \in \mathbb{Z}} \sum_{j=-\infty}^{-1} \mathbf{1}_{a^j K}(\xi + k) = \sum_{l=1}^{L} \sum_{j=1}^{\infty} \sum_{k \in \mathbb{Z}} |\hat{\psi}^l(a^j(\xi + k))|^2.$$

Suppose next that Ψ is not combined MSF. Then j_0 given by (3.7) is the smallest integer $j \geq 1$ such that a^j is rational.

Consider first the case when $j_0 = 1$, i.e., $a = p/q \in \mathbf{Q}$, where $\gcd(p, q) = 1$. For simplicity assume that $\Psi = \{\psi\}$. To see that V given by (3.5) is SI it suffices to show that for any $n \in \mathbb{Z}$, $\psi_{j,k_1}(x - n) \perp \psi_{m,k_2}(x)$, where $j < 0$, $m \geq 0$, $k_1, k_2 \in \mathbb{Z}$. Indeed, pick any $w \in \mathbb{Z}$ such that $w \equiv 0$ mod q^m and $w \equiv n$ mod p^{-j}. Therefore, $a^j(w - n), a^m w \in \mathbb{Z}$, and by a simple change of variables $x = y + w$,

$$\int_{\mathbb{R}} \psi_{j,k_1}(x - n)\overline{\psi_{m,k_2}(x)}dx$$

$$= a^{(j+m)/2} \int_{\mathbb{R}} \psi(a^j(x - n) - k_1)\overline{\psi(a^m x - k_2)}dx$$

$$= a^{(j+m)/2} \int_{\mathbb{R}} \psi(a^j y + a^j(w - n) - k_1)\overline{\psi(a^m y + a^m w - k_2)}dy$$

$$= \langle \psi_{j,-a^j(w-n)+k_1}, \psi_{m,-a^m w+k_2} \rangle = 0.$$

Consider next the case when $j_0 \geq 2$. Define a a^{j_0}-multiplicatively invariant set E by

$$E = \bigcup_{l=1}^{L} \bigcup_{j \in j_0 \mathbb{Z}} a^j (\operatorname{supp} \hat{\psi}^l). \tag{3.8}$$

By [5, Theorem 4.1], the sets $E, aE, \ldots, a^{j_0-1}E$ form a partition of $\mathbb{R}$ (modulo sets of measure zero). Therefore, Ψ is a multiwavelet associated with the rational dilation a^{j_0} for $\check{L}^2(E)$. Therefore, by the argument above, V given by (3.5) is SI. Likewise, the space

$$\check{L}^2(E) \ominus V = \overline{\operatorname{span}}\{\psi^l_{jj_0,k} : j \geq 0, k \in \mathbb{Z}, \ l = 1, \ldots, L\}$$

is SI. Let $X^q(\Psi)$ denote the quasi-affine system generated by Ψ and associated with the rational dilation $\tilde{a} := a^{j_0} = p/q$. We claim that

$$V = \overline{\operatorname{span}}\{\tilde{\psi}^l_{j,k} : j < 0, k \in \mathbb{Z}, \ l = 1, \ldots, L\}.$$

Indeed, the inclusion "$\subset$" is trivial. To see "$\supset$", take any $j < 0, k \in \mathbb{Z}$ and notice that

$$\tilde{\psi}^l_{j,k}(x) = \frac{p^j}{q^{j/2}} \psi^l(\tilde{a}^j x - p^j k) = \frac{p^j}{q^{j/2}} \psi^l(\tilde{a}^j(x - l_1) - l_2),$$

where $l_1, l_2 \in \mathbb{Z}$ are such that $k = q^{-j}l_1 + p^{-j}l_2$. Since V is SI, this shows $\tilde{\psi}^l_{j,k} \in V$. Analogously we can show that

$$\check{L}^2(E) \ominus V = \overline{\operatorname{span}}\{\tilde{\psi}^l_{j,k} : j \geq 0, k \in \mathbb{Z}, \ l = 1, \ldots, L\}.$$

By Theorem 2.2, the quasi-affine system $X^q(\Psi)$ forms a tight frame with constant 1 for $\check{L}^2(E)$. Therefore, $\{\tilde{\psi}^l_{j,k} : j < 0, k \in \mathbb{Z}, \ l = 1, \ldots, L\}$ forms a tight frame with constant 1 for V. To compute the dimension function of V, notice that

$$\{\tilde{\psi}^l_{j,k} : j < 0, k \in \mathbb{Z}, \ l = 1, \ldots, L\} = E(\tilde{\Phi}),$$

where $\tilde{\Phi} = \{\tilde{\psi}^l_{j,d} : j < 0, d \in D_j, \ l = 1, \ldots, L\}$ and D_j are representatives of distinct cosets of $\mathbb{Z}/(q^{-j}\mathbb{Z})$. By Proposition 3.1,

$$\dim_V(\xi) = \sum_{l=1}^{L} \sum_{j<0} \sum_{k \in \mathbb{Z}} \sum_{d \in D_j} |\widehat{\tilde{\psi}^l_{j,d_j}}(\xi + k)|^2$$

$$= \sum_{l=1}^{L} \sum_{j<0} \sum_{k \in \mathbb{Z}} \sum_{d \in D_j} q^j |\hat{\psi}^l(\tilde{a}^{-j}(\xi + k))|^2 = \sum_{l=1}^{L} \sum_{j=1}^{\infty} \sum_{k \in \mathbb{Z}} |\hat{\psi}^l(a^{jj_0}(\xi + k))|^2.$$

This shows (3.6) and completes the proof of Theorem 3.2. $\quad\square$

Next we will show that Theorem 3.2 is the best possible result in the sense that the shift invariance of the core space V_0 given by (3.2) necessarily implies that either the dilation factor a is rational or Ψ is combined MSF. In addition, Theorem 5.4 will show that there indeed exist wavelets for which the core space V_0 is not SI.

Theorem 3.3. *Suppose Ψ is a multiwavelet associated with an irrational dilation factor a, $|a| > 1$. If the core space V_0 given by (3.2) is SI, then necessarily V_0 is SI under all real translations. In particular, Ψ is a combined MSF multiwavelet.*

Proof: It suffices to consider only a's such that $a^j \in \mathbf{Q}$ for some $j \geq 2$. Let $j_0 = j_0(a, \Psi)$ be the same as in (3.7). Clearly,

$$V_0 = V \oplus D_a V \oplus \ldots \oplus D_{a^{j_0 - 1}} V, \qquad (3.9)$$

where V is given by (3.5) and $D_a f(x) = \sqrt{|a|} f(ax)$. Since $D_{a^j} V \subset \check{L}^2(a^j E)$, where E is given by (3.8), and $E, aE, \ldots, a^{j_0 - 1} E$ are pairwise disjoint

$$S(V_0) = S(V) \oplus S(D_a V) \oplus \ldots \oplus S(D_{a^{j_0 - 1}} V).$$

In particular, $S(D_a V) = D_a V$, i.e., $D_a V$ is SI. Since, $D_a V$ is also SI with respect to translates by the lattice $1/a\mathbb{Z}$ and $a \notin \mathbf{Q}$, $D_a V$ is SI with respect to all real translations. Therefore, V_0 is also SI with respect to all real translations, i.e., $V_0 = \check{L}^2(S)$ for some measurable $S \subset \mathbb{R}$. Therefore,

$$\overline{\mathrm{span}}\{\tilde{\psi}_{0,k}^l : k \in \mathbb{Z}, \ l = 1, \ldots, L\} = V_1 \ominus V_0 = \check{L}^2(aS \setminus S),$$

and Ψ is combined MSF. $\square$

The proof of Theorem 3.3 shows that even though the core space V_0 in general is not SI, it can be decomposed as an orthogonal sum of a certain number of SI spaces with respect to different lattices of $\mathbb{R}$. Indeed, suppose that Ψ is a multiwavelet associated with the dilation a and $j_0 = j_0(a, \Psi) > 1$, where j_0 is the same as in (3.7). By (3.9),

$$V_0 = \bigoplus_{m=0}^{j_0 - 1} V^{(m)}, \qquad \text{where } V^{(m)} := D_{a^m} V, \qquad (3.10)$$

and V is given by (3.6). Clearly, the space $V^{(m)}$ is SI with respect to the lattice $a^{-m}\mathbb{Z}$, and hence it is meaningful to talk about its dimension function. Hence, at least formally, we can talk about the dimension function of V_0 as the sum of the dimension functions of its components $V^{(m)}$, $m = 0, \ldots, j_0 - 1$, see (3.13). We will use the following variant of Proposition 3.1 which again can serve as a definition of the dimension function of a SI space W with respect to a general lattice $b\mathbb{Z}$.

Proposition 3.4. *Suppose $b > 0$ and $W \subset L^2(\mathbb{R})$ is a SI space with respect to the lattice $b\mathbb{Z}$, i.e., $f \in W$ implies that $T_{kb}f \in W$ for any $k \in \mathbb{Z}$. Then $D_b W$ is SI (with respect to $\mathbb{Z}$), and the dimension function of W is a $1/b$-periodic function satisfying*

$$\dim_W^{b\mathbb{Z}}(\xi) := \dim \operatorname{span}\{(\hat{\varphi}(\xi + \frac{k}{b}))_{k \in \mathbb{Z}} : \varphi \in \Phi\} = \dim_{D_b W}(b\xi). \quad (3.11)$$

Finally, we are ready to define the wavelet dimension function for arbitrary real dilation factors.

Definition 3.5. *Suppose $\Psi = \{\psi^1, \dots, \psi^L\} \subset L^2(\mathbb{R})$ is a multiwavelet associated with the real dilation factor a, $|a| > 1$. Let $j_0 = j_0(a, \Psi)$ be given by (3.7). Define the wavelet dimension function of Ψ as*

$$D_\Psi(\xi) := \sum_{l=1}^{L} \sum_{j=1}^{\infty} \sum_{k \in \mathbb{Z}} |\hat{\psi}^l(a^j(\xi + a^{j - \lfloor j/j_0 \rfloor j_0}k))|^2. \quad (3.12)$$

The following result justifies the above definition and shows the connection of D_Ψ with the core space V_0 of a multiwavelet Ψ.

Theorem 3.6. *Suppose $\Psi = \{\psi^1, \dots, \psi^L\} \subset L^2(\mathbb{R})$ is a multiwavelet associated with the real dilation factor a, $|a| > 1$. Let V_0 be the core space of Ψ given by (3.2), and suppose V is defined by (3.6) and $j_0 = j_0(a, \Psi)$ by (3.7). Then (3.10) holds and*

$$\sum_{m=0}^{j_0 - 1} \dim_{V^{(m)}}^{a^{-m}\mathbb{Z}}(\xi) = D_\Psi(\xi) \qquad \text{for a.e. } \xi \in \mathbb{R}, \quad (3.13)$$

where $V^{(m)} = D_{a^m}V$. In particular, $D_\Psi(\xi)$ given by (3.12) is almost everywhere integer valued.

Proof: By Theorem 3.2 and Proposition 3.4,

$$\dim_{V^{(m)}}^{a^{-m}\mathbb{Z}}(\xi) = \dim_{D_{a^{-m}}V^{(m)}}(a^{-m}\xi)$$

$$= \dim_V(a^{-m}\xi) = \sum_{l=1}^{L} \sum_{j=1}^{\infty} \sum_{k \in \mathbb{Z}} |\hat{\psi}^l(a^{jj_0 - m}(\xi + a^m k))|^2.$$

Summing the above over $m = 0, \dots, j_0 - 1$, we immediately obtain (3.13). $\square$

§4. Applications of the Wavelet Dimension Function

The wavelet dimension function is a very powerful tool which enables us to show many properties of multiwavelets which are well-localized both in time and frequency. For the purposes of this work we say that $\psi \in L^2(\mathbb{R})$ is well-localized in time and frequency if it satisfies the condition $(\Re^0)$ below, see [16, §7.6].

Definition 4.1. *We say that a function $\psi \in L^2(\mathbb{R})$ satisfies condition $(\Re^0)$ if there exist $c, \delta > 0$ such that*

$$|\hat{\psi}(\xi)| \text{ is continuous on } \mathbb{R},$$
$$|\hat{\psi}(\xi)| \leq c|\xi|^{-1/2-\delta} \qquad \text{for all } \xi \in \mathbb{R}. \tag{4.1}$$

We say that a collection $\Psi = \{\psi^1, \ldots, \psi^L\}$ satisfies $(\Re^0)$ if each ψ^l satisfies $(\Re^0)$.

Daubechies in her book [13] has asked whether there are any well-localized wavelets with irrational dilations. This question was answered by the first author [5] by showing that well-localized multiwavelets can only exist for rational dilation factors. We will give an alternative proof of this result using the dimension function techniques, see Theorem 4.4. On the other hand, Auscher [1] has constructed Meyer-type wavelets for any rational dilation factor. Therefore, his construction is sharp in the sense that it cannot be extended to irrational dilation factors. However, it is not clear what is the minimal size L of a well-localized multiwavelet Ψ associated with rational non-integral dilation factor a. Auscher [2] has shown that if a is an integer then L has to be a multiple of $|a| - 1$. Theorem 4.3 extends this result to rational dilation factors.

Lemma 4.2. *Suppose $\psi \in L^2(\mathbb{R})$ satisfies $(\Re^0)$ and $|a| > 1$. Then $D_\psi(\xi)$ given by*

$$D_\psi(\xi) = \sum_{j=1}^{\infty} \sum_{k \in \mathbb{Z}} |\hat{\psi}(a^j(\xi+k))|^2$$

is a 1-periodic continuous function on $\mathbb{R} \setminus \mathbb{Z}$.

Proof: Notice first that $s(\xi) = \sum_{j=1}^{\infty} |\hat{\psi}(a^j\xi)|^2$ is continuous everywhere except possibly at the origin and that $s(\xi) \leq c/(|a|-1)|\xi|^{-1-2\delta}$. Therefore, $\sum_{k \in \mathbb{Z}} s(\xi+k)$ is 1-periodic and continuous on $\mathbb{R} \setminus \mathbb{Z}$. $\square$

Theorem 4.3. *Suppose $\Psi = \{\psi^1, \ldots, \psi^L\} \subset L^2(\mathbb{R})$ is a multiwavelet associated with the rational dilation factor $a = p/q$, where $\gcd(p, q) = 1$. If Ψ is well-localized in time and frequency, then L is a multiple of $|p| - |q|$.*

Proof: By Lemma 4.2, the dimension function of Ψ given by

$$D_\Psi(\xi) = \sum_{l=1}^{L} \sum_{j=1}^{\infty} \sum_{k \in \mathbb{Z}} |\hat{\psi}^l(a^j(\xi + k))|^2$$

is continuous on $\mathbb{R} \setminus \mathbb{Z}$. By Theorem 3.2, $D_\Psi(\xi)$ is integer valued for a.e. ξ, hence $D_\Psi(\xi) = M$ for some non-negative integer M. Since

$$\int_0^1 D_\Psi(\xi)d\xi = L/(|a| - 1) = L|q|/(|p| - |q|) = M,$$

L has to be a multiple of $|p| - |q|$, because $\gcd(|q|, |p| - |q|) = 1$. $\quad\square$

Theorem 4.3 indicates that it would be extremely difficult to have well-localized wavelet bases for irrational dilations. Indeed, as we approach an irrational dilation a by a sequence of rational dilation factors $(p_n/q_n)_{n \in \mathbb{N}}$, we see that the size of well-localized multiwavelets associated with p_n/q_n has to be a multiple of $|p_n| - |q_n|$, and therefore it must grow to infinity as $n \to \infty$. This heuristic argument is made precise in the proof of Theorem 4.4. The following proof is a variant of [5, Theorem 4.1].

Theorem 4.4. *There are no multiwavelets Ψ for irrational dilation factors which are well-localized in the sense of the condition $(\Re^0)$.*

Proof: Let $|a| > 1$ be irrational. It suffices to consider only the case when $j_0 = j_0(a, \Psi)$ given by (3.7) is > 1. Otherwise, Ψ is combined MSF and at least one of $\hat{\psi}^l$'s is not continuous, see [6].

Assume that Ψ satisfies $(\Re^0)$. Let V be the SI space given by (3.5). By Theorem 3.2 and Lemma 4.2, $\dim_V(\xi)$ is 1-periodic and continuous on $\mathbb{R} \setminus \mathbb{Z}$. Since $\dim_V(\xi)$ is integer valued, hence $\dim_V(\xi) = M$ for some $M \in \mathbb{N}$. Let

$$s(\xi) = \sum_{l=1}^{L} \sum_{j=1}^{\infty} |\hat{\psi}(a^{jj_0}\xi)|^2, \quad t(\xi) = \sum_{l=1}^{L} \sum_{j=1}^{\infty} \sum_{k \in \mathbb{Z} \setminus \{0\}} |\hat{\psi}(a^{jj_0}(\xi + k))|^2.$$

By Lemma 4.2, $s(\xi)$ is continuous on $\mathbb{R} \setminus \{0\}$, $t(\xi)$ is continuous on $(\mathbb{R} \setminus \mathbb{Z}) \cup \{0\}$, and $\dim_V(\xi) = s(\xi) + t(\xi) = M$. Let

$$E = \bigcup_{l=1}^{L} \sum_{j \in \mathbb{Z}} a^{jj_0}(\text{supp } \hat{\psi}^l).$$

By the orthogonality argument, see [5, Lemma 2.2], the sets $E, aE, \ldots,$ $a^{j_0-1}E$ are pairwise disjoint. Moreover, by [5, Corollary 3.3] the following Calderón formula holds:

$$\sum_{l=1}^{L} \sum_{j \in \mathbb{Z}} |\hat{\psi}^l(a^{jj_0}\xi)|^2 = \mathbf{1}_E(\xi), \qquad \text{for a.e. } \xi \in \mathbb{R}. \tag{4.2}$$

Therefore, for a.e. $\xi \in E$, we have

$$
\begin{aligned}
0 + t(0) &= \lim_{j \to -\infty} (s(a^{jj_0+1}\xi) + t(a^{jj_0+1}\xi)) = M \\
&= \lim_{j \to -\infty} (s(a^{jj_0}\xi) + t(a^{jj_0}\xi)) = 1 + t(0),
\end{aligned}
$$

since $a\xi \notin E$ by (4.2). This is a contradiction. $\square$

As an immediate corollary of Theorems 3.2, 4.3 and 4.4, we have that well-localized multiwavelets always come from an MRA possibly of higher multiplicity.

Corollary 4.5. *Suppose* $\Psi = \{\psi^1, \ldots, \psi^L\} \subset L^2(\mathbb{R})$ *is a well-localized multiwavelet associated with a real dilation factor* a, $|a| > 1$. *Then the dilation factor* a *is rational and* $d = L/(|a| - 1)$ *is an integer. Moreover,* Ψ *is associated with an MRA* $(V_j)_{j \in \mathbb{Z}}$ *of multiplicity* d, *i.e., there exists* $\Phi = \{\phi^1, \ldots, \phi^d\} \subset V_0$, *such that* $E(\Phi)$ *is an orthonormal basis of the core space* V_0 *given by (3.2).*

In particular, we obtain the following simple corollary of Theorems 4.3 and 4.4.

Corollary 4.6. *If a wavelet* ψ *associated with a real dilation factor* a *is well-localized in time and frequency, then* $a = \pm(q+1)/q$ *for some integer* $q \geq 1$.

Another consequence of the wavelet dimension function technique is the converse to Corollary 4.5 which says that only for rational dilation factors, multiwavelets (not necessarily well-localized) can be associated with an MRA (possibly of higher multiplicity). This has been already observed (without proof) by Auscher [1]. In fact, Theorem 4.7 states that among all (single) wavelets with arbitrary *real* dilation factors, only dyadic (or negative dyadic) wavelets can be associated with an MRA.

Theorem 4.7. *Suppose a multiwavelet* $\Psi = \{\psi^1, \ldots, \psi^L\} \subset L^2(\mathbb{R})$ *with a real dilation factor* a, $|a| > 1$, *is associated with an MRA* $(V_j)_{j \in \mathbb{Z}}$ *of multiplicity* d, *i.e., there exists* $\Phi = \{\phi^1, \ldots, \phi^d\} \subset V_0$, *such that* $E(\Phi)$ *is an orthonormal basis of the core space* V_0. *Then necessarily* $|a| = 1 + L/d$.

Proof: Since V_0 is SI, by Theorem 3.3, either Ψ is combined MSF or a is rational. In either case, by Theorem 3.2,

$$
d = \int_0^1 \dim_{V_0}(\xi)d\xi = \int_0^1 D_\Psi(\xi)d\xi = L/(|a| - 1),
$$

since $\dim_{V_0}(\xi) = d$ for a.e. $\xi \in \mathbb{R}$. Therefore, $L = (|a| - 1)d$. $\square$

Finally, we can address the question of the existence of well-localized wavelets for the Hardy space $H^2(\mathbb{R})$ which was originally posed by Meyer [17] in the case of dyadic wavelets. Auscher [2] has given a negative answer in the case of integer dilation factors. However, the non-integer case has not been addressed until recently. Using the same methods as in [5] one can show that there are no well-localized wavelets for $H^2(\mathbb{R})$ with irrational dilation factors. However, to show the same for rational dilations, it appears that one has to use the wavelet dimension function for rational dilation factors introduced in Section 3. The proof will then follow along the same lines as Auscher's original argument [2], see also [16, §7, Theorem 6.20].

Theorem 4.8. *There are no well-localized (in the sense of the condition* $(\Re^0)$*) multiwavelets* Ψ *for the Hardy space* $H^2(\mathbb{R})$ *and any real dilation factor.*

Proof: By [2,5] it remains to consider only the case of a rational dilation factor a. By Theorem 2.3 with $E = (0, \infty)$ and an $H^2(\mathbb{R})$ variant of Theorem 3.2, the space V_0 given by (3.2) is SI and its dimension function D_Ψ is given by (3.4).

Assume that Ψ satisfies $(\Re^0)$. By Lemma 4.2, $D_\Psi(\xi)$ is 1-periodic and continuous on $\mathbb{R} \setminus \mathbb{Z}$. Since $D_\Psi(\xi)$ is integer valued for a.e. ξ, hence $D_\Psi(\xi) = M$ for some $M \in \mathbb{N}$. Let

$$s(\xi) = \sum_{l=1}^{L} \sum_{j=1}^{\infty} |\hat{\psi}(a^j\xi)|^2, \quad t(\xi) = \sum_{l=1}^{L} \sum_{j=1}^{\infty} \sum_{k \in \mathbb{Z} \setminus \{0\}} |\hat{\psi}(a^j(\xi+k))|^2.$$

By Lemma 4.2, $s(\xi)$ is continuous on $\mathbb{R} \setminus \{0\}$, $t(\xi)$ is continuous on $(\mathbb{R} \setminus \mathbb{Z}) \cup \{0\}$, and $D_\Psi(\xi) = s(\xi) + t(\xi) = M$. By Theorem 1.3,

$$\sum_{l=1}^{L} \sum_{j \in \mathbb{Z}} |\hat{\psi}^l(a^j\xi)|^2 = \mathbf{1}_{(0,\infty)}(\xi) \qquad \text{for a.e. } \xi \in \mathbb{R}.$$

Therefore, for a.e. $\xi > 0$,

$$0 + t(0) = \lim_{j \to -\infty} \left(s(-a^j\xi) + t(-a^j\xi) \right)$$

$$= M = \lim_{j \to -\infty} \left(s(a^j\xi) + t(a^j\xi) \right) = 1 + t(0),$$

which is a contradiction. $\square$

§5. Dilations Admitting non-MSF Wavelets

In this section we characterize dilations that admit non-MSF wavelets. Since the arguments employed in this section work beyond the one-dimensional setting we consider higher dimensional wavelets. We prove that in the case of $n \times n$ expansive matrices A, there exists an A-wavelet which is not an MSF wavelet if and only if there exists an integer $j \neq 0$ such that $A^{*j}(\mathbb{Z}^n) \cap \mathbb{Z}^n \neq \emptyset$. We begin by recalling several definitions and theorems that will be useful in the sequel.

We say that a set E tiles $\mathbb{R}^n$ by translations if $\{E + k : k \in \mathbb{Z}^n\}$ is a partition of $\mathbb{R}^n$ and by A^*-dilations if $\{A^{*j}(E) : j \in \mathbb{Z}\}$ is a partition of $\mathbb{R}^n$. An MSF wavelet is a wavelet ψ such that the support of $\hat{\psi}$ has a minimal Lebesgue measure; in other words, $|\hat{\psi}|$ is the indicator function of a set. The support of an MSF wavelet (in the Fourier domain) is called an A-wavelet set, and is characterized by the fact that it tiles $\mathbb{R}^n$ by translations and A^*-dilations. That is,

$$
\begin{aligned}
\sum_{j \in \mathbb{Z}} \mathbf{1}_E(A^{*j}\xi) = 1, \qquad \text{for a.e. } \xi \in \mathbb{R}^n, \\
\sum_{k \in \mathbb{Z}^n} \mathbf{1}_E(\xi + k) = 1, \qquad \text{for a.e. } \xi \in \mathbb{R}^n,
\end{aligned}
\tag{5.1}
$$

where E is the support of $\hat{\psi}$. Note that this characterization does not depend on A being expansive or integer valued.

Assume now that A is expansive. Fix $C = [-1/2, 1/2]^n$, and suppose O is a bounded set which is bounded away from the origin and tiles $\mathbb{R}^n$ by A^*-dilations. We define the translation projection τ and the dilation projection d as

$$
\tau(E) = \cup_{k \in \mathbb{Z}^n}((E + k) \cap C), \qquad d(E) = \cup_{j \in \mathbb{Z}}(A^{*j}(E) \cap O).
$$

The following theorem gives a sufficient condition for sets to be contained in wavelet sets, which is a higher dimensional analogue of [22, Theorem 1.1]. The idea of the proof of Theorem 5.1 is to employ the iterative procedure as in [22], which can be intuitively described as "filling up arbitrary large pieces" of C while "filling up arbitrary small pieces" of O, and vice versa.

Theorem 5.1. *Let $E \subset \mathbb{R}^n$ and A be an expansive matrix. Then E is a subset of an A-wavelet set provided that:*

(i) $E \cap (E + k) = \emptyset$ *for all* $k \in \mathbb{Z}^n$,

(ii) $A^{*j}(E) \cap E = \emptyset$ *for all* $j \in \mathbb{Z}$,

(iii) *there exists an* $\epsilon_0 > 0$ *such that* $B_{\epsilon_0}(0) \cap \tau(E) = \emptyset$, *and*

(iv) $O \setminus d(E)$ *has a non-empty interior.*

Proof: We will make use of the following two elementary facts which are a consequence of A^* being an expansive matrix:

$$\forall \epsilon > 0 \ \exists M(\epsilon) \in \mathbb{Z}, \delta(\epsilon) > 0 : \quad A^{*M(\epsilon)}(O) \subset \big(B_\epsilon(0) \setminus B_{\delta(\epsilon)}(0)\big), \quad (5.2)$$

$$\forall S \subset O \text{ with non-empty interior } \exists k(S) \in \mathbb{Z}^n : \quad d(C + k(S)) \subset S. \ (5.3)$$

By (iv), we can write $O \setminus d(E)$ as the disjoint union of sets $\{E_i\}_{i=1}^\infty$, each of which has non-empty interior. We define by induction a decreasing sequence $(\epsilon_i)_{i=1}^\infty$ of positive numbers, a sequence $(F_i)_{i=1}^\infty$ of subsets of C, and a sequence $(G_i)_{i=1}^\infty$ of subsets of $\mathbb{R}^n$.

We use (5.2) with $\epsilon = \epsilon_0$, where ϵ_0 is the same as in (iii), and (5.3) with $S = E_1$ to define $\epsilon_1 = \delta(\epsilon_0)$,

$$F_1 = \emptyset, \quad G_1 = \Big(C \setminus \big(\tau(E) \cup B_{\epsilon_1}(0)\big)\Big) + k(E_1).$$

Given $\epsilon_1, \dots, \epsilon_i, F_1, \dots, F_i, G_1, \dots, G_i$, we use (5.2) with $\epsilon = \epsilon_i$ and (5.3) with $S = E_{i+1}$ to define $\epsilon_{i+1} = \delta(\epsilon_i)$,

$$F_{i+1} = A^{*M(\epsilon_i)}(E_i \setminus d(G_i)),$$

$$G_{i+1} = \Big(C \setminus \tau\Big(E \cup \bigcup_{j=1}^{i+1} F_j \cup \bigcup_{j=1}^{i} G_j \cup B_{\epsilon_{i+1}}(0)\Big)\Big) + k(E_{i+1}).$$

A simple induction argument shows that the dilation projections $d(E)$, $d(F_1)$, $d(G_1)$, $d(F_2)$, $d(G_2)$, ... are mutually disjoint. Moreover, for any $i \in \mathbb{N}$, $\bigcup_{j=1}^{i+1} E_j = d\big((\bigcup_{j=1}^{i+1} F_j) \cup (\bigcup_{j=1}^{i} G_j)\big)$. Likewise, the translation projections $\tau(E)$, $\tau(F_1)$, $\tau(G_1)$, $\tau(F_2)$, $\tau(G_2)$, ... are mutually disjoint and $C \setminus B_{\epsilon_i}(0) = \tau\big(E \cup (\bigcup_{j=1}^i F_j) \cup (\bigcup_{j=1}^i G_j)\big)$. Therefore, $E \cup (\bigcup_{i=1}^\infty F_i) \cup (\bigcup_{i=1}^\infty G_i)$ is a wavelet set containing E by (5.1). $\square$

We will also use the following two elementary facts.

Lemma 5.2. *Let $\mathcal{F}$ be a countable collection of continuous functions. Suppose that $\mathcal{F} = \{f_m : m \in \mathbb{N}\}$. For each $m \in \mathbb{N}$, let Y_m be the fixed points for f_m, and let $Y = \bigcup_{m \in \mathbb{N}} Y_m$. If (a) $y \notin Y$ and (b) there is a $\gamma > 0$ such that $f_m(B_\gamma(y)) \cap B_\gamma(y) = \emptyset$ for all but finitely many m's, then there is an $\epsilon > 0$ such that $f_m(B_\epsilon(y)) \cap B_\epsilon(y) = \emptyset$ for all $m \in \mathbb{N}$.*

Proof: Assume the contrary. Then, for every integer $k \geq 1$ there is an $m_k \in \mathbb{N}$ such that $f_{m_k}(B_{1/k}(y)) \cap B_{1/k}(y) \neq \emptyset$. By condition (b), m_k is bounded, so by passing to a subsequence, we can assume $m_k = m$ for all k. But, this implies by continuity that $f_m(y) = y$, contradicting condition (a). $\square$

Lemma 5.3. *Suppose A is an expansive matrix, $k_0 \in \mathbb{Z}^n \setminus \{0\}$, and $p \in \mathbb{Z} \setminus \{0\}$. Let Y be the collection of fixed points of $\mathcal{F} \cup \mathcal{G}$, where*

$$\mathcal{F} = \{f_c(x) = A^{*-p}x - c : c \in \mathbb{Z}^n\}, \quad \mathcal{G} = \{g_j(x) = A^{*-j}x - k_0 : j \in \mathbb{Z}\}.$$

Then for any $y \notin \overline{Y}$, there exists $\epsilon > 0$ such that

$$f_c(B_\epsilon(y)) \cap B_\epsilon(y) = g_j(B_\epsilon(y)) \cap B_\epsilon(y) = \emptyset \qquad \text{for all } c \in \mathbb{Z}^n, j \in \mathbb{Z}. \tag{5.4}$$

Proof: Let T_0, T_1 be the collection of points in $\mathbb{R}^n$ that are fixed by some function in $\mathcal{F}$, $\mathcal{G}$, respectively. Clearly, $T_0 = (A^{*-p} - I)^{-1}\mathbb{Z}^n$ is a lattice in $\mathbb{R}^n$ and $T_1 = \{(A^{*-j} - I)^{-1}k_0 : j \in \mathbb{Z}\}$ is a set with accumulation points 0 and $-k_0$. Suppose $y \notin \overline{Y} = \overline{T_0 \cup T_1 \cup \{0, -k_0\}}$. Let $\gamma = \min(\||y\||, \||y + k_0\||)/2$. Clearly, $f_c(B_\gamma(y)) \cap B_\gamma(y) = \emptyset$ for all but finitely many $c \in \mathbb{Z}^n$. We claim that $g_j(B_\gamma(y)) \cap B_\gamma(y) = \emptyset$ for all but finitely many $j \in \mathbb{Z}$. Indeed, $g_j(B_\gamma(y)) \cap B_\gamma(y) = (A^{*-j}(B_\gamma(y)) - k_0) \cap B_\gamma(y) = \emptyset$ for sufficiently large $j > 0$. Likewise $B_\gamma(y) \cap g_j^{-1}(B_\gamma(y)) = B_\gamma(y) \cap A^{*j}(B_\gamma(y) + k_0) = \emptyset$ for sufficiently small $j < 0$. Therefore, by Lemma 5.2, there exists $\epsilon > 0$ such that (5.4) holds. $\square$

We are now ready to present our main result.

Theorem 5.4. *Let A be an expansive $n \times n$ matrix. Then there is an A-wavelet ψ which is not MSF if and only if there exists a $p \in \mathbb{Z} \setminus \{0\}$ such that $A^{*p}(\mathbb{Z}^n) \cap \mathbb{Z}^n \neq \{0\}$.*

Proof: The forward direction was proven in [10] in the one-dimensional case and in [6] for the n-dimensional case. To see the reverse direction, let $k_0 \neq 0$ be an element of $A^{*p}(\mathbb{Z}^n) \cap \mathbb{Z}^n$, and denote $k_1 = A^{*-p}k_0$, which is also in $\mathbb{Z}^n$. We show that there exists a measurable set $I \subset \mathbb{R}^n$ such that the following four conditions hold:

$$\tau(I) \cap \tau(A^{*-p}(I)) = \emptyset, \tag{5.5}$$

$$d(I) \cap d(I + k_0) = \emptyset, \tag{5.6}$$

$$I \cup (A^{*-p}(I) + k_1) \text{ is contained in a wavelet set, and} \tag{5.7}$$

$$|I| > 0. \tag{5.8}$$

Indeed, let Y be the collection of fixed points of $\mathcal{F} \cup \mathcal{G}$ as in Lemma 5.3. Choose any $y \notin \overline{Y} \cup \mathbb{Z}^n \cup A^{*p}\mathbb{Z}^n$. By Lemma 5.3, there is an $\epsilon > 0$ such that (5.4) holds, i.e., (5.5) and (5.6) hold for the set $I = B_\epsilon(y)$. It is now clear that if we take $\epsilon' < \epsilon$ to be small enough and $I = B_{\epsilon'}(y)$, the hypotheses in Theorem 5.1 are satisfied for the set $I \cup (A^{*-p}(I) + k_1)$. Indeed, (i) and (ii) follow from (5.6) and (5.7), whereas (iii) and (iv) are a consequence of $y \notin \mathbb{Z}^n \cap A^{*p}\mathbb{Z}^n$. Hence, equations (5.5)–(5.8) above are satisfied.

Let W be a wavelet set containing $I \cup (A^{*-p}(I) + k_1)$, and define $T = W \setminus (I \cup (A^{*-p}(I) + k_1))$. Now, we define ψ by

$$\hat{\psi}(\xi) = \begin{cases} 1/\sqrt{2} & \text{for } \xi \in I \cup A^{*-p}I \cup (I + k_0) \\ -1/\sqrt{2} & \text{for } \xi \in A^{*-p}I + k_1 \\ 1 & \text{for } \xi \in T \\ 0 & \text{otherwise.} \end{cases}$$

To see that $\|\hat{\psi}\| = 1$, note that for almost every ξ exactly one of the following conditions holds:

(a) $\xi + w \in T$ for precisely one $w \in \mathbb{Z}^n$,

(b) $\xi + w \in I$ and $\xi + w + k_0 \in I + k_0$ for precisely one $w \in \mathbb{Z}^n$,

(c) $\xi + w \in A^{*-p}I$ and $\xi + w + k_1 \in A^{*-p}I + k_1$ for precisely one $w \in \mathbb{Z}^n$.

Thus, for a.e. ξ, $\sum_{w \in \mathbb{Z}^n} |\hat{\psi}(\xi + w)|^2 = 1$ from which it follows that $\|\hat{\psi}\| = 1$.

We turn now to showing that (1.2) is satisfied by $\hat{\psi}$, i.e., ψ is an A-wavelet. For $\alpha = 0$, an argument similar to the above paragraph works and shows $\sum_{j \in \mathbb{Z}} |\hat{\psi}(A^{*j}\xi)|^2 = 1$ for a.e. ξ. Indeed, note that for almost every ξ, exactly one of the following conditions holds:

(a') $A^{*j}\xi \in T$ for precisely one $j \in \mathbb{Z}$,

(b') $A^{*j}\xi \in I$ and $A^{*j-p}\xi \in A^{*-p}I$ for precisely one $j \in \mathbb{Z}$,

(c') $A^{*j}\xi \in I + k_0$ and $A^{*j-p}\xi \in A^{*-p}I + k_1$ for precisely one $j \in \mathbb{Z}$.

For $0 \neq \alpha \in \Lambda$, where Λ represents the set of A-adic vectors and is given by (1.3), the only possibility that the sum in (1.2) would not be zero is when there exist $(j, m) \in \mathbb{Z} \times \mathbb{Z}^n$, $\alpha = A^{*-j}m$ such that both $A^{*j}\xi$ and $A^{*j}\xi + m$ are in the support of $\hat{\psi}$. In this case, it is clear by (5.5), (5.7) and the definition of ψ, that either $m = k_0$ or $m = k_1$. Now, in the first case we have that $A^{*j}\xi \in I$ and $A^{*j}\xi + k_0 \in I + k_0$ if and only if $A^{*j-p}\xi \in A^{*-p}I$ and $A^{*j-p}\xi + k_1 \in A^{*-p}I + k_1$, where $(j - p, k_1) \in \mathbb{Z} \times \mathbb{Z}^n$, $\alpha = A^{*-j+p}k_1$. Hence, the sum in (1.2) is $(1/\sqrt{2})(1/\sqrt{2}) + (1/\sqrt{2})(-1/\sqrt{2}) = 0$. The second case follows similarly showing (1.2). By Theorem 1.4, ψ is an A-wavelet. $\square$

We note here that Theorem 5.4 does not provide a characterization of all matrices which yield non-MSF wavelets. Indeed, there is not yet even a full characterization of matrices which yield wavelets. Dai, Larson and the second author [12] showed that each expansive matrix admits an MSF wavelet, but that is not a necessary condition.

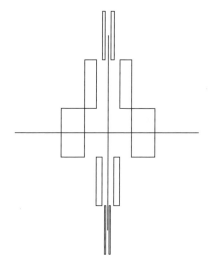

Fig. 1. Support of MSF wavelet for non-expansive dilation.

Example 5.5. *Let A be the (non-expansive) matrix $A = \begin{pmatrix} 2 & 0 \\ 0 & 1 \end{pmatrix}$. Then there exists an A-wavelet.*

Proof: A function of the form $\hat{\psi} = \mathbf{1}_E$ is an A-wavelet if and only if (5.1) holds. In other words, we need that $\{A^{*j}(E) : j \in \mathbb{Z}\}$ and $\{E+k : k \in \mathbb{Z}\}$ form partitions of $\mathbb{R}^2$. Let $k : \{0, 1, 2, 3, \ldots\} \to \mathbb{Z}$ be a bijection, and let

$$E_j = \big([-1/2^j, -1/2^{j+1}] \cup [1/2^{j+1}, 1/2^j]\big) \times \big([-1/2, 1/2] + k(j)\big).$$

Then ψ given by $\hat{\psi} = \mathbf{1}_E$, where $E = \cup_{j \geq 0} E_j$, is a wavelet. $\square$

For example, taking $k(2m + 1) = m + 1$ and $k(2m) = -m$ yields Figure 1, which is easily verified as the support of an MSF wavelet.

Finally, we turn to providing an explicit example of the type mentioned in Section 3. More examples of this type for other dilation factors can be easily constructed with the help of Theorem 5.1.

Example 5.6. *A $\sqrt{2}$-wavelet exists for which $D_\Psi(\xi)$ given by (3.4) is not integer valued, and the core space V_0 given by (3.2) is not shift invariant.*

Proof: Let $I = [\frac{1}{3\sqrt{2}}, 1/3]$. Then, $I \cup (2I - 2)$ is contained in a wavelet set T by Theorem 5.1. So, as in the proof of Theorem 5.4,

$$\hat{\psi}(\xi) = \begin{cases} 1/\sqrt{2} & \text{for } \xi \in I \cup 2I \cup (I - 1) \\ -1/\sqrt{2} & \text{for } \xi \in 2I - 2 \\ 1 & \text{for } \xi \in T \\ 0 & \text{otherwise} \end{cases}$$

is the Fourier transform of a $\sqrt{2}$-wavelet. By a straightforward calculation $D_\Psi(\xi) = \sum_{k\in\mathbb{Z}} \sum_{j\geq 1} |\hat{\psi}(2^{j/2}(\xi+k))|^2 \in \mathbb{Z} + 1/2$ for $\xi \in (\sqrt{2}I \cup \sqrt{2}(I-1)) + \mathbb{Z}$, and $D_\Psi(\xi) \in \mathbb{Z}$ otherwise. As in the proof of Theorem 3.3, one can show that the core space V_0 is not shift invariant, since $j_0(\sqrt{2}, \psi)$ given by (3.7) is 2.

Acknowledgments. The authors wish to thank Qing Gu and David Larson for useful discussions concerning the material in Section 5.

References

1. Auscher, P., Wavelet bases for $L^2(\mathbb{R})$ with rational dilation factor, in *Wavelets and their Applications*, Jones and Bartlett, Boston, 1992, 439–451.

2. Auscher, P., Solution of two problems on wavelets, J. Geom. Anal. **5** (1995), 181–236.

3. de Boor, C., R. A. DeVore, and A. Ron, The structure of finitely generated shift-invariant spaces in $L_2(\mathbb{R}^d)$, J. Funct. Anal. **119** (1994), 37–78.

4. Baggett, L. W., An abstract interpretation of wavelet dimension function using group representations, J. Funct. Anal. **173** (2000), 1–20.

5. Bownik, M., On a problem of Daubechies, Constr. Approx., to appear.

6. Bownik, M., Combined MSF multiwavelets, J. Fourier Anal. Appl., to appear.

7. Bownik, M., The structure of shift invariant subspaces of $L^2(\mathbb{R}^n)$, J. Funct. Anal. **177** (2000), 282–309.

8. Bownik, M., Z. Rzeszotnik, and D. Speegle, A characterization of dimension functions of wavelets, Appl. Comput. Harmon. Anal. **10** (2001), 71–92.

9. Chui, C. K., W. Czaja, M. Maggioni and G. Weiss, Characterization of general tight wavelet frames with matrix dilations and tightness preserving oversampling, J. Fourier Anal. Appl., to appear.

10. Chui, C. K., and X. Shi, Orthonormal wavelets and tight frames with arbitrary real dilations, Appl. Comput. Harmon. Anal. **9** (2000), 243–264.

11. Chui, C. K., X. Shi, and J. Stöckler, Affine frames, quasi-affine frames, and their duals, Adv. Comput. Math. **8** (1998), 1–17.

12. Dai, X., D. Larson, D. Speegle, Wavelet sets in $\mathbb{R}^n$, J. Fourier Anal. Appl. **3** (1997), no. 4, 451–456.

13. Daubechies, I. *Ten Lectures on Wavelets*, SIAM, Philadelphia, 1992.

14. Gu, Q., On interpolation families of wavelet sets, Proc. Amer. Math. Soc. **128** (2000), no. 10, 2973–2979.

15. Hernández, E., X. Wang, and G. Weiss, Smoothing minimally supported frequency wavelets. I. J. Fourier Anal. Appl. **2** (1996), no. 4, 329–340.

16. Hernández, E. and G. Weiss, *A First Course on Wavelets*, Studies in Advanced Mathematics, CRC Press, Boca Raton, FL, 1996.

17. Meyer, Y. *Wavelets and Operators*, Cambridge University Press, Cambridge (1992).

18. Ron, A., and Z. Shen, Frames and stable bases for shift-invariant subspaces of $L_2(\mathbb{R}^d)$, Canad. J. Math. **47** (1995), 1051–1094.

19. Ron, A., and Z. Shen, Affine systems in $L_2(\mathbb{R}^d)$: the analysis of the analysis operator, J. Funct. Anal. **148** (1997), 408–447.

20. Ron, A., and Z. Shen, The wavelet dimension function is the trace function of a shift-invariant system, preprint, 2001.

21. Rzeszotnik, Z., and D. Speegle, On wavelets interpolated from a pair of wavelet sets, Proc. Amer. Math. Soc., to appear.

22. D. Speegle, The s-elementary wavelets are path-connected, Proc. Amer. Math. Soc., **127** (1999), no. 1, 223–233.

Marcin Bownik
Department of Mathematics
University of Michigan
525 East University Ave.
Ann Arbor, MI 48109–1109
marbow@umich.edu

Darrin Speegle
Department of Mathematics
Saint Louis University
221 N. Grand Boulevard
Saint Louis, MO 63103
speegled@slu.edu

New Ties between Computational Harmonic Analysis and Approximation Theory

Emmanuel J. Candès

Abstract. Is the connection between approximation theory and harmonic analysis genuine? This question may seem a little provocative, especially in light of the recent literature about the significant interactions between wavelet analysis and nonlinear approximation. However, this connection might just as well be an accident as it merely seems to be limited to wavelets. For instance, do we know of any other constructions giving important insights into central problems in approximation theory? The work surveyed in this paper suggests that the connection between applied harmonic analysis and approximation theory is much larger than that wavelets or Fourier analysis imply. We introduce some fundamentally new systems for representing functions, and will show how well those new constructions connect with central problems in approximation theory. Especially, we will explore the subject of ridge function approximation and the problem of representing or approximating objects with spatial inhomogeneities.

§1. Introduction

This paper is the companion to the lecture delivered at the 10th Conference on Approximation Theory on March 28th, 2001. Rather than being about a specific problem, this is merely an account of some ideas that have been part of my intellectual life for about 5 years. Many of these ideas formed as I was a Ph. D. student in the Department of Statistics at Stanford University. David L. Donoho served as my advisor and this paper owes a lot to his intellectual generosity.

This article is loosely organized around a central theme, namely, the connection between computational harmonic analysis (CHA) and approximation theory (AT). The rapid development of wavelets in the eighties together with the key observation that wavelets can be construed as

Approximation Theory X: Wavelets, Splines, and Applications   87
Charles K. Chui, Larry L. Schumaker, and Joachim Stöckler (eds.), pp. 87–153.
Copyright ℗ 2002 by Vanderbilt University Press, Nashville, TN.
ISBN 0-8265-1416-2.

the building blocks of many functional classes and approximation spaces [20,48] fueled an explosive literature on this subject, see [29] for a wonderful survey. The importance of this literature was amplified by the fact that most of the ideas discussed bear great practical significance for signal and image processing. By now, we are all familiar with wavelet success stories such as their inclusion in JPEG 2000, the new still-picture compression standard, or the celebrated wavelet shrinkage for removing noise from signals or images.

1.1 An accidental connection?

By the late nineties, the connection between wavelet and approximation theory had been explored rather extensively. A concern –at least to the author– was whether the connection between CHA and approximation theory was genuine. Indeed, there was no evidence at the time of any other CHA construction giving important insights into central problems in approximation theory. In other words, was this tie between wavelets and approximation theory merely an accident?

From the CHA viewpoint, the domain of expertise of wavelets was clearly understood. At the same time, the concern that wavelets were perhaps not the answer to every problem in approximation theory developed. For instance, wavelets offer poor representations of two-dimensional objects that are singular along curves. An example, which to my knowledge should be attributed to Yves Meyer, shows that in two dimensions, nonlinear wavelet approximations of the indicator function of the unit disk converge 'slowly.' We used the word slowly because one can exhibit other methods of approximation such as using superpositions of indicators of triangles with all possible shapes which converge at a much faster rate. In the introduction to this lecture, Ron DeVore addressed this issue and perhaps best summarized that feeling when he said that "wavelets had hit a wall." Implications are clear. The challenge is to find new tools in CHA for problems that wavelets are not able to address efficiently. We should acknowledge, however, that this is an act of faith in which we express the belief that there are many possible CHA constructions beyond wavelets. Against this optimism, we often heard in conferences that not only wavelets but CHA had run its course and that we need ideas outside of CHA. The recent work of Mallat and his collaborators on bandelets [40] indeed explores other directions.

Conversely, from the approximation theory viewpoint, wavelets brought a remarkable sense of closure to an important line of research about spline and rational approximation as well as related topics in mathematical analysis such as applied functional analysis and interpolation theory. For instance, wavelet, free-knot spline or rational function approximation are, in some sense, optimal over the scale of Besov spaces and give the same speed of convergence.

The scope of approximation theory, however, is of course much larger than wavelet or spline approximation. In particular, nonlinear approximation received an increasing degree of attention, perhaps motivated by the close relationship existing between nonlinear approximation and the subject of data compression. We shall now describe a typical set of problems in this area.

1.2 Nonlinear approximation

Given a dictionary of generally overcomplete elements $\mathcal{D} = \{g_\lambda, \lambda \in \Lambda\}$, we are interested in the best m-term approximation

$$\inf_{f_m \in \Sigma_m} \|f - f_m\|_{L_2}, \tag{1.1}$$

where Σ_m is the set of objects that are linear combinations of at most m elements of $\mathcal{D}$

$$\Sigma_m = \{g, \ g = \sum_{i=1}^m \alpha_i g_{\lambda_i}\}. \tag{1.2}$$

Note that Σ_m is not a linear subspace. If g and h are both in Σ_m, then $f + g$ is generally not a member of Σ_m but rather, of the larger set Σ_{2m}. Let a target function f be given. How do we construct the best or near-best m-term approximation to f? What do we know about the degree of approximation

$$d_m(f, \mathcal{D}) \equiv \inf_{f_m \in \Sigma_m} \|f - f_m\|_{L_2} \quad \text{as} \quad m \to \infty? \tag{1.3}$$

Next, instead of being interested in the performance for a specific target f, one might study, instead, the quality of the approximation of a functional class $\mathcal{F}$ by the dictionary $\mathcal{D}$. For example, what do we know about

$$d_m(\mathcal{F}, \mathcal{D}) \equiv \sup_{f \in \mathcal{F}} d_m(f, \mathcal{D})? \tag{1.4}$$

There are also interesting characterization issues. For instance, given a dictionary $\mathcal{D}$ can we characterize those functions that can be approximated at a given rate by members of Σ_m? And conversely, given a functional class $\mathcal{F}$, how do we design a dictionary which, in some sense, is best for approximating its elements? Problems of this nature, whether they are about the characterization of approximation spaces or about the existence of constructive procedures for obtaining near-best m term approximations, are usually hopelessly difficult. There is a lack of universal principles which the author finds a little discouraging. Indeed, each new problem calls for a lot of hard work and progress are generally at the expense of a lot of hard and cumbersome analysis.

1.3 Our claim

I claim that there is much more to CHA than wavelets and related systems. Indeed, the work surveyed in this paper will introduce a wealth of new systems, and we will show how well those new constructions connect with central problems in approximation theory. Further, we suggest that the connection between CHA and approximation theory is much larger than what wavelets imply. We begin our exploration of this renewed connection with the problem of approximating with superpositions of ridge functions.

1.4 Ridge functions

In the seventies, Logan and Shepp [43] coined the terminology ridge function to designate functions of the form $g(k \cdot x) = g(k_1 x_1 + k_2 x_2 + \ldots + k_d x_d)$. In other words, a ridge function is a multivariate function constant on the hyperplanes $k \cdot x = t$ where t is a real valued constant. Ridge functions are also known under the name of planar waves. Over the last twenty years or so, ridge functions have appeared rather frequently in the scientific literature.

Computerized tomography. First, ridge functions play an important role in the literature of computerized tomography. Logan and Shepp [43] considered the problem of reconstructing a two-dimensional function $f(x)$ from its projections $(Rf(\cdot, \theta_1)$, $Rf(\cdot, \theta_2), \ldots, Rf(\cdot, \theta_k))$ for fixed and distinct directions $\theta_1, \ldots, \theta_k$ which belong to $\mathbf{S}^1$, the unit sphere of $\mathbb{R}^2$. Here Rf is the Radon transform

$$Rf(t, \theta) = \int_{\theta \cdot x = t} f(x) \, dx, \quad t \in \mathbb{R}, \ \theta \in \mathbf{S}^1. \quad (1.5)$$

This problem is, indeed, a mathematical idealization of several imaging problems [18] as in tomography. Define the set of 'candidates' as all functions whose Radon transform match the given projections for every θ_j, $j = 1, \ldots, k$. We use the method of regularization and seek that object f^* with minimum L_2 norm among all possible candidates. The result proved by Logan and Shepp is that f^* is a superposition of at most k ridge functions.

Statistics. In statistics, ridge functions were introduced to overcome the adverse effect of the curse of dimensionality. A central problem there is that of estimating an unknown regression surface given data $(x_i, y_i)_{i=1}^N$, where x_i is a d-dimensional input variable and y_i is a real-valued output response and the model

$$y_i = f(x_i) + \epsilon_i; \quad (1.6)$$

here, ϵ is a stochastic and noisy contribution which is assumed to have zero-mean so that the problem is to estimate the conditional mean of y given x

as $f(x) = E(y|x)$. Friedman and Stuetzle [31] suggest approximating the unknown regression function f by a sum of ridge functions

$$f(x) \sim \sum_{j=1}^{m} g_j(k_j \cdot x),$$

where the k_j's are vectors of unit length, i.e. $\|k_j\| = 1$. In its abstract version, the approximation process operates in a stepwise and greedy fashion. At stage m, it augments the fit f_{m-1} by adding a ridge function $g_m(k_m \cdot x)$, where k_m and g_m are chosen so that $g_m(k_m \cdot x)$ best approximates the residuals $f(x) - f_{m-1}(x)$.

For completeness, we should stress that the original paper presented a principle for fitting finite linear combinations of ridge functions in the sampling case. Given data $(x_i, y_i)_{i=1}^{N}$, the procedure is analogous to that described above. At stage m, the fit f_{m-1} is augmented by adding a ridge function $g_j(k_j \cdot x)$ obtained as follows: calculate the residuals of the $(m-1)$th fit $r_i = x_i - \sum_{j=1}^{m-1} g_j(k_j \cdot x_i)$; and for a fixed direction a, plot the residuals r_i against $k \cdot x_i$; fit a smooth curve g and choose the best direction a, so as to minimize the residuals sum of squares $\sum_i (r_i - g(k \cdot x_i))^2$. The algorithm stops when the improvement is small.

Partial differential equations. Planar waves appear frequently in the study of partial differential equations and especially in certain hyperbolic problems [35]. To give a flavor of this connection, we follow [33] and consider the differential operator with constant coefficients

$$L = \sum_{\alpha} a_\alpha D^\alpha,$$

where D is either one of the partial derivatives $\partial_1, \partial_2, \ldots, \partial_d$. Suppose we want to solve

$$Lu = f, \tag{1.7}$$

where f is a nice object, namely, f belongs to the Schwartz class $\mathcal{S}(\mathbb{R}^d)$. Start with the equation

$$Lu = f_\theta, \tag{1.8}$$

where $\theta \in \mathbf{S}^{d-1}$ and $f_\theta(x) = f_\theta(x \cdot \theta)$ is a ridge function. Then, looking for solutions of the form $u_\theta(x) = u_\theta(x \cdot \theta)$ amounts to solving an *ordinary* differential equation. If we can synthesize the right hand-side of (1.7) as a superposition of ridge functions, then under some conditions, we would be able to write down a solution to (1.7) as a superposition of ridge functions as well. For completeness, such synthesizing principles are possible by means of the Radon transform.

As an example, consider the Cauchy problem for the wave equation with constant coefficients

$$\Delta u = \frac{\partial^2 u}{\partial t^2}, \quad u(x,0) = f_0(x), \ u_t(x,0) = f_1(x),$$

where f_0 and f_1 are given smooth functions. Then one has available an explicit formula which expresses the Cauchy problem as a continuous superposition of ridge functions [33]. Note also that for $h \in C^2$ and $\theta \in \mathbf{S}^{d-1}$, the ridge function

$$v^{\pm}(x,t) = h(x \cdot \theta \pm t)$$

satisfies $\Delta v = (\partial^2/\partial t^2)v$.

Neural networks. There is no real definition of a neural network. However, the most commonly studied neural network is the one hidden-layer feedforward neural network which is a name given to a function constructed by the rule

$$f_m = \sum_{j=1}^{m} \alpha_j \sigma(k_j \cdot x - b_j); \tag{1.9}$$

the α_j and b_j are scalars and the k_j are d-dimensional vectors. A one hidden-layer feedforward neural network is then a superposition of m ridge functions which are often called **neurons**. The activation function σ is usually **sigmoidal**, a terminology which means that σ is bounded and monotone with e.g. $\lim_{t \to -\infty} \sigma(t) = 0$ and $\lim_{t \to \infty} \sigma(t) = 1$. Popular choices for σ are the Heavyside $\sigma(t) = 1_{\{t>0\}}$ and the logistic function $\sigma(t) = 1/(1 + e^{-t})$.

Approximation theory. Many of the problems in the neural nets methodology are of an approximation theoretic nature, hence the significant place of ridge functions in the literature of approximation theory. Throughout the remainder of this paper, $\mathcal{D}_{NN}$ will denote the collection of neurons (neural network dictionary)

$$\mathcal{D}_{NN} = \{\sigma(a \cdot x - b), \ k \in \mathbb{R}^d, b \in \mathbb{R}\}. \tag{1.10}$$

The most fundamental questions in this field are a recast of those formulated in a preceding paragraph. For instance, one can ask about the capabilities of this dictionary. As an example, we may be curious about the speed of convergence of finite linear combination of neurons to a given target f as the number of neurons n is increasing and tending to infinity. Or about how well does $\mathcal{D}_{NN}$ approximate a given functional class $\mathcal{F}$. In addition, a central question is about the existence of very concrete procedures with good approximation properties. Later sections will review some results in this area but in truth, very little is known about neural network as a form of approximation. We quote from Petrushev [53] "It is surprising, therefore, that the most fundamental questions concerning the efficiency of approximation by ridge functions are unanswered." There seems to be a general consensus about this, see Pinkus [54] who writes "there is very, very little known about the degree of approximation by ridge functions."

We should emphasize that in order to talk about the speed of convergence, one should, of course, better verify that $\mathcal{D}_{NN}$ is, in some sense, complete. Various completeness results are known for neural networks [13,17,32,34]. These results are fairly recent and we introduce two of them. First, Cybenko [42] shows that the span of $\mathcal{D}_{NN}$ is dense in $C(K)$ for any compact set $K \subset \mathbb{R}^d$, for any continuous σ obeying $\lim_{t \to -\infty} \sigma(t) = 0$ and $\lim_{t \to \infty} \sigma(t) = 1$. This means that for every function $f \in C(K)$ and every $\epsilon > 0$, one can find g in the span of $\mathcal{D}_{NN}$ with the property

$$\sup_K |f(x) - g(x)| < \epsilon.$$

Note that this result gives completeness in other spaces such as in $L_2(K)$ since $C(K)$ is dense in $L_2(K)$. Second, Leshno, Lin, Pinkus and Schocken [42] obtained a rather definitive result concerning the density property of of $\mathcal{D}_{NN}$. They proved that continuous activation σ yields the density in $C(K)$ if, and only if, σ is not a polynomial.

Many mathematical problems in the field of neural networks are of an approximation-theoretic nature. We shall first discuss the problem of constructing neural net approximations, and then review what is known about their degree of approximation.

1.5 Construction of neural networks

Fundamental questions remain open about the computational efficiency of neural networks. First and foremost, it is unclear how to construct neural networks. A previous article [4] pointed out this major conceptual weakness, and our exposition will closely follow the argument presented in that paper. The problem here is that we do not really know how to represent a multivariate function as a superpositions of neurons $\sigma(k \cdot x - b)$. This is in stark contrast with some other areas of approximation theory such as polynomial, Fourier or wavelet approximations.

Because of the lack of synthesizing principle, any approximation procedure amounts to minimizing highly multimodal error surfaces. For instance, to find the best m-term approximation

$$f_m(x) = \sum_{j=1}^{m} \alpha_j \sigma(k_j \cdot x - b_j),$$

one has to solve the following optimization problem

$$\inf \ \|f - f_m\|_{L_2}$$

where the infimum ranges over the coefficients $(\alpha_j)_{j=1}^m$ and the network parameters $(k_j)_{j=1}^m$ and $(b_j)_{j=1}^m$. Solving such problems belong to the realm of dreams. We quote Barron [16]: "There is no known algorithm... Gradient search and its variants produce a local optimum of dubious scientific merit." There is more. In a practical setting where one is given sampled data $(x_i, y_i)_{i=1}^N$, Vu [62] showed that finding the minimum of

$$\sum_i (y_i - f(x_i))^2$$

or even an *approximate* minimum is NP-hard, as soon as $m \geq 2$. Vu improved upon the pioneering work of Jones [37]; the title of Jones' article, "The computational intractability of training sigmoidal neural networks" surely drives our point home. The aim of this line of research is to show that it is impossible to design algorithms running in polynomial time that would produce 'accurate estimates' (the exact formulation is that this problem is NP-hard and it is a conjecture that NP-hard problems cannot be solved in polynomial time).

A possibly more reasonable method of approximation is the greedy algorithm as discussed above. This algorithm and its variants synthesize the approximation f_m through a greedy stepwise addition of terms; begin with $f_0(x) = 0$, the relaxed greedy algorithm inductively defines for each $i = 1, \ldots, m$

$$f_i = \alpha^* f_{i-1} + (1 - \alpha^*)\sigma(k^* \cdot x - b^*), \qquad (1.11)$$

where (α^*, k^*, b^*) are solutions of the optimization problem

$$\min_{0 \leq \alpha \leq 1} \ \min_{(k,b) \in \mathbf{R}^n \times \mathbf{R}} \ \|f - \alpha f_{i-1} + (1 - \alpha)\sigma(k \cdot x - b)\|_2. \qquad (1.12)$$

At stage i, the algorithm updates the current approximation f_{i-1} with a convex combination involving f_{i-1} and a new term, a neuron $\sigma(k \cdot x - b)$, that results in the largest decrease in approximation error (1.12). As we pointed out, this strategy sounds more concrete, but is still subject to some rather serious objections.

- *Algorithm?* At each stage, there are many feasible choices (α_i, k_i, b_i), and the minimization (1.12) involves a nonlinear search over those parameters. This is a nonconvex problem, and to the author's knowledge there is no obvious practical algorithm for solving (1.12). (In a discrete setting, there is work showing that the number of local minima may be bounded below by $C \cdot N^d$, where N is the sample size and *d* the dimension of the space [1]). In a realistic implementation, one would need to discretize the set of parameters to perform a search. How fine does the discretization of the network parameters need to be?

- *Stability.* The greedy approach does not yield stable decompositions. A small perturbation of the input function *f* will typically produce radically different parameter values. In other words, these parameters do not have any reliable scientific meaning.

- *Efficiency.* The work of DeVore and Temlyakov proves that the greedy algorithm obeys very weak approximation bounds even when good approximations exist. To be more concrete, it is possible to synthesize a target function as a superposition of only two elements of the neural net dictionary and prove that the greedy algorithm, producing a sequence of *m*-term approximations, will converge at the rate $O(m^{-1/2})$ and not faster. And the target is only a superposition of two terms!

  These weak properties are well-known to engineers and statisticians. In statistics, this says that stepwise regression may be severely inefficient for model selection. In signal processing, where the greedy algorithm is also known under the name of "matching pursuit," it is known that the inability to look ahead may cause initial errors the algorithm will keep on trying to correct. Chen et al. [14,15] give a sequence of rather spectacular computational examples in this direction.

We say "that there is no obvious practical algorithm for solving (1.12)" in light of the scientific standards in common use in numerical optimization. There, the word algorithm has a very precise meaning, namely, that of a procedure which, in a given number of steps, gives the minimum guaranteed or an approximate minimum with a ticket which says how far we are from the minimum, see the literature on Linear Programming (LP) or Semi-Definite Programming (SDP), for example.

It is not the author's intention to sound sweepingly negative. As a matter of fact, the neural network methodology is used quite successfully in many applications. The point here is that, by and large, this methodology seems removed from mathematical and scientific standards. Success is often measured in terms of how well does a particular methodology perform on a specific example rather than on the scientific underpinning it provides. This is not a criticism, merely a fact. The neural network methodology is a guiding principle and researchers in this field are happy trying everything that works well in practice. However, true scientific progress has to do with a better understanding of these mathematical models.

1.6 Approximation by ridge functions

In this section, we will consider the degree of approximation with n-term approximations from $\mathcal{D}_{NN}$. For each n, put

$$\Sigma_n = \Big\{ \sum_{j=1}^{n} \alpha_j \sigma(k_j \cdot x - b_j), \ \alpha_j \in \mathbb{R}, k_j \in \mathbb{R}^d, b_j \in \mathbb{R} \Big\}.$$

There is a body of work which tells us that ridge function or neural network approximation is as efficient as other means of approximation such as splines or polynomials for approximating classical smoothness classes which we now define.

We first introduce some notations. Let $\alpha = (\alpha_1, \ldots, \alpha_d)$ be a d-uple of nonnegative integers and D^α be the partial derivative $D^\alpha = \partial_1^{\alpha_1} \cdots + \partial_d^{\alpha_d}$. Put $|\alpha| = \alpha_1 + \ldots \alpha_d$. Further, we set m to be a nonnegative integer and Ω to be an open set of $\mathbb{R}^d$. The Sobolev space $W_p^m(\Omega)$ is the completion of $C^m(\Omega)$ with respect to the norm

$$\|f\|_{W_p^m} = \|f\|_{L_p(\Omega)} + \sum_{\alpha:|\alpha|=m} \|D^\alpha f\|_{L_p(\Omega)}, \quad p \in [1, \infty]. \qquad (1.13)$$

An object f belongs to the space W_p^m if it has finite Sobolev norm or, in other words, if f and all of its partial derivatives up to order m are in L_p. Interpolation theory allows the extension of Sobolev norms to the half-line $m \geq 0$ and we will omit this definition. Finally, define Sobolev balls $W_p^m(C)$ by

$$W_p^m(C) = \{f \in W_p^m, \|f\|_{W_p^m} \leq C\}$$

meaning that one get control of the size of f and its derivatives up to order m. Unless specified otherwise, we will take Ω to be the unit ball D of $\mathbb{R}^d$, $D = \{x, x_1^2 + \cdots + x_d^2 \leq 1\}$.

Mhaskar [49] proves the following result. Assume that the activation function σ is C^∞ and that σ is not a polynomial. Then for each $p \in [1, \infty]$ and $m \geq 1$

$$d_n(W_p^m(C), \mathcal{D}_{NN}; L_p) \equiv \sup_{f \in W_p^m(C)} \inf_{g \in \Sigma_n} \|f - g\|_{L_p} \leq K_{s,p} \cdot C \cdot n^{-m/d}. \tag{1.14}$$

There are converse results as well. For instance, letting σ be the logistic sigmoid, $\sigma(t) = 1/(1 + e^{-t})$, Maiorov and Meir [45] prove that

$$d_n(W_p^m(C), \mathcal{D}_{NN}; L_p) \geq K'_{s,p} \cdot C \cdot (n \log n)^{-m/d}. \qquad (1.15)$$

This shows that the degree of approximation $n^{-m/d}$ is, in some sense, optimal. In fact, [19] showed that the Sobolev balls cannot be approximated at a rate faster than $n^{-m/d}$ by any reasonable means of approximation. By 'reasonable,' we mean a method of approximation which depends continuously on f, see the above reference for details. Also, the exponent m/d is that appearing in the Kolmogorov ϵ-entropy, and in the minimax risk of estimation over Sobolev balls.

We would like to mention another interesting result due to Petrushev [53] which is an extension of an earlier work from DeVore, Oskolkov and Petrushev [21]. Set $I = [-1, 1]$. We let X_n be a linear space of univariate functions in $L_2(I)$ of dimension n, and let Θ_n be a finite subset of the unit sphere of $\mathbb{R}^d$. The collection of functions of the form

$$g(x) = \sum_{\theta \in \Theta_n} \alpha_\theta \rho_\theta(\theta \cdot x), \quad \rho_\theta \in X_n, \; \theta \in \Theta_n\}$$

is a linear space Y_n of dimension $\leq n \times \#\Theta_n$. We suppose that X_n obeys

$$\inf_{h \in X_n} \|g - h\|_{L_2(I)} \leq C_s \cdot n^{-s} \cdot \|g\|_{W_2^s(I)}. \tag{1.16}$$

Then Petrushev proves that for appropriately chosen sets Θ_n of cardinality $O(n^{d-1})$, we have

$$\inf_{g \in Y_n} \|f - g\|_{L_2(D)} \leq C_r \cdot n^{-r} \cdot \|g\|_{W_2^r(D)}, \quad r = s + (d-1)/2. \tag{1.17}$$

Note that the cardinality of Y_n obeys $\#Y_n = O(n^d)$. For instance we may construct X_n as the univariate space spanned by $\sigma(nt - k)$, $0 \leq k \leq n$ and assume that the activation function σ is chosen such that (1.16) holds. Then define Y_n as described above, i.e.

$$Y_n = \text{span}\{\sigma(n x \cdot \theta_n - k), \theta_n \in \Theta_n, 0 \leq k < n\}.$$

Then Y_n obeys (1.17). Petrushev remarks that there is "an unexpected gain of $(d-1)/2$ in the approximation order." To describe this phenomenon, consider the case where σ is the Heavyside $\sigma(t) = 1_{\{t > 0\}}$, and observe that for this choice of sigmoid, X_n obeys (1.16) for $s = 1$. Then Y_n approximates elements with $1 + (d-1)/2$ derivatives at the optimal rate, even though σ, and by the same token the elements of Σ_n, are discontinuous. We shall return to this point later. The point of Petrushev's findings is that the condition (1.16) is about the approximation of univariate functions which automatically translates into corresponding approximation properties for multivariate smoothness classes by superpositions of ridge functions.

We remark that (1.17) is a result about linear approximation which, in some sense, is an extension of Mhaskar's theorem (1.14) since it allows for very general X_n. On the other hand, however, it is only established for $p = 2$.

We would like to point the reader to other references in this direction of research such as [44] and [52] which give other results with means of approximation which are not continuous. See also, Pinkus [55] for a rapid tour of those references.

This line of research shows, essentially, that neural networks enjoy the same degree of approximation as polynomials over classical smooth-ness classes. In fact, polynomials play a central role in the proofs of the above results. For instance, the key observation behind Mhaskar's theo-rem (1.14) is that for every integer $n \geq m$ there exists a polynomial $\pi_n(f)$ of coordinatewise degree not exceeding n such that for every $f \in W_p^m$, we have

$$\|f - \pi_n(f)\|_{L_p} \leq C \cdot n^{-m} \cdot \|f\|_{W_p^m}.$$

The idea is that one can then approximate each monomial of $\pi_n(f)$ with finite differences of neurons, see [49]. There is a similar idea underlying the result (1.17) of Petrushev. The starting point is again to use polynomials to approximate elements of W_2^s, and to decompose those into 'ridge' poly-nomials. The profile of those 'ridge' polynomials are of course polynomial and one can use X_n to approximate those univariate polynomials.

The point of this research, however, is not very clear to the author. We have available a very concrete, stable and extraordinarily well-understood method of approximation by polynomials. Why should we prefer obscure methods of approximations by neural networks which are often unstable [49] and algorithmically nonconstructive? Isn't it possible to introduce function classes over which neural networks are arguably better than other means of approximation? In some sense, this is the spirit of the work of Barron we introduce next.

1.7 Barron's results

Barron [2] published a result about the degree of approximation of neural networks which had an enormous impact upon the community. Letting $\hat{f}$ be the Fourier transform of f, Barron introduces the class of functions

$$\mathcal{B}(C) = \left\{ f \in L_2([0,1]^d) : \int_{\mathbf{R}^d} |\xi||\hat{f}(\xi)| \, d\xi \leq C \right\}. \qquad (1.18)$$

(Here and throughout the remainder of the paper, the notation $|\cdot|$ stands for the Euclidean norm.) Barron shows that for this class, the relaxed greedy algorithm (1.11)–(1.12) produces a sequence of approximations with the property

$$\|f - f_n\|_2 \leq 2 \cdot C \cdot n^{-1/2}, \qquad (1.19)$$

where f_n is the output of the algorithm at stage n. One notices that the exponent of convergence does not depend upon the dimension d, which launched a curious discussion about the fact that neural networks defeated the curse of dimensionality.

The argument, here, is that Barron's class is included in the convex hull of the neural net dictionary. In other words, every $f \in \mathcal{B}(C)$ can be written as

$$f = \sum_j \alpha_j \sigma(k_j \cdot x - b_j), \quad \text{with} \quad \sum_j |a_j| \leq C. \qquad (1.20)$$

A stochastic argument which goes back to Maurey [56] then shows that if f belongs to the convex hull of $\{\sigma(k \cdot x - b)\}$, there exists a sequence of n-term approximations which converge at the rate $n^{-1/2}$. Further, in a remarkable paper, Jones [36] showed that the greedy algorithm converges at this same rate.

On the one hand, we already highlighted the lack of constructive character of the greedy algorithm which does not turn (1.19) into a very constructive result. On the other hand, the dictionary of orthogonal sinusoids is optimal for approximating elements of $\mathcal{B}(C)$, which is hardly surprising since the class $\mathcal{B}(C)$ is defined by means of the Fourier transform. We let $\mathcal{D}_F$ be the classical orthobasis $\{e^{i2\pi k \cdot x}, \ k \in \mathbb{Z}^d\}$, and let f_n^F be the approximation obtained by keeping only the n largest terms of the trigonometric series –ironically, trigonometric exponentials are ridge functions. We have

$$\sup_{f \in \mathcal{B}(C)} \|f - f_n^F\|_{L_2} \leq C \cdot n^{-1/2 - 1/d}. \qquad (1.21)$$

Roughly, this follows from the equivalence (f is compactly supported)

$$\int_{\mathbb{R}^d} |\xi| |\hat{f}(\xi)| \, d\xi \asymp \sum_{k \in \mathbb{Z}^d} |k| |\hat{f}(2\pi k)|,$$

which is a simple consequence of a famous theorem about the sampling of bandlimited functions due to Polỳa and Plancherel [57]. Therefore, $f \in \mathcal{B}(C)$ implies that the Fourier coefficients $c_k(f)$ of f obey

$$\sum_k |k| \, |c_k(f)| \leq C. \qquad (1.22)$$

The inequality (1.22) gives a bound on the decay coefficient sequence of f. Skipping the details and letting $|c(f)|_{(n)}$ be the nth largest entry in the sequence, (1.22) gives

$$\sum_{m > n} |c(f)|_{(m)}^2 \leq C \cdot n^{-(1+2/d)},$$

and, therefore, (1.21).

Actually, Makovoz [46] later improved the bound (1.19), and showed

$$d_n(\mathcal{B}(C), \mathcal{D}_{NN}) \leq C \cdot n^{-1/2 - 1/2d}. \tag{1.23}$$

This upperbound is still not as good as the rate provably obtained by trigonometric approximations (1.21).

In short, these results are not very satisfying as one may exhibit other well-established method of approximation, namely, the thresholding of Fourier series, with at least as good approximation properties.

1.8 Key issues

Neural networks are used everyday for approximation, prediction, pattern recognition, etc. in the applied sciences and engineering. This is a gigantic field; there are several journals in the field with several thousands of papers published on neural networks, many annual conferences and several dozens of textbooks on this subject. And yet, there seems to be no real theoretical basis, and also practical issues such as the construction of neural nets need to be addressed. From the viewpoint of approximation, there is a need to understand the properties of neural net expansions, to understand what they can and what they cannot do, and where they do well and where they do not. How and which type of functions should we approximate with ridge functions? How do we develop approximation bounds?

1.9 A different approach

The work surveyed in this paper suggests that there is a very different way to go about this problem. We develop transforms which allow the representation of arbitrary objects by superpositions of ridge functions. Those transforms can be used to construct stable approximations. An analogy may be helpful to illustrate this shift in emphasis [27].

In one dimension, consider the problem of approximating a function f by dilations and translations $\sigma(at - b)$, $a > 0, b \in \mathbb{R}$ of a single template σ. If σ is sigmoidal, this is far from being obvious. Now, substitute the sigmoid σ with an oscillatory profile ψ. Harmonic analysis tells us that this becomes a much better posed problem. Indeed, the wavelet transform allows the representation of rather arbitrary signals as superpositions of elements of the form $\psi(at - b)$. Moreover, one has available wavelet orthobases and fast algorithms which yield very concrete procedures for obtaining stable n-term approximations. This is especially relevant for practical applications. The philosophy is the same, namely, that of approximating objects with dilations and translations of a single (or a few) templates. Only the shape of this template has changed.

Just as the wavelet transform allows the representations of arbitrary functions with superpositions of dilations and translations of a single function, the ridgelet transform will allow the representations of multivariate objects with dilated, rotated and translated versions $\psi(au \cdot x - b)$ of a single ridge function, say, $\psi(x_1)$. Like in wavelet theory, there is both a discrete and a continuous transform which we now introduce.

§2. Ridgelets

This section introduces the ridgelet transforms and surveys some of their main properties. All of the forthcoming claims and results are proved in [4]. For now, $\hat{g}$ will denote the Fourier transform of g,

$$\hat{g}(\xi) = \int_{\mathbf{R}^d} f(x)e^{-ix\cdot\xi} \, dx. \tag{2.1}$$

2.1 The continuous ridgelet transform

In d dimensions, the ridgelet construction starts with a univariate function ψ satisfying an admissibility condition, namely,

$$K_\psi = \int |\hat{\psi}(\xi)|^2/|\xi|^d \, d\xi < \infty; \tag{2.2}$$

this condition says that ψ is oscillatory and has vanishing moments up to about $d/2$. Here, the number of vanishing moments grows linearly with the dimension of the space. Sigmoidal activation functions in use in the theory of neural networks are not admissible. A ridgelet is a function of the form

$$\frac{1}{a^{1/2}}\psi\left(\frac{u \cdot x - b}{a}\right), \tag{2.3}$$

where a and b are scalar parameters and u is a vector of unit length. Of course, a ridgelet is a ridge function and resembles a neuron but for the oscillatory behavior of the profile. A ridgelet has a scale a, an orientation u, and a location parameter b. Ridgelets are concentrated around hyperplanes: roughly speaking the ridgelet (2.3) is supported near the strip $\{x, |u \cdot x - b| \leq a\}$. Ridgelets are pictured in Fig. 1 for various values of these parameters.

Define a ridgelet coefficient as

$$\mathcal{R}_f(a, u, b) = \int f(x) \, a^{-1/2}\psi(\frac{u \cdot x - b}{a}) \, dx; \tag{2.4}$$

then for any $f \in L_1 \cap L_2(\mathbf{R}^d)$, we have

$$f(x) = \int \mathcal{R}_f(a, u, b)a^{-1/2}\psi(\frac{u \cdot x - b}{a}) \, d\mu(a, u, b), \tag{2.5}$$

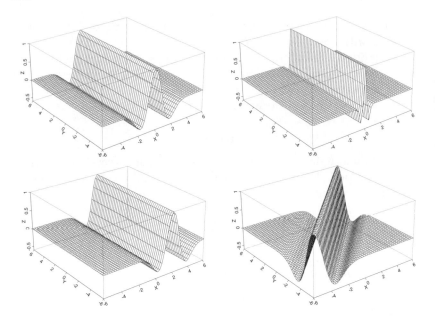

Fig. 1. Ridgelets.

where $d\mu(a, u, b) = da/a^{d+1} \, du \, db$ (du being the uniform measure on the sphere) which holds true if ψ is properly normalized, i.e. $K_\psi = 1/(2\pi)^{d-1}$ in (2.2). The formula (2.5) expresses the idea that one can represent any function as a superposition of these ridgelets. In fact, this formula has been independently discovered by two other researchers, namely, Murata [50] and Rubin [58]. The former is working on neural networks while the latter is working in the field of mathematical analysis. The interpretation of the formula is classical. For $f \in L_1 \cap L_2(\mathbb{R}^d)$, put

$$f_{\epsilon,\delta}(x) = \int_{\epsilon < a < \delta} \int_{\mathbf{S}^{d-1}} \int_{\mathbb{R}} \mathcal{R}_f(a, u, b) a^{-1/2} \psi\left(\frac{u \cdot x - b}{a}\right) d\mu(a, u, b);$$

$f_{\epsilon,\delta}$ is well-defined and obeys

$$\|f - f_{\epsilon,\delta}\|_{L_2} \to 0 \text{ as } \epsilon \to 0, \ \delta \to \infty.$$

Furthermore, the reproducing formula is stable as one has a Parseval relation

$$\|f\|_2^2 = \int |\mathcal{R}_f(a, u, b)|^2 d\mu(a, u, b). \tag{2.6}$$

Like in Fourier or wavelet analysis, this says that a perturbation of the function (resp. the coefficient sequence) has a well-controlled effect on the coefficient sequence (resp. the reconstructed object).

As in Littlewood-Paley or wavelet theory, one may want to discretize the scale on a a dyadic lattice. Let us choose a profile ψ obeying

$$\sum_{j \in \mathbb{Z}} \frac{|\hat{\psi}(2^{-j}\xi)|^2}{|2^{-j}\xi|^{d-1}} = K'_\psi, \qquad (2.7)$$

a condition which greatly resembles the admissibility condition (2.2) introduced earlier. Note that if one is given a function Ψ obeying

$$\sum_{j \in \mathbb{Z}} |\hat{\Psi}(2^{-j}\xi)|^2 = c$$

as in wavelet theory, ψ defined by $\hat{\psi}(\xi) = |\xi|^{(d-1)/2}\hat{\Psi}(\xi)$ will verify (2.7). Assume the special normalization $2K'_\psi = (2\pi)^{-(d-1)}$ in (2.7). Then

$$f(x) = \sum_{j \in \mathbb{Z}} 2^{jd} \int \mathcal{R}_f(2^{-j}, u, b) 2^{j/2} \psi(2^j(u \cdot x - b)) \, du \, db, \qquad (2.8)$$

where again the inequality holds in an L_2 sense; for $f \in L_1 \cap L_2(\mathbb{R}^d)$, the partial sums of the right-hand side are square integrable and converge to f in L_2.

Finally, as in wavelet theory, we may introduce some special coarse scale ridgelets. We choose a profile φ such that $\xi \in \mathbb{R}$

$$\frac{|\hat{\varphi}(\xi)|^2}{|\xi|^{d-1}} + \sum_{j \geq 0} \frac{|\hat{\psi}(2^{-j}\xi)|^2}{|2^{-j}\xi|^{d-1}} = K'_{\psi'} \qquad (2.9)$$

Note that the above equality implies $|\hat{\varphi}(\xi)|^2 \leq |\xi|^{d-1}$, which is very much *unlike* Littlewood-Paley or wavelet theory: our coarse scale ridgelets are also oscillating since $\hat{\varphi}$ must have some decay near the origin, that is, φ itself must have some vanishing moments.

Let (φ, ψ) be a pair obeying (2.9) with $2K'_\psi = (2\pi)^{-(d-1)}$. Then

$$f = \int \mathcal{R}_f^0(u, b)\varphi(u \cdot x - b) \, du \, db$$

$$+ \sum_{j \geq 0} 2^{jd} \int \mathcal{R}_f(2^{-j}, u, b) 2^{j/2} \psi(2^j(u \cdot x - b)) \, du \, db,$$

$$(2.10)$$

with

$$\mathcal{R}_f^0(u, b) = \int f(x)\varphi(u \cdot x - b) \, dx,$$

and $\mathcal{R}_f$ as before.

2.2 The discrete ridgelet transform

Similar to the continuous transform, there is a discrete transform. Let R be the triple (j, ℓ, k) where the indices run as follows

$$R \in \mathcal{R} := \{(j, \ell, k),\ j, k \in \mathbb{Z}, j \geq j_0, \ell \in \Lambda_j\},$$

and define the collection of discrete ridgelets

$$\psi_R(x) = 2^{j/2} \psi(2^j u_{j,\ell} \cdot x - k), \quad R \in \mathcal{R}. \tag{2.11}$$

Note that the range of the parameter ℓ is scale dependent as it depends on j. Ridgelets are directional and, here, the interesting aspect is the discretization of the directional variable u; this variable is sampled at increasing resolution so that at scale j, the discretized set is a net of nearly equispaced points at a distance of order 2^{-j}; a detailed exposition on the ridgelet construction is given in [4].

The key result is that the discrete collection of ridgelets $(\psi_R)_{R \in \mathcal{R}}$ is complete in $L_2[0, 1]^d$, and any function f can be reconstructed from the knowledge of its coefficients $(\langle f, \psi_R \rangle)_{R \in \mathcal{R}}$. (The notation $\langle \cdot, \cdot \rangle$ stands here and throughout this paper for the usual inner product of L_2: $\langle f, g \rangle = \int f(x) g(x) dx$.) There exist two constants A and B such that for any $f \in L_2[0, 1]^d$, we have

$$A \|f\|^2 \leq \sum_{R \in \mathcal{R}} |\langle f, \psi_R \rangle|^2 \leq B \|f\|^2. \tag{2.12}$$

The previous equation says that the datum of the ridgelet transform at the points $(a = 2^j, u = u_{j,\ell}, b = k2^{-j})_{(j,k,\ell) \in \mathcal{R}}$ suffices to reconstruct the function perfectly. In this sense, this is analogous to the Shannon sampling theorem for the reconstruction of bandlimited functions. Indeed, standard arguments show that there exists a dual collection $(\tilde{\psi}_R)_{R \in \mathcal{R}}$ with the property

$$f = \sum_{R \in \mathcal{R}} \langle f, \tilde{\psi}_R \rangle \psi_R = \sum_{R \in \mathcal{R}} \langle f, \psi_R \rangle \tilde{\psi}_R, \tag{2.13}$$

which gives perfect and stable reconstruction.

2.3 Frequency-side picture

In d-dimensions, ridgelets are localized around planes of codimension 1. To make things concrete, consider the situation in two dimensions where the discretization of the angular variable is as follows: we let $u_{j,\ell} = (\cos \theta_{j,\ell}, \sin \theta_{j,\ell})$ and for a fixed discretization step $\alpha > 0$, set

$$\theta_{j,\ell} = \alpha \pi 2^{-j} \ell, \quad \ell = 0, 1, \ldots, \lfloor 2^j/\alpha \rfloor.$$

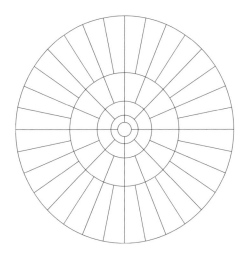

Fig. 2. Illustration of ridgelet sampling scheme in the frequency plane.

Ridgelets are oriented in the codirection $\theta_{j,\ell} = \ell/2^j$ and located near the line $x_1 \cos\theta_{j,\ell} + x_2 \sin\theta_{j,\ell} - k2^{-j} = 0$. Their width is about 2^{-j} so that we can think of ridgelets as a collection of fat lines at different scales.

The frequency-side picture also gives some nice insights about the organization of the ridgelet transform. In the Fourier domain, a ridge function $\rho(k \cdot x)$ is, in fact, supported on the radial line (λk), $\lambda \in \mathbb{R}$. Then, the Fourier transform of a ridgelet ψ_R (polar coordinates) is given by

$$(\hat{\psi}_R)(\lambda\cos\theta, \lambda\sin\theta) = \hat{\psi}_I(2^{-j}\lambda)\delta(\theta - \theta_{j,\ell}), \quad \lambda \in \mathbb{R},\ \theta \in [0, 2\pi), \quad (2.14)$$

where ψ_I is the dyadic wavelet $\psi_I(t) = 2^{j/2}\psi(2^j t - k)$, $I = (j, k)$. Recall the Fourier transform of ψ_I

$$\hat{\psi}_I(\lambda) = 2^{-j/2}\,\psi(2^{-j}\lambda)\,e^{ik2^{-j}}.$$

Therefore, in the frequency plane, ridgelets are supported near the dyadic segments represented on Fig. 2. The angular sampling principle is explicit; the number of ridgelet orientations is doubled from one dyadic corona to the next. We close this section by summarizing the main points of the discrete ridgelet transform.

- Like in Littlewood-Paley theory, divide the frequency domain into dyadic annuli $|x| \in [2^j, 2^{j+1})$.
- Sample each annulus with radial lines $\theta = \alpha\pi 2^{-j}\ell$.
- The angular resolution increases with the spatial resolution as illustrated in Fig. 2.

2.4 Connection with neural nets

The introduction described a popular approach – the greedy algorithm – to compute neural net approximations. At each step, one would need to solve an optimization problem of the form (1.12), and in any real implementation, one would probably need to restrict the search for a minimum over a grid. What are the properties of a restricted search? Is there a grid preserving the completeness property? If so, what is the proper spacing of this grid? In other words, what is the real complexity of the search (1.12)? In some sense, the discretization (2.11) gives a precise answer to these questions.

§3. Orthonormal Ridgelets

In dimension 2, Donoho [26] introduced a new orthonormal basis whose elements he called 'orthonormal ridgelets.' We quote from [10]: "Such a system can be defined as follows: let $(\psi_{j,k}(t) : j \in \mathcal{Z}, k \in \mathcal{Z})$ be an orthonormal basis of Meyer wavelets for $L^2(\mathbb{R})$ [41], and let $(w^0_{i_0,\ell}(\theta)$, $\ell = 0, \ldots, 2^{i_0} - 1;\ w^1_{i,\ell}(\theta),\ i \geq i_0,\ \ell = 0, \ldots, 2^i - 1)$ be an orthonormal basis for $L^2[0, 2\pi)$ made of periodized Lemarié scaling functions $w^0_{i_0,\ell}$ at level i_0 and periodized Meyer wavelets $w^1_{i,\ell}$ at levels $i \geq i_0$. (We suppose a particular normalization of these functions). Let $\hat{\psi}_{j,k}(\omega)$ denote the Fourier transform of $\psi_{j,k}(t)$, and define ridgelets $\rho_\lambda(x)$, $\lambda = (j, k; i, \ell, \varepsilon)$ as functions of $x \in \mathbb{R}^2$ using the frequency-domain definition

$$\hat{\rho}_\lambda(\xi) = |\xi|^{-\frac{1}{2}}(\hat{\psi}_{j,k}(|\xi|)w^\varepsilon_{i,\ell}(\theta) + \hat{\psi}_{j,k}(-|\xi|)w^\varepsilon_{i,\ell}(\theta + \pi))/2 . \qquad (3.1)$$

Here the indices run as follows: $j, k \in \mathbb{Z}$, $\ell = 0, \ldots, 2^{i-1} - 1$; $i \geq i_0$, $i \geq j$. Notice the restrictions on the range of ℓ and on i. Let λ denote the set of all such indices λ. It turns out that $(\rho_\lambda)_{\lambda \in \Lambda}$ is a complete orthonormal system for $L^2(\mathbb{R}^2)$."

Orthonormal ridgelets are *not* ridge functions, although there is a close connection between 'pure' and orthonormal ridgelets. Using a formulation emphasizing the resemblance with (3.1), the frequency representation of a pure ridgelet is given by

$$\hat{\psi}_{j,\ell,k}(\xi) = (\hat{\psi}_{j,k}(|\xi|)\delta(\theta - \alpha\pi 2^{-j}\ell) + \hat{\psi}_{j,k}(-|\xi|)\delta(\theta + \pi - \alpha\pi 2^{-j}\ell))/2. \quad (3.2)$$

In the ridgelet construction, the angular variable θ is uniformly sampled at each scale; the sampling step being proportional to the scale 2^{-j}. In contrast, the sampling idea is replaced by the wavelet transform for orthonormal ridgelets. The orthonormal ridgelet basis is a tensor basis in the polar coordinate system. It is properly renormalized by the Jacobian underlying the Cartesian to polar change of coordinates so that it yields an orthobasis of $L^2(\mathbb{R}^2)$. It is interesting to note that the restriction on the range, namely, $i \geq j$ in the definition (3.1), gives the same angular scaling as in the original ridgelet construction.

§4. Ridgelet Thresholding

Whereas the construction of neural networks involved a delicate stepwise construction of approximations, ridgelet analysis gives an *explicit and stable* formula for representing a function as a superposition of ridge functions. Further, the expansion (2.13) allows the construction of simple finite approximations using naive ideas like thresholding. For instance, we may truncate the exact series

$$f = \sum_R \theta_R \tilde{\psi}_R, \quad \theta_R = \langle f, \psi_R \rangle,$$

by extracting the terms corresponding to the n largest coefficients; that is,

$$f_n^R = \sum_\lambda \theta_R 1_{\{|\theta_R| \geq \delta\}} \tilde{\psi}_R, \tag{4.1}$$

where δ is chosen so that $\#\{R, |\theta_R| \geq \delta\} = n$, i.e. δ is about the nth largest entry of the sequence $|\theta|$ (in case of ties, i.e. $|\theta_R| = \delta$ for several R's, any subset will do).

We note that (4.1) is a simple, constructive and stable method of approximation, and we propose this strategy as an alternative to the ill-posed construction of neural networks. The philosophy, however, is the same; that of approximating multivariate objects by finite linear combinations of dilations, translations and rotations of a single ridge function.

§5. Ridgelets and Neural Networks

Ridgelet thresholding may seem like a very naive strategy. As a first impulse, one would imagine that it would not be able to match the approximation performance of possibly very abstract neural network approximation strategies. This section shows that, in some sense, this intuition is wrong. Ignoring boundary issues, we claim that there is no such function which is approximated at a faster rate, in an asymptotic sense, with sigmoidal feedforward neural networks than with ridgelet thresholding. We now detail this claim and refer the reader to [6] for further reference.

Assume σ is the logistic function $\sigma(t) = (1 + e^{-t})$, and consider a sequence $(f_n)_{n \geq 1}$ of feedforward neural networks

$$f_n^{NN} = \sum_{i=1}^n \alpha_i^n \sigma(k_i^n \cdot x - b_i^n) \tag{5.1}$$

with coefficients having polynomial growth, that is

$$\sup_i |\alpha^n| = O(n^\beta), \quad \text{for some} \quad \beta \geq 0. \tag{5.2}$$

We emphasize that the parameters of the network, namely the directions and locations, are allowed to change as the number of terms or neurons in the approximation increases. The restriction concerning the growth of the coefficients α^n (5.2) is very mild. It says that the coefficients have at most polynomial growth in the number of terms which prevents the consideration of very wild approximants that would practically be irrelevant, see the discussion in [6].

Theorem 5.1. *Let* $f \in L_2(\mathbb{R}^d)$ *be supported in the unit ball* D *of* $\mathbb{R}^d$, *and suppose that there is a sequence* (f_n) *of feedforward neural networks (5.1) obeying*

$$\|f - f_n^{NN}\|_{L_2(2D)} = O(n^{-r}).$$

Then the n-term approximations f_n^R *(4.1) obtained by simply thresholding the exact ridgelet series of* f *obeys*

$$\|f - f_n^R\|_{L_2(2D)} = O(n^{-r+\delta}), \quad \text{for any} \quad \delta > 0.$$

In short, ridgelets thresholding guarantees at least the same rate as the best neural net approximation.

We already argued that the best approximation f_n^{NN} is merely existential; we may think about it as an ideal approximation able to pick the best possible n directions $(k_1, k_2, \ldots, k_n)$. A remarkable feature of the ridgelet thresholding procedure is that it does not know anything about those directions and yet performs nearly equally as well. That is, ridgelet thresholding gives rates of approximation which rival those attainable by very abstract and complicated procedures.

Important Remark. We would like to point out that *this result is not limited to the use of the logistic function*. Actually, Theorem 5.1 still holds whether one uses a sigmoidal and monotone C^∞ activation function. It is also true if σ is C^∞ and vanishes at infinity, e.g. σ belongs to the Schwartz class. Moreover, σ may not be continuous. The theorem is still valid in the case where σ is the Heavyside, say.

We note that the formulation of Theorem 5.1 avoids discussing problems related to the geometry of the domain D, say. We mention, in passing, that other formulations of Theorem 5.1 are possible with other boundary conditions. Boundaries introduce, of course, artificial discontinuities which require a special treatment and, at the moment, we are not interested in studying those boundary effects. (Nearly everyone is aware of boundary issues. If by any chance, the reader is not familiar with these, consider the following situation. In two dimensions, we let f be a superb C^∞ function supported on the unit disk D. Then, wavelet orthobases of $L_2(\mathbb{R}^2)$ yield poor approximations of f although it is C^∞ on the disk. The problem is that the extension of f to the plane introduces a discontinuity along ∂D.)

Theorem 5.1 is proved in [6], and we will briefly outline the argument in a later section. A preliminary version of Theorem 5.1 was first introduced in [3].

§6. Ridgelets and Ridge Functions

At the heart of Theorem 5.1 lies a key property about the ridgelet representation of ridge functions which we now present.

6.1 Ridgelets and sigmoidal ridge functions

Let σ be a C^∞ sigmoidal function, and construct the ridge function

$$f(x) = \sigma(k \cdot x - b); \qquad (6.1)$$

f is not in any L_p and, therefore, the inner product between a ridge function and a ridgelet does not make much sense. We localize this ridge function near the unit disk with a multiplication by a smooth window w in C_0^∞. The key-property is that the sorted ridgelet coefficients $|\alpha(fw)|_{(n)}$ of $f\,w$ decay faster than any negative power of n. We establish that for any $p > 0$,

$$\sum_R |\alpha_R(fw)|^p \leq C_p, \qquad (6.2)$$

where the constant C_p may be chosen independently of the parameters k and b. It is well-known that the sparsity of the decomposition controls the quality of partial reconstructions, see [22] for example. Suppose that w is identically equal to one over the unit D and vanishes out of $2D$, and let f_n^R be the truncated ridgelet series of $f\,w$, keeping the n-largest terms in the ridgelet series. Then for any $s > 0$, we have

$$\|f - f_n^R\|_{L_2(D)} \leq C_s \cdot n^{-s}, \qquad (6.3)$$

where again the constant C_s is may be chosen independently of the parameters of the ridge function (6.1).

As we pointed out, the estimates (6.2)–(6.3) hold uniformly over k and b. As k varies, the ridge functions (6.1) exhibit very different behaviors. When $|k|$ is small, say $|k| \leq 1$, f is a very gentle function; its derivatives up to any order are well behaved. As $|k|$ increases, however, f becomes less smooth and in the limit is discontinuous, e.g.

$$\sigma(a \cdot (\theta \cdot x - b)) \to H(\theta \cdot x - b), \text{ as } a \to \infty, \quad \theta \in \mathbf{S}^{d-1}$$

where H is the Heavyside $H(t) = 1_{\{t>0\}}$. It is then interesting to note that the ridgelet sequence is equally sparse in either situation. First, ridgelets provide sparse representations of smooth functions, which explains (6.2) whenever the parameter $|k|$ is not too large. Then as $|k|$ increases, a singularity develops. In the ridgelet domain, however, very few coefficients actually feel this singularity just as in one dimension, only a few wavelets would feel a point singularity. This localization phenomenon is responsible for the sparsity of the sequence, see Section 8 and [7] for further reference.

6.2 Ridgelets and 'Ridge-Wavelets'

Theorem 5.1 is not limited to smooth sigmoidal activation functions. Likewise, the properties (6.1)–(6.3) hold for other profiles. For instance, let ψ^M be the father Meyer wavelet [48], and consider the Meyer ridge function

$$f(x) = a^{1/2}\psi^M(a\theta \cdot x - b), \quad a > 0, \theta \in \mathbf{S}^{d-1}, \quad b \in \mathbb{R}. \tag{6.4}$$

Recall that the Meyer ψ^M belongs to the Schwartz class and is bandlimited, i.e. its Fourier transform is compactly supported. Then, the ridgelet coefficient sequence of the ridge Meyer wavelet (6.4) also obeys the estimate (6.2).

The ℓ_p-summability of ridge-wavelets (6.4) is especially interesting because it says that ridgelets are nearly orthogonal. Indeed, (6.2) applied to a smooth and oscillatory profile guarantees that the Gramm matrix

$$T(R, R') := \langle \psi_R, \psi_{R'} \rangle_{L_2(w)} := \int \psi_R(x)\psi_{R'}(x)\, w(x)\, dx \tag{6.5}$$

is nearly diagonal; Specifically, suppose ψ belongs to the Schwartz class and has vanishing moments up to any order. Then

$$\sup_R \sum_{R'} |T(R, R')|^p \leq A_p. \tag{6.6}$$

This key property is proved in the Appendix. In other words, (6.6) shows that the rows are very sparse as they are uniformly bounded in ℓ_p for any $p > 0$.

The results collected so far bring up an interesting question which Theorem 5.1 partially addresses. The grail would be to show that if a function can be well-approximated by ridge-wavelets, or ridge-free-knots-splines, it can be well approximated by ridgelet thresholding. We now discuss some results in this direction.

Inspired by Donoho [27], we consider the ridge function

$$f_\theta = f(\theta \cdot x), \quad \theta \in \mathbf{S}^{d-1}, \tag{6.7}$$

and assume that the profile f is supported over the interval $[-1, 1]$, say. We might develop an ideal n-term approximation of f as follows: we use an orthobasis of wavelets (ψ_I) for approximating the profile and let f_n be an n-term wavelet approximation of the ridge profile; we construct $f_{n,\theta}$ by the rule

$$f_{n,\theta} = f_n(\theta \cdot x), \tag{6.8}$$

Assume now that nonlinear wavelet partial reconstructions of the profile obey

$$\|f - f_n\|_{L_2} \leq C \cdot n^{-s}, \tag{6.9}$$

for a constant C not depending on n. The point is that (6.9) is about one-dimensional functions and automatically gives the corresponding multivariate degree of approximation

$$\|f - f_{n,\theta}\|_{L_2(D)} \leq C \cdot n^{-s}. \tag{6.10}$$

Again, let w be a fixed window in $\mathcal{S}(\mathbb{R}^d)$ with the property that w is identically equal to one over the unit ball D and vanishes out of $2D$. Consider now the ridgelet approximation f_n^R obtained by thresholding the ridgelet expansion of $f\,w$, keeping the n-largest terms. Then f_n^R obeys

$$\|f - f_n^R\|_{L_2(D)} \leq C \cdot n^{-s}. \tag{6.11}$$

This is a remarkable property. Although thresholding does not know about the special direction θ, thresholding gives the rate (6.10) attainable by ideal procedures which know about θ.

6.3 Orthonormal ridgelets and ridge functions

Donoho [27] explored a similar situation in a slightly different setting. In two dimensions, consider the orthonormal ridgelet expansion

$$f_\theta = \sum_\lambda \alpha_\lambda \rho_\lambda$$

of the ridge function

$$f_\theta(x) = f(x_1 \cos \theta + x_2 \sin \theta), \quad \theta \in \mathbf{S}^1.$$

Let $n(\delta)$ be

$$n(\delta) = \sum_\Lambda 1_{\{|\alpha_\lambda|\|\rho_\lambda\|_{L_\infty(D)} > \delta\}},$$

and set

$$f_\delta = \sum_\Lambda \alpha_\lambda 1_{\{|\alpha_\lambda|\|\rho_\lambda\|_{L_\infty(D)} > \delta\}} \rho_\lambda.$$

To construct f_δ, we reorder the coefficients such that $|\alpha_\lambda|\|\rho_\lambda\|_{L_\infty(D)}$ is nonincreasing and keep the $n(\delta)$ largest. The reason for this special ordering is that we will be interested in L_∞ rather than L_2-errors of approximation. We thereby define a sequence of approximants $(\overline{f}_n)$ by letting $\overline{f}_{n(\delta)}$ be the $n(\delta)$ term approximation from f_θ. Suppose that f belongs to the homogeneous Besov space $\dot{B}^s_{p,p}$, with $s = 1/p$ and $0 < p < 1$, and vanishes at $\pm\infty$. Donoho showed

$$\|f_\theta - \overline{f}_n\|_{L_\infty(D)} \le C \|f\|_{\dot{B}^s_{p,p}} n^{-(s-1)}, \quad n = 1, 2, \ldots \qquad (6.12)$$

He compares this with an ideal ridge approximation knowing θ constructed as in the previous section. That is, one would use wavelets to construct an n-term approximation f_n of the profile and synthesize an approximation of f_θ with

$$f_{\theta,n} = f_n(x_1 \cos\theta + x_2 \sin\theta).$$

The error of the ideal 1-dimensional approximation obeys

$$\|f - f_n\|_{L_\infty([-1,1])} \le C \|f\|_{\dot{B}^s_{p,p}} n^{-(s-1)}, \quad s = 1/p.$$

We now have an isometry

$$\|f - f_n\|_{L_\infty([-1,1])} = \|f_\theta - f_{n,\theta}\|_{L_\infty(D)}$$

and, therefore, the ideal approximation $f_{n,\theta}$ obeys

$$\|f_\theta - f_{n,\theta}\|_{L_\infty(D)} \le C \|f\|_{\dot{B}^s_{p,p}} n^{-(s-1)}, \quad s = 1/p. \qquad (6.13)$$

Again, thresholding does not know the direction θ, and yet, does just as well as a method with full knoweledge of the structure of f_θ. The assumption about the profile f, namely $f \in \dot{B}^s_{p,p}$, says that, in some sense, the univariate rate $n^{-(s-1)}$ is optimal. Therefore, one cannot fundamentally improve the bound (6.13).

These results are of the same flavor as those presented in the last section. The difference here is that one uses an L_∞-norm to measure the degree of approximation as opposed to the L_2-norm. This is hardly a major distinction, however, as results similar to those presented in this section are likely to hold in L_2. The important point here is that the guiding principle is the same; that is, ridgelet series or orthoridgelet series of ridge-wavelets are very sparse.

6.4 Why does Theorem 5.1 hold?

In some sense, the results developed in this section are very special cases of Theorem 5.1. The theorem considers the case of ideal approximations which can carefully select the best n directions to approximate f. (For instance, if f were truly a superposition of finitely many ridge functions, the approximation might select those directions.) In Theorem 5.1, those directions are arbitrary whereas Sections 6.1 and 6.2 involved only a single direction.

Theorem 5.1 is a statement about the sparsity of α, the ridgelet coefficient sequence of f. Define the ℓ_p norm of an arbitrary sequence (a_n) by

$$\|a\|_{\ell_p}^p = \sum_n |a_n|^p. \tag{6.14}$$

Recall that if $|a|_{(m)}$ denotes the m largest entry in the sequence $(|a_n|)$, we have

$$\sum_{n>m} |a|_{(n)}^2 \leq K_p \cdot m^{-2r} \cdot \|a\|_{\ell_{p^*}}^2, \tag{6.15}$$

where r and p^* are related to each other via $1/p^* = r + 1/2$, e.g. compare with Lemma 1 in [22].

To prove the theorem it is then sufficient to establish that for any p such that $1/p < r + 1/2$, (r is the exponent appearing in the statement of Theorem 5.1), the ridgelet coefficient sequence α obeys

$$\|\alpha\|_{\ell_p} \leq C_p. \tag{6.16}$$

To see why this implies Theorem 5.1, observe that the frame property gives the following inequality which is classical and proved in [7], say: letting (a_λ) be a sequence in ℓ_2 we have

$$\|f\|_2^2 \leq A^{-1} \|a\|_{\ell_2}^2, \quad f = \sum_\lambda a_\lambda \tilde{\psi}_\lambda,$$

where A is the constant appearing on the left-hand side of (2.12). We apply this inequality to the rest $(f - f_n^R)$ and obtain

$$\|f - f_n^R\|_{L_2}^2 \leq A^{-1} \sum_{m>n} |\alpha(f)|_{(m)}^2,$$

which gives $\|f - f_n^R\|_{L_2}^2 = O(n^{-r'})$ for any $r' > r$ thanks to (6.16) and (6.15).

The strategy for proving our theorem will be to establish a key property about the sparsity of ridgelet coefficients of sigmoidal functions. For each $p > 0$, [6] proves (6.2) and, here, we simply sketch the idea behind this estimate. First, we show that the one-dimensional expansion of a neuron in a nice wavelet basis is very sparse. Specifically, we prove that

$$\sigma(at - b) = \sum_k b_k \varphi_k(t) + \sum_{j \geq 0} \sum_k a_{jk} \psi_{j,k}(t), \tag{6.17}$$

with for each $p > 0$,

$$\|b\|_{\ell_\infty} \leq C_\infty, \quad \|a\|_{\ell_p} \leq C_p.$$

Then, for $\theta \in \mathbf{S}^{d-1}$, we then decompose a neuron as a superposition of ridge-wavelets by the rule

$$\sigma(a\,\theta \cdot x - b) = \sum_k b_k \varphi_k(\theta \cdot x) + \sum_{j \geq 0} \sum_k a_{jk} \psi_{j,k}(\theta \cdot x).$$

The point is that ridgelet coefficients of 'ridge-wavelets' are very sparse. This can be quantified, and roughly speaking, this is the content of (6.6); indeed, the infinite matrices mapping the a_{jk}'s and b_k's into ridgelet coefficients α_R are nearly diagonal and preserve sparsity. A careful argument then gives (6.2).

To prove the theorem, we then start by writing f as the telescoping sum

$$f = \sum_{n \geq 0} g_n, \quad g_n = f_{2^n} - f_{2^{n-1}}. \tag{6.18}$$

where f_{2^n} is the best approximation using 2^n neurons as in (5.1). Each term g_n is a finite linear combination of at most $2^n + 2^{n-1}$ neurons whose coefficients obey (5.2) and by assumption, the L_2-norm of g_n converges to zero at the rate 2^{-nr}. An argument which uses these two special facts then shows

$$\|\alpha(g_n)\|_{\ell_p} \leq B^p \,\epsilon_n^p, \tag{6.19}$$

with

$$\sum_n \epsilon_n^p \leq 1, \quad 1/p < r + 1/2.$$

For $p \leq 1$, say, the p-triangle inequality gives

$$\|\alpha(f)\|_{\ell_p}^p \leq \sum_n \|\alpha(g_n)\|_{\ell_p}^p \leq B^p \,\epsilon_n^p \leq B^p, \tag{6.20}$$

which is what we seeked to establish.

§7. Ridgelets Analysis

7.1 CHA and Approximation Theory

Harmonic analysis is concerned with developing new ways for representing functions which may have great potential in approximation theory. Typical approximation theoretic questions are about how well one can approximate an object by finite linear combinations of given templates. Harmonic analysis, however, brings emphasis on a whole different set of issues. To name just a few, some of the key concepts are

- analysis,
- synthesis,
- stability, and
- discretization.

To be complete, one should add that much of the literature in CHA is also concerned with the development of rapid algorithms for computing these new representations. Those issues are perhaps not central in approximation theory. Nevertheless, this shift in emphasis helps in reformulating important approximation theoretic questions, as we have seen. It also suggests new ways of approaching these problems.

In addition, CHA provides some new tools which may very helpful in gaining a renewed mathematical understanding of approximation theoretic problems. As an example, the ridgelet transform proves to be very powerful for studying the capabilities and the limitations of ridge function approximations. In addition, this transform is also very helpful for identifying functional classes that are well approximated by ridge functions.

7.2 Examples

To illustrate our purpose, consider the following problem. In d-dimensions, let f be the indicator function of the unit ball

$$f(x) = 1_{\{|x| \leq 1\}}, \tag{7.1}$$

and let us ask about the rate of convergence of neural networks

$$d_n(f, D^{NN}) \text{ as } n \to \infty.$$

This may seem like a trivial question. After all, it is hard to think of a simpler object than f. In truth, quantifying –giving a sharp upper bound, say– the rate of decay of

$$d_n(f, \mathcal{D}_{NN}), \quad n \to \infty$$

is certainly not an easy task.

In some sense, ridgelet analysis solves this type of problem rather effortlessly. We simply need to calculate and quantify the size of the ridgelet coefficients of f. The degree of approximation is immediately read off the decay of the coefficient sequence. We will carry out this program on the example (7.1).

First, recall the definition of the Radon transform Rf of an integrable function f (see [18] for details)

$$Rf(u,t) = \int_{u \cdot x = t} f(x)\, dx, \quad u \in \mathbf{S}^{d-1},\, t \in \mathbb{R}. \tag{7.2}$$

One way to calculate ridgelet coefficients is to observe that the ridgelet transform is precisely the application of a 1-dimensional wavelet transform to the slices of the Radon transform where the angular variable u is held constant and t is varying [4]. Mathematically speaking, the ridgelet coefficient (2.4) can be expressed as

$$\mathcal{R}_f(a,u,b) = \int Rf(u,t)\, a^{-1/2} \psi\left(\frac{t-b}{a}\right) dt. \tag{7.3}$$

Loosely speaking, ridgelet analysis is some kind of wavelet analysis in the Radon domain.

A simple calculation then shows for f, the indicator of the unit ball, that

$$R_u f_\alpha(t) = c_d (1 - t^2)^{(d-1)/2}. \tag{7.4}$$

We now study the sparsity of the vector of ridgelet coefficients. Without loss of generality, we will take ridgelets with a profile ψ compactly supported in the spatial domain. We will assume further that ψ is R times differentiable and has vanishing moments through order D. Finally, let p^* be defined by $p^* = 2 - 2/d$. We show that the coefficient sequence (α_R) of f is in ℓ_p for any $p > p^*$ provided that $\min(R, D)$ is sufficiently large.

Define g by

$$g(t) = (1 - t^2)^{(d-1)/2},$$

and as usual let $\psi_{j,k}(t)$ denote the one dimensional wavelet $2^{j/2} \psi(2^j t - k)$. We have

$$|\langle g, \psi_{j,k} \rangle| \leq C\, 2^{-jd/2} \left(1 + (|k| - 2^j)\right)^{-2}. \tag{7.5}$$

The proof is a simple integration by parts and we omit it. Since $\alpha_R = c_d \langle g, \psi_{j,k} \rangle$, we have

$$|\alpha_R| \leq C\, 2^{-jd/2} \left(1 + (|k| - 2^j)\right)^{-2},$$

and, therefore, for $p > p^*$,

$$\sum_k |\alpha_R|^p \leq C\, 2^{-jdp/2} \sum_k \left(1 + (|k| - 2^j)\right)^{-2p} \leq C\, 2^{-jdp/2}.$$

We sum this inequality over the angular variable (the number of orientations is of the order $2^{j(d-1)}$), and obtain

$$\sum_{\ell}\sum_{k}|\alpha_R|^p \le C\, 2^{-jdp/2}\, 2^{j(d-1)}.$$

In short, $\sum_R |\alpha_R|^p$ is finite provided $dp/2 > d-1$, or equivalently $p > p^*$. In fact, careful bookkeeping also gives

$$\|\alpha\|_{w\ell_{p^*}} \le A_p, \tag{7.6}$$

where the weak-ℓ_p or Marcinkiewicz quasi-norm is defined as follows: let $|\theta|_{(n)}$ be the nth largest entry in the sequence $(|\theta_n|)$; we set

$$|\theta|_{w\ell_p} = \sup_{n>0} n^{1/p}|\theta|_{(n)}. \tag{7.7}$$

It is well known that inequality (6.15) holds with the weak-ℓ_p in place of the ℓ_{p^*} norm, and therefore, thresholding the ridgelet series gives

$$\|f - f_n^R\|_{L_2} \le C\, n^{-\frac{1}{2(d-1)}}. \tag{7.8}$$

The ridgelet sequence does not belong to any weak-ℓ_p space for $p < p^*$ and thus, the rate $n^{-\frac{1}{2(d-1)}}$ is the best one can hope for. In light of Theorem 5.1, this also gives a lower bound on the degree of approximation with neural networks, provided that σ is a smooth sigmoidal function.

We may generalize this example and consider different types of singularities. For instance, let

$$f_\alpha = (1 - |x|)_+^\alpha, \quad \alpha > -1/2;$$

$\alpha = 0$ is our previous example, $\alpha = 1$ says the first derivative is discontinuous at $|x| = 1$, etc. The condition $\alpha > -1/2$ ensures that f_α is square-integrable. Then, the ridgelet coefficient sequence of f_α is in $w\ell_{p^*(\alpha)}$ for $1/p^*(\alpha) - 1/2 = (\alpha + 1/2)/(d-1)$ and thresholding the ridgelet series gives

$$\|f - f_n^R\|_{L_2} \le C\, n^{-(\alpha+1/2)/(d-1)}. \tag{7.9}$$

We take another example. In two dimensions, let f now be the indicator function of the unit square

$$f(x) = 1_{[0,1]^2}(x).$$

How well can we approximate f by superpositions of ridge functions? Again, this is far from trivial. Simple calculations show that the ridgelet sequence obeys $\|\alpha\|_{w\ell_{2/3}} \le C$ and, therefore,

$$\|f - f_n^R\|_{L_2} \le C\, n^{-1}. \tag{7.10}$$

We might multiply examples of this kind at will but hope that the power of the ridgelet transform is now quite clear. In another direction, the ridgelet transforms helps identifying those functions which are well approximated by ridge functions, as we are about to see.

§8. Ridgelets and Linear Singularities

Traditional methods based on wavelets, Fourier series, local cosine transforms, or splines fail at efficiently representing objects which are discontinuous along lines in dimension 2, planes in dimension 3, and so on. To detail this claim, let us examine a very simple example. On $[0,1]^d$, suppose that we want to represent the simple object

$$f(x) = 1_{\{u \cdot x > t_0\}} \, g(x) \quad g \in \overline{W}_2^s([0,1]^d), \tag{8.1}$$

where $\overline{W}_2^s([0,1]^d)$ is the closure of $C_0^\infty([0,1]^d)$ with respect to the W_2^s-Sobolev norm. The object f is singular on the hyperplane $u \cdot x = t_0$ (u is a unit vector) but may be very smooth elsewhere. Suppose for instance that one wishes to represent this object in a wavelet basis. Then, the vector of wavelet coefficients is not sparse. In fact, the number of wavelet coefficients exceeding $1/n$ is greater than $c\,n^{2(1-1/d)}$. This immediately translates into lower bounds for nonlinear approximations. Letting f_n^W be the best n-term partial reconstruction of f, the L_2-squared error of such an approximation obeys

$$\|f - f_n^W\|_{L_2}^2 \geq c\,n^{-\frac{1}{2(d-1)}}. \tag{8.2}$$

This lower bound holds even when g is as nice as we want, i.e., $g \in C^\infty$. Of course, one could develop an approximation which would track the singularity $\{u \cdot x = t_0\}$ and partition the space into two halves along this hyperplane. One would then use specially adapted polynomials, splines, or wavelets to those half-spaces, and obtain a degree of approximation as if there were no singularity, i.e. of the order of $n^{-s/d}$. Hence, wavelet thresholding performs very badly vis a vis these ideal strategies.

Whereas the presence of the singularity had a dramatic effect on the sparsity of wavelet coefficients, *it does not ruin the sparsity of the ridgelet series*. Indeed, let us consider let α ($\alpha_R = \langle f, \psi_R \rangle$) denote the ridgelet coefficient sequence of f. Then, [7] shows the sequence α is sparse as if f were not singular in the sense that

$$\#\{i, \; s.t. \; |\alpha_i| \geq 1/n\} \leq C\,n^p \, \|g\|_{H^s}, \quad with \;\; 1/p = s/d + 1/2, \tag{8.3}$$

where the constant C does not depend on f. There is a direct consequence of this result. Consider the n-term obtained by naive thresholding. Then,

$$\|f - f_n^R\| \leq C\,n^{-s/d} \, \|g\|_{H^s}, \tag{8.4}$$

where, again, the constant C is independent of f. Just as wavelets are optimal for representing objects with point-like discontinuities, ridgelets provide optimally sparse representations of objects with discontinuities along hyperplanes. To my knowledge, there is not any other system with similar features.

The bound (8.4) also highlights remarkable spatial adaptivity. Ridgelet thresholding does not know whether or not there is a singularity and if there is one, where it is. In addition, ridgelet thresholding does not need to know the degree of smoothness s of smooth part of the object. And yet, ridgelet thresholding does as just as well as an ideal approximation which would know about the location of a possible singularity, and about the smoothness away from the singularity.

We would like to point out that in two dimensions, both results (8.3) and (8.4) continue to hold with orthonormal ridgelets in place of 'pure' ridgelets (2.11), see [7].

§9. Ridge Spaces

This section introduces a new scale of functional spaces which we believe are not studied in classical analysis. The aim here is to show that this new scale is closely related to ridge function approximation.

9.1 Definition

We start with a few classical definitions/conditions, and will assume that they hold throughout the remainder of this section:

(i) $\psi \in \mathcal{S}(\mathbb{R})$.

(ii) supp $\hat{\psi} \subset \{1/2 \leq |\xi| \leq 2\}$.

(iii) $|\hat{\psi}(\xi)| \geq c$ if $3/5 \leq |\xi| \leq 5/3$. These conditions are standard in Littlewood-Paley theory, see [30].

As before, we let ψ_j denote the univariate function $2^{j/2}\psi(2^j \cdot)$ and set $R_j(u, b)$ as

$$R_j(u, b) = \int f(x) \psi_j(u \cdot x - b) \, dx \quad u \in \mathbf{S}^{d-1}, b \in \mathbb{R}.$$

For $s \in \mathbb{R}$, $p, q > 0$, define

$$\|f\|_{\dot{R}^s_{p,q}} = \left(\sum_{j \in \mathbb{Z}} \left(2^{js} 2^{jd/2} \|R_j\|_{L_p} \right)^q \right)^{1/q}, \tag{9.1}$$

which is well defined for integrable functions, say. Here $\|R_j\|_{L_p}$ is of course the L_p-norm defined by $\|R_j\|_{L_p} = \left(\int_{\mathbf{S}^{d-1} \times \mathbf{R}} |R_j(u, b)|^p \, dudb \right)^{1/p}$. There is an obvious modification in the case where $q = \infty$. It is not difficult to see that these expressions are norms for $1 \leq p \leq \infty$, $1 \leq q \leq \infty$ and quasi norms in general. Note that it follows from the reproducing formula (2.8) that for $f \in L_1 \cap L_2$, $\|f\|_{\dot{R}^s_{p,q}} = 0$ if only if $f = 0$. Finally, one can show that replacing ψ with $\psi^{\#}$ obeying the properties of ψ listed introduced above yields equivalent norms or quasi-norms.

Sobolev spaces are defined by means of the Fourier transform. Similarly, Besov spaces may be defined by means of the wavelet transform. Likewise, what we have done here is merely to define a norm based on the ridgelet transform. The definition of $\|\cdot\|_{\dot{R}^s_{p,q}}$ bears much resemblance with Besov norms. Actually if $d = 1$, $\|\cdot\|_{\dot{R}^s_{p,q}}$ is the homogeneous Besov norm with the same indices. In higher dimensions, the quantities $\|\cdot\|_{\dot{R}^s_{p,q}}$ measure a very different behavior than that measured by Besov norms. We will not explore these differences here and simply point to [3] for further reference.

At first, the definition may seem rather internal. It is possible, however, to give an external characterization of these spaces, at least in the case where $p = q$. Let $Rf(u, t)$ be the Radon transform (7.2) of f. Then, Section 7 showed that

$$R_j(u, b) = (Rf(u, \cdot) * \tilde{\psi}_j)(b),$$

where $\tilde{\psi}_j(t) = \psi_j(-t)$ allowing an automatic substitution in (9.1). Now for $p = q$, this gives

$$\|f\|^p_{\dot{R}^s_{p,p}} = \int \left(\sum_j 2^{jsp} 2^{jdp/2} \|Rf(u, \cdot) * \tilde{\psi}_j\|^p_{L_p(\mathbf{R})} \right) du$$

$$\sim \int \|Rf(u, \cdot)\|^p_{\dot{B}^{s+(d-1)/2}_{p,p}} du, \tag{9.2}$$

where $\dot{B}^{s+(d-1)/2}_{p,p}$ stands for the usual one-dimensional homogeneous Besov norm. This interpretation makes clear that s is a smoothness parameter and that both parameters p, q serve to measure smoothness. Here, smoothness has to be understood in a non-classical way; we are not talking about the local behavior of a function but rather about its behavior near lines (or if one is in dimension $d > 2$, near hyperplanes). We observe that this characterization highlights an interesting aspect: the condition does not require any particular smoothness of the Radon transform along the directional variable u. As an aside comment, (9.2) also supports the claim that one obtains equivalent quantities for two ψ and $\psi^{\#}$ verifying the properties listed at the beginning of the section.

It is interesting to notice that for $p = q = 2$, $\|\cdot\|_{R^s_{2,2}}$ is an equivalent norm for the homogeneous Sobolev norm $\dot{W}^s_2$,

$$\|f\|^2_{\dot{W}^s_2} = \int_{\mathbf{R}^d} |\hat{f}(\xi)|^2 |\xi|^{2s} d\xi.$$

This follows from (9.2) and the fact that

$$\int \|Rf(u,\cdot)\|^2_{\dot{B}^{s+(d-1)/2}_{2,2}} \sim \|f\|^2_{\dot{W}^s_2},$$

see [51].

Definition 9.1. *Let D be the unit ball of $\mathbb{R}^d$, say. The space $\overline{R}^s_{p,q}(D)$ will denote the closure of $C_0^\infty(D)$ of f with respect to the homogeneous norm $\|\cdot\|_{\dot{R}^s_{p,q}}$.*

Other definitions, with other 'boundary conditions' are of course possible, see [3].

It is beyond the scope of this paper to explore the properties of the new spaces $\overline{R}^s_{p,q}$, such as investigating for instance embedding relationships or interpolation properties, etc. Instead, we will show that those new functional classes are optimally approximated by ridge functions and especially, by ridgelet thresholding.

We mention, however, that these functional classes may model objects with a special kind of inhomogeneity. For instance, consider a linearly mutilated object as in Section 8 $f_i(x) = 1_{\{u_i \cdot x - b_i \geq 0\}} g_i(x)$ with $g_i \in \overline{W}^s_2$. and linear combinations of these. Take the class $\mathcal{F}$ of those f which may be decomposed as a convex combination of our templates

$$f = \sum_i a_i f_i, \quad \sum |a_i| \leq 1, \text{ with } \|g_i\|_{W^2_s} \leq C.$$

It turns out that this intuitive class of objects nearly corresponds to a ball in one of our functional classes. In fact, [3] proves that there exist two constants such that

$$R^{(d+1)/2}_{1,1}(C_1) \subset \mathcal{F} \subset R^{(d+1)/2}_{1,\infty}(C_2). \tag{9.3}$$

Now, the bracketing classes are nearly the same which means that membership in $\mathcal{F}$ would be roughly equivalent to membership in a ball in $R^{(d+1)/2}_{1,q}$ ($1 \leq q \leq \infty$). Therefore, we should really think about these spaces as describing the kind of spatial inhomogeneities we introduced in Section 8.

9.2 Approximation

Let $\overline{R}^s_{p,q}(C)$ be the ball

$$\overline{R}^s_{p,q}(C) = \{f \in \overline{R}^s_{p,q}, \|f\|_{\dot{R}^s_{p,q}} \leq C\}. \tag{9.4}$$

We are going to characterize the degree of approximation of these functional classes in the L_2 metric. We give both an upper and a lower bound of approximation which are of the same order. We will consider a range of indices where $s > d(1/p - 1/2)$, as this guarantees that elements of $\overline{R}^s_{p,q}$ are square integrable, see [3].

Theorem 9.2. *Consider the class* $\overline{R}_{p,q}^{s}(C)$, *and assume* $s > d(1/p - 1/2)_{+}$. *Then*

- *For any "reasonable" dictionary* $\mathcal{D}$

$$d_n(\overline{R}_{p,q}^{s}(C), \mathcal{D}) \geq K_1 \, n^{-s/d}, \tag{9.5}$$

where the constant K_1 *depends at most on* s, p, q.
- *Ridgelet thresholding achieves the optimal rate i.e.*

$$\sup_{f \in \overline{R}_{p,q}^{s}(C)} \|f - f_n^R\|_{L_2(D)} \leq K_2 \, n^{-s/d}, \tag{9.6}$$

where again K_2 *might depend on* s, p, q.

Essentially, what the theorem says is that no other "reasonable" dictionary exists with better approximation estimates for the classes $\overline{R}_{p,q}^{s}(C)$ than what can be obtained via ridgelet thresholding. We must clarify, however, the meaning of the word "reasonable": when considering lower approximation bounds by finite linear combinations from dictionaries, we remark that we must only consider certain kinds of dictionaries. We quote from [28]. "If one allows infinite dictionaries (even discrete countable ones), we would then be considering dictionaries $\mathcal{D} = \{g_\lambda, \ \lambda \in \Lambda\}$ enumerating a dense subset of all common functional classes (including $\overline{R}_{p,q}^{s}(C).$), and which can perfectly reproduce any f with a singleton: $d_1(f, \mathcal{D}) = 0$". Thus when we say "reasonable dictionary," we have in mind that one considers only sequences of dictionaries whose size grow polynomially in the number of terms to be kept in the approximation (1.3).

There are other ways of looking at the lower bound. Let $\mathcal{F}$ be a compact set of functions in $L^2(D)$. We recall that the Kolmogorov ϵ-entropy $N(\epsilon, \mathcal{F})$ of the class $\mathcal{F}$ is the minimum number of bits that is required to specify any element f from $\mathcal{F}$ within an accuracy of ϵ. It may be defined as follows. Let $N(\epsilon, \mathcal{F})$ be the minimum number of L_2-balls of radius ϵ one needs to cover $\mathcal{F}$. Formally,

$$N(\epsilon, \mathcal{F}) = \min\{n : \ \exists (f_i)_{i=1}^n \text{ such that } \forall f \in \mathcal{F}, \ \inf_i \|f - f_i\| \leq \epsilon\};$$

the Kolmogorov entropy is

$$L(\epsilon, \mathcal{F}) = \lceil \log_2 N(\epsilon, \mathcal{F}) \rceil. \tag{9.7}$$

Then inequality (9.5) says that the Kolmogorov entropy of $\overline{R}_{p,q}^{s}(C)$ is bounded below by

$$L(\epsilon, \overline{R}_{p,q}^{s}(C)) \geq c \cdot \epsilon^{-1/(s/d)}.$$

It also gives correponding lower bounds on rates of estimation, see [5]. Finally, following [19], it is most likely that there is no method of approximation which depends continuously on f with better properties than (9.5).

The upper bound says that we have an asymptotically near-optimal procedure for binary encoding elements of $\overline{R}^s_{p,q}(C)$: let $L(\epsilon, \overline{R}^s_{p,q}(C))$ be the minimum number of bits necessary to store in a lossy encoding/decoding system in order to be sure that the decoded reconstruction of every $f \in \overline{R}^s_{p,q}(C)$ will be accurate to within ϵ (in an L_2 sense). Then, a coder-decoder based on simple uniform quantization (depending on ϵ) of the coefficients α_i followed by simple run length coding achieves both a distortion smaller than ϵ and a codelength that is optimal up to multiplicative factors like $\log(\epsilon^{-1})$.

9.3 About the lower bounds

The proof of the lower bound uses an argument which is rather classical in statistics and perhaps in rate distortion theory. The goal is to construct a "fat" hypercube which is embedded in the functional class you wish to approximate. More specifically, a result from [28] shows how the hypercube embedding limits the approximation error.

Theorem 9.3. *Suppose that the class $\mathcal{F}$ contains embedded hypercubes of dimension $n(\delta)$ and side δ, and that*

$$n(\delta) \geq K \, \delta^{-2/(2r+1)}, \qquad 0 < \delta < \delta_0.$$

Let $\mathcal{D}_k$ be a family of finite dictionaries indexed by $k = k_0, k_0 + 1, \ldots$ obeying the size estimate $\#\mathcal{D}_k \leq Bk^\beta$. Let $\pi(t)$ be a polynomial. Then

$$d_n(\mathcal{F}, \mathcal{D}_{\pi(n)}) \geq K' n^{-r}.$$

In our situation, the construction of embedded hypercubes involves properties of the ridgelet frame and is proved in [5]. Letting $\nu = (s, p, q)$. We construct a sequence of cubes $\mathcal{H}^\nu_j$ indexed by $j \geq j_0$ and dimension of the order of 2^{jd} such that

$$\mathcal{H}^\nu_j \subset \overline{R}^s_{p,q}(C), \quad \nu = (s, p, q),$$

and of sidelength $\delta_{j\nu}$ obeying the hypothesis of Theorem 9.3. If ridgelets were orthogonal, the vertices of the cube $\mathcal{H}^\nu_j$ would be properly renormalized ridgelets at scale j. The actual construction $\mathcal{H}^\nu_j$ involves the 'orthogonalization' of the ridgelet frame (so that the vertices are 'perturbed ridgelets') and is delicate. The argument is not reproduced here.

9.4 Proof of the upper bound

Let (ψ_R) be a nice ridgelet frame such that ψ is at least R times differentiable and has vanishing moments up to order D. We define a discrete norm on the coefficient sequence $\alpha_R = \langle f, \psi_R \rangle$ as follows

$$\|\alpha\|_{\dot{\mathbf{r}}^s_{p,q}} \equiv \left(\sum_j \left(2^{j\sigma} \left(\sum_{|R|=j} |\alpha_R|^p \right)^{1/p} \right)^q \right)^{1/q}, \quad \sigma = s + d(1/2 - 1/p). \tag{9.8}$$

Here, the notation $\sum_{|R|=j}$ means that the summation extends to all those coefficients at a fixed scale j. The norm (9.8) is, of course, the discrete analog of (9.1). Recall that we may interpret the frame coefficients as being the samples from the continuum of ridgelet coefficients, namely

$$\alpha_R = \mathcal{R}_f(2^{-j}, u_{j\ell}, k2^{-j}).$$

To have the discrete analogy, one would like to have a sampling theorem which says that the discrete Riemann sum is equivalent to the corresponding integral, i.e.

$$\|\alpha\|_{\dot{\mathbf{r}}^s_{p,q}} \sim \|f\|_{\dot{R}^s_{p,q}}$$

valid for elements of $\overline{R}^s_{p,q}$. Notice that we have already established this equivalence in the case where $s = 0$ and $p = q = 2$ since the frame property gives

$$\|\alpha\|_{\ell_2} \sim \|f\|^2_{L_2} \sim \|f\|^2_{\dot{R}^0_{2,2}}.$$

In fact, a slight modification of the argument underlying the proof of the frame property (2.12) [4] would give the same equivalence for arbitrary s, $s > 0$.

Lemma 9.4. *Let $f \in \overline{R}^s_{p,q}$. There is $r^*(s)$ so that, if we use a ridgelet frame with $\min(R, D) \geq r^*(s)$, then there is a constant C possibly depending on s, p, q such that*

$$\|\alpha\|_{\dot{\mathbf{r}}^s_{p,q}} \leq C \|f\|_{\dot{R}^s_{p,q}}. \tag{9.9}$$

Proof: It is simpler to prove the lemma for a ridgelet frame $(\psi_R)_{R \in \mathcal{R}}$ with a compactly supported profile ψ. In addition, to be compactly supported, we will assume that ψ has as many derivatives and vanishing moments as we want.

Further, we will take $\varphi^\#$ and $\psi^\#$ compactly supported and obeying (2.9) with $2K'_\psi = (2\pi)^{-(d-1)}$ so that the following identity holds:

$$f(x) = \int R^0(u, b) \varphi^\#(u \cdot x - b) \, du \, db$$
$$+ \sum_{j \geq 0} 2^{jd} \int R^1_j(u, b) 2^{j/2} \psi^\#(2^j(u \cdot x - b)) \, du \, db,$$

with

$$R^0(u, b) = \int f(x)\varphi^{\#}(u \cdot x - b)\, dx,$$

$$R_j^1(u, b) = \int f(x)2^{j/2}\psi^{\#}(2^j(u \cdot x - b))\, dx.$$

We recall that the reproducing formula is valid for square integrable and compactly supported functions, say.

Inspired by the discrete ridgelet transform, we further decompose our reproducing formula. We keep the same notations as for the ridgelet frames, and recall that R indexes the triples (j, ℓ, k) and $u_{j,\ell}$ is the collection of sampled orientations at scale j. At each scale j, we introduce a Voronoi partition of the sphere, and let $S_{j,\ell}$ be those directions u on $\mathbf{S}^d$ which are closer to $u_{j,\ell}$ (Euclidean distance) than any other sampled orientations $u_{j,\ell'}$. We then define the 'cell' Q_R by

$$Q_R = S_{j,\ell} \otimes [k2^{-j}, (k+1)2^{-j}].$$

With these notations, we set

$$f = \sum_{R \in \mathcal{R}^0} m_R^0 + \sum_{R \in \mathcal{R}} m_R^1,$$

where $\mathcal{R}^0 = \{R \in \mathcal{R}, |R| = 0\}$ and

$$m_R^0 = \int_{Q_R} R^0(u, b)\varphi^{\#}(u \cdot x - b)\, du\, db$$

$$m_R^1 = 2^{jd} \int_{Q_R} R_j(u, b)2^{j/2}\psi^{\#}(2^j(u \cdot x - b))\, du\, db.$$

The reader will object that we have not defined cells allowing a proper definition of m_R^0; we take the same cells as those in m_R^1 for $j = 0$.

Now, let w be a C^∞ window identically equal to one over D and vanishing outside of $2D$. We recall that $f = fw$ since f is compactly supported in D. For notational convenience, we let $\mathcal{R}^1$ be a copy of $\mathcal{R}$ and write

$$\alpha_{R'} = \sum_{\varepsilon \in \{0,1\}} \sum_{R \in \mathcal{R}^\varepsilon} \langle \psi_{R'}, m_R^\varepsilon w \rangle.$$

We introduce some notations. Define

$$T_1(R', R) = \sup_{Q_R} \left| \int_{\mathbf{R}^d} 2^{j/2}\psi^{\#}(2^j(u \cdot x - b))\, \psi_{R'}(x)w(x)\, dx \right|. \tag{9.10}$$

and

$$\beta_{R'}^1 = 2^{jd} \int_{Q_R} |R_j^1(u, b)|\, du\, db, \tag{9.11}$$

and similarly for T_0 and β_R^0. Note that per scale, there is only a finite number of β_R^1's which are possibly nonzero because we have chosen a pair $(\varphi^\#, \psi^\#)$ where both are compactly supported; at scale j, β_R^1 vanishes whenever $k > c2^{-j}$ where the constant c is a function of the support of $\psi^\#$. A similar conclusion applies to β_R^0. As a consequence, the number of nonzero β_R^0's is at most $O(1)$ and that of nonzero β_R^1's is at most 2^{jd} per scale.

Observe now that

$$|\langle \psi_{R'}, m_R^\varepsilon w \rangle| \leq T_\varepsilon(R', R)\,\beta_R^\varepsilon.$$

This follows from

$$|\langle \psi_{R'}, m_R^1 w \rangle| =$$

$$= \left| 2^{jd} \int_{Q_R} \int_{\mathbf{R}} R_j^1(u,b) 2^{j/2} \psi^\#(2^j(u \cdot x - b))\, \psi_{R'}(x) w(x)\, dx\, du\, db \right|$$

$$\leq 2^{jd} \int_{Q_R} |R_j^1(u,b)| \left| \int_{\mathbf{R}} 2^{j/2} \psi^\#(2^j(u \cdot x - b)) \psi_{R'}(x) w(x)\, dx \right|\, du\, db,$$

and similarly for $\varepsilon = 0$. In short,

$$|\alpha_R| \leq \sum_\varepsilon \sum_{R' \in \mathcal{R}^\varepsilon} T_\varepsilon(R, R')\beta_{R'}^\varepsilon.$$

Lemma 9.5. *Let* $1 \leq p, q \leq \infty$ *and* $s > d(1/2 - 1/p)$. *The operators* T_0 *and* T_1 *obey*

$$\|T_0\beta^0\|_{\dot{\mathbf{r}}_{p,q}^s} \leq A_0 \,\|\beta^0\|_{\ell_p}, \quad \|T_1\beta^1\|_{\dot{\mathbf{r}}_{p,q}^s} \leq A_1 \,\|\beta^1\|_{\dot{\mathbf{r}}_{p,q}^s}.$$

This lemma is at the heart of the proof of (9.9) and we postpone its demonstration. It implies

$$\|\alpha\|_{\dot{\mathbf{r}}_{p,q}^s} \leq A_0\|\beta^0\|_{\ell_p} + A_1\|\beta^1\|_{\dot{\mathbf{r}}_{p,q}^s},$$

and (9.9) follows from the claim that for $f \in \overline{R}_{p,q}^s$

$$\|\beta^0\|_{\ell_p} + \|\beta^1\|_{\dot{\mathbf{r}}_{p,q}^s} \leq A \,\|f\|_{\dot{R}_{p,q}^s}. \tag{9.12}$$

We shall now argue about (9.12).

Recall the general probability inequality

$$\int_\Omega |g(x)|\, \mu(dx) \leq \mu(\Omega)^{1-1/p} \left(\int_\Omega |g(x)|^p\, \mu(dx) \right)^{1/p}.$$

We apply this inequality and obtain

$$|\beta_R^1| \leq C \left(2^{jd} \int_{Q_R} |R_j^1(u,b)|^p \, du \, db \right)^{1/p}.$$

This follows from the ridgelet discretization which gives

$$\sup_R |Q_R| \leq C \, 2^{-jd}.$$

Therefore,

$$\sum_{|R|=j} |\beta_R^1|^p \leq C \, 2^{jd} \int_{\mathbf{S}^{d-1}} \int_{\mathbf{R}} |R_j^1(u,b)|^p \, du \, db.$$

Similarly

$$\sum_{R \in \mathcal{R}^0} |\beta_R^0|^p \leq C \int_{\mathbf{S}^{d-1}} \int_{\mathbf{R}} |R_j^0(u,b)|^p \, du \, db.$$

Hence, we have proved that

$$\|\beta^0\|_{\ell_p} + \|\beta^1\|_{\dot{\mathbf{r}}_{p,q}^s} \leq C \left(\|R^0\|_{L_p} + \left(\sum_{j \geq 0} (2^{js} 2^{jd/2} \|R_j\|_{L_p})^q \right)^{1/q} \right).$$

$$(9.13)$$

The right-hand side of the above inequality is the definition (9.1) but for the coarse scale variation where the sum over the negative indices j is replaced by the single term $\|R^0\|_{L_p}$. In fact, the right-hand side of (9.13) may be taken as the definition of the inhomogeneous $R_{p,q}^s$-norm. For $f \in \overline{R}_{p,q}^s$, we have

$$\|R^0\|_{L_p} + \left(\sum_{j \geq 0} \left(2^{js} 2^{jd/2} \|R_j\|_{L_p} \right)^q \right)^{1/q} \sim \|f\|_{\dot{R}_{p,q}^s} \qquad (9.14)$$

where the $\sim$ symbol has the meaning of norm equivalence $\dot{R}_{p,q}^s$-norm. This norm equivalence is proved in [3], and holds for we defined $\overline{R}_{p,q}^s$ as the closure of $C_0^\infty(D)$ with respect to the . The proof is straightforward and is totally analogous to the following situation: for $f \in \overline{W}_2^s$ (see Section 8 for the definition of $\overline{W}_2^s$), we have

$$\|f\|_{W_2^s} = \|f\|_{L_2} + \|f\|_{\dot{W}_2^s} \sim \|f\|_{\dot{W}_2^s}.$$

(The reader familiar with Besov norms will also know that if one defines Besov spaces over the unit ball $\overline{B}_{p,q}^s$ by the closure of $C_0^\infty(D)$ with respect to the homogeneous Besov norm, say, then over that space, the homogeneous and inhomogeneous Besov norms are equivalent.) Of course, (9.14) finishes the proof of Lemma 9.4. $\square$

Proof of Lemma 9.5. We have

$$T_1(R', R) = \sup_{(u,b) \in Q_R} K(j, u, b; j', u_{R'}, b_{R'}),$$

with

$$K(j, u, b; j', u', b') =$$
$$= \left| \int_{\mathbf{R}^d} 2^{j/2} \psi^\#(2^j(u \cdot x - b)) 2^{j'/2} \psi(2^{j'}(u' \cdot x - b')) w(x)\, dx \right|.$$

Recall that $Q_R = S_{u_R} \otimes [k2^{-j}, (k+1)2^{-j}]$ where the diameter of S_{u_R} does not exceed $C\, 2^{-j}$. Now, Lemma 14.1 below gives an upper bound on the quantity $K(j, u, b; j', u', b')$. We then apply this lemma and take the supremum of the right hand-side of (14.8) over the cell Q_R. It is easy to verify that this gives an upper estimate as in (14.13), i.e. with the notations of the Appendix,

$$T_1(R', R) \le C 2^{-(j'+j)(n+1/2)} \delta_j^{2n+1}(\ell, \ell') \left(1 + |k'2^{-j'}|\right)^{-m}$$
$$\left(1 + \delta_j(\ell, \ell')|k'2^{-j'} \cos\theta - k2^{-j}|\right)^{-m}.$$

We first show the lemma at 'the corners', namely for $p, q \in \{1, \infty\}$. We will then establish the general case by interpolation. We let $\alpha^\varepsilon = T_\varepsilon \beta^\varepsilon$. Suppose first that $p = 1$. Then

$$\sum_{|R|=j} |\alpha_R^1| \le \sum_{R'} \sum_R T_1(R, R') |\beta_{R'}^1|.$$

The Appendix establishes (14.18)

$$\sum_{|R|=j} |T(R; R')| \le C\, 2^{-|j-j'|((n+1/2)-d)},$$

where $n \le \min(R, D)$. Therefore,

$$\sum_{|R|=j} |\alpha_R^1| \le C \sum_{R'} 2^{-|j-j'|((n+1/2)-d)} |\beta_{R'}^1|$$
$$= C \sum_{j'} 2^{-|j-j'|((n+1/2)-d)} \sum_{|R'|=j'} |\beta_{R'}^1|.$$

Letting

$$a_j = 2^{js}2^{-jd/2} \sum_{|R|=j} |\alpha_R^1|, \quad b_j = 2^{js}2^{-jd/2} \sum_{|R|=j} |\beta_R^1|,$$

we have

$$a_j \leq C \sum_{j'} \epsilon_{j,j'} \, b_{j'}, \quad \epsilon_{j,j'} = 2^{-|j-j'|((n+1/2)-d)} \, 2^{(j-j')s}.$$

For $q = \infty$ we have

$$a_j \leq C \sup_{j' \geq 0} b_{j'} \sum_{j' \geq 0} \epsilon_{j,j'} \leq B \, \|b\|_{\ell_\infty},$$

provided that n is large enough so that $(n + 1/2) > d$. For $q \leq 1$ we have (again $(n + 1/2) > d$)

$$\sum_j \epsilon_{j,j'}^q \leq B_q^q,$$

and, therefore, the q-triangle inequality gives

$$\sum_{j \geq 0} a_j^q \leq \sum_{j \geq 0} \sum_{j' \in \mathbb{Z}} \epsilon_{j,j'}^q b_{j'}^q \leq B_q^q \, \|b\|_{\ell_q}^q.$$

The case for $p = \infty$ is nearly identical. We have

$$\sum_{|R'|=j'} T(R, R') \leq C \, 2^{-|j-j'|((n+1/2)-d)},$$

and, therefore,

$$|\alpha_R^1| \leq \sum_{j'} 2^{-|j-j'|((n+1/2)-d)} \sup_{|R'|=j'} |\beta_{R'}^1|.$$

Letting

$$a_j = 2^{js}2^{jd/2} \sup_{|R|=j} |\alpha_R^1|, \quad b_j = 2^{js}2^{jd/2} \sup_{|R|=j} |\beta_R^1|,$$

gives

$$a_j \leq C \sum_{j'} \epsilon_{j,j'} b_{j'}, \quad \epsilon_{j,j'} = 2^{-|j-j'|(n-(s+d/2))}.$$

We use the exact same argument as before. The ℓ_q summability for any $q > 0$ of $\epsilon_{j,j'}$ with respect to either j or j' (provided $n > s + d/2$) gives

$$\|a\|_{\ell_q} \leq C \, \|b\|_{\ell_q}.$$

We have now proved the property for $p, q \in \{1, \infty\}$.

We finish the proof by interpolation. The sequence norm $\dot{\mathbf{r}}_{p,q}^{s}$ is a weighted norm of the type $\ell_q(\ell_p)$. We write

$$\alpha^1 = T_1\beta^1,$$

and as we observed earlier, at scale j the coefficients α_R^1 and β_R^1 vanish for $k \geq c \cdot 2^j$. Therefore, at each scale, there is a set of indices of cardinality at most $O(2^{jd})$ which can contribute possibly nonzero coefficients. This shows that there is, of course, a full analogy with the Besov scale which is also a weighted space of the type $\ell_q(\ell_p)$ with 2^{jd} coefficients per scale. Therefore, the sequence spaces $\dot{\mathbf{r}}_{p,q}^{s}$ are clearly interpolation spaces with the same interpolation properties as for the d-dimensional Besov scale. The boundedness of T_1 in the $\dot{\mathbf{r}}_{p,q}^{s}$ norm for arbitrary $1 \leq p, q \leq \infty$ follows.

The story is nearly identical for T_0. The operator T_0 obeys the estimates developed in the Appendix, namely (14.22). Further, the Appendix (14.23) shows that

$$\sum_{|R'|=j'} \sum_{|R|=0} |T_0(R', R)|^p \leq C \, 2^{-j'((n+1/2)p-d)},$$

where n is as large as we want provided we have enough regularity and vanishing moments. Now, using the same arguments as before, the previous estimates on T_0 show that

$$\|T_0\beta^0\|_{\dot{\mathbf{r}}_{p,q}^{s}} \leq C \, \|\beta^0\|_{\ell_p}, \quad p, q \in \{1, \infty\}.$$

Interpolation allows the extension of the above inequality to arbitrary parameters $1 \leq p, q \leq \infty$. $\square$

Corollary 9.6. *Assume $s > d(1/p-1/2)_+$ and let p^* be defined by $1/p^* = s/d + 1/2$. Then for f in $\overline{R}_{p,q}^{s}(D)$ we have*

$$\|\alpha\|_{w\ell_{p^*}} \leq C \, \|f\|_{\dot{R}_{p,q}^{s}(D)}, \tag{9.16}$$

where $w\ell_{p^*}$ is the weak-ℓ_{p^*} quasi-norm (7.7).

Proof: We just showed that $\|\alpha\|_{\dot{\mathbf{r}}_{p,q}^{s}} \leq C \, \|f\|_{\dot{R}_{p,q}^{s}(D)}$, and we have

$$\|\alpha\|_{w\ell_{p^*}} \leq C \, \|\alpha\|_{\dot{\mathbf{r}}_{p,q}^{s}}$$

for $s > d(1/p - 1/2)_+$. This follows from a similar statement about the weak-ℓ_p boundedness of wavelet coefficient sequences taken from d-dimensional Besov spaces since $r_{p,q}^{s}$ and $b_{p,q}^{s}$-Besov balls have exactly the same structure. The proof then consists of a minor adaptation of a result in [22]. We do not reproduce it here. $\square$

Of course, (9.16) gives that ridgelet thresholding satisfies

$$\|f - f_n^R\| \leq C \, n^{-s/d} \, \|f\|_{\dot{R}_{p,q}^{s}(D)},$$

which proves (9.6), the second part of Theorem 9.2.

§10. Beyond Ridgelets

10.1 Limitations of ridge function approximation

Ridgelet analysis may give very precise information about the degree of approximation by finite linear combination of ridge functions. As we have seen, we may use the ridgelet transform as a fundamental tool to identify those target functions that are well-suited or ill-suited for ridge function approximation. Objects with curved singularities are in the latter class.

We recall that Section 7 developed the following sharp lower bound: letting f be the indicator function of the unit ball, we have

$$\|f - f_n^R\| \sim n^{-1/(2(d-1))}.$$

This is a general phenomenon. If f is singular along a smooth manifold of codimension 1, we cannot, in general, hope for better rates than $n^{-1/(2(d-1))}$. (By the way, it is remarkable that d-dimensional wavelets give the same degree of approximation, i.e. $n^{-1/(2(d-1))}$.) In two dimensions, this says that wavelets and ridgelets are inefficient for representing edges as the rate is only of the order $n^{-1/2}$. We are entitled to say that they are inefficient because of the existence of other approximation strategies with provably better performances, as we are about to see.

Finding nonadaptive and optimally sparse representations of objects with curved singularities has been a special concern of mine; this is our next stop. The beautiful thing is that although ridgelets fail at representing efficiently smooth singularities, they will be the cornerstone of refined transforms which will not. In short, they pave the way to more sophisticated constructions.

10.2 CHA and curved singularities

It is well-known that, for objects with discontinuities, wavelets offer an improvement on traditional representations like sinusoids, but wavelets are far from optimal.

To make this concrete, consider approximating such an f from the best n-terms in a Fourier expansion adapted to $[-\pi, \pi]^2$ (say). The squared error of such an n-term expansion $\tilde{f}_n^F$ would obey

$$\|f - \tilde{f}_n^F\|_2^2 \asymp n^{-1/2}, \quad n \to \infty. \tag{10.1}$$

For comparison, consider an approximation $\tilde{f}_n^W$ from the best n-terms in a wavelet expansion; then

$$\|f - \tilde{f}_n^W\|_2^2 \asymp n^{-1}, \quad n \to \infty, \tag{10.2}$$

which is considerably better. However, from [28,25,33] we know that there exist dictionaries of (nonorthogonal) elements, and procedures for selecting from those dictionaries that will yield m-term approximations obeying

$$\|f - \tilde{f}_n^D\|_2^2 \asymp n^{-2}, \quad n \to \infty. \tag{10.3}$$

The next section constructs tight frames of curvelets. These frames are based on the ridgelet transform and provide optimal representations of C^2 objects smooth away from C^2 curves. Indeed, thresholding the curvelet expansion gives (10.3) up to a logarithmic factor.

§11. Curvelet Construction

We now briefly discuss the curvelet frame; for more details, see [8]. The construction combines several ingredients, which we briefly review:

- *Ridgelets*, a method of analysis very suitable for objects which are discontinuous across straight lines.

- *Multiscale Ridgelets*, a pyramid of analyzing elements which consists of ridgelets renormalized and transported to a wide range of scales and locations.

- *Bandpass Filtering*, a method of separating an object into a series of disjoint scales.

Sections 2 and 3 introduced ridgelets, and we first briefly discuss multiscale ridgelets and bandpass filtering. We then describe the combination of these three components. There is a difference between this construction and the one given in [8] at large scales.

11.1 Multiscale ridgelets

Think of ortho ridgelets as objects which have a "length" of about 1 and a "width" which can be arbitrarily fine. The multiscale ridgelet system renormalizes and transports such objects, so that one has a system of elements at all lengths and all finer widths.

The construction begins with a smooth partition of energy function w with $w(x_1, x_2) \geq 0$, $w \in C_0^\infty([-1,1]^2)$ obeying $\sum_{k_1,k_2} w^2(x_1 - k_1, x_2 - k_2) \equiv 1$. Define a transport operator associated with index Q indicating a dyadic square $Q = (s, k_1, k_2)$ of the form $[k_1/2^s, (k_1+1)/2^s) \times [k_2/2^s, (k_2+1)/2^s)$, by $(T_Q f)(x_1, x_2) = f(2^s x_1 - k_1, 2^s x_2 - k_2)$. The Multiscale Ridgelet with index $\mu = (Q, \lambda)$ is then

$$\psi_\mu = 2^s \cdot T_Q(w \cdot \rho_\lambda)$$

In short, one transports the normalized, windowed orthoridgelet.

Letting $\mathcal{Q}_f$ denote the dyadic squares of side 2^{-s}, we can define the subcollection of Monoscale Ridgelets at scale s:

$$\mathcal{M}_s = \{(Q, \lambda) : Q \in \mathcal{Q}_s, \lambda \in \Lambda\}.$$

It is immediate from the orthonormality of the ridgelets that each system of monoscale ridgelets makes a tight frame, in particular obeying the Parseval relation

$$\sum_{\mu \in \mathcal{M}_s} \langle \psi_\mu, f \rangle^2 = \|f\|_{L^2}^2.$$

It follows that the dictionary of multiscale ridgelets at all scales, indexed by

$$\mathcal{M} = \cup_{s \geq 1} \mathcal{M}_s,$$

is not frameable, as we have energy blow-up:

$$\sum_{\mu \in \mathcal{M}} \langle \psi_\mu, f \rangle^2 = \infty. \tag{11.1}$$

The Multiscale Ridgelets dictionary is simply too massive to form a good analyzing set. It lacks inter-scale orthogonality – $\psi_{(Q,\lambda)}$ is not typically orthogonal to $\psi_{(Q',\lambda')}$ if Q and Q' are squares at different scales and overlapping locations. In analyzing a function using this dictionary, the repeated interactions with all different scales causes energy blow-up (11.1).

The construction of curvelets solves this problem by in effect disallowing the full richness of the Multiscale Ridgelets dictionary. Instead of allowing all different combinations of 'lengths' and 'widths', we allow only those where *width* $\approx$ *length*^{2}.

11.2 Subband filtering

Our remedy to the 'energy blow-up' (11.1) is to decompose f into subbands using standard filterbank ideas. Then we assign one specific monoscale dictionary $\mathcal{M}_f$ to analyze one specific (and specially chosen) subband.

We define coronae of frequencies $|\xi| \in [2^{2s}, 2^{2s+2}]$, and subband filters D_s extracting components of f in the indicated subbands; a filter P_0 deals with frequencies $|\xi| \leq 1$. The filters decompose the energy exactly into subbands:

$$\|f\|_2^2 = \|P_0 f\|_2^2 + \sum_s \|D_s f\|_2^2.$$

The construction of such operators is standard [61]; the coronization oriented around powers 2^{2s} is nonstandard – and essential for us. Explicitly, we build a sequence of filters Φ_0 and $\Psi_{2s} = 2^{4s}\Psi(2^{2s}\cdot)$, $s = 0, 1, 2, \ldots$ with the following properties: Φ_0 is a lowpass filter concentrated near frequencies $|\xi| \leq 1$; Ψ_{2s} is bandpass, concentrated near $|\xi| \in [2^{2s}, 2^{2s+2}]$; and we have

$$|\hat{\Phi}_0(\xi)|^2 + \sum_{s \geq 0} |\hat{\Psi}(2^{-2s}\xi)|^2 = 1, \quad \forall \xi.$$

Hence, D_s is simply the convolution operator $D_s f = \Psi_{2s} * f$.

11.3 Definition of curvelet transform

Assembling the above ingredients, we are able to sketch the definition of the Curvelet transform. We let M' consist of M merged with the collection of integral triples (s, k_1, k_2, e) where $s \leq 0$, $e \in \{0, 1\}$, indexing all dyadic squares in the plane of side $2^s > 1$.

The curvelet transform is a map $L^2(\mathbb{R}^2) \mapsto \ell^2(M')$, yielding Curvelet coefficients $(\alpha_\mu : \mu \in M')$. These come in two types. At **coarse scales** we have wavelet coefficients:

$$\alpha_\mu = \langle W_{s,k_1,k_2,e}, P_0 f \rangle, \qquad \mu = (s, k_1, k_2) \in M' \backslash M,$$

where each $W_{s,k_1,k_2,e}$ is a Meyer wavelet, while at **fine scale** we have Multiscale Ridgelet coefficients of the bandpass filtered object:

$$\alpha_\mu = \langle D_s f, \psi_\mu \rangle, \qquad \mu \in M_s, s = 1, 2, \ldots.$$

Note well that for $s > 0$, each coefficient associated to scale 2^{-s} derives from the subband filtered version of $f - D_s f$ – and not from f. Several properties are immediate;

• Tight Frame:

$$\|f\|_2^2 = \sum_{\mu \in M'} |\alpha_\mu|^2.$$

• Existence of Coefficient Representers (Frame Elements): There are $\gamma_\mu \in L^2(\mathbb{R}^2)$ so that

$$\alpha_\mu \equiv \langle f, \gamma_\mu \rangle.$$

• L^2 Reconstruction Formula:

$$f = \sum_{\mu \in M'} \langle f, \gamma_\mu \rangle \gamma_\mu.$$

• Formula for Frame Elements: for $s \leq 0$, $\gamma_\mu = P_0 \phi_{s,k_1,k_2}$, while for $s > 0$,

$$\gamma_\mu = D_s \psi_\mu, \qquad \mu \in \mathcal{Q}_s. \tag{11.2}$$

In short, fine-scale curvelets are obtained by bandpass filtering of Multiscale Ridgelets coefficients where the **passband** is rigidly linked to the **scale** of spatial localization.

- Anisotropy Scaling Law: By linking the filter passband $|\xi| \approx 2^{2s}$ to the scale of spatial localization 2^{-s} imposes that: (1) most curvelets are negligible in norm (most multiscale ridgelets do not survive the bandpass filtering D_s); (2) the nonnegligible curvelets obey *length* $\approx 2^{-s}$ while *width* $\approx 2^{-2s}$. In short, the system obeys approximately the scaling relationship

$$width \approx length^2.$$

Note: it is at this last step that our 2^{2s} coronization scheme comes fully into play.

- Oscillatory Nature. Both for $s > 0$ and $s \leq 0$, each frame element has a Fourier transform supported in an annulus away from 0.

§12. Curvelets and Curved Singularities

12.1 Functions which are C^2 away from C^2 edges

We now formally specify a class of objects with discontinuities along edges; our notation and exposition are taken from [25,28,23]; related models were introduced some time ago in the mathematical statistics literature by [38,39]. It is clear that nothing in the arguments below would depend on the specific assumptions we make here, but the precision allows us to make our arguments uniform over classes of such objects.

A star-shaped set $B \subset [0,1]^2$ has an origin $b_0 \in [0,1]^2$ from which every point of B is 'visible'; i.e. such that the line segment $\{((1-t)b_0 + tb : t \in [0,1]\} \subset B$ whenever $b \in B$. This geometrical regularity is useful; it forces very simple interactions of the boundary with dyadic squares at sufficiently fine scales. We use this to guarantee that 'sufficiently fine' has a uniform meaning for every B of interest.

We define $\mathsf{Star}^2(A)$, a class of star-shaped sets with 2-smooth boundaries, by imposing regularity on the boundaries using a kind of polar coordinate system. Let $\rho(\theta) : [0, 2\pi) \to [0,1]$ be a radius function and $b_0 = (x_{1,0}, x_{2,0})$ be an origin with respect to which the set of interest in star-shaped. Define $\Delta_1(x) = x_1 - x_{1,0}$ and $\Delta_2(x) = x_2 - x_{2,0}$; then define functions $\theta(x_1, x_2)$ and $r(x_1, x_2)$ by

$$\theta = \tan^{-1}(-\Delta_2/\Delta_1); \qquad r = ((\Delta_1)^2 + (\Delta_2)^2)^{1/2}.$$

For a starshaped set, we have $(x_1, x_2) \in B$ iff $0 \leq r \leq \rho(\theta)$. In particular, the boundary ∂B is given by the curve

$$\beta(\theta) = (\rho(\theta)\cos(\theta) + x_{1,0}, \rho(\theta)\sin(\theta) + x_{2,0}). \qquad (12.1)$$

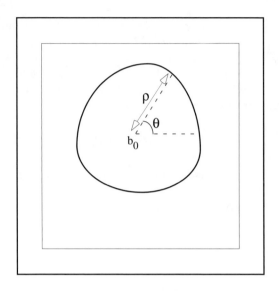

Fig. 3. Typical star-shaped set, and associated notation.

Fig. 3 gives a graphical indication of some of the objects just described. The class $\mathsf{Star}^2(A)$ of interest to us can now be defined by

$$\mathsf{Star}^2(A) = \{B : B \subset [\frac{1}{10}, \frac{9}{10}]^2, \; \frac{1}{10} \leq \rho(\theta) \leq \frac{1}{2}, \; \theta \in [0, 2\pi),$$
$$\rho \in \mathsf{Hölder}^2(A)\}.$$

Here the condition $\rho \in \mathsf{Hölder}^2(A)$ means that ρ is continuously differentiable and

$$|\rho'(\theta) - \rho'(\theta')| \leq A \cdot |\theta - \theta'|, \quad \theta, \theta' \in [0, 2\pi).$$

The actual objects of interest to us are functions which are twice continuously differentiable except for discontinuities along edges ∂B of starshaped sets. We define $C_0^2(A)$ to be the collection of twice continuously differentiable functions supported strictly inside $[0, 1]^2$.

Definition 12.1. *Let $\mathcal{E}^2(A)$ denote the collection of functions f on $\mathbb{R}^2$ which are supported in the square $[0, 1]^2$ and obey*

$$f = f_1 + f_2 \cdot 1_B \tag{12.2}$$

where $B \in \mathsf{Star}^2(A)$, and each $f_i \in C_0^2(A)$. We speak of $\mathcal{E}^2(A)$ as consisting of functions which are C^2 away from a C^2 edge.

12.2 Near-optimality of curvelet expansions

The point is that the curvelet coefficient sequence $(c_\mu)_{\mu \in M}$ of an object f that is C^2 away from a C^2 edge, is in some sense, as sparse as if f were not singular.

Theorem 12.2. *Let $\mathcal{E}^2(A)$ be the collection (12.2) of objects which are C^2 away from a C^2 curve. There exists a constant C such that for every $f \in \mathcal{E}^2(A)$, the curvelet coefficient sequence $(c_\mu)_{\mu \in M}$ of f obeys*

$$\#\{\mu, , |c_\mu| \geq \epsilon\} \leq C \log(\epsilon^{-1}) \, \epsilon^{-2/3}. \tag{12.3}$$

There is a natural companion to this theorem. Let f_m be the m-term approximation of f obtained by extracting from the curvelet series

$$f = \sum_\mu \langle f, \gamma_\mu \rangle \gamma_\mu,$$

the terms corresponding to the m largest coefficients. Then,

Theorem 12.3. *Under the assumptions of Theorem 12.2, we have*

$$\|f - f_m\|_{L_2}^2 \leq C \, m^{-2} \, (\log m)^3. \tag{12.4}$$

The proof of Theorem 12.3 is very technical and over 35 pages long. We cannot possibly give an idea of the argument here.

Comparing with the results (10.1)–(10.3) we see that m-term approximations in the curvelet frame are almost rate optimal, and in fact perform far better than m-term sinusoid or wavelet approximations, in an asymptotic sense.

12.3 Significance

I believe that Theorem 12.3 is very significant. In fact, it comes as a little surprise. Finding optimal representations for smooth objects singular along smooth edges has been a long standing problem in CHA. Some expressed the belief that the way to go about singularities was to use adaptive bases as in [25], i.e. bases that would depend upon the object to be approximated, see the discussion in [9]. In other words, the existence of orthobases, frames, or tight frames yielding approximation rates similar to (12.4) by naive thresholding was thought to be highly doubtful. In some sense, Theorem 12.3 disproves this conjecture.

Beyond Theorem 12.3, the curvelet transform introduces a new data structure or, mathematically speaking, fundamentally new linear functionals. It has great potential in various fields outside of approximation theory, such as statistical estimation, mathematical analysis, partial differential equations, scientific computing, and data processing [59,11,60,12].

§13. Discussion

13.1 Renewed understanding

In this paper, we presented a whole set of new ideas and showed how one can deploy these ideas to tackle central problems in approximation theory.

First, we replaced a delicate and rather ill-posed problem of constructing ridge function approximations by a well-posed and constructive procedure, namely, the thresholding of ridgelet expansions. We proved that ridgelet thresholding is a viable substitute for traditional neural network approximations, as not only, it is more constructive, but it also rivals —at least asymptotically— rates attainable by very abstract approximation neural net procedures.

Second, we were able to identify those objects that are well approximated by ridgelet thresholding. We proved that ridgelet provide optimally sparse representations of smooth objects with discontinuities along hyperplanes of codimension 1. We introduced a new scale of functional spaces which have an intimate relationship with ridge-wavelet, ridge-free-knot splines or ridgelet approximations. We explained why, in some sense, this relationship is similar to the role that the Besov scale plays vis a vis free-knot splines or wavelet approximations.

13.2 Multiscale geometric analysis (MGA)

At the same time, ridgelet analysis gives decisive insights about the limitations and capabilities of ridge function approximation. For instance, we proved that ridgelets are suboptimal for approximating objects with curved inhomogeneities. In this manuscipt, we used those limitations as a motivation for further CHA constructions such as the curvelet transform.

We would like to emphasize that there is a wealth of new multiscale constructions which use the ridgelet transform as a core component. In addition to the curvelet transform, we would like to mention the possibility of constructing ridgelet packets which would involve the joint recursive dyadic splitting of both time and frequency. It is not the scope of this paper to discuss such constructions. However, the reader familiar with ideas such as Recursive Dyadic Partitioning on the one hand, and the curvelet pyramid on the other, will see that one can obtain a pyramid of windowed and filtered ridgelets at all lengths, and widths and that fast algorithms for searching sparse decompositions are very likely to exist.

In another direction, [24] explores a natural extension of the ideas developed here, namely, the construction of k-plane ridgelets in spaces of arbitrary dimension d.

There are undoubtedly many other possible multiscale constructions and, in some sense, it is only the beginning of a new subject we thought

–together with David Donoho– about calling MGA as in *Multiscale Geometric Analysis*. This is a rather unexplored territory with many research opportunities both at the theoretical and practical level.

13.3 Other connections between approximation theory and CHA

There are other opportunities for building bridges between approximation theory and CHA. We close this paper by describing one potential connection which is especially dear to my heart.

It seems of interest to bring together concepts of approximation and statistical theory with those of time-frequency analysis. There are interesting issues such as: how well can we approximate or estimate certain classes of rapidly oscillating functions where the oscillation rate may be changing over time? Can we construct approximations or estimators that will match the best theoretical performances? Time-frequency analysis provides a natural framework for studying these problems, but the tools to address them remain, I believe, to be constructed.

In time-frequency analysis, signals of primary interest oscillate rapidly and their frequency of oscillations is also rapidly changing over time. Those signals commonly referred to as chirps take the general form

$$f(t) = A(t)\cos(N\phi(t)); \qquad (13.2)$$

here N is a (large) base frequency, $\phi(t)$ is time-varying and the amplitude $A(t)$ is slowly varying.

The method of choice for representing chirps is to use Gabor frames of the form

$$g_{m,n}(t) = w((t - mL)/L)\, e^{i2\pi n L t}, \quad n, m \in \mathbb{Z};$$

that is smoothly windowed sinusoids. This representation breaks the time-frequency plane into congruent rectangles and assigns one basis function to each rectangle. Schematically, Gabor approximations correspond to piecewise approximation of the instantaneous frequency. In a sense, Gabor frames are as good for approximating smooth chirps as wavelets are for approximating objects with edges. This brings up a central question: is there a nonadaptive representation which provides optimally sparse decompositions of chirps just as curvelets do for objects with smooth edges? To put it bluntly: is there something beyond Gabor? A mathematical result in this direction – either positive or negative – would certainly be very significant.

13.4 Closing

During my Ph.D. at Stanford, I recall that on several occasions I asked David Donoho the reason why he was not writing a book about the connections between harmonic analysis, statistical estimation, approximation theory and data compression. At the time (1998), he was drafting a manuscript entitled "Harmonic analysis and data compression" [29] for a special issue of *IEEE Trans. Inform. Theory* in the honor of the 50th birthday of Shannon's seminal paper on communication theory. He answered rather vaguely.

Two years later, he obviously had a very different attitude about this as he gave ten lectures about this same topic at the CBMS-NSF Regional Conference 2000 in Applied Mathematics. Much of those lectures were about ridgelets and curvelets. As if the picture suddenly became more complete, and much richer...

§14. Appendix

14.1 Fundamental estimates

The purpose of this technical section is to show that the kernel

$$T(R, R') = \langle \psi_R, \psi_{R'} \rangle_{L_2(w)}, \quad R, R' \in \mathcal{R} \tag{14.1}$$

is "almost diagonal." The fact that $\langle \psi_R, \psi_{R'} \rangle_{L_2(w)}$ decays rapidly as the "distance" between the indices (R, R') increases is a crucial fact of our analysis.

Let f_u and $g_{u'}$ be two ridge functions given by $f_u(x) = f(u \cdot x)$ and $g_{u'}(x) = g(u' \cdot x)$, respectively. Of course, $\langle f_u, g_{u'} \rangle$ makes no sense, as the ridge functions are not square-integrable. We then take a fixed window w in $\mathcal{S}(\mathbb{R}^d)$ and look at the inner product with respect to the signed measure w. We set

$$\langle f_u, g_{u'} \rangle_{L_2(w)} = \int f(u \cdot x) g(u' \cdot x) w(x) \, dx, \tag{14.2}$$

and derive a simpler expression for this quantity. Let Q be an orthogonal change of coordinates ($x = Qx'$) such that

$$u' \cdot x = x'_1 \quad \text{and} \quad u \cdot x = (u \cdot u')x'_1 - \sqrt{1 - (u \cdot u')^2} \, x'_2,$$

and put

$$v = u \cdot u' / \sqrt{1 - (u \cdot u')^2}; \tag{14.3}$$

letting θ be the angle between the unit vectors u and u' gives

$$\cos \theta = u \cdot u', \quad v = \cos \theta / |\sin \theta|. \tag{14.4}$$

With these notations, the inner product (14.2) can be rewritten as

$$\langle f_u, g_{u'} \rangle_{L_2(w)} = \int f(x_1' \cos\theta - x_2' |\sin\theta|) \, g(x_1') \, w(Qx') \, dx_1' dx_2' \dots dx_d'.$$
(14.5)

Define $w_Q(x_1, x_2)$ by $\int w(Qx) \, dx_3 \dots dx_d$. Then,

$$\langle f_u, g_{u'} \rangle_{L_2(w)} = \int f(x_1 \cos\theta - x_2 |\sin\theta|) \, g(x_1) \, w_Q(x_1, x_2) \, dx_1 dx_2. \quad (14.6)$$

Examine now the partial derivatives of w_Q. Let D_1 be the partial derivative with respect to the first variable, i.e. x_1, and similarly for D_2. It is trivial to see that w_Q belongs to $\mathcal{S}(\mathbb{R}^2)$. Moreover, for any n_1, n_2, and each $m > 0$, there is a constant C depending upon w, n_1, n_2, and m such that

$$\sup_Q |D_1^{n_1} D_2^{n_2} w_Q(x_1, x_2)| \le C \, (1 + |x_1| + |x_2|)^{-2m}$$
$$\le C \, (1 + |x_1|)^{-m}(1 + |x_2|)^{-m}. \quad (14.7)$$

In this section we will assume that ψ satisfies a few standard conditions: namely, ψ is R times differentiable and for every nonnegative integer m there is a constant C (depending on m) so that

$$|(D^n \psi)(t)| \le C(1 + |t|)^{-m}$$

for some constant C (depending only upon m and n). Further, we will suppose that ψ has vanishing moments through order D. The next lemma gives an upper bound on the inner product

$$K(j, u, b; j', u', b') := \int \psi_j(u \cdot x - b)\psi_{j'}(u' \cdot x - b')w(x) \, dx,$$

where $\psi_j(t) = 2^{j/2}\psi(t)$.

Lemma 14.1. *Assume $j' \ge j$ and suppose $n < \min(R, D)$. Then, for each $m \ge 0$, there is a constant C (depending on n and m) so that*

$$|K(j, u, b; j', u', b')| \le C \, 2^{-(j'+j)(n+1/2)} \, \delta_j^{2n+1}(\theta) \, T_j(u, b; u', b'); \quad (14.8)$$

here, θ is the angle between u and u' and

$$\delta_j(\theta) = \min(2^j, |\sin\theta|^{-1})$$

and

$$T_j(u, b; u', b') = (1 + |b'|)^{-m} \, (1 + \delta_j(\theta)|b' \cos\theta - b|)^{-m}.$$

Proof: Equation (14.6) gives

$$K(j, u, b; j', u', b')$$
$$= \int_{\mathbf{R}^2} \psi_{j'}(x_1 - b')\psi_j(x_1 \cos\theta - x_2|\sin\theta| - b)w_Q(x_1, x_2)\, dx_1 dx_2.$$

Suppose first that $|\sin\theta| \geq 2^{-j}$ so that $\delta_j(\theta) = |\sin\theta|^{-1}$. The change of variables $x_1' = x_1, x_2' = x_1 \cos\theta - x_2|\sin\theta|$ allows rewriting the kernel as

$$K(j, u, b; j', u', b')$$
$$= \int_{\mathbf{R}^2} \psi_{j'}(x_1' - b')\,\psi_j(x_2' - b)\, w_Q(x_1', |\sin\theta|^{-1}(x_1' \cos\theta - x_2'))\, dx_1' dx_2'.$$

Now let $\tilde{w}_Q$ be the function defined by $\tilde{w}_Q(x_1, x_2) = w_Q(x_1, |\sin\theta|^{-1}(x_1 \cos\theta - x_2))$. For each pair of integers n_1, n_2. We check that $\tilde{w}_Q$ obeys

$$|D_1^{n_1} D_2^{n_2} \tilde{w}_Q)(x_1, x_2)| = C\, |\sin\theta|^{-(n_1+n_2)}$$
$$(1 + |x_1| + |\sin\theta|^{-1}|x_1 \cos\theta - x_2|)^{-2m},$$
$$(14.9)$$

where again, the constant C does not depend on Q. This property is directly inherited from (14.7) and we omit the proof. Let n be an integer. By assumption, the wavelet ψ is of class $\mathcal{C}^R$ for $R > n$ and has at least n vanishing moments. It follows from standard wavelet estimates that one can find a constant C such that

$$|K(j, u, b; j', u', b')| \leq C\, 2^{-(j'+j)(n+1/2)}|\sin\theta|^{-(2n+1)}$$
$$(1 + |b'| + |\sin\theta|^{-1}|b' \cos\theta - b|)^{-2m}.$$
$$(14.10)$$

The proof is essentially an integration by parts. We have

$$K(j, u, b; j', u', b') = \langle \psi_{j'}(\cdot - b') \otimes \psi_j(\cdot - b), \tilde{w}_Q \rangle$$
$$= \langle D_1^{-n}\psi_{j'}(\cdot - b') \otimes D_2^{-n}\psi_j(\cdot - b), D_1^n D_2^n \tilde{w}_Q \rangle.$$

Now the inequality follows from the estimate about the size of the partial derivatives of $\tilde{w}_Q$ and the localization properties of $D^{-n}\psi$ as for each n and m, there is a constant C such that

$$|D^{-n}\psi_j(t)| \leq C\, 2^{-j(n-1/2)}\, (1 + 2^j|t|)^{-m}.$$

The point is that the bulk of the mass of $|D_1^{-n}\psi_{j'}(\cdot - b') \otimes D_2^{-n}\psi_j(\cdot - b)|$ is concentrated near the rectangle $[b' \pm c2^{-j'}] \times [b \pm c2^{-j}]$. The upper bound on the function $D_1^n D_2^n \tilde{w}_Q$ is slowly varying over this effective support since $|\sin \theta|^{-1} \leq 2^j$. The bound (14.10) follows from the fact that

$$\|D_1^{-n} \psi_{j'}(\cdot - b') \otimes D_2^{-n} \psi_j(\cdot - b)\|_{L_1(\mathbf{R}^2)} \leq 2^{-(j+j')/2} \|\psi\|_{L_1(\mathbf{R}^2)}$$

and the fact that over the effective support, the upper bound is about $C\left(1 + |b'| + |\sin\theta|^{-1}|b'\cos\theta - b|\right)^{-2m}$. This reasoning may be turned into a rigorous argument and the calculations giving (14.10) are absolutely standard, see [47] for instance.

The inequality (14.10) together with

$$(1+|b'|+|\sin\theta|^{-1}|b'\cos\theta - b|)^{-2m} \leq (1+|b'|)^{-m}(1+|\sin\theta|^{-1}|b'\cos\theta - b|)^{-m}$$

give the result for $\delta_j(\theta) = |\sin\theta|^{-1}$.

The argument is a little different when the ridgelets are nearly parallel; that is, when $|\sin\theta| \leq 2^{-j}$ or equivalently $\delta_j(\theta) = 2^j$. We apply Fubini's theorem and express the inner product as

$$K(j, u, b; j', u', b') = \int \psi_{j'}(t - b')\, g(t)\, dt,$$

where

$$g(t) = \int_{\mathbf{R}} \psi_j(t\cos\theta - t'\sin\theta - b) w_Q(t, t')\, dt'.$$

The function g is smooth and well localized. On the one hand, for any n there is a constant C such that for any $\ell \leq n$

$$\left| \frac{d^\ell}{dt^\ell} \psi_j(t\cos\theta - t'\sin\theta - b) \right| \leq C\, 2^{j(\ell+1/2)}(1 + 2^j|t\cos\theta - t'\sin\theta - b|)^{-m}.$$

On the other,

$$\left| \frac{\partial^{n-\ell}}{\partial t^{n-\ell}} w(t, t') \right| \leq C\,(1 + |t| + |t'|)^{-2m} \leq C\,(1 + |t|)^{-m}(1 + |t'|)^{-m}.$$

Now, an integration by parts gives

$$\left| \frac{d^n}{dt^n} g(t) \right| \leq C\, 2^{j(n+1/2)}(1 + |t|)^{-m}(1 + 2^j|t\cos\theta - b|)^{-m}. \qquad (14.12)$$

Again, we apply classical wavelet analysis techniques for bounding the coefficients of a smooth and well-localized function. The argument is essentially an integration by parts as explained above. We write

$$K(j, u, b; j', u', b') = \langle D^{-n}\psi_{j'}(\cdot - b'), D^n g \rangle,$$

and use (14.12) and (14.11) to obtain

$$|K(j, u, b; j', u', b')| = \left| \int \psi_{j'}(t - b') g(t)\, dt \right|$$
$$\le C 2^{-(j'-j)(n+\frac{1}{2})} (1 + |b'|)^{-m} (1 + 2^j |b' \cos\theta - b|)^{-m}.$$

This last inequality is the content of (14.8) when $|\sin\theta| \le 2^{-j}$. The lemma is proved. $\square$

For $j' \ge j$, Lemma 14.1 gives the upper bound on the entries of the Gramm matrix (14.1)

$$|T(R, R')| \le C 2^{-(j'+j)(n+1/2)} \delta_j^{2n+1}(\ell, \ell')\, L(R, R'), \tag{14.13}$$

with

$$L(R, R') = \left(1 + |k' 2^{-j'}|\right)^{-m} \left(1 + \delta_j(\ell, \ell') |k' 2^{-j'} \cos\theta - k 2^{-j}|\right)^{-m}. \tag{14.14}$$

Here, δ_j is as before, namely,

$$\delta_j(\ell, \ell') = \min(2^j, 1/|\sin\theta|)$$

where θ is the angle between $u_{j,\ell}$ and $u_{j',\ell'}$. The estimate for $j' < j$ is obtained by symmetry.

Lemma 14.2. *Suppose ψ is C^∞ and has vanishing moments up to any order. Then for any $p > 0$, we have*

$$\sup_{R'} \sum_R |T(R, R')|^p \le A_p. \tag{14.15}$$

(If $r = \min(R, D)$ is finite, then (14.15) holds for $p > p^*$ with $(r+1/2)p^* = d$.)

Proof: First, for $j' \ge j$, (14.14) gives

$$\sum_k |L(R, R')|^p \le \sum_k \left(1 + \delta_j(\ell, \ell') |k' 2^{-j'} \cos\theta - k 2^{-j}|\right)^{-mp}$$
$$\le C 2^j |\delta_j(\ell, \ell')|^{-1}, \tag{14.16}$$

provided that $mp > 1$. Second, for $j > j'$ and m large enough so that $mp > 1$, we have

$$\sum_k |L(R, R')|^p$$
$$= \sum_k \left(1 + |k 2^{-j}|\right)^{-mp} \left(1 + \delta_{j'}(\ell, \ell') |k 2^{-j} \cos\theta - k' 2^{-j'}|\right)^{-mp}$$
$$\le C 2^j \min(1, (|\delta_{j'}(\ell, \ell') \cos\theta|^{-1}).$$

In particular, this last upper bound gives

$$\sum_k |L(R,R')|^p \leq \begin{cases} C\,2^j, & |\cos\theta| < 1/\sqrt{2} \\ C\,2^j\,|\delta_{j'}(\ell,\ell')|, & |\cos\theta| \geq 1/\sqrt{2}. \end{cases} \tag{14.17}$$

We now sum the inequalities (14.16)–(14.17) over the angular variable. To develop upper bound, we shall use a little lemma whose proof is postponed.

Lemma 14.3. *Let $\beta > d - 1$. Then,*

$$\sum_\ell |\delta_j(\ell,\ell')|^\beta \leq C\,2^{j\beta}.$$

and, for $j' \leq j$

$$\sum_\ell |\delta_{j'}(\ell,\ell')|^\beta \leq C\,2^{j'\beta}\,2^{(j-j')(d-1)}.$$

Suppose $j' \geq j$. The bound (14.16) yields

$$\sum_k |T(R,R')|^p \leq C\,2^{-(j+j')(n+1/2)p}\,2^j\,|\delta_j(\ell,\ell')|^{(2n+1)p-1}.$$

For $(2n+1)p > d$, Lemma 14.3 gives

$$\sum_\ell |\delta_j(\ell,\ell')|^{(2n+1)p-1} \leq C\,2^{j((2n+1)p-1)},$$

and, therefore,

$$\sum_\ell \sum_k |T(R,R')|^p \leq C\,2^{-(j+j')(n+1/2)p}\,2^{j(2n+1)p} = C\,2^{-(j'-j)(n+1/2)p}.$$

Let us turn our attention to the case $j' \leq j$. For $|\cos\theta| > 1/\sqrt{2}$, the bound (14.17) yields

$$\sum_k |T(R,R')|^p \leq C\,2^{-(j+j')(n+1/2)p}\,2^j\,|\delta_{j'}(\ell,\ell')|^{(2n+1)p-1}.$$

For $(2n+1)p > d$, Lemma 14.3 now gives

$$\sum_\ell |\delta_{j'}(\ell,\ell')|^{(2n+1)p-1} \leq C\,2^{j'((2n+1)p-1)}\,2^{(j-j')(d-1)},$$

and, therefore,

$$\sum_{\ell:|\cos\theta|>1/\sqrt{2}}\sum_{k}|T(R,R')|^p \leq C\,2^{-(j+j')(n+1/2)p}\,2^{j'((2n+1)p}\,2^{(j-j')d}$$

$$= C\,2^{-(j-j')((n+1/2)p-d)}.$$

Moreover, For $|\cos\theta| \geq 1/\sqrt{2}$, the bound (14.17) yields

$$\sum_{k}|T(R,R')|^p \leq C\,2^{-(j+j')(n+1/2)p}\,2^{j}\,|\delta_{j'}(\ell,\ell')|^{(2n+1)p}.$$

Observe that in the range $|\cos\theta| \geq 1/\sqrt{2}$, we have $|\delta_{j'}(\ell,\ell')| \leq \sqrt{2}$ since $\delta_{j'}(\ell,\ell') \leq 1/|\sin\theta|$. The last inequality then becomes

$$\sum_{k}|T(R,R')|^p \leq C\,2^{-(j+j')(n+1/2)p}\,2^{j},$$

which is independent of ℓ. At scale j, the total number of orientations is $O(2^{j(d-1)})$ and, hence,

$$\sum_{\ell:|\cos\theta|\leq 1/\sqrt{2}}\sum_{k}|T(R,R')|^p \leq C\,2^{-(j+j')(n+1/2)p}\,2^{jd}.$$

Take $2(n+1/2)p > d$, then for $j' \geq 0$, we have $2^{j'(2n+1)p-d)} \geq 1$ which gives

$$\sum_{\ell:|\cos\theta|\leq 1/\sqrt{2}}\sum_{k}|T(R,R')|^p \leq C\,2^{-(j+j')(n+1/2)p}\,2^{jd}\,2^{j'(2n+1)p-d)}$$

$$= C\,2^{-(j-j')((n+1/2)p-d)}.$$

Collecting our results, we showed that for $j' \leq j$

$$\sum_{\ell:|\cos\theta|>1/\sqrt{2}}\sum_{k}|T(R,R')|^p \leq C\,2^{-(j-j')((n+1/2)p-d)}.$$

To summarize, we developed the following upper bound:

$$\sum_{|R|=j}|T(R,R')|^p \leq \begin{cases} C2^{-|j-j'|((n+1/2)p-d)}, & j' \leq j \\ C2^{-|j-j'|(n+1/2)p}, & j' \geq j. \end{cases} \qquad (14.18)$$

It is now clear that choosing n such that $(n+1/2)p > d$ gives

$$\sum_{R}|T(R,R')|^p \leq C,$$

where C is independent of R'. This finishes the proof of the lemma. $\square$

Proof of Lemma 14.3: Let u_0 be a fixed direction and let θ be the angle between u_0 and a running point u on the sphere. Define

$$\begin{aligned}
\Lambda_0 &= \{u, |\sin\theta| \geq 1/2\}, \\
\Lambda_m &= \{u, 2^{-(m+1)} \leq |\sin\theta| < 2^{-m}\}, \quad m = 1, 2, \dots
\end{aligned} \tag{14.19}$$

and let $\Lambda_{j,m}$ denote the set of indices obeying

$$\Lambda_{j,m} := \{\ell, \, u_{j,\ell} \in \Lambda_m\} \tag{14.20}$$

with the convention that for $m = j$, we will say that $\ell \in \Lambda_{j,m}$ if $u_{j,\ell} \in \Lambda_m$, for some $m \geq j$.

We recall that at scale j, the set of discrete angular variables $\{u_{j,\ell}, \ell \in \Lambda_j\}$ consists of points approximately uniformly distributed on the sphere. In particular, for $m = 0, 1, \dots, j$, we have

$$\# \Lambda_{j,m} \leq C \left(1 + 2^{j(d-1)} \mu(\Lambda_m)\right), \tag{14.21}$$

where μ is the uniform probability measure on the unit sphere of $\mathbb{R}^d$. In other words, the number of points falling in the set Λ_m, is essentially bounded — up to a multiplicative constant — by the total number of sampled points on the sphere times the area of the set Λ_m. As one can see, this says that the points $u_{j,\ell}$ are approximately equi-distributed on the sphere.

To calculate $\mu(\Lambda_m)$, we introduce the spherical coordinates defined by $x_1 = \cos\theta_1$, $x_2 = \sin\theta_1 \cos\theta_2$, $\dots$, $x_d = \sin\theta_1 \sin\theta_2 \dots \sin\theta_{d-1}$, $0 \leq \theta_1, \dots$, $\theta_{d-2} \leq \pi$, $0 \leq \theta_{d-1} < 2\pi$. Let θ_1 be the angle between the fixed orientation u_0 and the running point u. With these notations, the distribution of θ_1 is proportional to $(\sin\theta_1)^{d-2} d\theta_1$, where the normalizing constant guarantees that the density integrates up to one. Then,

$$\mu(\Lambda_m) = c_d \int_{2^{-(m+1)} \leq |\sin\theta| \leq 2^{-m}} |\sin\theta|^{d-2} \, d\theta.$$

This last equation gives the upper bound

$$\mu(\Lambda_m) \leq C \, 2^{-m(d-1)},$$

which also holds for the special case $m = 0$. Therefore, from (14.21) we deduce a bound on the cardinality of $\Lambda_{j,m}$:

$$\# \Lambda_{j,m} \leq C(1 + 2^{(j-m)(d-1)}), \quad m = 0, 1, 2, \dots, j.$$

Observe that the above inequality is also valid in the special case where $m \in \{0, j\}$.

After these preliminaries, we are in a position to prove the lemma. Let $\beta > d - 1$. Over the subset of indices such that $\{\ell : |\sin \theta| \geq 2^{-j}\}$, we have

$$\sum_{\ell: |\sin \theta| \geq 2^{-j}} |\delta_j(\ell, \ell')|^\beta \leq \sum_{m=0}^{j-1} 2^{(m+1)\beta} \, \#\Lambda_{j,m}$$

$$\leq C \sum_{m=0}^{j-1} 2^{(m+1)\beta} (1 + 2^{(j-m)(d-1)}) \leq C \, 2^{j\beta}.$$

And for those ℓ's such that $|\sin \theta| \leq 2^{-j}$, we have

$$\sum_{\ell: |\sin \theta| \leq 2^{-j}} |\delta_j(\ell, \ell')|^\beta \leq 2^{j\beta} \, \#\Lambda_{j,j} \leq C \, 2^{j\beta}.$$

The last two inequalities yield the first part of the lemma.

Next, assume $j' \leq j$. Over the subset of indices such that $\{\ell : |\sin \theta| \geq 2^{-j'}\}$, we have

$$\sum_{\ell: |\sin \theta| \geq 2^{-j'}} |\delta_{j'}(\ell, \ell')|^\beta \leq \sum_{m=0}^{j'-1} 2^{(m+1)\beta} \, \#\Lambda_{j,m} \leq 2^{j'\beta} \, 2^{(j-j')(d-1)}.$$

And for those ℓ's such that $|\sin \theta| \leq 2^{-j'}$, we have

$$\sum_{\ell: |\sin \theta| \leq 2^{-j'}} |\delta_{j'}(\ell, \ell')|^\beta \leq 2^{j'\beta} \sum_{m=j'}^{j} \#\Lambda_{j,m} \leq C \, 2^{j'\beta} \, 2^{(j-j')(d-1)}.$$

The last two inequalities yield the second part of the lemma. $\square$

We close this section by proving similar estimates for the inner product

$$K_0(u, b; j', u', b') := \int \phi(u \cdot x - b) \psi_{j'}(u' \cdot x - b') w(x) \, dx,$$

where φ is a coarse scale function as in Section 9. Let φ be R times differentiable such that, for each $m \geq 0$, its derivatives up to order R obey

$$|(D^n \varphi)(t)| \leq C(1 + |t|)^{-m}.$$

The same arguments as in the proof of Lemma 14.8 show that

$$|K_0(u, b; j', u', b')| \leq C \, 2^{-j'(n+1/2)}$$
$$(1 + |b'|)^{-m} \, (1 + |b' \cos \theta - b|)^{-m}. \tag{14.22}$$

In particular, this last inequality gives $(j' \geq 0)$

$$|T_0(R', R)| \leq C \, 2^{-j'(n+1/2)} \left(1 + |k'2^{-j'}|\right)^{-m} \left(1 + |k'2^{-j'}\cos\theta - k|\right)^{-m},$$

with T_0 as in Section 9. Then, for $m > 1/p$, we have

$$\sum_k |T_0(R', R)|^p \leq C \, 2^{-j'(n+1/2)p} \left(1 + |k'2^{-j'}|\right)^{-mp},$$

and further

$$\sum_k \sum_{k'} |T_0(R', R)|^p \leq C \, 2^{-j(n+1/2)p} \, 2^{j'}.$$

We sum the previous inequalities over the angular variables. At scale j', the total number of orientations is $O(2^{j'(d-1)})$ and, hence,

$$\sum_{|R'|=2^{j'}} \sum_{|R|=0} |T_0(R', R)|^p \leq C \, 2^{-j'(n+1/2)p} \, 2^{j'd}. \tag{14.23}$$

It is clear that for $(n + 1/2)p > d$, T_0 is maps bounded ℓ_∞ sequences into ℓ_p bounded sequences, say.

Acknowledgments. This work was supported by an Alfred P. Sloan Fellowship. I would like to dedicate this paper to all the people who have helped me in my young career, and especially to David L. Donoho.

References

1. Auer, P., M. Hebster, and M. K. Warmuth, Exponentially many local minima for single neurons, in *Advances in Neural Information Processing Systems*, D. S. Touretzky, M. C. Mozer, and M. E. Hasselmo (eds.), volume 8, The MIT Press, 1996, 316–322.

2. Barron, A. R., Universal approximation bounds for superpositions of a sigmoidal function, IEEE Trans. Inform. Theory **39** (1993), 930–945.

3. Candès, E. J., Ridgelets: theory and applications, PhD thesis, Department of Statistics, Stanford University, 1998.

4. Candès, E. J., Harmonic analysis of neural networks, Appl. Comput. Harmonic Anal. **6** (1999), 197–218.

5. Candès, E. J., Ridgelets: Estimating with ridge functions, Technical report, Department of Statistics, Stanford University, 1999. Submitted for publication.

6. Candès, E. J., Ridgelets and sigmoidal neural networks, Submitted for publication, see `http://www.acm.caltech.edu/~emmanuel/publi cations.html`, 2001.

7. Candès, E. J., Ridgelets and the representation of mutilated Sobolev functions, SIAM J. Math. Anal. **33** (2001), 347–368.

8. Candès, E. J., and D. L. Donoho. Curvelets, manuscript, see `http://www-stat.stanford.edu/donoho/Reports/1998/curvelets.zip`, 1999.

9. Candès, E. J., and D. L. Donoho, Curvelets—a surprisingly effective nonadaptive representation for objects with edges, in *Curve and Surface Fitting: Saint-Malo 1999*, Albert Cohen, Christophe Rabut, and Larry L. Schumaker (eds.), Vanderbilt University Press, Nashville, 2000, 105–120.

10. Candès, E. J., and D. L. Donoho, Ridgelets: the key to higher dimensional intermittency?, Phil. Trans. R. Soc. Lond. A. **357** (1999), 2495–2509.

11. Candès, E. J., and D. L. Donoho, Recovering edges in ill-posed inverse problems: Optimality of curvelet frames, Technical report, Department of Statistics, Stanford University, 2000, to appear Ann. Statist.

12. Candès, E. J., and D. L. Donoho. Curvelets and curvilinear integrals, J. Approx. Theory **113** (2001), 59–90.

13. Carroll, S. M., and B. W. Dickinson, Construction of neural nets using the Radon tranform, in *Proceedings of the IEEE 1989 International Joint Conference on Neural Networks*, IEEE, New York, 1989, 661–667.

14. Chen, S. S., Basis Pursuit, PhD thesis, Department of Statistics, Stanford University, 1995.

15. Chen, S. S., D. L. Donoho, and M. A. Saunders, Atomic decomposition by basis pursuit, SIAM J. Sci. Comput. **20** (1999), 33–61.

16. Cheng, B., and D. M. Titterington, Neural networks: a review from a statistical perspective. With comments and a rejoinder by the authors, Stat. Sci. **9** (1994), 2–54.

17. Cybenko, G., Approximation by superpositions of a sigmoidal function, Math. Control Signals Systems **2** (1989), 303–314.

18. Deans, S. R., *The Radon Transform and Some of Its Applications*, John Wiley & Sons, 1983.

19. DeVore, R. A., R. Howard, and C. A. Micchelli, Optimal nonlinear approximation, Manuscripta Mathematica **63** (1989), 469–478.

20. DeVore, R. A., B. Jawerth, and V. Popov, Compression of wavelet decompositions, Amer. J. Math. **114** (1992), 737–785.

21. DeVore, R. A., K. I. Oskolkov, and P. P. Petrushev, Approximation by ridge functions and neural networks, Ann. Numer. Math. **4** (1997), 261–287. The heritage of P. L. Chebyshev: a Festschrift in honor of the 70th birthday of T. J. Rivlin.

22. Donoho, D. L., Unconditional bases are optimal bases for data compression and for statistical estimation, Appl. Comput. Harmonic Anal. **1** (1993), 100–115.

23. Donoho, D. L., Sparse components analysis and optimal atomic decomposition, Technical report, Department of Statistics, Stanford University, 1998.

24. Donoho, D. L., Tight frames of k-plane ridgelets and the problem of representing objects that are smooth away from d-dimensional singularities in $\mathbb{R}^n$, Proc. Natl. Acad. Sci. USA **96** (1999), 1828–1833.

25. Donoho, D. L., Wedgelets: nearly minimax estimation of edges, Ann. Statist, **27** (1999), 859–897.

26. Donoho, D. L., Orthonormal ridgelets and linear singularities, SIAM J. Math. Anal. **31** (2000), 1062–1099.

27. Donoho, D. L., Ridge functions and orthonormal ridgelets, J. Approx. Theory **111** (2001), 143–179.

28. Donoho, D. L., and I. M. Johnstone, Empirical atomic decomposition, manuscript, 1995.

29. Donoho, D. L., M. Vetterli, R. A. DeVore, and I. Daubechies, Data compression and harmonic analysis, IEEE Trans. Inform. Theory **44** (1998), 2435–2476.

30. M. Frazier, B. Jawerth, and G. Weiss, *Littlewood-Paley Theory and the Study of Function Spaces*, volume 79 of NSF-CBMS Regional Conf. Ser. in Mathematics, American Math. Soc., Providence, RI, 1991.

31. Friedman, J. H., and W. Stuetzle, Projection pursuit regression. J. Amer. Statist. Assoc. **76** (1981), 817–823.

32. Funahashi, K., On the approximate realization of continuous mapping by neural networks, Neural Networks **2** (1989), 183–192.

33. Helgason, S., *The Radon Transform*, Birkhäuser, Boston, second edition, 1999.

34. Hornik, K., M. Stinchcombe, and H. White, Multilayer feedforward networks are universal approximators, Neural Networks **2** (1989), 359–366.

35. John, F., *Plane Waves and Spherical Means Applied to Partial Differential Equations*, Interscience Publishers, Inc., New York, 1955.

36. Jones, L. K., A simple lemma on greedy approximation in Hilbert space and convergence rates for projection pursuit regression and neural network training, Ann. Statist. **20** (1992), 608–613.

37. Jones, L. K., The computational intractability of training sigmoidal neural networks, IEEE Transactions on Information Theory **43** (1997), 167–173.

38. Khas'minskii, R. Z., and V. S. Lebedev, On the properties of parametric estimators for areas of a discontinuous image, Problems Control Inform. Theory, 1990.

39. Korostelev, A. P., and A. B. Tsybakov, *Minimax Theory of Image Reconstruction,* volume 82 of Lecture Notes in Statistics, Springer-Verlag, New York, 1993.

40. Le Pennec, E., and S. Mallat, Image compression with geometrical wavelets, in *Proceedings of International Conference on Image Processing* (ICIP) 2000.

41. Lemarié, P. G., and Y. Meyer, Ondelettes et bases Hilbertiennes, Rev. Mat. Iberoamericana **2** (1986), 1–18.

42. Leshno, M., V. Y. Lin, A. Pinkus, and S. Schoken, Multilayer feedforward networks with a nonpolynomial activation function can approximate any function, Neural Networks **6** (1993), 861–867.

43. Logan, B. F., and L. A. Shepp, Optimal reconstruction of a function from its projections, Duke Math. J. **42** (1975), 645–659.

44. Maiorov, V., and A. Pinkus, Lower bounds for approximation by MLP neural networks, Neurocomputing **25** (1999), 81–91.

45. Maiorov, V. E., and R. Meir, On the near optimality of the stochastic approximation of smooth functions by neural networks, Adv. Comput. Math. **13** (2000), 79–103.

46. Makovoz, Y., Random approximants and neural networks, J. Approx. Theory **85** (1996), 98–109.

47. Meyer, Y., *Ondelettes et Opérateurs: II. Opérateurs de Calderón Zygmund,* Hermann, 1990.

48. Meyer, Y., *Wavelets and Operators,* Cambridge University Press, 1992.

49. Mhaskar, H. N., Neural networks for optimal approximation of smooth and analytic functions, Neural Computation **8** (1996), 164–177.

50. Murata, N., An integral representation of functions using three-layered networks and their approximation bounds, Neural Networks **9** (1996), 947–956.

51. Natterer, F., *The Mathematics of Computerized Tomography,* B. G. Teubner, John Wiley & Sons, 1986.

52. Oskolkov, K., Chebyshev—Fourier analysis and optimal quadrature formulas, Proc. Steklov Math. Inst. **219** (1997), 269–285. (in Russian).

53. Petrushev, P. P., Approximation by ridge functions and neural networks, SIAM J. Math. Anal. **30** (1999), 155–189.

54. Pinkus, A., Approximating by ridge functions, in *Surface Fitting and Multiresolution Methods*, A. Le Méhauté, C. Rabut, and L. L. Schumaker (eds.), Vanderbilt University Press, Nashville, 1997, 279–292.

55. Pinkus, A., Approximation theory of the MLP model in neural networks, Acta Numerica **8** (1999), 143–196.

56. Pisier, G., Remarques sur un résultat non publié de B. Maurey, in *Séminaire d'Analyse Fonctionelle, 1980–1981*, Centre des Mathématiques, Ecole Polytechnique, Palaiseau, France.

57. Plancherel, M., and G. Pólya, Fonctions entières et intégrales de Fourier multiples, Commentarii Math. Helv. **10** (1938), 110–163.

58. Rubin, B., The Calderón reproducing formula, windowed X-ray transforms and Radon transforms in L^p-spaces, J. Fourier Anal. Appl. **4** (1998), 175–197.

59. Starck, J. L., E. J. Candès, and D. L. Donoho, The curvelet transform for image denoising, IEEE Transactions on Image Processing, 2000, to appear.

60. Starck, J. L., E. J. Candès, and D. L. Donoho, Very high quality image restoration, in *Wavelet Applications in Signal and Image Processing IX*, A. Aldroubi, A. F. Laine, and M. A. Unser (eds.), Proc. SPIE 4478, 2001.

61. Vetterli, M., and J. Kovacevic, *Wavelets and Subband Coding*, Prentice Hall, Englewood Cliffs, NJ, 1995.

62. Vu, V. H., On the infeasibility of training neural networks with small mean-squared error, IEEE Transactions on Information Theory **44** (1998), 2892–2900.

Emmanuel J. Candès
Applied and Computational Mathematics
California Institute of Technology
Pasadena, California 91125
emmanuel@acm.caltech.edu
http://www.acm.caltech.edu/~emmanuel

Approximation of the Eigenvalues of Regular Fourth Order Sturm-Liouville Problems Using Interpolation Theory

B. Chanane

Abstract. Sampling theory has been used recently to compute the eigenvalues of second order Sturm-Liouville (SL) problems. We demonstrate that the method is capable of approximating the eigenvalues of fourth order SL problems. We show that transforms of the solution of the associated initial value problem and of its derivatives are analytic in a strip containing the real axis and satisfy a condition which allows them to be recovered from their samples very accurately using the sinc series approximation. Then, the zeros of the reconstructed boundary function are seen as the fourth roots of the sought eigenvalues.

§1. Introduction

The spectral theory of differential operators is well developed (see references). However, when it comes to computing the spectral data much remains to be done. In the second order case, a number of computer codes have been produced SLEIGN2, SLEDGE, etc. These packages are based on piecewise constant approximation of the coefficients in the Sturm-Liouville (SL) problem, and on the well-known shooting method. Sampling theory has been used successfully to compute the eigenvalues of second order regular SL problems with Dirichlet boundary conditions in [4], and with general separable boundary conditions in [7]. In [8], we provided high order approximations to the eigenvalues and provided a very sharp estimate. In [11], we considered the coupled SL problem, and showed how to effectively compute the double eigenvalues. Several numerical examples have been worked out to show the effectiveness of the method. We have

Approximation Theory X: Wavelets, Splines, and Applications   155
Charles K. Chui, Larry L. Schumaker, and Joachim Stöckler (eds.), pp. 155–166.
Copyright ⊖ 2002 by Vanderbilt University Press, Nashville, TN.
ISBN 0-8265-1416-2.

considered the singular case in [5] and [6] and the random case in [12]. In 1997 Greenberg and Marletta [20] released a software package dealing with the computation of eigenvalues of fourth order Sturm-Liouville problems. The underlying theory was presented in [19]. It is based on the approximation of the coefficients in the differential equation and a suitable zero counting algorithm. It is remarkable to note at this point that this code named SLEUTH (Sturm-Liouville Eigenvalues Using Theta matrices) is the only package available dealing with fourth order problems.

Recently, Chanane [9,10], using the concepts of Fliess series and iterated integrals, introduced a novel series representation for the boundary function associated with Sturm-Liouville problems of order two and four. The zeros of this boundary function are the eigenvalues of the problem. A few examples were provided and the results were in agreement with the output of SLEIGN2 in the second order case and SLEUTH in the fourth order case.

We shall see in this paper how we can compute the eigenvalues of fourth order Sturm-Liouville problems using interpolation.

§2. Main Result

Consider the fourth order Sturm-Liouville problem

$$y^{(4)} - (s(x)y^{(1)})^{(1)} + q(x)y = \lambda y, \qquad x \in (0,1), \tag{1}$$

with the general separated self-adjoint boundary conditions [19]

$$\begin{cases} A_1 u + A_2 v = 0 & \text{at } x = 0 \\ B_1 u + B_2 v = 0 & \text{at } x = 1, \end{cases} \tag{2}$$

where

$$u = (u_1, u_2)' , v = (v_1, v_2)'$$

and

$$u_1 = y , u_2 = y' , v_1 = sy' - y''' , v_2 = y'' \tag{3}$$

are the quasiderivatives, A_1 , A_2 , B_1 , B_2 are 2×2 matrices such that $A_1 A_2^T = A_2 A_1^T$, $B_1 B_2^T = B_2 B_1^T$ and the 2×4 matrices $M = (A_1|A_2)$, $N = (B_1|B_2)$ have rank 2.

We shall assume that q and s are in $C^0(0,1)$ and $C^1(0,1)$ respectively. Under these assumptions, it is well known that the eigenvalues are bounded from below and can be ordered as $\lambda_0 \leq \lambda_1 \leq \lambda_2 \leq \lambda_3 \leq \cdots$, $\lambda_n \to +\infty$ as $n \to \infty$ and each eigenvalue has multiplicity at most 2, see [14,19,25]. We are interested in computing the positive eigenvalues, the negative ones can be computed using a simple shift in λ, so let $\lambda = \mu^4$.

We shall associate to (1) an initial condition as follows. Let $z = (u_1, u_2, v_1, v_2)' = (z_1, z_2, z_3, z_4)'$, let M_1, M_2, M_3, M_4 denote the columns of M, i.e., $M = (M_1|M_2|M_3|M_4)$. We have $Mz = 0$ at $x = 0$. Since the rank of M is 2, there exists a permutation σ_0 of $1, 2, 3, 4$ for which the $\det(M_{\sigma_0(1)}|M_{\sigma_0(2)}) \neq 0$. Let $M_{\sigma_0} = (M_{\sigma_0(1)}|M_{\sigma_0(2)})|M_{\sigma_0(3)}|M_{\sigma_0(4)})$ and $z_{\sigma_0} = (z_{\sigma_0(1)}, z_{\sigma_0(2)}, z_{\sigma_0(3)}, z_{\sigma_0(4)})'$. From $M_{\sigma_0} z_{\sigma_0} = 0$ at $x = 0$, we get at once

$$(M_{\sigma_0(1)}|M_{\sigma_0(2)}) \begin{pmatrix} z_{\sigma_0(1)} \\ z_{\sigma_0(2)} \end{pmatrix} + (M_{\sigma_0(3)}|M_{\sigma_0(4)}) \begin{pmatrix} z_{\sigma_0(3)} \\ z_{\sigma_0(4)} \end{pmatrix} = \begin{pmatrix} 0 \\ 0 \end{pmatrix}. \quad (4)$$

Thus,

$$\begin{pmatrix} z_{\sigma_0(1)} \\ z_{\sigma_0(2)} \end{pmatrix} = -(M_{\sigma_0(1)}|M_{\sigma_0(2)})^{-1}(M_{\sigma_0(3)}|M_{\sigma_0(4)}) \begin{pmatrix} z_{\sigma_0(3)} \\ z_{\sigma_0(4)} \end{pmatrix}, \quad (5)$$

which yields

$$z = \begin{pmatrix} -(M_{\sigma_0(1)}|M_{\sigma_0(2)})^{-1}(M_{\sigma_0(3)}|M_{\sigma_0(4)}) \begin{pmatrix} z_{\sigma_0(3)} \\ z_{\sigma_0(4)} \end{pmatrix} \\ z_{\sigma_0(3)} \\ z_{\sigma_0(4)} \end{pmatrix}_{\sigma_0^{-1}} \quad \text{at } x = 0, \quad (6)$$

where σ_0^{-1} denotes the inverse permutation; i.e., we recover the original ordering of the components of z.

The initial condition we were looking for is taken to be

$$z(0) = z^0 = \begin{pmatrix} -(M_{\sigma_0(1)}|M_{\sigma_0(2)})^{-1}(M_{\sigma_0(3)}|M_{\sigma_0(4)}) \begin{pmatrix} 1 \\ \alpha \end{pmatrix} \\ 1 \\ \alpha \end{pmatrix}_{\sigma_0^{-1}}. \quad (7)$$

Thus we are normalizing the eigenfunctions using $z_{\sigma_0(3)} = 1$ at $x = 0$. We shall decompose z^0 as $z^0 = \hat{z}_1^0 + \alpha \hat{z}_2^0$, where $\hat{z}_1^0$ and $\hat{z}_2^0$ are free from α, that is

$$\hat{z}_1^0 = \begin{pmatrix} -(M_{\sigma_0(1)}|M_{\sigma_0(2)})^{-1} M_{\sigma_0(3)} \\ 1 \\ 0 \end{pmatrix}_{\sigma_0^{-1}},$$

$$\hat{z}_2^0 = \begin{pmatrix} -(M_{\sigma_0(1)}|M_{\sigma_0(2)})^{-1} M_{\sigma_0(4)} \\ 0 \\ 1 \end{pmatrix}_{\sigma_0^{-1}}.$$

If $\hat{z}_1$ and $\hat{z}_2$ denote the solutions of (1) with the initial conditions $\hat{z}_1(0) = \hat{z}_1^0$ and $\hat{z}_2(0) = \hat{z}_2^0$ respectively, then $\hat{z} = \hat{z}_1 + \alpha \hat{z}_2$ is the solution satisfying $\hat{z}(0) = z^0$, so that $N\hat{z}(x, \mu^4, \alpha) = \binom{0}{0}$ at $x = 1$ becomes $N\hat{z}_1 + \alpha N\hat{z}_2 = \binom{0}{0}$ at $x = 1$ from which we get the eigenvalues as the fourth power of the zeroes of the boundary function $B(\mu) = \det(C(\mu))$, where $C(\mu) =$

$(N\widehat{z_1}|N\widehat{z_2})$ when $x = 1$. Note that the initial condition $\widehat{z_1}(0) = \widehat{z_1^0}$ (say) gives rise to an initial condition for y, y', y'', y''' as $y(0) = u_1^0$, $y'(0) = u_2^0$, $y''(0) = v_2^0$ and $y'''(0) = s(0)u_2^0 - v_1^0$. To obtain the eigenfunction (s) belonging to the eigenvalue $\overline{\mu}^4$, we proceed as follows. If $\dim \text{Ker}(C(\overline{\mu})) = 1$, then we associate to this eigenvalue one independent eigenfunction with $\alpha = \delta/\gamma$, where $(\gamma, \delta)^T$ is an eigenvector of $C(\overline{\mu})$. If $\dim \text{Ker}(C(\overline{\mu})) = 2$ then we associate to this eigenvalue two linearly independent eigenfunctions with $\alpha = \delta_1/\gamma_1$ and $\alpha = \delta_2/\gamma_2$ where $(\gamma_1, \delta_1)^T$ and $(\gamma_2, \delta_2)^T$ are two linearly independent eigenvectors of $C(\overline{\mu})$. We therefore have a way to check the multiplicity of the eigenvalue whether it is simple or double.

Going back to (1), we write the equation as an integral equation with $\lambda = \mu^4$. We start with

$$y^{(4)} - \mu^4 y = p(x, \mu) = (s(x)y')' - q(x)y, \tag{8}$$

from which we get

$$y(x, \mu^4) = \varphi_4(x, \mu)y(0, \mu^4) + \varphi_3(x, \mu)y'(0, \mu^4) + \varphi_2(x, \mu)y''(0, \mu^4)$$
$$+ \varphi_1(x, \mu)y'''(0, \mu^4) + \int_0^x \varphi_1(x - t, \mu)p(t, \mu)dt, \tag{9}$$

where

$$\begin{cases} \varphi_1(x, \mu) = \dfrac{\sinh \mu x - \sin \mu x}{2\mu^3} \\[2mm] \varphi_2(x, \mu) = \dfrac{\cosh \mu x - \cos \mu x}{2\mu^2} = \varphi_1'(x, \mu) \\[2mm] \varphi_3(x, \mu) = \dfrac{\sinh \mu x + \sin \mu x}{2\mu} = \varphi_2'(x, \mu) \\[2mm] \varphi_4(x, \mu) = \dfrac{\cosh \mu x + \cos \mu x}{2} = \varphi_3'(x, \mu). \end{cases} \tag{10}$$

Note that $\varphi_4'(x, \mu) = \mu^4 \varphi_1(x, \mu)$, where the differentiation is with respect to x. Using (10), the last term in (9) can be integrated by parts to yield

$$\int_0^x \varphi_1(x - t, \mu)p(t, \mu)dt$$
$$= \int_0^x \varphi_1(x - t, \mu)\left\{(s(t)y'(t, \mu^4))' - q(t)y(t, \mu^4)\right\}dt$$
$$= \left[\varphi_1(x - t, \mu)s(t)y'(t, \mu^4)\right]_{t=0}^x + \int_0^x \varphi_2(x - t, \mu)s(t)y'(t, \mu^4)dt$$
$$- \int_0^x \varphi_1(x - t, \mu)q(t)y(t, \mu^4)dt$$

$$= -\varphi_1(x,\mu)s(0)y'(0,\mu^4) + \left[\varphi_2(x-t,\mu)s(t)y(t,\mu^4)\right]_{t=0}^{x}$$

$$- \int_0^x \left\{\varphi_2(x-t,\mu)s'(t) - \varphi_3(x-t,\mu)s(t) + \varphi_1(x-t,\mu)q(t)\right\}y(t,\mu^4)dt$$

$$= -\varphi_1(x,\mu)s(0)y'(0,\mu^4) - \varphi_2(x,\mu)s(0)y(0,\mu^4)$$

$$+ \int_0^x \left\{-\varphi_2(x-t,\mu)s'(t) + \varphi_3(x-t,\mu)s(t) - \varphi_1(x-t,\mu)q(t)\right\}y(t,\mu^4)dt.$$

Thus,

$$y(x,\mu^4) = \Psi_0(x,\mu) + \int_0^x K(x,t,\mu)y(t,\mu^4)dt, \tag{11}$$

where

$$\Psi_0(x,\mu) = \{\varphi_4(x,\mu) - \varphi_2(x,\mu)s(0)\}\,y(0,\mu^4) + \\ \{\varphi_3(x,\mu) - \varphi_1(x,\mu)s(0)\}\,y'(0,\mu^4) + \tag{12} \\ \varphi_2(x,\mu)y''(0,\mu^4) + \varphi_1(x,\mu)y'''(0,\mu^4),$$

and

$$K(x,t,\mu) = -\varphi_1(x-t,\mu)q(t) - \varphi_2(x-t,\mu)s'(t) + \varphi_3(x-t,\mu)s(t). \tag{13}$$

We shall need the following lemmata which are instrumental in the sequel.

Lemma 1. *The following estimates hold for all $z \in \mathbf{C}$,*

$$|\sinh z - \sin z| \le \frac{a\,|z|^3}{1+|z|^3}\exp|z|$$

$$|\cosh z - \cos z| \le \frac{|z|^2}{1+|z|^2}\exp|z| \tag{14}$$

$$|\sinh z + \sin z| \le \frac{|z|}{1+|z|}\exp|z|$$

$$|\cosh z + \cos z| \le 2\exp|z|,$$

where a is some constant (we may take $a = 0.53$).

Proof: To prove the first inequality, we proceed as follows. For small $x \in \mathbb{R}$, $|\sinh x - \sin x| \le |x|^3$, and for large real x, $|\sinh x - \sin x| \le \exp|x|$. Thus for all $x \in \mathbb{R}$, $|\sinh x - \sin x| \le a\,|x|^3 \exp|x| / (1 + |x|^3)$ for some constant $a > 0$. If x is replaced by $z \in \mathbf{C}$, then $|\sinh z - \sin z|$ grows like $\exp|z|$ which leads to the result. The other inequalities can be proved in a similar way. $\square$

Lemma 2. *The kernel $K(x,t,\mu)$ satisfies*

$$\left|\frac{\partial^\alpha K}{\partial x^\alpha}(x,t,\mu)\right| \le c_\alpha(1+|\mu|)^{\alpha-1}\exp\left\{(x-t)\,|\mu|\right\}, \qquad (15)$$

for $\alpha = 0,1,2,3$ with absolute constants c_α that depend only on q and s.

Proof: We shall give the proof only for $\alpha = 0$; the other cases are dealt with in a similar way. From (13) and using (14), we have

$$
\begin{aligned}
|K(x&,t,\mu)| \\
&\le |\varphi_1(x-t,\mu)|\,|q(t)| + |\varphi_2(x-t,\mu)|\,|s'(t)| + |\varphi_3(x-t,\mu)|\,|s(t)| \\
&\le \frac{1}{2}(x-t)^3\frac{ae^{(x-t)|\mu|}}{1+(x-t)^3\,|\mu|^3}||q||_\infty + \frac{1}{2}(x-t)^2\frac{e^{(x-t)|\mu|}}{1+(x-t)^2\,|\mu|^2}||s'||_\infty \\
&\quad + \frac{1}{2}(x-t)\frac{e^{(x-t)|\mu|}}{1+(x-t)\,|\mu|}||s||_\infty \\
&\le \frac{1}{2}\left\{\frac{a}{1+|\mu|^3}||q||_\infty + \frac{1}{1+|\mu|^2}||s'||_\infty + \frac{1}{1+|\mu|}||s||_\infty\right\}e^{(x-t)|\mu|} \\
&\le \frac{c_0}{1+|\mu|}e^{(x-t)|\mu|}
\end{aligned}
$$

for some constant $c_0 > 0$. We have used the fact that $\frac{(x-t)^m}{1+(x-t)^m|\mu|^m}$ is increasing in $(x-t)$. $\square$

Lemma 3. *For some constant $b > 0$,*

$$|\Psi_0(x,\mu)| \le b\exp\left\{x\,|\mu|\right\}. \qquad (16)$$

Proof: From (12), we obtain

$$
\begin{aligned}
|\Psi_0(x,\mu)| &\le \{|\varphi_4(x,\mu)| + |\varphi_2(x,\mu)|\,|s(0)|\}\,|y(0,\mu^4)| \\
&\quad + \{|\varphi_3(x,\mu)| + |\varphi_1(x,\mu)|\,|s(0)|\}\,|y'(0,\mu^4)| \\
&\quad + |\varphi_2(x,\mu)|\,|y''(0,\mu^4)| + |\varphi_1(x,\mu)|\,|y'''(0,\mu^4)| \\
&\le \frac{1}{2}\left\{2\exp\left\{x\,|\mu|\right\} + x^2\frac{\exp\left\{x\,|\mu|\right\}}{1+x^2\,|\mu|^2}\,|s(0)|\right\}\,|y(0,\mu^4)| \\
&\quad + \frac{1}{2}\left\{x\frac{\exp\left\{x\,|\mu|\right\}}{1+x\,|\mu|} + x^3\frac{a\exp\left\{x\,|\mu|\right\}}{1+x^3\,|\mu|^3}\,|s(0)|\right\}\,|y'(0,\mu^4)| \\
&\quad + \frac{1}{2}x^2\frac{\exp\left\{x\,|\mu|\right\}}{1+x^2\,|\mu|^2}\,|y''(0,\mu^4)| + \frac{1}{2}x^3\frac{a\exp\left\{x\,|\mu|\right\}}{1+x^3\,|\mu|^3}\,|y'''(0,\mu^4)|
\end{aligned}
$$

$$\leq \frac{1}{2} \left\{ 2 + \frac{1}{1 + |\mu|^2} |s(0)| \right\} |y(0, \mu^4)| \exp\{x |\mu|\}$$

$$+ \frac{1}{2} \left\{ \frac{1}{1 + |\mu|} + \frac{a}{1 + |\mu|^3} |s(0)| \right\} |y'(0, \mu^4)| \exp\{x |\mu|\}$$

$$+ \frac{1}{2} \frac{\exp\{x |\mu|\}}{1 + |\mu|^2} |y''(0, \mu^4)| + \frac{1}{2} \frac{a \exp\{x |\mu|\}}{1 + |\mu|^3} |y'''(0, \mu^4)|$$

$$\leq b \exp\{x |\mu|\}$$

for some constant b. $\square$

Let

$$\Psi_n(x, \mu) = \sum_{j=0}^{n} (\mathcal{K}^j \Psi_0)(x, \mu),$$

where $\mathcal{K}$ is the operator defined by $(\mathcal{K}\Psi)(x, \mu) = \int_0^x K(x, t, \mu)\Psi(t, \mu)dt$.

Lemma 4. *For $y_{-2+i}(x, \mu) = y(x, \mu^4) - \Psi_i(x, \mu)$, $0 \leq i \leq 3$, we have*

$$|y_{-2+i}(x, \mu)| \leq \frac{d_i}{(1 + |\mu|)^{i+1}} \exp\{x |\mu|\} \tag{17}$$

with absolute constants d_i that depend only on q, s and the initial values $y^j(0)$, $0 \leq j \leq 3$.

Proof: (17) is true for $i = 0$. Indeed, from (11) we get

$$|y_{-2}(x, \mu)| \leq \int_0^x |K(x, t, \mu)| |\Psi_0(t, \mu)| dt + \int_0^x |K(x, t, \mu)| |y_{-2}(t, \mu)| dt,$$

and using (15) and (16), we obtain

$$|y_{-2}(x, \mu)| \leq \frac{c_0 b}{1 + |\mu|} \int_0^x \exp\{x |\mu|\} dt$$

$$+ \int_0^x c_0 \exp\{(x - t) |\mu|\} |y_{-2}(t, \mu)| dt.$$

Multiplying by $\exp\{-x |\mu|\}$ before using Gronwall's lemma, and multiplying back by $\exp\{x |\mu|\}$ leads to

$$|y_{-2}(x, \mu)| \leq \frac{c_0 b e^{c_0}}{1 + |\mu|} \exp\{x |\mu|\} = \frac{d_0}{1 + |\mu|} \exp\{x |\mu|\},$$

for some constant $d_0 > 0$. That (17) is true for $i = 1, 2, 3$ follows immediately from the fact that $y_{-2+i} = \mathcal{K}^i y_{-2}$ and the case $i = 0$. $\square$

Lemma 5. *Let* $y_2(x, \mu) = y_1'(x, \mu)$, $y_3(x, \mu) = y_2'(x, \mu) - s(x)y_0(x, \mu)$, $y_4(x, \mu) = y_3'(x, \mu) + s'(x)y_0(x, \mu)$, *then there exist absolute constants* e_i, $0 \leq i \leq 2$, *that only depend on* s, q *and the initial values* $y^{(j)}(0)$, $0 \leq j \leq 3$, *such that*

$$|y_{2+i}(x, \mu)| \leq \frac{e_i}{\{1 + |\mu|\}^{3-i}} \exp\{x\,|\mu|\}, \qquad i = 0, 1, 2. \tag{18}$$

Proof: Since $y_1 = \mathcal{K}y_0$, we have

$$y_2(x, \mu) = \int_0^x \frac{\partial K}{\partial x}(x, t, \mu)y_0(t, \mu)dt, \tag{19}$$

so that

$$|y_2(x, \mu)| \leq \int_0^x \left|\frac{\partial K}{\partial x}(x, t, \mu)\right| |y_0(t, \mu)|\, dt$$

$$\leq \frac{c_1 d_0}{\{1 + |\mu|\}^3} \exp\{x\,|\mu|\} = \frac{e_0}{\{1 + |\mu|\}^3} \exp\{x\,|\mu|\}.$$

So, (18) is true for $i = 0$. Differentiating (19) two more times, using the definitions of y_3 and y_4 and the estimates (15) and (17), we obtain that (18) is true for $i = 1, 2$. $\square$

Following Stenger [27], let $D_d = \{w \in C : |\text{Im}\, w| < d\}$, for some $d > 0$. We introduce the following notation.

Definition 1. *Let* $1 \leq p \leq \infty$, *and let* $H^p(D_d)$ *denote the family of all functions analytic in* D_d *such that if* $D_d(\epsilon)$ *is defined for* $0 < \epsilon < 1$ *by*

$$D_d(\epsilon) = \{z \in C : |Re\, z| < 1/\epsilon,\ |Im\, z| < d(1 - \epsilon)\},$$

then $N_p(f, D_d) < \infty$, *with*

$$N_p(f, D_d) = \begin{cases} \lim\limits_{\epsilon \to 0} \left(\int_{\partial D_d(\epsilon)} |f(z)|^p |dz|\right)^{1/p}, & if\ 1 \leq p < \infty, \\ \lim\limits_{\epsilon \to 0} \sup\limits_{z \in D_d(\epsilon)} |f(z)|, & if\ p = \infty. \end{cases}$$

Let $S(k, h)(x) = \sin[\pi(x - kh)/h]/[\pi(x - kh)/h]$, let $f \in H^p(D_d)$, $h > 0$, let N be a positive integer, and let us set

$$C(f, h) = \sum_{k=-\infty}^{\infty} f(kh)S(k, h), \qquad C_N(f, h) = \sum_{k=-N}^{N} f(kh)S(k, h),$$

$$E(f, h) = f - C(f, h), \qquad E_N(f, h) = f - C_N(f, h).$$

Consider the Paley-Wiener space

$$\mathrm{PW}_{\pi/h} = \{f \text{ analytic in } \mathbb{C} : |f(z)| \le A e^{\pi |\mathrm{Im}\ z|/h}, \int_{-\infty}^{\infty} |f(t)|^2 dt < \infty\}.$$

It is well known that if $f \in \mathrm{PW}_{\pi/h}$, then $C(f, h) = f$. While this is no longer true for $f \in H^p(D_d)$, the function $C(f, h)$ provides us, nevertheless, with a very accurate approximation of f (see [27]). Now we are ready to state our main theorem.

Theorem 1. *The functions* $y_i(x, \mu)/\cosh \mu x$, $i = 1, \dots, 4$, *are analytic in the strip* D_d *(*$0 < d \le \pi/2$*) and are in* $H^2(D_d)$ *for each* x *as a function of* μ. *Furthermore,*

$$|y_i(x, \mu)/\cosh \mu x| \le \frac{f_i}{\{1 + |\mu|\}^{5-i}} \exp\{x\,|Im\mu|\}\ ,\quad i = 1, \dots, 4,$$

for some constant f_i, $i = 1, \dots, 4$.

Proof: It is enough to note that $|\cosh z| > \frac{\cos^2 d}{4} \exp\{|\mathrm{Re}\ z|\}$ for z in the strip D_d ($0 < d \le \pi/2$) and make use of the estimates above. Now, it is not hard to see that $y_i(1, \mu)/\cosh \mu$ is in $H^2(D_d)$ as a function of μ for $i = 1, \dots, 4$. Therefore we can reconstruct them from their values at $\mu_k = k\pi$, $k \in \mathbb{Z}$. That is,

$$y_i(1, \mu) \simeq \cosh \mu \sum_{k=-\infty}^{k=\infty} \frac{y_i(1, k\pi)}{\cosh k\pi} \frac{\sin(\mu - k\pi)}{\mu - k\pi}, \tag{20}$$

for $i = 1, \dots, 4$. In practice, we consider their approximations by finite sums as

$$y_{i,N}(1, \mu) \simeq \cosh \mu \sum_{k=-N}^{k=N} \frac{y_i(1, k\pi)}{\cosh k\pi} \frac{\sin(\mu - k\pi)}{\mu - k\pi}, \quad i = 1, \dots, 4, \tag{21}$$

for some given integer N large enough. Thus, $y^{(i-1)}(1, \mu^4)$ can be approximated by $y_N^{(i-1)}(1, \mu^4)$, for $i = 1, \dots, 4$, and

$$
\begin{aligned}
y_N(1, \mu^4) &= y_{1,N}(1, \mu) + \Psi_3(1, \mu) \\
y_N'(1, \mu^4) &= y_{2,N}(1, \mu) + \Psi_3'(1, \mu) \\
y_N^{(2)}(1, \mu^4) &= y_{3,N}(1, \mu) + \Psi_3''(1, \mu) + s(1)(y_N(1, \mu^4) - \Psi_2(1, \mu)) \\
y_N^{(3)}(1, \mu^4) &= y_{4,N}(1, \mu) + \Psi_3'''(1, \mu) + s'(1)(y_N'(1, \mu^4) - \Psi_2'(1, \mu)).
\end{aligned}
\tag{22}
$$

Now, $z_1(1, \mu) = y(1, \mu^4)$, $z_2(1, \mu) = y'(1, \mu^4)$, $z_3(1, \mu) = s(1)y'(1, \mu^4) - y'''(1, \mu^4)$ and $z_4(1, \mu) = y''(1, \mu^4)$ can be approximated by $z_{1,N}(1, \mu) = y_N(1, \mu^4)$, $z_{2,N}(1, \mu) = y'_N(1, \mu^4)$, $z_{3,N}(1, \mu) = s(1)y'_N(1, \mu^4) - y'''_N(1, \mu^4)$ and $z_{4,N}(1, \mu) = y''_N(1, \mu^4)$. Thus, we have the approximation

$$B(\mu) \simeq B_N(\mu) = \det(C_N(\mu)),$$

where $C_N(\mu) = (N\widehat{z}_{1,N}|N\widehat{z}_{2,N})$. $\square$

§3. Conclusion

In this paper we have considered the computation of eigenvalues of fourth order Sturm-Liouville problems. We have shown that although transforms of the solution of the associated initial value problem and of its derivatives are not in a Paley-Wiener space (not even entire), they are in $H^2(D_d)$, a fact which allowed them to be recovered from their samples very accurately using the sinc series approximation. Having reconstructed the boundary function, the eigenvalues are seen as the fourth powers of its zeros. An error analysis of the method will be presented in a forthcoming paper together with extensive numerical computations. Comparison with SLEUTH and the author's method based on Fliess series will also be presented.

Acknowledgments. The author is pleased to acknowledge the support of K.F.U.P.M. and thank an anonymous referee for his thorough review and pertinent suggestions.

References

1. Algazin, S. D., Calculating the eigenvalues of ordinary differential equations, Comp. Math. Phys. **35**, No.4 (1995), 477–482.

2. Bailey, P. B., W. N. Everitt and A. Zettl, Computing eigenvalues of singular Sturm-Liouville problems, in *Results in Mathematics* Vol. 20, Birkhäuser Verlag, Basel, 1991.

3. Bailey, P. B., M. K. Gordon and L. F. Shampine, Automatic solution of the Sturm-Liouville problem, ACM Trans. Math. Software **4** (1978), 193–208.

4. Boumenir, A., and B. Chanane, Eigenvalues of S-L systems using sampling theory, Applicable Analysis **62** (1996), 323–334.

5. Boumenir, A., and B. Chanane, Computing eigenvalues of Sturm-Liouville systems of Bessel type, Proceedings of the Edinburgh Mathematical Society **42** (1999), 257–265.

6. Boumenir, A., and B. Chanane, Computing the negative eigenvalues of singular Sturm-Liouville problems, IMA. J. Numer. Anal. **21** (2001) No.2, 489–501.

7. Chanane, B., Computing eigenvalues of regular Sturm-Liouville problems, Applied Math. Letters **12** (1999), 119–125.

8. Chanane, B., High order approximations of the eigenvalues of regular Sturm-Liouville problems, J. Math. Anal. and Appl. **226** (1998), 121–129.

9. Chanane, B., Eigenvalues of Sturm-Liouville problems using Fliess series, Applicable Analysis **69** (1998), 233–238.

10. Chanane, B., Eigenvalues of fourth order Sturm-Liouville problems using Fliess series, J. Comput. and Appl. Math. **96** (1998), 91–97.

11. Chanane, B., High order approximations of the eigenvalues of Sturm-Liouville problems with coupled self-adjoint boundary conditions, Applicable Analysis, to appear.

12. Chanane, B., On a class of random Sturm-Liouville problems, Intern. J. Appl. Math. **8** (2002) No. 2, 171–182.

13. Chen, Xi, The shooting method for solving eigenvalue problems, J. Math. Anal. and Appl. **203** (1996), 435–450.

14. Coddington, E. A., and N. Levinson, *Theory of Ordinary Differential Equations*, Mc Graw-Hill Co., New York, 1955.

15. Fliess, M., Fonctionelles causalles nonlineaires et indeterminées non commutatives, Bull. Soc. Math. France **109** (1981), 3–40.

16. Everitt, W. N., On the transformation theory of ordinary second order linear symmetric differential expressions, Czechoslovak Math. J. **32**, 107 (1982), 275–305.

17. Fulton, C. T., and S. A. Pruess, Mathematical software for Sturm-Liouville problems, ACM Trans. Math. Software **19** (1993), 360–376.

18. Fulton, C. T., and S. A. Pruess, Eigenvalue and eigenfunction asymptotics for regular Sturm-Liouville problems, J. Math. Anal. Appl. **188** (1994), 297–340.

19. Greenberg, L., and M. Marletta, Oscillation theory and numerical solution of fourth order Sturm-Liouville problems, I.M.A. J. Numer. Anal. (1995), 319–356.

20. Greenberg, L., and M. Marletta, The code SLEUTH for solving fourth order Sturm-Liouville problems, 1997.

21. Hargrave, B. A., Numerical approximation of eigenvalues of Sturm-Liouville systems, J. Computational Phys. **20** (1976), 381–396.

22. Harris, B. J., Asymptotics of eigenvalues for regular Sturm-Liouville problems, J. Math. Anal. and Appl. **183** (1994), 25–36.

23. Kong, Q., and A. Zettl, Eigenvalues of regular Sturm-Liouville problems, J. Diff. Eqs. **131** (1996), 1–19.

24. Marletta, M., and J. D. Pryce, Automatic solution of Sturm-Liouville problems using the Pruess method, J. Computational and Applied Mathematics **39** (1992), 57–78.

25. Naimark, M. A., *Linear Differential Operators II*, G.G. Harrap & Co. Ltd, 1968.

26. Pryce, J. D., *Numerical Solution of Sturm-Liouville problems*, Oxford Science Publications, Clarendon Press, 1993.

27. Stenger, F., *Numerical Methods Based on Sinc and Analytic Functions*, Springer-Verlag, 1993.

28. Zayed, A. I., *Advances in Shannon's Sampling Theory*, CRC Press, 1993.

B. Chanane
Department of Mathematical Sciences
K. F. U. P. M., Dhahran 31261
Saudi Arabia
chanane@kfupm.edu.sa

Introduction to Synthetic Aperture Radar (SAR) and SAR Interferometry

Margaret Cheney

Abstract. This paper presents a tutorial on the foundations of Synthetic Aperture Radar. The paper shows how a simple antenna model can be used together with a linearized scattering approximation to predict the received signal. The paper then outlines the conventional image formation process and shows how two images from different flight passes can be used to obtain topographic information.

§1. Introduction and Relation to Approximation Theory

Approximation theory can be thought of as the reconstruction of a function from incomplete information. This description applies to many inverse problems and imaging problems, in which one generally wants to determine properties of an object from remote measurements. A good example is Synthetic Aperture Radar (SAR) imaging, a very successful imaging technique that has been developed in the engineering community and which has received little attention from mathematicians.

In conventional strip-mode SAR imaging, a plane or satellite carrying an antenna flies along a straight track, which we will assume is in the direction of the x_2 axis. The antenna emits pulses of electromagnetic radiation in a directed beam perpendicular to the flight track (i.e., in the x_1 direction). These waves scatter off the terrain, and the scattered waves are detected with the same antenna. The received signals are then used to produce an image of the terrain. (See Figure 1.)

The data depend on two variables, namely time and position along the x_2 axis, so we expect to be able to reconstruct a function of two variables.

Approximation Theory X: Wavelets, Splines, and Applications   167
Charles K. Chui, Larry L. Schumaker, and Joachim Stöckler (eds.), pp. 167–177.
Copyright ⊖ 2002 by Vanderbilt University Press, Nashville, TN.
ISBN 0-8265-1416-2.

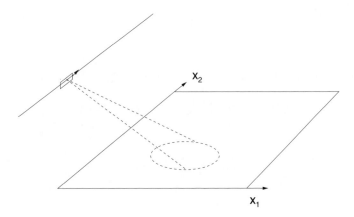

Fig. 1. Geometry of a SAR system.

§2. The Mathematical Model

A model for the wave propagation

The correct model for radar is of course Maxwell's equations, but the simpler scalar wave equation is commonly used:

$$\left(\nabla^2 - \frac{1}{c^2(x)}\partial_t^2\right) U(t, x) = 0. \tag{1}$$

This is the equation satisfied by each component of the electric and magnetic fields in free space, and is thus a good model for the wave propagation in dry air. When the electromagnetic waves interact with the ground, their polarization is certainly affected, but if the SAR system does not measure this polarization, then (1) is an adequate model.

We assume that the earth is roughly situated at the plane $x_3 = 0$, and that for $x_3 > 0$, the wave speed is $c(x) = c_0$, the speed of light in vacuum (a good approximation for dry air).

A model for the field from an antenna

In free space, the field $G_0(t, x)$ at x, t due to a delta function point source at the origin at time zero is given [15] by

$$G_0(t, x) = \frac{\delta(t - |x|/c_0)}{4\pi|x|}, \tag{2}$$

and satisfies the equation

$$\left(\nabla^2 - c_0^{-2}\partial_t^2\right) G_0(t, x) = -\delta(t)\delta(x). \tag{3}$$

The antenna, however, is not a point source $\delta(x)$; typical SAR satellite antennas are rectangular arrays measuring roughly 10m by 1m. Moreover, the delta function $\delta(t)$ on the right side of (3) is not a good model for the signal sent to the antenna; typically a *chirp* of the form $P(t) = \exp(i\alpha t^2)$ (over a limited time interval) is used. The field U^{in} emanating from the antenna then satisfies an equation of the form (3) with $\delta(t)$ replaced by the waveform $P(t)$, and $\delta(x)$ replaced by the current distribution $J(x)$ over the antenna.

The field emanating from the antenna then satisfies

$$\left(\nabla^2 - c_0^{-2}\partial_t^2\right) U^{in}(t, x) = -P(t)J_s(x), \tag{4}$$

so that

$$U^{in}(t, x) = G_0 * (PJ) = \int \frac{P(t - |x - y|/c_0)}{4\pi|x - y|} J(y)dy, \tag{5}$$

where the star denotes convolution in t and x.

We write P in terms of its Fourier transform $\tilde{P}$:

$$P(t) = \int e^{-i\omega t}\tilde{P}(\omega)d\omega. \tag{6}$$

In all SAR systems, $\tilde{P}(\omega)$ is negligible outside an interval which we call the frequency band. An important special case arises when the effective support is a narrow band (i.e., interval) centered at the carrier frequency ω_0; in this case it is useful to write

$$P(t) = A(t)e^{i\omega_0 t}, \tag{7}$$

where A is a slowly varying amplitude that is allowed to be complex. All the present satellite systems are narrowband systems for which (7) is most useful, but some of the airborne systems are broadband systems for which (7) is inappropriate and (6) is more useful. Using (6) in (5) gives us

$$U^{in}(t, x) = \int \int \frac{e^{-i\omega(t - |x - y|/c_0)}}{4\pi|x - y|}\tilde{P}(\omega)d\omega J(y)dy. \tag{8}$$

Next we assume that the antenna is small compared with the distance from the antenna to the ground. We denote the center of the antenna by y^0; thus a point on the antenna can be written $y = y^0 + q$, where q is a vector from the center of the antenna to a point on the antenna. In this notation, the assumption that the scattering point x is far from the antenna can be expressed $|q| << |x - y^0|$. For such x, we can write

$$|x - y| = |x - y^0| - \widehat{(x - y^0)} \cdot q + O(q^2/|x - y^0|), \tag{9}$$

where the hat denotes a unit vector. We use the expansion (9) in (8) to obtain

$$U^{in}(t,x) \approx \int\int \frac{e^{-i\omega(t-|x-y^0|)}}{4\pi|x-y^0|} e^{-i\omega(\widehat{x-y^0})\cdot q} \tilde{P}(\omega) d\omega J(y^0+q) dq \tag{10}$$
$$\approx \int \frac{e^{-i\omega(t-|x-y^0|)}}{4\pi|x-y^0|} \tilde{P}(\omega)\tilde{J}(\omega,x,y^0) d\omega,$$

where we have written

$$\tilde{J}(\omega,x,y^0) = \int e^{-i\omega(\widehat{x-y^0})\cdot q} J(y^0+q) dq.$$

This Fourier transform of the current density gives the antenna beam pattern in the far-field at each fixed frequency. A typical example is the case in which J is constant over a rectangular antenna. The antenna beam pattern $\tilde{J}$ is then a product of sinc functions, so that the antenna directs a beam perpendicular to the face of the antenna. At a frequency ω, the width of this beam is roughly $2\lambda/L$, where $\lambda = c_0/(2\pi\omega)$ is the wavelength and L is the length of the antenna.

Generally, the antenna beam pattern is nearly independent of frequency over the frequency band of $\tilde{P}(\omega)$; indeed, a great deal of work goes into designing antennas for which this is the case. Antennas whose beam patterns are nearly constant over a wide frequency band are called **broadband antennas**. For such antennas, (10) becomes

$$U^{in}(t,x) \approx \frac{P(t-|x-y^0|)}{4\pi|x-y^0|} \tilde{J}(\omega_0,x,y^0).$$

A linearized scattering model

Since we are ignoring polarization effects, the current density J is scalar and the model we use for wave propagation, including the source, is

$$\left(\nabla^2 - c_0^{-2}\partial_t^2\right) U(t,x) = -P(t)J_s(x). \tag{11}$$

We write $U = U^{in} + U^{sc}$ in (11) and use (4) to obtain

$$\left(\nabla^2 - c_0^{-2}\partial_t^2\right) U^{sc}(t,x) = -V(x)\partial_t^2 U(t,x), \tag{12}$$

where

$$V(x) = \frac{1}{c_0^2} - \frac{1}{c^2(x)}.$$

Most SAR systems operate at frequencies for which the waves do not penetrate appreciably into the earth. Thus we can approximate V by

$V(x) = V(x')\delta(x_3)$, where $x' = (x_1, x_2)$. We can write (12) as an integral equation

$$U^{sc}(t, x) = \int \int G_0(t - \tau, x - z)V(z)\partial_\tau^2 U(\tau, z)d\tau dz. \qquad (13)$$

The single-scattering or **Born approximation** is to replace the full field U on the right side of (13) by the incident field U^{in}:

$$U^{sc}(t, x) \approx \int \int G_0(t - \tau, x - z)V(z)\partial_\tau^2 U^{in}(\tau, z)d\tau dz. \qquad (14)$$

The value of this approximation is that it *linearizes* the inverse problem: the product of unknowns VU is replaced by the product of the unknown V with the known incident field.

With (2) and the expression (10) for the incident field from an antenna at position y^0, (14) becomes

$$
\begin{aligned}
U_{y^0}^{sc}(t, x) &\approx \int \int \frac{\delta(t - \tau - |x - z|/c_0)}{4\pi|x - z|} V(z) \\
&\quad \cdot \int \frac{e^{-i\omega(\tau - |z - y^0|/c_0)}}{4\pi|z - y^0|} \tilde{P}(\omega)\tilde{J}(\omega, z, y^0)\omega^2 d\omega d\tau dz \\
&\approx \int \int \frac{e^{-i\omega((t - (|x - z| + |z - y^0|)/c_0)}}{(4\pi)^2|x - z||z - y^0|} \tilde{P}(\omega)\tilde{J}(\omega, z, y^0)\omega^2 d\omega \; V(z)dz.
\end{aligned}
$$

At the center of the antenna,

$$U_{y^0}^{sc}(t, y^0) \approx \int \int \frac{e^{-i\omega(t - 2|z - y^0|/c_0)}}{(4\pi)^2|z - y^0|^2} \tilde{P}(\omega)\tilde{J}(\omega, z, y^0)\omega^2 d\omega \; V(z)dz.$$

In practice, measurements are not simply made at the center of the antenna; instead what is measured is the integral of the field over the whole antenna. This gives rise to another antenna pattern which, in most cases, is the same as the transmission antenna beam pattern. In this case, the expression for the signal measured at antenna location y is

$$S(t, y) = U_y^{sc}(t, y) \approx \int \int \frac{e^{-i\omega(t - 2|z - y|/c_0)}}{(4\pi)^2|z - y|^2} \tilde{P}(\omega)\tilde{J}^2(\omega, z, y)\omega^2 d\omega \; V(z)dz. \qquad (15)$$

For a narrowband system, we have

$$S_N(t, y) \approx \int \frac{P(t - 2|z - y|/c_0)}{(4\pi)^2|z - y|^2} \tilde{J}^2(\omega_0, z, y)\omega_0^2 \; V(z)dz. \qquad (16)$$

In (15) we are implicitly making the *start-stop* approximation, i.e., we are assuming that the antenna is stationary while it is transmitting and receiving. This is a good approximation because the antenna speed is so much slower than the speed of propagation of the electromagnetic signals.

From knowledge of the signal S for a large interval in t and for y along a line, we want to determine V.

§3. Formation of the Image

The idea underlying SAR reconstruction algorithms is to apply a matched filter to the measured signal (15). Applying a matched filter means integrating the signal against a shifted copy of the complex conjugate of the transmitted signal:

$$I(x) = \int \int \overline{P(t - 2|x - y|/c_0)} S(t, y) dt dy. \tag{17}$$

For a narrowband system, the term matched filter is a natural one: taking $V(z) = \delta(z - x)$ in (16) shows that the measured signal from a point scatterer at position x is proportional to $P(t - 2|x - y|/c_0)$. Thus in (17) we are "matching" the received signal to the signal that would have been received from a point scatterer at x. The "matching" process is done by taking the inner product; we expect that the integral (17) will be a maximum when the signal $S(t, y)$ is proportional to $P(t - 2|x - y|/c_0)$. Under certain hypotheses on the noise, it is known that a matched filter gives the optimal signal-to-noise ratio in the class of linear filters [3,12].

The matched filter operation (17) is also called backprojection because it corresponds to summing all the signals to which a scatterer at x could have contributed. This amounts to summing over all spheres passing through the point x. Summing over the antenna locations y corresponds to forming a synthetic aperture, and, roughly speaking, corresponds to synthesizing the data one would have obtained from an extremely long antenna. Using (15) in (17) results in

$$I(x) \approx \int \int \overline{P(t - 2|x - y|/c_0)}$$
$$\cdot \int \int \frac{e^{-i\omega(t-2|z-y|/c_0)}}{(4\pi)^2 |z-y|^2} \tilde{P}(\omega) \tilde{J}^2(\omega, z, y) \omega^2 d\omega \, V(z) dz dt dy$$
$$\approx \int \int \int \frac{e^{-i\omega(2(|x-y|-|z-y|)/c_0)}}{(4\pi|z-y|)^2} |\tilde{P}(\omega)|^2 \tilde{J}^2(\omega, z, y) \omega^2 d\omega dy \, V(z) dz.$$
$$\tag{18}$$

The last equation of (18) can be written as

$$I(x) = \int W(x, z) V(z) dz,$$

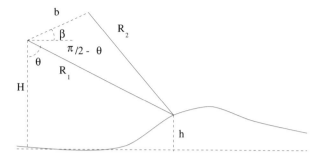

Fig. 2. Geometry for stereometry.

where

$$W(x, z) = \int \int \frac{e^{-i\omega(2(|x-y|-|z-y|)/c_0)}}{(4\pi|z-y|)^2} |\tilde{P}(\omega)|^2 \tilde{J}^2(\omega(\widehat{z-y}))\omega^2 d\omega dy$$

is the **point spread function** of the imaging system. This term arises because $W(x, z)$ is the image that would result from a point scatterer (i.e., delta function) located at the point z.

We see from a large-ω stationary phase analysis of W that the main contribution to W does indeed arise from the point $x = z$. An analysis of the resolution, which can be found in many references (for example [2,4,17]), shows that resolution in the range (i.e., distance from the antenna track) is determined by the bandwidth of the signal. The resolution in the along-track direction is determined by the effective aperture, which is the distance the antenna travels while keeping a given point in the beam. For narrowband systems, it turns out that the along-track resolution is independent of range and wavelength, and is better for small antennas.

§4. Determination of Ground Topography

In this section, instead of assuming the earth is flat, we assume that it has a varying elevation h, which we want to determine. In other words, we take the ground reflectivity function V to be of the form $V(z) = V(z')\delta(z_3 - h(z'))$, where $z' = (z_1, z_2)$.

If two images are made from two separate, known flight tracks, then in principle the elevation of an object that appears in both images can be determined by stereometry [8].

Stereometry

Suppose the first flight track is flown at height H, and the second is flown a distance b and angle of elevation β from the first. If the same object can be identified in both images, then its distances (ranges) R_1 and R_2 from the first and second flight track, respectively, can be determined. (See Figure 2.)

From R_1, R_2, b, and β, we can determine the angle of elevation θ from the law of cosines:

$$
\begin{aligned}
R_2^2 &= R_1^2 + b^2 - 2bR_1 \cos(\beta + \frac{\pi}{2} - \theta) \\
&= R_1^2 + b^2 + 2bR_1 \sin(\theta - \beta),
\end{aligned}
\tag{19}
$$

and from knowledge of θ, the object's elevation h can be determined from $h = H - R_1 \cos \theta$.

The difficulty with this method is twofold. First, it requires that common objects be identified in both images; this often requires human intervention. Second, the process is very sensitive to errors in the determination of the range difference $\Delta R = R_2 - R_1$. To see this, we use the chain rule

$$
\frac{dh}{d(\Delta R)} = \frac{dh}{d\theta} \frac{d\theta}{d(\Delta R)} = R_1 \sin \theta \frac{d\theta}{d(\Delta R)}
\tag{20}
$$

and calculate the derivative $d\theta/d(\Delta R)$ implicitly from (19) with $R_2 = R_1 + \Delta R$:

$$
\frac{d\theta}{d(\Delta R)} = -\frac{R_1 + \Delta R}{bR_1 \cos(\theta - \beta)}.
$$

Using this in (20) gives

$$
\frac{dh}{d(\Delta R)} = \frac{-(R_1 + \Delta R) \sin \theta}{b \cos(\theta - \beta)} \approx \frac{-R_2}{b}.
\tag{21}
$$

For satellite-borne SAR systems, the ratio R_2/b is very large; for the ERS-1 SAR, for example, R_1 and R_2 are both on the order of 800 km, and the baseline b is about 100 m [8]. Thus in estimating the ground height, range errors are magnified by a factor of 8000; the range resolution ΔR, however, is about 10 m, which means that the uncertainty in the estimated height is about 8 km, which is clearly unacceptable. It is for this reason that many SAR systems use instead an interferometric method to estimate ΔR and thus find the ground topography.

Interferometry

The interferometric technique applies to narrow-band SAR systems, for which (16) applies. Using (7) in (16) results in

$$
S_N(t, y) \approx \int \frac{e^{-i\omega_0(t - 2|z - y|/c_0)}}{(4\pi)^2 |z - y|^2} A(t - 2|z - y|/c_0) \tilde{J}^2(\omega_0(\widehat{z - y})) \omega_0^2 V(z) dz.
\tag{22}
$$

The reflectivity function V we assume to be of the form $V(z) = V(z_T)$ $\delta(z_3 - h(z_T))$ where we have written $z_T = (z', 0)$ and $z = z_T + h\hat{e}_3$. (See

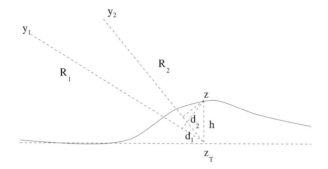

Fig. 3. Geometry for interferometry.

Figure 3.) We assume also that the height h is much less than the distance from the antenna to the ground, so that

$$|z - y| = |z_T + h\hat{e}_3 - y| = |z_T - y| - \widehat{z_T - y} \cdot h\hat{e}_3 + O(h^2/|z_T - y|). \quad (23)$$

We use this expansion in the expression (22) for the received signal, obtaining

$$S_N(t, y) \approx \int \frac{e^{-i\omega_0(t - 2|z_T - y|/c_0)}}{(4\pi)^2 |z_T - y|^2} A(t - 2|z_T - y|/c_0)\omega_0^2 \cdot$$
$$\tilde{J}^2(\omega_0(\widehat{z_T - y}))e^{2i\omega_0 d(z_T)/c_0} V(z_T) dz_T,$$

where we have written $d(z_T) = \widehat{z_T - y} \cdot h\hat{e}_3$ and where we have used the fact that A is slowly varying. We see that S_N corresponds to the signal that would have been obtained from a flat earth with complex reflectivity function $e^{2i\omega_0 d(z_T)/c_0} V(z_T)$. The image formation process of Section 3 will thus result in a complex-valued image, in which the phase at each point contains information about the height of the ground at that point.

This information is extracted as follows. Two separate, known flight tracks are each used to produce a complex-valued image, in which the pixel at location z_T represents the quantity $e^{2i\omega_0 d(z_T)/c_0} V(z_T)$. In the jth image, this pixel occurs at range $r_j = |y_j - z_T|$; the true range is $R_j = r_j - d_j \approx |y_j - z|$. The two images are *co-registered* to make $r_1 = r_2$. The complex conjugate of one image is then multiplied by the other, resulting in a complex image whose phase at z_T is $2\omega_0(d_1(z_T) - d_2(z_T))/c_0$; from this, the range difference $d_1 - d_2$ is found (after solving a phase unwrapping problem). Then (19) can be used with $R_2 = R_1 + (d_1 - d_2)$.

The determination of $\Delta R = d_1 - d_2$ can be done with great accuracy; for example, to millimeter accuracy in the case of the ERS-1 satellite [8]. The estimate (21) then results in an uncertainty in h of the order of meters.

Acknowledgments. This work was partially supported by the National Science Foundation Engineering Research Centers Program under award number EEC-9986821 and by the Focus Groups in Mathematical Sciences Program.

References

1. Bamler, R. and P. Hartl, Synthetic aperture radar interferometry, Inverse Problems **14** (1998), R1–R54.

2. Cheney, M., A mathematical tutorial on Synthetic Aperture Radar, SIAM Review **43** (2001), 301–312.

3. Curlander, J. C. and R. N. McDonough, *Synthetic Aperture Radar*, Wiley, New York, 1991.

4. Cutrona, L. J., Synthetic Aperture Radar, in *Radar Handbook*, second edition, ed. M. Skolnik, McGraw-Hill, New York, 1990.

5. Edde, B., *Radar: Principles, Technology, Applications*, Prentice Hall, New York, 1993.

6. Elachi, C., *Spaceborne Radar Remote Sensing: Applications and Techniques*, IEEE Press, New York, 1987.

7. Guyenne, T.-D., ed., *Engineering Achievements of ERS-1*, European Space Agency SP-1197/III, ESA Publications Division, ESTEC, Noordwij, The Netherlands, 1977.

8. Franceschetti, G. and R. Lanari, *Synthetic Aperture Radar Processing*, CRC Press, New York, 1999.

9. Langenberg, K. J, M. Brandfass, K. Mayer, T. Kreutter, A. Brüll, P. Felinger, D. Huo, Principles of microwave imaging and inverse scattering, EARSeL Advances in Remote Sensing, **2** (1993), 163–186.

10. Langenberg, K. J., Applied inverse problems, in *Basic Methods of Tomography and Inverse Problems*, ed. P.C. Sabatier, Adam Hilger, Bristol, 1987.

11. Natterer, F., *The Mathematics of Computerized Tomography*, Wiley, New York, 1986.

12. North, D. O., An analysis of the factors which determine signal/noise discrimination in pulsed-carrier systems, Proc. IEEE **51** (1963), 1016–1027 (reprint of RCA Technical Report PTR-6C, June 25, 1943).

13. Oppenheim, A. V. and R. W. Shafer, *Digital Signal Processing*, Prentice-Hall, Englewood Cliffs, New Jersey, 1975.

14. Soumekh, M., *Synthetic Aperture Radar Signal Processing with MATLAB Algorithms*, Wiley, New York, 1999.

15. Treves, F., *Basic Linear Partial Differential Equations*, Academic Press, New York, 1975.

16. Therrien, C. W., *Discrete Random Signals and Statistical Signal Processing*, Prentice Hall, Englewood Cliffs, New Jersey, 1992.

17. Ulander, L. M. H., and H. Hellsten, A new formula for SAR spatial resolution, AEÜ Int. J. Electron. Commun. **50** (1996) no. 2, 117–121.

18. Ulander, L. M. H., and P.-O. Frölund, Ultra-wideband SAR interferometry, IEEE Trans. on Geoscience and Remote Sensing, **36** no. 5, September 1998, 1540–1550.

19. Ziomek, L. J., *Underwater Acoustics: A Linear Systems Theory Approach*, Academic Press, Orlando, 1985.

Margaret Cheney
Department of Mathematical Sciences
Rensselaer Polytechnic Institute
Troy, NY 12180 USA
cheney@rpi.edu
http://www.rpi.edu/~cheney

Supports of Multivariate Refinable Functions

Hoi Ling Cheung and Canqin Tang

Abstract. We study supports of multivariate refinable functions associated with general dilation matrices M. In the univariate case, when $M = 2$, the support is always a closed interval. When $M > 2$, supports of refinable functions are more complicated and were investigated in [1] by means of attractors of iterated function systems. In this paper, we extend our study to the multivariate case with M a general dilation matrix. Under the assumption of local linear independence, it is shown that the support of a refinable function is the attractor of an iterated function system. A characterization of self-affine tilings is presented in terms of the existence of L_2 solutions of refinement equations.

§1. Introduction

We consider refinement equations of the form

$$\phi(x) = \sum_{\alpha \in \mathbb{Z}^s} a(\alpha)\phi(Mx - \alpha), \qquad (1.1)$$

where ϕ is on the s-dimensional Euclidean space $\mathbb{R}^s$, a is a finitely supported sequence of $s \times s$ matrices called the **refinement mask**, and M is an $s \times s$ integer matrix such that $\lim_{n \to \infty} M^{-n} = 0$ called the **dilation matrix**. Solutions of (1.1) are called refinable functions.

The purpose of this paper is to study the support of a compactly supported refinable function ϕ by means of the attractor of an iterated function system. A connection between self-affine tilings and L_2 solutions of refinement equations without assuming the basic sum rule is pointed out, which leads to further problems.

In the univariate case, it is well known that for $M = 2$, the support of a compactly supported distributional solution of (1.1) exactly equals the interval $[0, N]$ if $a(0), a(N) \neq 0$. For M an integer greater than (or equal

Approximation Theory X: Wavelets, Splines, and Applications   179
Charles K. Chui, Larry L. Schumaker, and Joachim Stöckler (eds.), pp. 179–185.
Copyright Ⓔ 2002 by Vanderbilt University Press, Nashville, TN.
ISBN 0-8265-1416-2.

to) 2, we have the inclusion $\text{supp}\phi \subset [0, \frac{N}{M-1}]$. However, things are more difficult in the multivariate case. In Section 2, we show that the support is the attractor of an iterated function system under the assumption of local linear independence. In Section 3, we mention a connection between multivariate refinable functions and self-affine tilings.

§2. Supports Studied by Means of IFS Attractors

Iterated function systems have been widely studied for computer graphics applications and for fractal image compression. The attractors of an iterated function system represent the property of self-similarity since it is made of a union of reduced copies of itself. Thus, a natural way to study the supports of multivariate refinable functions is by means of the attractors of iterated function systems.

Let D be the support of the mask a:

$$D = \{\alpha \in \mathbb{Z}^s : a(\alpha) \neq 0\}.$$

From the refinement equation (1.1), we have the inclusion

$$\text{supp}\phi \subset \bigcup_{\alpha \in D} (M^{-1}\alpha + M^{-1}\text{supp}\phi).$$

An iterated function system corresponding to a set of contractive maps $\{S_\alpha\}_{\alpha \in D}$ on $\mathbb{R}^s$ is given by

$$S_\alpha x = M^{-1}(x + \alpha), \qquad x \in \mathbb{R}^s.$$

It is shown in [6] that there exists a unique compact subset Ω of $\mathbb{R}^s$ such that

$$\Omega = \bigcup_{\alpha \in D} S_\alpha(\Omega).$$

This set is called the **attractor of the iterated function system** $\{S_\alpha\}_{\alpha \in D}$. In our situation, this is

$$T(M, D) := \left\{ \sum_{\alpha=1}^{\infty} M^{-\alpha}\varepsilon_\alpha : \varepsilon_\alpha \in D \right\}.$$

Define the set-valued map S by

$$S(\Gamma) = \bigcup_{\alpha \in D} (M^{-1}\alpha + M^{-1}\Gamma).$$

Also define the iterated sequence

$$S^{(n+1)}(\Gamma) = S(S^n(\Gamma)), \qquad n = 1, 2, \ldots.$$

Then for any nonempty bounded subset $\Gamma \subset \mathbb{R}^s$,

$$\lim_{n \to \infty} S^n(\Gamma) = T(M, D)$$

in the Hausdorff metric [6]. This tells us that supp$\phi \subset T(M, D)$.

Under the assumption of local linear independence, we show that the support of ϕ is exactly equal to $T(M, D)$. We say that a set F of functions is locally linearly independent if, for any nonempty open subset G of $\mathbb{R}^s$, the set

$$\{f|_G : f \in F, G \cap \text{supp} f \neq \emptyset\}$$

is linearly independent.

Theorem 1. *Let ϕ be a nontrivial compactly supported integrable solution of (1.1), and suppose the support of the mask a is D. Assume that the shifts $\{\phi(x - \alpha)\}_{\alpha \in \mathbb{Z}^s}$ of ϕ are locally linearly independent. Then*

$$\text{supp}\, \phi = T(M, D).$$

Proof: We use the uniqueness of the attractor. Here $A = B$ means that two sets A and B are equal up to a set of measure zero. Since $T(M, D)$ is the (unique) attractor of $\{S_\alpha\}_{\alpha \in D}$, we need to show that $S(\text{supp}\, \phi) = \text{supp}\, \phi$. The refinement equation (1.1) implies supp$\phi \subset S(\text{supp}\, \phi)$ up to a set of measure zero. Then it is sufficient to verify that $S(\text{supp}\, \phi) \subset \text{supp}\, \phi$, i.e., $M^{-1}(\text{supp}\, \phi + \alpha) \subset \text{supp}\, \phi$ for each $\alpha \in D$. This is equivalent to

$$\mathbb{R}^s \setminus (\text{supp}\, \phi) \subset \mathbb{R}^s \setminus M^{-1}(\text{supp}\, \phi + \alpha).$$

Let G be the open set $\mathbb{R}^s \setminus (\text{supp}\, \phi)$. By the refinement equation (1.1),

$$\phi(x) = a(\alpha)\phi(Mx - \alpha) + \sum_{\beta \in D \setminus \{\alpha\}} a(\beta)\phi(Mx - \beta).$$

Now $\phi|_G$ is identically zero and $a(\alpha) \neq 0$, and the local linear independence of the shifts of ϕ tells us that

$$\phi(M \cdot -\alpha)|_G = 0.$$

Hence, supp$(\phi(M \cdot -\alpha)) \cap G = [M^{-1}(\text{supp}\, \phi + \alpha)] \cap G = \emptyset$.

Consider the complement of $M^{-1}(\text{supp}\, \phi + \alpha)$. Then

$$G = \mathbb{R}^s \setminus (\text{supp}\, \phi) \subset \mathbb{R}^s \setminus M^{-1}(\text{supp}\, \phi + \alpha).$$

This proves that $S(\text{supp}\, \phi) = \text{supp}\, \phi$. Since ϕ is nontrivial, suppϕ is a nonempty compact subset of $\mathbb{R}^s$. Therefore, it must be the attractor $T(M, D)$. $\square$

The inclusion supp$\phi \subset T(M, D)$ holds even without assuming the local linear independence. The following example is taken from [5].

Example 1. Let a be a mask supported on $[-2, 2] \times [-1, 1]$ and given by

$$\begin{bmatrix} \frac{-1}{16} & 0 & \frac{1}{8} & 0 & \frac{-1}{16} \\ 0 & \frac{1}{2} & \mathbf{1} & \frac{1}{2} & 0 \\ \frac{-1}{16} & 0 & \frac{1}{8} & 0 & \frac{-1}{16} \end{bmatrix},$$

where the origin is distinguished by boldface type. Then

$$D = \{(-2, 1), (0, 1), (2, 1), (-1, 0), (0, 0), (1, 0), (-2, -1), (0, -1), (2, -1)\}.$$

Consider the refinement equation

$$\phi(x) = \sum_{\alpha \in \mathbb{Z}^2} a(\alpha)\phi(Mx - \alpha), \qquad x \in \mathbb{R}^2,$$

where $M = \begin{bmatrix} 1 & -1 \\ 1 & 1 \end{bmatrix}$. We first compute the set $\left\{ \sum_{j=1}^{4} M^{-j}\varepsilon_j : \varepsilon_j \in D \right\}$. It is observed that the set is contained in

$$\left\{ \frac{1}{4}\alpha : \alpha = (\alpha_1, \alpha_2) \in \mathbb{Z}^2, -17 \le \alpha_1 + \alpha_2 \le 17, -19 \le \alpha_1 - \alpha_2 \le 19 \right\}.$$

Observe that $M^{-4} = (\frac{-1}{4})I$. Then we know that $T(M, D)$, and hence supp ϕ, is contained in the set

$$\left\{ x = (x_1, x_2) \in \mathbb{R}^2 : \frac{-17}{3} \le x_1 + x_2 \le \frac{17}{3}, \frac{-19}{3} \le x_1 - x_2 \le \frac{19}{3} \right\}.$$

The equality given in Theorem 1 does not hold in general without the assumption of local linear independence, even in the univariate case. The following is such an example.

Example 2. Let $s = 2$, $M = \begin{bmatrix} 3 & 0 \\ 0 & 3 \end{bmatrix}$ and a be supported on $[0, 8]^2$ with $a(\alpha_1, \alpha_2) = -1$ when exactly one of α_1 and α_2 is in $\{3, 4, 5\}$, and $a(\alpha_1, \alpha_2) = 1$ for other (α_1, α_2) in $[0, 8]^2$. Then the compactly supported integrable solution ϕ to the refinement equation (1.1) with $\hat{\phi}(0) = 1$ is

$$\phi = \frac{1}{4}\chi_{[0,1)^2} + \frac{1}{4}\chi_{[0,1)\times[3,4)} + \frac{1}{4}\chi_{[3,4)\times[0,1)} + \frac{1}{4}\chi_{[3,4)^2}.$$

The shifts of ϕ are locally linearly dependent and supp$\phi = [0, 1]^2 \cup ([0, 1] \times [3, 4]) \cup ([3, 4] \times [0, 1]) \cup [3, 4]^2$, while $T(M, D) = [0, 4]^2$.

The concept of local linear independence of refinable functions has been considered for various purposes in the wavelet literature. To construct wavelets on the interval, Meyer [11] and Lemarié [10] investigated linear independence of the shifts of a refinable function over the unit interval $[0, 1]$. In the univariate case, it is shown that the local linear independence of 2-refinable functions is equivalent to the global linear independence [8, 12, 13]. The situation is totally different for the refinable functions with dilation factor $M > 2$, see the examples [2, 3, 4]. The characterization of local linear independence by means of the mask in this case was presented by Cheung, Tang and Zhou [1]. In the multivariate case, Goodman, Jia and Zhou [4] gave a complete characterization for the local linear independence of refinable vectors of functions with dilation matrix $M = 2I$. However, how to characterize the local linear independence for general M is a nontrivial problem.

§3. Multivariate Refinable Functions and Self-affine Tilings

As an interesting family of refinable functions, the characteristic functions of self-affine tilings can be studied by methods for refinable functions. Self-affine tilings are defined in terms of dilation matrices and digit sets. A digit set D associated with the dilation matrix M is a complete set of representatives of distinct cosets of the quotient group $\mathbb{Z}^s/M\mathbb{Z}^s$. If $|det(M)| = m$ for some integer $m > 1$, then D is a subset of $\mathbb{Z}^s$ containing m elements. In fact, the attractor $T(M, D)$ is called a self-affine tiling if its Lebesgue measure is positive and the measure of its boundary is zero. If $T(M, D)$ is not a self-affine tiling, $\chi_{T(M,D)}$ is supported on a set of measure zero, hence equals zero as a function in L_2. Self-affine tilings have been studied in a variety of contexts in the literature.

An example of an attractor which is not a self-affine tiling arises with $M = 3$ and $D = \{0, 1, 4\}$. How to characterize the self-affine tiling property by means of the digit set D remains an open problem for general M, even in the univariate case, although the case $M = p^m$ for some prime integer p has been handled by Kenyon, and Lagarias and Wang [9]. This problem has an equivalent form in terms of refinement equations as follows.

Theorem 2. *Let $D \subset \mathbb{Z}^s$ and $\#D = m$. Then $T(M, D)$ is a self-affine tiling if and only if the (unique) compactly supported distributional solution ϕ with $\hat{\phi}(0) = 1$ to the refinement equation*

$$\phi(x) = \sum_{\alpha \in D} \phi(Mx - \alpha) \tag{3.1}$$

lies in $L_2(\mathbb{R}^s)$.

Proof: *Necessity.* By the relation $S(T(M, D)) = T(M, D)$, we know that the characteristic function $\phi := \chi_{T(M,D)}$ of the set $T(M, D)$ satisfies

$$\phi(x) \le \sum_{\alpha \in D} \phi(Mx - \alpha).$$

Since the measure of $T(M, D)$ is $\hat{\phi}(0) > 0$, integrating both sides tells us that the equality holds. Hence (3.1) has a nontrivial L_2 solution

$$\frac{\chi_{T(M,D)}}{\operatorname{meas} T(M, D)}.$$

Sufficiency. Suppose the normalized solution ϕ of (5.1) is in $L_2(\mathbb{R}^s)$. Then $\operatorname{supp}\phi$ has positive measure. Hence $T(M, D)$, which contains $\operatorname{supp}\phi$, has positive measure. $\square$

Thus, as in [15], we may use the method for the existence of L_2 solutions to refinement equations (without assuming stability or the basic sum rule, see [7]) to check the self-affine tilings by means of a finite nonnegative matrix for a given digit set. However, how to characterize the self-affine tiling property by means of some nice number theory properties of the digit set remains unknown, and is an interesting problem for further investigation.

Acknowledgments. The authors would like to thank D. X. Zhou for his support. Research was supported by City University of Hong Kong under Grant #7001029 and by Research Grant Council of Hong Kong under Grant #9040463.

References

1. H. L. Cheung, C. Q. Tang and D. X. Zhou, Supports of locally linearly independent M-refinable functions, attractors of iterated function systems and tilings, Adv. Comput. Math., to appear.

2. X. R. Dai, D. R. Huang, and Q. Y. Sun, Some properties of five-coefficient refinement equation, Arch. Math. **66** (1996), 299–309.

3. I. Daubechies and J. C. Lagarias, Two-scale difference equations: II. Local regularity, infinite products of matrices and fractals, SIAM J. Math. Anal. **23** (1992), 1031–1079.

4. T. N. T. Goodman, R. Q. Jia and D. X. Zhou, Local linear independence of refinable vectors of functions, Proc. Royal Soc. Edinburgh **130** (2000), 813–826.

5. B. Han and R. Q. Jia, Quincunx fundamental refinable functions and quincunx biorthogonal wavelets, Math. Comp., to appear.

6. J. E. Hutchinson, Fractals and self similarity, Indiana Univ. Math. J. **30** (1981), 713–747.

7. R. Q. Jia, K. S. Lau, and D. X. Zhou, L_p solutions of refinement equations, J. Fourier Anal. Appl. **7** (2001), 144–169.

8. R. Q. Jia and J. Z. Wang, Stability and linear independence associated with wavelet decompositions, Proc. Amer. Math. Soc. **117** (1993), 1115–1124.

9. J. C. Lagarias and Y. Wang, Integral self-affine tiles in $\mathbb{R}^n$ I. standard and nonstandard digit sets, J. London Math. Soc. **54** (1996), 161–179.

10. P. G. Lemarié, Fonctions à support compact dans les analyses multi-résolutions, Revista Mathemática Iberoamericana **7** (1991), 157–182.

11. Y. Meyer, Ondelettes sur l'intervalle, Revista Mathemática Iberoamericana **7** (1991), 115–133.

12. A. Ron, Characterizations of linear independence and stability of the shifts of a univariate refinable function in terms of its refinement mask, CMS TSR # 93-3, University of Wisconsin-Madison.

13. J. Z. Wang, On local linear independence of refinable distributions, preprint.

14. D. X. Zhou, Stability of refinable functions, multiresolution analysis and Haar bases, SIAM J. Math. Anal. **27** (1996), 891–904.

15. D. X. Zhou, Self-similar lattice tilings and subdivision schemes, SIAM J. Math. Anal. **33** (2001), 1–15.

Hoi Ling Cheung
Department of Mathematics
City University of Hong Kong
Kowloon
Hong Kong
96473695@plink.cityu.edu.hk

Canqin Tang
Department of Mathematics
Changde Teacher's College
Hunan
P. R. China
and
Institute of Applied Mathematics
Hunan University
Changsha, Hunan
P. R. China
tangcq2000@yahoo.com.cn

Tight Frames with Maximum Vanishing Moments and Minimum Support

Charles K. Chui, Wenjie He, and Joachim Stöckler

Abstract. The introduction of vanishing moment recovery (VMR) functions in our recent work (also called "fundamental functions" in an independent paper by Daubechies, Han, Ron, and Shen) modifies the so-called "unitary extension principle" to allow the construction of compactly supported affine frames with any desirable order of vanishing moments up to the order of polynomial reproduction of the given associated compactly supported scaling function. The objective of this paper is to unify and extend certain tight-frame results in the two papers mentioned above, with primary focus on the investigation of tight frame generators with minimum supports. In particular, a computational scheme to be described as an algorithm is developed for constructing such minimum-supported tight frame generators. An example is included as an illustration of this algorithm.

§1. Introduction

The parametric representation of curves and surfaces in terms of B-splines, and more generally NURBS, is a standard method in computer-aided design and manufacturing (CAD/CAM). Local support, variation-diminishing properties, and fast computational methods for B-splines and NURBS constitute some of the most important features for the selection of these basis functions for the CAD/CAM industry standards. However, only during the past 15 years, the properties of B-spline multiresolution analysis and spline wavelets entered into the picture, and it was clear from the very beginning that there does not exist an L_2-orthonormal basis of continuous spline-wavelets with compact support. This led to the idea of using a semi-orthogonal spline-wavelet basis with local support for the synthesis (the basis functions of the parametric representation) and dual basis functions of global support, but exponential decay, for the analysis (the dual functionals rendering the coefficients in the representation),

Approximation Theory X: Wavelets, Splines, and Applications
Charles K. Chui, Larry L. Schumaker, and Joachim Stöckler (eds.), pp. 187–206.
Copyright © 2002 by Vanderbilt University Press, Nashville, TN.
ISBN 0-8265-1416-2.

although for better performance, change-of-bases is recommended in order to use the same compactly supported spline-wavelets both for analysis and for synthesis (see [1]). Of course, semi-orthogonal wavelets and their duals have the maximum order of vanishing moments in the sense that they annihilate all polynomials of degree $L - 1$, when L^{th} order B-spline functions are used to construct the spline-wavelets.

To avoid the need of change-of-bases but still use compactly supported spline-wavelets for both analysis and synthesis, a more general approach to multiresolution representation of curves and surfaces is offered by tight affine frames of $L_2 = L_2(\mathbb{R})$, which also lead to stable parametric representations. In our present work, we discuss the shift-invariant setting in L_2. Extension to bounded intervals where B-splines with non-uniform knot sequences provide a local basis, is studied in a paper under preparation.

For the L^2 setting, tight affine frames are generated by functions $\psi_i \in L_2$ by shifts and dilations, such that the family

$$\Psi := \{\psi_{i;j,k} = 2^{j/2}\psi_i(2^j \cdot -k); \ j \in \mathbb{Z}, \ k \in \mathbb{Z}, \ 1 \le i \le n\}$$

satisfies

$$A\|f\|^2 = \sum_{i=1}^{n} \sum_{j,k \in \mathbb{Z}} |\langle f, \psi_{i;j,k}\rangle|^2, \qquad f \in L_2(\mathbb{R}), \tag{1}$$

where the constant $A > 0$, called the tight frame constant, does not depend on f. If $A = 1$ we say that Ψ is a normalized tight frame. Here, normalization is achieved simply by dividing each ψ_i by $A^{1/2}$. In order to simplify notation, we also call the functions $\{\psi_i : 1 \le i \le n\}$ tight frame generators of Ψ. If the functions ψ_i are finite linear combinations of B-splines, they can be employed as the multiresolution synthesis and analysis tool for parametric spline curves and tensor-product surfaces. In addition to having local support, the tight frame generators should also exhibit L vanishing moments for the purpose of providing an effective analysis tool.

In the following, we consider a more general setting in our discussion of the construction and characterization of tight frame generators. Let $\phi \in L_2$ be a refinable function with compact support, satisfying a refinement equation

$$\phi(x) = \sum_{k=M}^{N} p_k \phi(2x - k), \tag{2}$$

where p_k are real coefficients and ϕ is the normalized solution of (2) with $\hat{\phi}(0) = 1$ such that the corresponding Laurent polynomial

$$P(z) = \frac{1}{2} \sum_{k=M}^{N} p_k z^k, \tag{3}$$

called the two-scale symbol of ϕ, satisfies $P(-1) = 0$. The integer $L \geq 1$ is so chosen that $P(z) = (1 + z)^L P_0(z)$, and P_0 is a Laurent polynomial with $P_0(-1) \neq 0$. It is known [2,7] that the spaces V_j that are the closure of the span of the $2^{-j}\mathbb{Z}$-shifts of $\phi(2^j \cdot)$, generate a multiresolution approximation

$$\{0\} \leftarrow \cdots \subset V_{-1} \subset V_0 \subset V_1 \subset \cdots \rightarrow L_2(\mathbb{R}).$$

Without giving away much generality, we assume throughout that ϕ is a minimally supported refinable function in V_0, and therefore by the results in [8], the integer shifts of ϕ constitute a Riesz basis of V_0. (Note that this conclusion does not generalize to higher dimensions.) The characterization and construction of (minimally supported) tight frame generators $\psi_i \in V_1$, $1 \leq i \leq n$, that have L_0 vanishing moments, where $1 \leq L_0 \leq L$ is arbitrary, is our main concern in this paper. The functions ψ_i can be defined by

$$\hat{\psi}_i(\omega) = Q_i(z)\hat{\phi}(\omega/2), \qquad z := e^{-i\omega/2},$$

where Q_i, $1 \leq i \leq n$, are Laurent polynomials with real coefficients such that

$$Q_i(z) = (1 - z)^{L_0} q_i(z), \qquad 1 \leq i \leq n, \tag{4}$$

with new Laurent polynomials q_i. The factor $(1-z)^{L_0}$ entails the vanishing moment property of ψ_i.

A characterization of all orthonormal wavelets ψ (without the necessity of being defined by an MRA) appears in the book [6], and its extention to tight frame generators $\{\psi_i\}$ was given later by others such as [5,12]. Tight frame generators in the given setting of an MRA (i.e. associated with a compactly supported scaling function) can be characterized by the identities

$$S(z^2)P(z)P(1/z) + \sum_{i=1}^{N} Q_i(z)Q_i(1/z) = S(z), \tag{5}$$

$$S(z^2)P(z)P(-1/z) + \sum_{i=1}^{N} Q_i(z)Q_i(-1/z) = 0, \tag{6}$$

which must be satisfied for all $z \in \mathbb{C} \setminus \{0\}$. Here, the function S must be a Laurent polynomial with $S(1) = 1$ and $S(z) \geq 0$ on the unit circle (to be denoted by $\mathbb{T}$ throughout this paper). Moreover, the functions ψ_i have vanishing moments of order L_0, *i.e.*,

$$\int_{\mathbb{R}} x^\ell \psi_i(x) dx = 0, \qquad 0 \leq \ell < L_0, \tag{7}$$

if and only if

$$S(z)\Phi(z) - 1 = \mathcal{O}(|z - 1|^{2L_0}) \qquad \text{near } z = 1, \tag{8}$$

where $\Phi(z) = \sum_k c_k z^k$ is the autocorrelation symbol of ϕ defined by the coefficients $c_k = \int_{\mathbf{R}} \phi(x)\phi(x + k)dx$. The last two statements were mostly developed in [2], [4]. Minor improvements for tight frames are made in the present paper in Theorems 1 and 2. We coined S as a vanishing moment recovery (VMR) function in [2]; the term "fundamental function" is employed in [4]. Well-known properties of the autocorrelation function of cardinal B-splines show that $L_0 > 1$ cannot be achieved, if ϕ is a cardinal B-spline and S in (5)–(6) is the constant function 1. This was implicitly assumed in the construction of tight frame generators in [12], however, and one of the tight frame generators of their construction by a matrix extension method has only one vanishing moment.

The outline of this paper is as follows. A review of some results in [2], [4] with minor improvement is given in Section 2, where the characterization by (5)–(8) is developed. The natural question that addresses to whether or not for any refinable (stable) function ϕ, there exists a VMR function S such that $L_0 = L$ vanishing moments can be obtained for all compactly supported generators ψ_i, was answered to be positive in [2] (see Theorem 3). In Section 3 we extend the technique in [2] for the construction of tight frame generators with two functions ψ_1, ψ_2, in order to find all such generators that have minimum support. Although the derivation looks rather technical, the outcome is a simple algorithm which is given at the end of Section 3.

§2. Results for Tight Frames

In this section, we restrict our discussion to the results of the authors on univariate tight frames in [2] and of Daubechies et al. in [4], particularly in characterization and construction of tight frame generators.

Theorem 1. *Let ϕ be a refinable function with compact support and two-scale symbol P as in (3), and assume that the shifts of ϕ are stable. Let Q_i, $1 \leq i \leq n$, be Laurent polynomials with real coefficients. Then the functions ψ_i defined by $\hat{\psi}_i(2\xi) = Q_i(e^{-i\xi})\hat{\phi}(\xi)$ are tight frame generators, with frame constant 1, if and only if there exists a Laurent polynomial S with real coefficients, $S(1) = 1$, $S(z) \geq 0$ on $\mathbb{T}$, that satisfies (5)–(6).*

Theorem 1 is a reformulation of a result in [2], under the additional hypothesis of stability of the shifts of ϕ. Note that identity (5) implies that $Q_i(1) = 0$ holds for all $1 \leq i \leq n$. This condition is necessary and sufficient for the boundedness of the series on the right-hand side of (1).

The characterization of tight frame generators with L_0 vanishing moments in (7) can also be given. The sufficiency of property (8) of the

Laurent polynomial S was shown in [2]. Our subsequent discussion in the present paper shows that (8) is also necessary.

Theorem 2. *Let $1 \leq L_0 \leq L$ and let the assumptions of Theorem 1 on ϕ and Q_i, $1 \leq i \leq N$, be satisfied. Then the functions ψ_i are tight frame generators, with frame constant 1 and L_0 vanishing moments, if and only if there exists a Laurent polynomial S with real coefficients, $S(1) = 1$, $S(z) \geq 0$ on $\mathbb{T}$, that satisfies (5)–(6) and (8).*

Proof: We recall from [1] that the autocorrelation symbol Φ satisfies $\Phi(1) = 1$ and

$$|P(z)|^2 \Phi(z) + |P(-z)|^2 \Phi(-z) = \Phi(z^2), \quad z \in \mathbb{T}. \tag{9}$$

Consequently, we obtain

$$\frac{1}{\Phi(z)} - \frac{|P(z)|^2}{\Phi(z^2)} = \mathcal{O}(|z-1|^{2L}) \qquad \text{near } z = 1. \tag{10}$$

Let $1 \leq L_0 \leq L$, where L is the multiplicity of the zero $z = -1$ of P. All Laurent polynomials Q_i, $1 \leq i \leq n$, have the form (4), as a consequence of (5), if and only if

$$S(z) - S(z^2)|P(z)|^2 = \mathcal{O}(|z-1|^{2L_0}) \qquad \text{near } z = 1. \tag{11}$$

If we insert (10), we obtain the equivalent relation

$$S(z) - \frac{1}{\Phi(z)} - |P(z)|^2 \left[S(z^2) - \frac{1}{\Phi(z^2)} \right] = \mathcal{O}(|z-1|^{2L_0}).$$

By analyticity of S and Φ near 1 and $P(1) = 1$, the previous relation is equivalent to

$$S(z) - \frac{1}{\Phi(z)} = \mathcal{O}(|z-1|^{2L_0}),$$

which, in turn, is equivalent to (8). If we combine this result with Theorem 1, we obtain the result of Theorem 2. $\square$

There are many ways to find a Laurent polynomial S that satisfies (8). By symmetry of $\Phi(e^{i\xi})$ around 0, there exists a unique solution S of (8) that has the form

$$S(z) = \sum_{k=0}^{L_0-1} s_k(z^k + z^{-k}). \tag{12}$$

The coefficients s_k of this solution are real, so that S is real on the unit circle and has minimum degree (possibly less than $2(L_0-1)$). The function

S defined in this way may not be a VMR function, however. In addition
to (8), S must be chosen such that identities (5)–(6) admit Laurent polynomial solutions for Q_i, $1 \leq i \leq n$. By rewriting these identities with an
argument $-z$ for z, we have the equivalent condition

$$
\mathcal{M}(z) := \begin{pmatrix} S(z) - S(z^2)P(z)P(1/z) & -S(z^2)P(1/z)P(-z) \\ -S(z^2)P(z)P(-1/z) & S(-z) - S(z^2)P(-z)P(-1/z) \end{pmatrix}
$$

$$
= \begin{pmatrix} Q_1(1/z) & \cdots & Q_n(1/z) \\ Q_1(-1/z) & \cdots & Q_n(-1/z) \end{pmatrix} \begin{pmatrix} Q_1(z) & Q_1(-z) \\ \vdots & \vdots \\ Q_n(z) & Q_n(-z) \end{pmatrix}, \qquad (13)
$$

which we use as a substitute for (5)–(6) from now on. Apparently, the
VMR function S must be so chosen that, in addition to (8), the matrix
$\mathcal{M}(z)$ is positive semi-definite for all $z \in \mathbb{T}$. This is accomplished in [2]
(see Theorem 5) as follows.

Theorem 3. *Let the assumptions on ϕ in Theorem 1 be satisfied. Then
there exists a Laurent polynomial S with real coefficients, $S(1) = 1$, $S(z) >
0$ for all $z \in \mathbb{T}$, such that (8) is satisfied and $\mathcal{M}(z)$ in (13) is positive
semidefinite for all $z \in \mathbb{T}$.*

We only mention the main ideas of the proof of this result. First, we
make use of spectral properties of the transfer operator

$$
T_{|P|^2}(f)(z^2) = |P(z)|^2 f(z) + |P(-z)|^2 f(-z), \qquad z \in \mathbb{T},
$$

which are developed in [9]. This leads to the construction of a Laurent
polynomial R that is an eigenfunction of $T_{|P|^2}$ with respect to an eigenvalue
$0 < \lambda < 1$, satisfies $R = \mathcal{O}(|1 - z|^{2L_0})$ and $R(z) > 0$ on $\mathbb{T} \setminus \{1\}$. The
existence of R, which is not shown in [9], can be deduced from a study of
positivity and irreducibility of the restriction of $T_{|P|^2}$ to certain invariant
subspaces. The notations in [11] prove to be very useful for this discussion.
Then, we construct a Laurent polynomial S with real coefficients, for which

$$
\frac{1}{\Phi + \beta R} \leq S \leq \frac{1}{\Phi + R}, \qquad 1 < \beta < 1/\lambda,
$$

is satisfied on $\mathbb{T}$. This problem is solved by trigonometric approximation
with interpolatory constraints at $z = 1$. The upper and lower bounds for
$1/S$ are inserted in

$$
T_{|P|^2}(1/S) \leq T_{|P|^2}(\Phi + \beta R) = \Phi + \beta \lambda R \leq \Phi + R \leq 1/S, \qquad (14)
$$

where the equality sign is justified by (9). Finally, the matrix $\mathcal{M}$ is shown
to be positive semi-definite on $\mathbb{T}$, which is a consequence of

$$
\frac{S(z) - S(z^2)|P(z)|^2}{S(z)S(z^2)} \geq (\mathrm{id} - T_{|P|^2})(1/S) = \frac{\det \mathcal{M}(z^2)}{S(z)S(-z)S(z^2)},
$$

where the term in the middle is nonnegative by (14).

A similar result as stated in Theorem 3 for refinable functions ϕ with dilation factor $M \geq 2$ was recently obtained in our work [3]. The proof does not provide precise bounds for the degree of S. For the special case of cardinal B-splines, however, a precise analysis of the minimum degree Laurent polynomial S in (12) is given in [4]. Moreover, it can be shown that their result implies that the matrix $\mathcal{M}$ in (13) is positive semi-definite on $\mathbb{T}$. The same result is confirmed for low order B-splines ($L \leq 4$) in [2].

§3. Tight Frames with Two Generators and Minimum Support

From now on we assume that the VMR function S is given such that (8) holds and the matrix $\mathcal{M}$ in (13) is positive semi-definite on $\mathbb{T}$. Moreover, S is supposed to have minimum degree, whenever we discuss factorizations of $\mathcal{M}$ with minimum degree Laurent polynomials. We show, after performing three consecutive transformations (15), (21), and (27) of this matrix, that a factorization of $\mathcal{M}$ in (13) exists where we need only two Laurent polynomials Q_1 and Q_2. Upper bounds for the degree of Q_1 and Q_2 can be derived from Theorem 4. These bounds are experimentally found to be sharp for tight frame generators from B-spline MRA of low order. In this way, tight frame generators $\psi_1, \psi_2 \in V_1$ with L_0 vanishing moments are constructed. Special emphasis is given to the aspect of finding minimum degree Laurent polynomials Q_1, Q_2 in (13), because they define tight frame generators with minimum support. This is the reason why we keep track of the degree of all Laurent polynomials that are involved in the construction, and why certain transformations of $\mathcal{M}$ are chosen. Although this produces some notational overhead, there is a simple algorithm at the end of this section which summarizes all steps for the construction of tight frame generators with minimum support in a compact way.

Before we begin with the construction, we need to agree on the meaning of comparing size of support of function pairs (ψ_1, ψ_2). In the following, $|I|$ denotes the length of an interval I. For a function f on $\mathbb{R}$, we let $I(f)$ be the smallest interval that contains the support of f. We define a partial ordering as follows. The support of the pair (ψ_1, ψ_2) is larger than the support of $(\tilde{\psi}_1, \tilde{\psi}_2)$ if one of the two conditions is satisfied: (1) $|I(\psi_1)| > |I(\tilde{\psi}_1)|$ and $|I(\psi_2)| \geq |I(\tilde{\psi}_2)|$, or (2) $|I(\psi_1)| = |I(\tilde{\psi}_1)|$ and $|I(\psi_2)| > |I(\tilde{\psi}_2)|$. Similarly, we define a partial ordering for pairs of Laurent polynomials. The pair (Q_1, Q_2) has larger degree than the pair $(\tilde{Q}_1, \tilde{Q}_2)$ if either $\deg(Q_1) > \deg(\tilde{Q}_1)$ and $\deg(Q_2) \geq \deg(\tilde{Q}_2)$, or $\deg(Q_1) = \deg(\tilde{Q}_1)$ and $\deg(Q_2) > \deg(\tilde{Q}_2)$. Here, the degree of a Laurent polynomial $Q(z) = \sum_{k=m}^{n} a_k z^k$ is defined to be $n - m$, if $a_m a_n \neq 0$. The space of all Laurent polynomials Q as above is denoted by $L[m : n]$. First we recall some important definitions and agree on the notations. The two-scale symbol of ϕ is given by

$$P(z) = \left(\frac{1+z}{2}\right)^L P_0(z) = \sum_{k=-M_P}^{N_P} p_k z^k, \qquad p_{-M_P} p_{N_P} \neq 0.$$

The VMR function S is a Laurent polynomial with real coefficients, which is real on $\mathbb{T}$. Therefore, it has the expansion

$$S(z) = \sum_{k=0}^{N_S} s_k(z^k + z^{-k}), \qquad s_{N_S} \neq 0.$$

The first transformation of the matrix $\mathcal{M}$ is performed by making use of (4) in order to replace identity (13) with

$$
\begin{aligned}
\mathcal{M}_0(z) &= \begin{pmatrix} X(z) & Y(z) \\ Y(-z) & X(-z) \end{pmatrix} \\
&= \begin{pmatrix} q_1(1/z) & q_2(1/z) \\ q_1(-1/z) & q_2(-1/z) \end{pmatrix} \begin{pmatrix} q_1(z) & q_1(-z) \\ q_2(z) & q_2(-z) \end{pmatrix},
\end{aligned}
\tag{15}
$$

where q_1, q_2 are (minimum degree) Laurent polynomials and

$$
\begin{aligned}
X(z) &:= \frac{S(z) - S(z^2)P(z)P(1/z)}{(1-z)^{L_0}(1-1/z)^{L_0}}, \\
Y(z) &:= \frac{-S(z^2)P(1/z)P(-z)}{(1+z)^{L_0}(1-1/z)^{L_0}}.
\end{aligned}
\tag{16}
$$

By the assumptions on S and P, the functions X and Y are Laurent polynomials that have the form

$$X(z) = \sum_{k=0}^{N_X} x_k(z^k + z^{-k}), \qquad Y(z) = \sum_{k=0}^{N_X} y_k(z^k + (-1)^k z^{-k}), \tag{17}$$

where $N_X = 2N_S + M_P + N_P - L_0$ and the leading coefficients satisfy $x_{N_X} = \pm y_{N_X} \neq 0$.

The determinant

$$\det \mathcal{M}_0(z) =$$
$$\frac{S(z)S(-z) - S(z^2)(S(-z)P(z)P(1/z) + S(z)P(-z)P(-1/z))}{((1-z)(1-1/z)(1+z)(1+1/z))^{L_0}}$$

is an even Laurent polynomial which is nonnegative on $\mathbb{T}$. We assume that it does not vanish identically. Then we define

$$\Delta(z) = \sum_{k=0}^{N_\Delta} \delta_k(z^k + z^{-k}), \qquad \delta_{N_\Delta} \neq 0, \tag{18}$$

such that $4\Delta(z^2) = \det \mathcal{M}_0(z)$. Note that

$$N_X - N_\Delta \geq N_S + M_P + N_P - K \geq \begin{cases} 1, & \text{if } N_S + M_P + N_P < 3, \\ 2, & \text{if } N_S + M_P + N_P \geq 3, \end{cases} \tag{19}$$

where K is the largest integer such that the coefficient of z^{2K} in the expansion of $S(-z)P(z)P(1/z) + S(z)P(-z)P(-1/z)$ is nonzero. (If this Laurent polynomial equals zero, we let $K = 0$.)

It is suitable to present the following result at this point, although the proof of the existence can only be given later. The second part of the theorem is clear, since $X(z) = q_1(z)q_1(1/z) + q_2(z)q_2(1/z)$ by (15).

Theorem 4. *With X, Y, and Δ from above, let $\mu := \min(2, N_X - N_\Delta)$. Then there exist Laurent polynomials q_1, q_2 with real coefficients and $\deg q_1 = N_X$, $\deg q_2 \leq N_X - \mu$ which satisfy (15). Moreover, no pair of Laurent polynomials (q_1, q_2) with $\deg q_1 < N_X$ and $\deg q_2 < N_X$ can satisfy (15).*

We continue with our preparations for the construction of a factorization of $\mathcal{M}$, from which the existence part of Theorem 4 will be deduced. As the second transformation of $\mathcal{M}$, we perform a polyphase decomposition in order to decouple the entries of the matrix $\mathcal{M}_0$ in (15). If we let

$$\begin{aligned}\nu = 0, \quad T(z) &= \frac{1}{2}\begin{pmatrix} 1 & 1 \\ z & -z \end{pmatrix}, \qquad \text{if } x_{N_X} = y_{N_X}, \\ \nu = 1, \quad T(z) &= \frac{1}{2}\begin{pmatrix} 1/z & -1/z \\ 1 & 1 \end{pmatrix}, \qquad \text{if } x_{N_X} = -y_{N_X},\end{aligned} \tag{20}$$

we can define Laurent polynomials A, B, and C by

$$\begin{pmatrix} A(z^2) & B(z^2) \\ B(1/z^2) & C(z^2) \end{pmatrix} = T(z)\begin{pmatrix} X(z) & Y(z) \\ X(-z) & Y(-z) \end{pmatrix} T^t(1/z). \tag{21}$$

Then q_1, q_2 satisfy (15) if and only if

$$q_i(z) = z^{-\nu + 2\ell_i}[u_i(z^2) + zv_i(z^2)], \qquad i = 1, 2, \tag{22}$$

where ℓ_i is an integer, the lowest order monomial of $[u_i(z^2) + zv_i(z^2)]$ is either 1 or z, and

$$\begin{pmatrix} A(z) & B(z) \\ B(1/z) & C(z) \end{pmatrix} = \begin{pmatrix} u_1(1/z) & u_2(1/z) \\ v_1(1/z) & v_2(1/z) \end{pmatrix}\begin{pmatrix} u_1(z) & v_1(z) \\ u_2(z) & v_2(z) \end{pmatrix}. \tag{23}$$

We also note that Δ in (18) is the determinant of the matrix on the left-hand side of (23). Therefore, $d = u_1v_2 - u_2v_1$ satisfies the relation

$$d(z)d(1/z) = \Delta(z) = A(z)C(z) - B(z)B(1/z). \tag{24}$$

For later reference we display the representations of the new Laurent polynomials

$$A(z) = \sum_{k=0}^{N_A} a_k(z^k + z^{-k}), \quad a_k = \frac{x_{2k} + (-1)^\nu y_{2k}}{2},$$

$$C(z) = \sum_{k=0}^{N_C} c_k(z^k + z^{-k}), \quad c_k = \frac{x_{2k} - (-1)^\nu y_{2k}}{2}, \tag{25}$$

$$B(z) = \sum_{k=-M_B}^{N_B} b_k z^k, \quad b_k = \frac{x_{2k+1} - (-1)^\nu y_{2k+1}}{2},$$

where we let $y_{-k} = -y_k$ for odd k in the last expression, see (17). Here, N_A, N_C, M_B, and N_B are the largest integers for which the summands are nonzero. As a consequence of the choice of the parameter $\nu \in \{0, 1\}$ in (20), we obtain that

$$N_A = N_X/2, \qquad M_B, N_B + 1, N_C + 1 \le N_X/2, \quad \text{if } N_X \text{ is even,}$$
$$M_B = (N_X + 1)/2, \quad N_A, N_C, N_B + 1 \le (N_X - 1)/2, \quad \text{if } N_X \text{ is odd.} \tag{26}$$

The following observation summarizes the link between solutions of (15) and (23).

Proposition 5. *Let N_X be given as in (17), and let N be the largest integer less than or equal to $(N_X + 1)/2$. If (q_1, q_2) is a solution of (15) with $\deg q_i \le N_X$, $i = 1, 2$, then the polyphase components u_1, u_2, v_1, v_2 of q_1 and q_2, as defined in (22), are elements of $L[0 : N]$ and define a factorization (23).*

Proof: If q_1 and q_2 satisfy the assumptions of the proposition, we obtain from (22) that $z^{\nu - 2\ell_i} q_i \in L[0 : N_X + 1]$. Hence, the polyphase components u_i, v_i in (22), $i = 1, 2$, are elements of $L[0 : N]$. It is clear from the construction that they define a factorization (23). $\square$

Remark. The value of the last proposition for the construction of tight frame generators with minimum support can be seen as follows. In order to find *all* minimum degree solutions (q_1, q_2) of (15), where $\deg q_i \le N_X$, $i = 1, 2$, the search should be extended over all polyphase components $u_i, v_i \in L[0 : N]$. An algorithm that simplifies this search to an elementary problem of linear algebra is developed next.

Our main task remains to show the existence of the factorization of the positive semi-definite matrix on the left-hand side of (23). We have developed in [2] a method for the factorization of such matrices under certain constraints on the Laurent polynomial entries. Therefore, we arrange for a "generic" form of the matrix by the following third transformation. This transformation turns out to be more suitable for the discussion of minimum degree factorizations than the transformation in Lemma 4 of [2].

Lemma 6. *Assume that there is no common zero in $\mathbb{C} \setminus \{0\}$ of all four Laurent polynomials A, B, C, and $B(1/z)$. Then for almost every $r \in \mathbb{R}$, the Laurent polynomials $\widetilde{A}(z), \widetilde{B}(z)$ defined by*

$$\begin{pmatrix} \widetilde{A}(z) & \widetilde{B}(z) \\ \widetilde{B}(1/z) & C(z) \end{pmatrix} = \begin{pmatrix} 1 & r \\ 0 & 1 \end{pmatrix} \begin{pmatrix} A(z) & B(z) \\ B(1/z) & C(z) \end{pmatrix} \begin{pmatrix} 1 & 0 \\ r & 1 \end{pmatrix} \qquad (27)$$

have the form (25), with parameters $N_{\widetilde{A}} = \max(N_A, N_C, M_B, N_B)$, $M_{\widetilde{B}} = \max(M_B, N_C)$, $N_{\widetilde{B}} = \max(N_B, N_C)$, and $\widetilde{A}, \widetilde{B}$ have no common zeros in $\mathbb{C} \setminus \{0\}$.

Proof: The determinant of the matrix on the left hand side of (27) is the same, for every $r \in \mathbb{R}$; indeed, it agrees with Δ in (18). Consequently, only the roots of Δ are possible candidates for common zeros of

$$\widetilde{A}(z) = A(z) + r(B(z) + B(1/z)) + r^2 C(z), \qquad \widetilde{B}(z) = B(z) + rC(z).$$

For a fixed root z_0 of Δ, there is at most one value of $r \in \mathbb{R}$ such that $\widetilde{A}(z_0) = \widetilde{B}(z_0) = 0$, except for the case where $A(z_0) = B(z_0) = B(1/z_0) = C(z_0) = 0$. This shows that for at most $2N_\Delta$ values of r, the Laurent polynomials $\widetilde{A}$ and $\widetilde{B}$ can have common zeros, given the assumption on A, B, and C of the lemma. Evidently, almost all values of r yield $N_{\widetilde{A}} = \max(N_A, M_B, N_B, N_C)$. The other relations are clear as well. This completes the proof of the lemma. $\square$

Remark. It follows from the invertibility of T in (21) that all four Laurent polynomials A, B, $B(1/z)$, and C have a common zero $z_0 \neq 0$, if and only if $X(z_0) = X(-z_0) = Y(z_0) = Y(-z_0) = 0$. In the literature on stability of refinable functions, a common zero of X and $X(-z)$ is often called a symmetric zero of X. Therefore, the assumption of Lemma 6 is equivalent to the condition that X and Y have no common symmetric zeros. In our experiments we never encountered a case where this condition is violated.

We have now completed the three transformations of the matrix $\mathcal{M}$, that are the combination of (15), (21), and (27). Note that $N := N_{\widetilde{A}}$ is the largest number among $\{N_A, M_B, N_B, N_C\}$. In view of (26), this is the largest integer less than or equal to $(N_X + 1)/2$. Since all transformations

are invertible, we find a one-to-one correspondence between all Laurent
polynomial solutions $(\tilde{u}_1, \tilde{u}_2, \tilde{v}_1, \tilde{v}_2)$ of the factorization problem

$$\begin{pmatrix} \widetilde{A}(z) & \widetilde{B}(z) \\ \widetilde{B}(1/z) & C(z) \end{pmatrix} = \begin{pmatrix} \tilde{u}_1(1/z) & \tilde{u}_2(1/z) \\ \tilde{v}_1(1/z) & \tilde{v}_2(1/z) \end{pmatrix} \begin{pmatrix} \tilde{u}_1(z) & \tilde{v}_1(z) \\ \tilde{u}_2(z) & \tilde{v}_2(z) \end{pmatrix} \quad (28)$$

and the Laurent polynomial solutions (q_1, q_2) of (15), which gives

$$q_i(z) = z^{2\ell_i - \nu}(\tilde{u}_i(z^2) + (z - r)\tilde{v}_i(z^2)), \quad i = 1, 2. \quad (29)$$

If we compare (29) with (22) and apply Proposition 5, we obtain

$$\tilde{u}_1, \tilde{u}_2, \tilde{v}_1, \tilde{v}_2 \in L[0 : N], \qquad N = N_{\tilde{A}}, \quad (30)$$

if q_1, q_2 have degree at most N_X. It is this "a priori" specification of
the degree of the polynomials $\tilde{u}_1, \tilde{u}_2, \tilde{v}_1, \tilde{v}_2$, which renders the following
procedure for the construction of solutions feasible. Our main tool is a
link between all factorizations of the type (28) under the constraints (30)
and the system of equations

$$\widetilde{B}(z)\tilde{u}_1(z) - d(z)\tilde{u}_2(1/z) - \widetilde{A}(z)\tilde{v}_1(z) = 0, \quad (31)$$

$$\widetilde{B}(z)\tilde{u}_2(z) + d(z)\tilde{u}_1(1/z) - \widetilde{A}(z)\tilde{v}_2(z) = 0, \quad (32)$$

$$\tilde{u}_1^2(1) + \tilde{u}_2^2(1) = \widetilde{A}(1), \quad (33)$$

where d is a Laurent polynomial that satisfies (24). Recall that d can
be viewed as a "square root" of the determinant Δ of the matrix on the
left-hand side of (28). Note that (33) is merely a normalizing condition,
as long as u_1 and u_2 do not vanish simultaneously at 1.

The following result was shown in Theorem 4 of [2]. In order to
simplify notations, we leave out the tilde signs, which we then must do in
(31)–(33) and (28) as well. The result is reformulated in order to expose
the main tools needed for the proof of Theorem 4 and our new algorithm at
the end of this section. The proof in [2] is based on arguments concerning
elementary algebraic properties of Laurent polynomials and arguments
from linear algebra and is omitted here.

Theorem 7. *Let* $\begin{pmatrix} A(z) & B(z) \\ B(1/z) & C(z) \end{pmatrix}$ *be a matrix of Laurent polynomials*

$$A, B, C \in L[-N : N], \quad A(z) = \sum_{k=0}^{N} a_k(z^k + z^{-k}) \quad \text{with } a_N \neq 0.$$

Suppose that the matrix is positive semi-definite on $\mathbb{T}$, *and the determinant* Δ *does not vanish identically. Then*

(i) *If polynomials* $u_1, u_2, v_1, v_2 \in L[0:N]$ *satisfy (28), then the equations (31)–(33), with* $d := u_1 v_2 - u_2 v_1$, *are satisfied.*

Let $d \in L[0:2N]$ *be a polynomial that satisfies* $d(z)d(1/z) = \Delta(z)$, *and assume, in addition to the aforementioned assumptions, that* A *and* B *have no common zeros in* $\mathbb{C} \setminus \{0\}$. *Moreover,*

(ii) *If* $u_1, u_2, v_1 \in L[0:N]$ *define a nontrivial solution of (31), then there exists a polynomial* $v_2 \in L[0:N]$ *such that* u_1, u_2, v_2 *is a nontrivial solution of (32).*

(iii) *If* $u_1, u_2, v_1, v_2 \in L[0:N]$ *define a nontrivial solution of (31)–(32), then there is a constant* $c > 0$ *such that* cu_1, cu_2, cv_1, cv_2 *define a factorization (28).*

Let us discuss the meaning of Theorem 7 for our construction of minimum degree solutions (q_1, q_2) of (15). Parts (i) and (iii) of the theorem, together with (30), confirm that it is sufficient to inspect all solutions of (31)–(33) that are polynomials in $L[0:N]$, in order to find all solutions (q_1, q_2) whose degree does not exceed N_X. The parameter d in the equations (31)–(32) must vary over all polynomials in $L[0:2N]$ which have real coefficients and constitute a root of the determinant Δ. Therefore, d has the form

$$d(z) = z^\ell \sum_{k=0}^{N_\Delta} d_k z^k, \tag{34}$$

where $0 \leq \ell \leq 2N - N_\Delta$ and the real coefficients d_k determine all possible solutions of $d(z)d(1/z) = \Delta(z)$. There exist many methods in order to obtain these polynomials, *e.g.* spectral factorization as proposed in [10].

The second part of Theorem 7 implies a considerable reduction of the amount of work for solving (31)–(32). The following method makes use of elementary linear algebra. We insert the unknown coefficients of $\tilde{u}_1, \tilde{u}_2, \tilde{v}_1, \tilde{v}_2$, which we denote by $\tilde{u}_{1,k}$ etc., into two vectors

$$\begin{aligned}
\vec{x} &= (\tilde{v}_{1,0}, \ldots, \tilde{v}_{1,N}, \tilde{u}_{1,0}, \tilde{u}_{2,0}, \ldots, \tilde{u}_{1,N}, \tilde{u}_{2,N})^t, \\
\vec{y} &= (\tilde{v}_{2,0}, \ldots, \tilde{v}_{2,N}, \tilde{u}_{2,0}, -\tilde{u}_{1,0}, \ldots, \tilde{u}_{2,N}, -\tilde{u}_{1,N})^t.
\end{aligned} \tag{35}$$

Equation (31) can then be written as a linear homogeneous system $Z\vec{x} = 0$ by expanding the left hand side of (31) into a Laurent series and setting the coefficient of z^j, $-N \leq j \leq 2N$, to zero. Z is the corresponding real matrix with $3N + 1$ rows and $3N + 3$ columns. Likewise, equation (32) can be written as $Z\vec{y} = 0$ with the same matrix Z. Obviously, each of the systems $Z\vec{x} = 0$ and $Z\vec{y} = 0$ admits nontrivial solutions, since the number of unknowns in each system exceeds the number of equations by 2. Thanks to part (ii) of Theorem 7, the seemingly overdetermined system $(Z\vec{x} = 0, Z\vec{y} = 0)$, which has $6N + 2$ equations for $4N + 4$ unknowns, has as solutions all quadruplets $\tilde{u}_1, \tilde{u}_2, \tilde{v}_1, \tilde{v}_2$ where $(\tilde{u}_1, \tilde{u}_2, \tilde{v}_1)$ is a solution of the system $Z\vec{x} = 0$. This effect can also be read off the reduced row echelon form of Z, which is part (i) of the following lemma. Other features such as zero entries of the solution vectors $\vec{x}$ and $\vec{y}$ can also be determined from simple properties of $\widetilde{A}$, $\widetilde{B}$, and d, see parts (ii) and (iii) of the lemma. For ease of notation we skip the tilde sign again.

Lemma 8. *Assume that the Laurent polynomials A, B, C in (25) and d in (34) are given where $N := N_A \geq \max(N_C, M_B, N_B)$ and $d \in L[\ell : \ell + N_\Delta] \subset L[0 : 2N]$. Furthermore, assume that A and B have no common zeros in $\mathbb{C} \setminus \{0\}$. Then the matrix Z from above has full rank. Its first $3N + 1$ columns are linearly independent, and its reduced row echelon form is given by*

$$\begin{pmatrix} 1 & & & g_1 & h_1 \\ & \ddots & & \vdots & \vdots \\ & & 1 & g_{3N+1} & h_{3N+1} \end{pmatrix}. \tag{36}$$

All solutions of the linear systems $Z\vec{x} = 0$ and $Z\vec{y} = 0$ are given by

$$\vec{x} = u_{1,N}\alpha - u_{2,N}\beta, \qquad \vec{y} = u_{1,N}\beta + u_{2,N}\alpha, \tag{37}$$

where

$$\alpha = (-g_1, \ldots, -g_{3N+1}, 1, 0)^t, \qquad \beta = (h_1, \ldots, h_{3N+1}, 0, -1)^t \tag{38}$$

and $u_{1,N}$, $u_{2,N}$ are free parameters. Moreover, the entries in the last two columns in (36) satisfy

(i) $g_i = h_{i+1}$ *and* $g_{i+1} = -h_i$ *for all* $i = N + 2, N + 4, \ldots, 3N$,

(ii) $g_i = h_i = 0$ *for all* $1 \leq i \leq \min(N - M_B, \ell)$ *and* $\max(N_B, \ell + N_\Delta - N) + 2 \leq i \leq N + 1$,

(iii) $h_{N_B+1} = 0$, *if* $N_\Delta < N + N_B - \ell$.

Proof: The combined system $Z\vec{x} = 0$, $Z\vec{y} = 0$ is consistent, as mentioned above. Let $r := 3N + 1$. If the first r columns of Z are linearly dependent, there exists a nontrivial solution of $Z\vec{x} = 0$ with $u_{1,N} = u_{2,N} = 0$. Hence, equation (31) has a nontrivial solution (u_1, u_2, v_1) where the degree of u_1 and u_2 is at most $N - 1$. By (ii) and (iii) in Theorem 7, the Laurent polynomial A of degree $2N$ divides the nonzero Laurent polynomial $u_1(z)u_1(1/z) + u_2(z)u_2(1/z)$ of degree less than $2N - 1$. This is a contradiction. Therefore, the first r columns of Z must be linearly independent. This implies that Z has full rank and its reduced row echelon form is given by (36). The vectors α, β in (38) are a basis of the solution set of $Z\vec{x} = 0$. This leads to the form of the solutions for $\vec{x}$ and $\vec{y}$ in (37). Moreover, since the coefficients $u_{1,k}$, $u_{2,k}$ appear in the definition of both vectors $\vec{x}$ and $\vec{y}$, part (i) of the lemma follows by inserting the values $u_{1,N} = 1$ and $u_{2,N} = 0$ in (37).

For parts (ii) and (iii) of the lemma, only the system $Z\vec{x} = 0$ needs to be analyzed. In order to prove part (ii), we define $s = \min(-M_B, \ell - N)$ and inspect the homogeneous equations for the coefficients of z^j in (31), where $-N \leq j < s$. Only the product $A(z)v_1(z)$ contributes to these coefficients. This leads to $s + N$ equations of the form

$$-\sum_{k=0}^{j+N} a_{N+k-j}v_{1,k} = 0, \quad -N \leq j \leq s - 1,$$

with unknowns $v_{1,0}, \ldots, v_{1,s+N-1}$. These equations are part of the linear system $Z\vec{x} = 0$. Since the matrix of this subsystem is invertible, every solution of $Z\vec{x} = 0$ must have $v_{1,0} = \cdots = v_{1,s+N-1} = 0$. Consequently, both solutions α and β in (38) must have zeros in rows i where $1 \leq i \leq s + N$, and this gives the first part of (ii). The second part of (ii) follows in an analogous way by inspecting the coefficients of z^j in (31) for $\max(N_B + N, \ell + N_\Delta) + 1 \leq j \leq 2N$.

Similarly, for part (iii), we inspect the coefficient of z^{N+N_B} in (31). If we have $\ell + N_\Delta < N_B + N$, then the coefficient of z^j, $j = N_B + N$, in (31) vanishes if and only if

$$b_{N_B}u_{1,N} - \sum_{k=N_B}^{N} a_{j-k}v_{1,k} = 0. \tag{39}$$

By part (ii), this can be simplified to $b_{N_B}u_{1,N} - a_N v_{1,N_B} = 0$. Consequently, the solution β in (38) with $u_{1,N} = 0$ and $u_{2,N} = -1$ must have $v_{1,N_B} = h_{N_B+1} = 0$. We have thus shown (iii). $\square$

Parts (ii) and (iii) of the Lemma 8 are essential for the proof of Theorem 4. Before we enter the proof, we specify the following consequence of the lemma.

Lemma 9. *Let the assumptions of Lemma 8 be satisfied. Assume that $B \in L[-N : N - 1]$ and $d \in L[0 : 2N - \mu]$, where $\mu \in \{1, 2\}$. Then there exists a nontrivial solution (u_1, u_2, v_1, v_2) of equations (31)–(33), such that*

$$u_1 \in L[0 : N], \quad u_2 \in L[0 : N - 1],$$
$$v_1 \in L[0 : N - 1], \quad v_2 \in L[0 : N - \mu].$$

Proof: The assumptions on B and d imply that $N_B < N_A = N$ and $\ell + N_\Delta - N \leq N - \mu$, where $\mu \in \{1, 2\}$. This gives $\max(N_B, \ell + N_\Delta - N) + 2 \leq N + 1$ in part (ii) of Lemma 8. It follows that $g_{N+1} = h_{N+1} = 0$. If $\mu = 2$ and $N_B \leq N - 2$, another application of (ii) gives $g_N = h_N = 0$. If $\mu = 2$ and $N_B = N - 1$, however, then Lemma 8(iii) implies that $h_N = 0$. The parameters $u_{1,N} = 1$ and $u_{2,N} = 0$ in (37) define the solution

(u_1, u_2, v_1, v_2) with coefficients $v_{1,N} = -g_{N+1} = 0$ and $v_{2,N} = h_{N+1} = 0$. This is the assertion of Lemma 9 for $\mu = 1$. In the case where $\mu = 2$ we obtain that $v_{2,N-1} = h_N = 0$, which again proves the assertion of the lemma. □

Finally, we are in a position to give the proof of Theorem 4.

Proof of Theorem 4: We have mentioned before that (15) cannot hold if q_1 and q_2 have degree less than N_X. Therefore, let us turn to the existence of Laurent polynomials q_1 and q_2 with the degree constraints $\deg q_1 = N_X$ and $\deg q_2 \leq N_X - \mu$. We first deal with the case where X and Y have no common symmetric zeros.

Let A, B, C be the Laurent polynomials in (21) and r be chosen as in Lemma 6. Then we obtain the matrix (27), where the Laurent polynomials $\widetilde{A}$, $\widetilde{B}$ have no common zeros and $N_{\widetilde{A}} = \max(N_A, N_B, M_B, N_C)$. Recall from (26) and (27) that if N_X is even, we have

$$N_{\widetilde{A}} = N_A = N_X/2 > N_C, \quad M_{\widetilde{B}} \leq N_X/2, \quad N_{\widetilde{B}} \leq N_X/2 - 1.$$

If N_X is odd, we obtain

$$N_{\widetilde{A}} = M_{\widetilde{B}} = M_B = (N_X + 1)/2, \quad N_A, N_{\widetilde{B}}, N_C \leq (N_X - 1)/2.$$

In both cases we conclude that $M_{\widetilde{B}} \leq N_{\widetilde{A}}$ and $N_{\widetilde{B}}, N_C < N_{\widetilde{A}}$. Furthermore, we have

$$N_\Delta \leq N_X - \mu \leq 2N_{\widetilde{A}} - \ell - \mu,$$

where we let $\ell = 0$, if N_X is even, and $\ell = 1$, if N_X is odd.

Let $N = N_{\widetilde{A}}$ as in Lemma 8. The determinant Δ in (18) remains unchanged by the transformation (27). We choose d in (34), such that $d(z)d(1/z) = \Delta(z)$ and $d \in L[\ell : \ell + N_\Delta]$, with $\ell \in \{0, 1\}$ from above. Then the result of Lemma 9 assures, that there exists a nontrivial solution $(\tilde{u}_1, \tilde{u}_2, \tilde{v}_1, \tilde{v}_2)$ of equations (31)–(33), where

$$\begin{aligned} \tilde{u}_1 \in L[0 : N], \quad \tilde{u}_2 \in L[0 : N - 1], \\ \tilde{v}_1 \in L[0 : N - 1], \quad \tilde{v}_2 \in L[0 : N - \mu]. \end{aligned} \tag{40}$$

A nontrivial solution of (31)–(33), where the original Laurent polynomials A and B in (21) are inserted, is obtained by

$$(u_1, u_2, v_1, v_2) = (\tilde{u}_1 - r\tilde{v}_1, \tilde{u}_2 - r\tilde{v}_2, \tilde{v}_1, \tilde{v}_2). \tag{41}$$

If N_X is even, (40) and (41) combined with $N = N_X/2$ lead to

$$\begin{aligned} u_1 \in L[0 : \tfrac{N_X}{2}], \quad u_2 \in L[0 : \tfrac{N_X - 2}{2}], \\ v_1 \in L[0 : \tfrac{N_X - 2}{2}], \quad v_2 \in L[0 : \tfrac{N_X - 2\mu}{2}]. \end{aligned}$$

The corresponding Laurent polynomials q_1 and q_2 in (22) satisfy

$$q_1 \in L[2\ell_1 - \nu : 2\ell_1 - \nu + N_X], \quad q_2 \in L[2\ell_2 - \nu : 2\ell_2 - \nu + N_X - \mu].$$

This shows the existence result of the theorem, if N_X is even. If N_X is odd, the choice of $\ell = 1$ in (34) has the effect that the homogeneous equations

$$\tilde{b}_{-N}\tilde{u}_{i,0} - \tilde{a}_{-N}\tilde{v}_{i,0} = 0, \qquad i = 1, 2, \tag{42}$$

follow from (31) (for $i = 1$) and (32) (for $i = 2$). Here, $\tilde{b}_{-N}$, $\tilde{a}_{-N}$ denote the coefficients of the monomial z^{-N} in the Laurent polynomials $\widetilde{B}$ and $\widetilde{A}$. The definition of $\widetilde{A}$ and $\widetilde{B}$ in (27) combined with $N = M_B > N_A, N_B, N_C$ give $\tilde{b}_{-N} = b_{-N}$ and $\tilde{a}_{-N} = rb_{-N}$. Therefore, the solutions of (42) satisfy $\tilde{u}_{i,0} - r\tilde{v}_{i,0} = 0$, $i = 1, 2$. This, together with (40), (41), leads to

$$u_1 \in L[1 : \tfrac{N_X+1}{2}], \quad u_2 \in L[1 : \tfrac{N_X-1}{2}],$$
$$v_1 \in L[0 : \tfrac{N_X-1}{2}], \quad v_2 \in L[0 : \tfrac{N_X+1-2\mu}{2}].$$

Inserting these polynomials into (22) gives

$$q_1 \in L[2\ell_1 - \nu + 1 : 2\ell_1 - \nu + N_X + 1],$$
$$q_2 \in L[2\ell_2 - \nu + 1 : 2\ell_2 - \nu + N_X + 1 - \mu].$$

This completes the proof of the theorem under the additional constraint that X and Y have no common symmetric zeros.

If X and Y have a common symmetric zero $z_0 \in \mathbb{C} \setminus \{0\}$, then $1/z_0$ must also be a common symmetric zero, since $X(z) = X(1/z)$ and $Y(-z) = Y(1/z)$ hold by (17). Moreover, $\overline{z_0}$ and $1/\overline{z_0}$ are common symmetric zeros as well, since X and Y have real coefficients. Let z_k, $1 \leq k \leq 2\kappa$, denote all common symmetric zeros of X and Y. Since the matrix in (15) is positive semi-definite on $\mathbb{T}$, each zero $z_k \in \mathbb{T}$ must have even multiplicity. We can therefore order the zeros such that $z_k z_{k+\kappa} = 1$, $1 \leq k \leq \kappa$, and $\prod_{k=1}^{\kappa}(z - z_k)$ is a polynomial with real coefficients. The division of X and Y by all factors $(z - z_k)(1/z - z_k)$, $1 \leq k \leq \kappa$, leads to new Laurent polynomials $\widetilde{X}$, $\widetilde{Y}$ that have real coefficients and the same form (17) as X and Y, where $N_{\widetilde{X}} = N_X - \kappa$. The parameter N_Δ of the determinant is reduced by 2κ. Therefore, the parameter μ in the theorem remains 2 or increases from 1 to 2. An application of the result proved so far leads to Laurent polynomials $\tilde{q}_i$, $i = 1, 2$, that satisfy (15) (with $\widetilde{X}$ and $\widetilde{Y}$ instead of X and Y) and have $\deg \tilde{q}_1 = N_{\widetilde{X}}$, $\deg \tilde{q}_2 = \leq N_{\widetilde{X}} - 2$. Multiplication by all factors $(z - z_k)$, $1 \leq k \leq \kappa$, gives Laurent polynomials q_1, q_2 with real coefficients that satisfy the assertion of the theorem. $\square$

We summarize the computational method for finding the minimum degree solutions q_1, q_2 of (15), whose degree does not exceed N_X, in the following algorithm.

Algorithm. *Assume that X and Y are given as in (16)–(17), and that X and Y have no common symmetric roots. Then minimum degree solutions of (15) whose degree does not exceed N_X are found by the following procedure:*

1) *Let N denote the largest integer less than or equal to $(N_X + 1)/2$. Choose ν as in (20) and compute the Laurent polynomials A, B, C in (21).*

2) *Choose $r \in \mathbb{R}$ such that the assumptions of Lemma 8 for the modified Laurent polynomials $\widetilde{A}$, $\widetilde{B}$, C are satisfied. Almost every $r \in \mathbb{R}$ is suitable.*

3) *For each choice of d in (34), where $d \in L[0 : 2N]$ and $d(z)d(1/z) = \Delta(z)$:*

   3a) *Compute the matrix Z in the matrix representation of equation (31), using the vector $\vec{x}$ in (35) as unknowns. (Columns of Z repeat the coefficient sequences of $-\widetilde{A}$, $\widetilde{B}$, and the reverse coefficient sequence of $-d$, each padded with zeros on top and/or bottom.)*

   3b) *Find the reduced row echelon form of Z.*

   3c) *Among all nontrivial choices of parameters $\tilde{u}_{1,N}$, $\tilde{u}_{2,N}$ for the computation of $\vec{x}$ and $\vec{y}$ in (37), find those for which the pair*

$$q_1(z) = z^{-\nu}(\tilde{u}_1(z^2) + (z - r)\tilde{v}_1(z^2)),$$
$$q_2(z) = z^{-\nu}(\tilde{u}_2(z^2) + (z - r)\tilde{v}_2(z^2))$$

   has minimum degree.

4) *Multiply the minimum degree solution(s) (q_1, q_2) found in step 3 by a positive constant so as to satisfy $q_1^2(1) + q_2^2(1) = X(1)$.*

The result is a pair (q_1, q_2) with $\deg q_1 = N_X$, $\deg q_2 \leq N_X - \mu$, where $\mu = \min(2, N_X - N_\Delta)$.

If X and Y have common symmetric zeros, the conclusions (ii) and (iii) of Theorem 7 may fail. Part (i) of that theorem remains valid, however. This implies the necessity of equations (31)–(33). N is chosen to be the maximum of N_A, M_B, N_B, N_C. Therefore, in the absence of conclusion (ii) of the theorem, the combined system ($Z\vec{x} = 0$, $Z\vec{y} = 0$) should be solved. Nontrivial solutions exist by Theorem 4. They must be cross-checked, however, with (28), since the conclusion (iii) of Theorem 7 may fail as well. This search would produce the minimum degree solutions (q_1, q_2) of (15). On the other hand, it may be much simpler to work with the reduced Laurent polynomials $\widetilde{X}, \widetilde{Y}$, which are X and Y, respectively, divided by linear factors that contain symmetric zeros as in parts of the proof of Theorem 4. This will produce solutions (q_1, q_2) of (15), where q_1 and q_2 have as common zeros half of the common symmetric zeros of X and Y. (The other half appears in $q_1(1/z)$ and $q_2(1/z)$ in (15).) Their degrees satisfy the inequalities in Theorem 4. It is not clear, however, that the minimum degree solution (q_1, q_2) of (15) can be obtained in this way.

The following example serves as an illustration of our algorithm.

Example 1. Find Laurent polynomials (q_1, q_2) of minimum degree such that (15) holds, where

$$X(z) = 5(z^6 + z^{-6}) + 14(z^5 + z^{-5}) + 26(z^4 + z^{-4}) + 28(z^3 + z^{-3}) + 49(z^2 + z^{-2}) + 74(z + z^{-1}) + 122,$$

$$Y(z) = 5(z^6 + z^{-6}) + 6(z^5 + z^{-5}) + 10(z^4 + z^{-4}) + 14(z^3 + z^{-3}) + 45(z^2 + z^{-2}) + 16(z + z^{-1}) + 40.$$

Neither X nor Y have symmetric zeros in $\mathbb{C}$. Therefore, the algorithm can be started with $N_X = 6$, $N = 3$ and $\nu = 0$ in (20). The polyphase decomposition reveals

$$A(z) = 5(z^3 + z^{-3}) + 18(z^2 + z^{-2}) + 47(z + z^{-1}) + 81,$$
$$C(z) = 8(z^2 + z^{-2}) + 2(z + z^{-1}) + 41,$$
$$B(z) = 4z^{-3} + 7z^{-2} + 29z^{-1} + 45 + 21z + 10z^2$$

and

$$\Delta(z) = 325 - 150(z + z^{-1}), \qquad N_\Delta = 1.$$

Hence, we obtain $\mu = 2$ in Theorem 4 and know that solutions (q_1, q_2) exist with $\deg q_1 = 6$ and $\deg q_2 \leq 4$. Step 2 of the algorithm can be skipped ($r = 0$ is suitable in Lemma 6). Since Δ has two roots $z_1 = 3/2$, $z_2 = 2/3$, there are 12 choices for $d \in L[0:6]$ in step 3 of the algorithm. They are given by

$$d(z) = z^\ell(15z - 10), \quad d(z) = z^\ell(10z - 15), \quad 0 \leq \ell \leq 5.$$

We demonstrate step 3 of the algorithm for $d(z) = 15z - 10$. Equation (31) leads to a linear system of 10 equations for 12 unknowns

$$\vec{x} = (v_{1,0}, \ldots, v_{1,3}, u_{1,0}, u_{2,0}, \ldots, u_{1,3}, u_{2,3})^t.$$

The last two columns of the reduced row-echelon form of Z are

$$\begin{pmatrix} -4 & 1 & -2 & 0 & -5 & -5 & -3 & -4 & -1 & -2 \\ 2 & 4 & 0 & 0 & 5 & -5 & 4 & 3 & 2 & -1 \end{pmatrix}^t.$$

The patterns described in Lemma 8 can be recognized. The choice $u_{1,N} = 1$, $u_{2,N} = 0$ leads to minimum degree solutions

$$q_1(z) = (5 + 3z^2 + z^4 + z^6) + z(4 - z^2 + 2z^4),$$
$$q_2(z) = (5 + 4z^2 + 2z^4) + z(2 + 4z^2).$$

Since $q_1^2(1) + q_2^2(1) = 514 = X(1)$, q_1 and q_2 possess the correct normalization. We verified that the other cases do not lead to solutions with smaller degree.

Acknowledgment. This research was supported by NSF Grants CCR-9988289 and CCR-0098331 and ARO Grant DAAD19-00-1-0512.

References

1. Chui, C. K., *An Introduction to Wavelets*, Academic Press, Boston, 1992.

2. Chui, C. K., W. He and J. Stöckler, Compactly supported tight and sibling frames with maximum vanishing moments, Appl. Comput. Harmonic Anal., to appear.

3. Chui, C. K., W. He, J. Stöckler and Q. Sun, Compactly supported tight affine frames with integer dilations and maximum vanishing moments, J. Fourier Anal. and Appl., to appear.

4. Daubechies, I., B. Han, A. Ron and Z. Shen, Framelets: MRA-based constructions of wavelet frames, Appl. Comput. Harmonic Anal., to appear.

5. Han, B., On dual wavelet tight frames, Appl. Comput. Harmon. Anal. **4** (1997), 380–413.

6. Hernández, E., and G. Weiss, *A First Course on Wavelets*, CRC Press, Boca Raton, 1996.

7. Jia, R.-Q., and Z. Shen, Multiresolution and wavelets, Proc. Edinburgh Math. Soc. **37** (1994), 271–300.

8. Jia, R.-Q., and J. Z. Wang, Stability and linear independence associated with wavelet decompositions, Proc. Amer. Math. Soc. **117** (1993), 1115–1124.

9. Lawton, W., S. L. Lee and Z. Shen, Stability and orthonormality of multivariable refinable functions, SIAM J. Math. Anal. **28** (1997), 999–1014.

10. Pólya, G., and G. Szegö, *Problems and Theorems in Analysis II*, (reprint ed.), Springer, Heidelberg, 1998.

11. Rheinboldt, W. C., and J. S. Vandergraft, A simple approach to the Perron-Frobenius theory for positive operators on general partially-order finite-dimensional linear spaces, Math. Comp. **27** (1973), 139–145.

12. Ron, A., and Z. W. Shen, Affine systems in $L_2(R^d)$: the analysis of the analysis operator, J. Funct. Anal. **148** (1997), 408–447.

Charles K. Chui and Wenjie He
University of Missouri–St. Louis
Department of Mathematics
St.Louis, MO 63121-4499

Joachim Stöckler
Universität Dortmund
Fachbereich Mathematik
44221 Dortmund, Germany

cchui@stat.stanford.edu   he@neptune.cs.umsl.edu
joachim.stoeckler@math.uni-dortmund.de

Nonstationary Wavelets and Refinement Sequences of Nonuniform B-splines

Charles K. Chui and Jian-ao Lian

Abstract. This paper is devoted to the study of spline refinement and semi-orthogonal wavelets with arbitrary knots, with emphasis on explicit formulas and analytical formulation. For instance, refinement sequences (or masks) are formulated as divided differences with respect to the coarse knots of some truncated polynomials in terms of the fine knots, and B-wavelets are expressed as linear combinations of mth order derivatives of $2m^{\text{th}}$ order B-splines. Illustrative examples are used to demonstrate the elegance of these formulations.

§1. Introduction

For nonuniform simple knot sequences on bounded intervals, Chui and De Villiers [3] considered formulation of decomposition and reconstruction matrices for B-splines and B-wavelets as well as their "duality principle," where one new knot is inserted between two consecutive knots to form the finer knot sequence. To emphasize the generalization from the study of classical cardinal spline wavelets, we will call the minimally supported B-spline wavelets with general knots B-wavelets. Here, general knots mean arbitrary and possibly multiple knots, and that arbitrarily many refinement knots are allowed. B-wavelets were first studied by Lyche and Mørken in [5]. Recently, Lyche, Mørken and Quak [6] extended this study. However, formulation of the decomposition and reconstruction matrices were not discussed in detail.

As usual, to study the multiresolution structure of B-splines with general knots, their **refinement sequences**, which are generalizations of the two-scale sequences for the cardinal spline setting, are of primary importance. For this reason, we will investigate, in this paper, an *explicit algorithm* for calculating the refinement sequences. When the knots are

Approximation Theory X: Wavelets, Splines, and Applications   207
Charles K. Chui, Larry L. Schumaker, and Joachim Stöckler (eds.), pp. 207–229.
Copyright Ⓒ 2002 by Vanderbilt University Press, Nashville, TN.
ISBN 0-8265-1416-2.

nonuniform but simple, an *explicit* expression of these sequences will be explored, and for demonstrative purposes, the corresponding B-wavelets will also be revisited.

To facilitate our presentation, we begin with the necessary notations. Let m be a positive integer that represents the order of spline functions with knot sequences $\mathbf{t}_j = \{t_{j,k}\}$, $j \in \mathbb{Z}$, that satisfy

$$\mathbf{t}_j : \quad \cdots \leq t_{j,-1} \leq t_{j,0} \leq t_{j,1} \leq \cdots, \tag{1}$$

with

$$t_{j,k} < t_{j,k+m}, \quad j, k \in \mathbb{Z}, \tag{2}$$

$$\cdots \subset \mathbf{t}_{-1} \subset \mathbf{t}_0 \subset \mathbf{t}_1 \subset \cdots. \tag{3}$$

Of particular interest are simple and interlaced knot sequences, i.e.,

$$t_{j+1,2k} := t_{j,k}, \tag{4}$$

$$t_{j+1,2k+1} \in (t_{j,k}, t_{j,k+1}), \quad k \in \mathbb{Z}, \tag{5}$$

with typical example $t_{j+1,2k+1} = (t_{j,k} + t_{j,k+1})/2$. In general, new knots can be arbitrarily inserted to $\mathbf{t}_j$ to define $\mathbf{t}_{j+1}$ as long as (2) is satisfied. Let $M_{m,\mathbf{t}_j,k}$ be the mth order normalized B-splines defined by $m+1$ knots $t_{j,k}, \ldots, t_{j,k+m}$, $j, k \in \mathbb{Z}$, i.e.,

$$M_{m,\mathbf{t}_j,k}(t) := (-1)^m (t_{j,k+m} - t_{j,k}) [t_{j,k}, t_{j,k+1}, \ldots, t_{j,k+m}] (t - \bullet)_+^{m-1}. \tag{6}$$

Let $\{p_{m,j,k;\ell}\}_{\ell \in \mathbb{Z}}$ be the **refinement sequences** of the B-splines $M_{m,\mathbf{t}_j,k}$, i.e.,

$$M_{m,\mathbf{t}_j,k}(t) = \sum_{\ell \in \mathbb{Z}} p_{m,j,k;\ell} M_{m,\mathbf{t}_{j+1},\ell}(t), \qquad k \in \mathbb{Z}. \tag{7}$$

For the cardinal (i.e., equally spaced) setting, these sequences are also called **two-scale sequences**.

We are concerned with the calculation and possible explicit formulation of the $p_{m,j,k;\ell}$'s. With general knots, the $p_{m,j,k;\ell}$'s will be established in Section 2. For arbitrary nonuniform but *simple* knots, explicit formulations will be given in Section 3. For practical applications, B-wavelets on a finite interval $[a, b]$ are required. Chui and De Villiers (see Theorem 3.1 and Algorithm 5.1 in [3]) developed an efficient algorithm for constructing the finite reconstruction matrix P whose entries are given by these refinement sequences.

Certainly, the construction matrix P can also be computed efficiently by the Oslo algorithm [4]. On the other hand, B-spline wavelets with equally spaced knots on a finite interval $[a, b]$ can be simply viewed as B-wavelets induced by B-splines with multiple knots at the endpoints a and b. We will formulate B-wavelets on $[a, b]$ in Section 4. Reconstruction and decomposition matrices will be revisited in Section 5. An illustrative example with details will constitute Section 6. B-wavelets with nonuniform but simple interior knots on $[a, b]$ will be also revisited in Section 7, with an illustrative example given in Section 8.

§2. Refinement Sequences of B-splines with Arbitrary Knots

Let $\mathcal{S}_{m,j}$, $j \in \mathbb{Z}$, be the linear spaces of splines of order m and with knot sequences $\mathbf{t}_j$, i.e.,

$$\mathcal{S}_{m,j} := \mathrm{Clos}_{L^2} \mathrm{span}\{M_{m,\mathbf{t}_j,k}(\cdot) : k \in \mathbb{Z}\}, \quad j \in \mathbb{Z}.$$

It is clear from $\mathbf{t}_j \subset \mathbf{t}_{j+1}$ that

$$\mathcal{S}_{m,j} \subset \mathcal{S}_{m,j+1}, \quad j \in \mathbb{Z}.$$

It is also well-known that, with $t_{j,k} = k/2^j$, $j, k \in \mathbb{Z}$, the classical m^{th} order cardinal B-spline $N_m = M_{m,\mathbf{t}_0,0}$ satisfies

$$N_m(t) = \sum_{\ell=0}^{m} \frac{1}{2^{m-1}} \binom{m}{\ell} N_m(2t - \ell).$$

In other words, in terms of $M_{m,\mathbf{t}_j,k}(t) = N_m(2^j t - k)$,

$$M_{m,\mathbf{t}_0,0}(t) = \sum_{\ell \in \mathbb{Z}} p_{m,0,0;\ell} M_{m,\mathbf{t}_1,\ell}(t),$$

with

$$p_{m,0,0;\ell} = \frac{1}{2^m} \binom{m}{\ell}, \quad \ell = 0, \ldots, m,$$

and

$$M_{m,\mathbf{t}_j,k}(t) = \sum_{\ell \in \mathbb{Z}} p_{m,j,k;\ell} M_{m,\mathbf{t}_{j+1},\ell}(t),$$

with

$$p_{m,j,k;\ell} = p_{m,0,0;\ell-2k} = \frac{1}{2^{m-1}} \binom{m}{\ell - 2k}, \quad j, k, \ell \in \mathbb{Z}.$$

However, we are concerned with refinement sequences $\{p_{m,j,k;\ell}\}_{\ell \in \mathbb{Z}}$ of B-splines $M_{m,\mathbf{t}_j,k}$ in (7) with general (i.e., *arbitrary* nonuniform) knot sequences $\mathbf{t}_j$ satisfying (1)–(3). Precisely, we have the following.

Theorem 2.1. *The refinement sequences $p_{m,j,k;\ell}$ in (7) are given by*

$$p_{m,j,k;\ell} = (-1)^m \left(t_{j,k+m} - t_{j,k} \right) [t_{j,k}, \ldots, t_{j,k+m}] \Theta_{j+1,\ell}, \tag{8}$$

with

$$\Theta_{j+1,\ell}(s) = (\tau_{j+1,\ell} - s)_+^0 \prod_{r=1}^{m-1} (t_{j+1,\ell+r} - s), \tag{9}$$

where $\tau_{j+1,\ell} \in (t_{j+1,\ell}, t_{j+1,\ell+m})$, $\ell \in \mathbb{Z}$.

Proof: First, $[t_{j,k}, \ldots, t_{j,k+m}]\, g$ can be written as (page 46, [8])

$$[t_{j,k}, \ldots, t_{j,k+m}]\, g = \sum_{r=k}^{k+m} d_{j,k;r}\, g^{(\ell_{j,k;r})}(t_{j,r}), \tag{10}$$

where

$$\ell_{j,k;r} = \max\{s : r - s \geq k, t_{j,r-s} = t_{j,r}\}, \quad r = k, \ldots, k+m. \tag{11}$$

Secondly, recall the quasi-interpolant (page 178, [2]) Q_j on $\mathcal{S}_{m,j}$,

$$Q_j g := \sum_{k \in \mathbb{Z}} (\lambda_{j,k}\, g) M_{m,\mathbf{t}_j,k}(t), \tag{12}$$

where

$$\lambda_{j,k}\, g = \sum_{r=0}^{m-1} (-1)^{m-1} \Omega_{j,k}^{(m-1-r)}(\tau_{j,k})\, D^r g(\tau_{j,k}), \tag{13}$$

$$\Omega_{j,k}(t) = \frac{1}{(m-1)!}(t_{j,k+1} - t) \cdots (t_{j,k+m-1} - t), \tag{14}$$

$$\tau_{j,k} \in (t_{j,k}, t_{j,k+m}). \tag{15}$$

It is well-known that $\lambda_{j,k}$ are the dual functionals for the B-splines and satisfy

$$\lambda_{j,k}\, M_{m,\mathbf{t}_j,\ell} = \delta_{k,\ell}, \quad \ell \in \mathbb{Z},$$

and that the quasi-interpolant Q_j reproduces $\mathcal{S}_{m,j}$. Hence,

$$(Q_{j+1}\, M_{m,\mathbf{t}_j,k})(t) = M_{m,\mathbf{t}_j,k}(t)$$
$$= \sum_{\ell \in \mathbb{Z}} \left(\lambda_{j+1,\ell} M_{m,\mathbf{t}_j,k} \right) M_{m,\mathbf{t}_{j+1},\ell}(t),$$

so that

$$p_{m,j,k;\ell} = \lambda_{j+1,\ell}\, M_{m,\mathbf{t}_j,k}, \quad j, k, \ell \in \mathbb{Z}. \tag{16}$$

Thirdly, with $\lambda_{j,k}$, $\Omega_{j,k}$, and $\tau_{j,k}$ in (12)–(15), we see that

$$\lambda_{j,k}\, (\bullet - s)_+^{m-1} = (m-1)!\, (\tau_{j,k} - s)_+^0\, \Omega_{j,k}(s). \tag{17}$$

Hence, from (16), (6), and (10)–(11), we have

$$p_{m,j,k;\ell} =$$

$$= (-1)^m\, (t_{j,k+m} - t_{j,k}) \sum_{r=k}^{k+m} d_{j,k;r} \left\{ D_s^{\ell_{j,k;r}} \left[\lambda_{j+1,\ell}\, (\bullet - s)_+^{m-1} \right] \right\} \Big|_{s=t_{j,r}}$$

$$= (-1)^m\, (t_{j,k+m} - t_{j,k}) \sum_{r=k}^{k+m} d_{j,k;r} \left\{ D_s^{\ell_{j,k;r}} \Theta_{j+1,\ell}(s) \right\} \Big|_{s=t_{j,r}}$$

$$= (-1)^m\, (t_{j,k+m} - t_{j,k})[t_{j,k}, \ldots, t_{j,k+m}]\, \Theta_{j+1,\ell},$$

where the identity

$$\lambda_{j+1,\ell} \left[D_s^{\ell_j,k;r} (\bullet - s)_+^{m-1} \right] = D_s^{\ell_j,k;r} \left[\lambda_{j+1,\ell} (\bullet - s)_+^{m-1} \right]$$

and (17) have been used in the first and second equalities, respectively.
□

We end this section by including the following remarks.

Remark 2.2. It is worthwhile to mention another way to consider $p_{m,j,k;\ell}$ in (7). It follows from the **polar forms** [cf., e.g., 7 and 9] of B-splines that $p_{m,j,k;\ell}$ are also given by

$$p_{m,j,k;\ell} = \mathcal{M}_{m,\mathbf{t}_j,k}(t_{j+1,\ell+1}, \ldots, t_{j+1,\ell+m-1}),$$

where $\mathcal{M}_{m,\mathbf{t}_j,k}$ is the polar form of the B-spline $M_{m,\mathbf{t}_j,k}$. In other words, $\mathcal{M}_{m,\mathbf{t}_j,k}(u_1, \ldots, u_{m-1})$ is the *symmetric $(m-1)$-affine* function that satisfies

$$M_{m,\mathbf{t}_j,k}(t) = \mathcal{M}_{m,\mathbf{t}_j,k}(t, \ldots, t).$$

See [9] for the recurrence relation of $\mathcal{M}_{m,\mathbf{t}_j,k}$. Theorem 2.1 reconfirms another expression for $\mathcal{M}_{m,\mathbf{t}_j,k}$, namely [1],

$$\mathcal{M}_{m,\mathbf{t}_j,k}(t_{j+1,\ell+1}, \ldots, t_{j+1,\ell+m-1}) = \lambda_{j+1,\ell} \, M_{m,\mathbf{t}_j,k}.$$

Remark 2.3. Observe that $\lambda_{j+1,\ell}, \ell \in \mathbb{Z}$, are linear functionals on $\mathcal{S}_{m,j+1}$, that satisfy $\lambda_{j+1,k} \, M_{m,\mathbf{t}_{j+1},\ell} = \delta_{k,\ell}$, all $\ell \in \mathbb{Z}$. However, $\lambda_{j+1,k} \, M_{m,\mathbf{t}_j,\ell} \neq \delta_{k,\ell}$ (unless $\mathbf{t}_{j+1} = \mathbf{t}_j$).

Remark 2.4. The sample points $\tau_{j+1,\ell}$ are *chosen* in $(t_{j+1,\ell}, t_{j+1,\ell+m})$. Also, $\lambda_{j+1,\ell} \, g$ is a function of $\tau_{j+1,\ell}$, in terms of the values of g and up to the $(m-1)^{\text{st}}$-order derivatives of g at $\tau_{j+1,\ell}$. However, $p_{m,j,k;\ell}$'s are *independent of* $\tau_{j+1,\ell}$.

Remark 2.5. If the formulation

$$M_{m,\mathbf{t}_j,k}(t) = (t_{j,k+m} - t_{j,k}) \, [t_{j,k}, t_{j,k+1}, \ldots, t_{j,k+m}] \, (\bullet - t)_+^{m-1}$$

is used, then the explicit formula for $p_{m,j,k;\ell}$ in (8)–(9) becomes

$$p_{m,j,k;\ell} = (t_{j,k+m} - t_{j,k})[t_{j,k}, \ldots, t_{j,k+m}] \widetilde{\Theta}_{j+1,\ell}, \qquad (18)$$

with

$$\widetilde{\Theta}_{j+1,\ell}(s) = (s - \tau_{j+1,\ell})_+^0 \prod_{r=1}^{m-1} (t_{j+1,\ell+r} - s), \qquad \ell \in \mathbb{Z}. \qquad (19)$$

Hence, $p_{m,j,k;\ell} = 0$ follows from (8)–(9) and (18)–(19) provided $\tau_{j+1,\ell} \notin (t_{j,k}, t_{j,k+m})$.

Remark 2.6. The formulation of $p_{m,j,k;\ell}$ in (8)–(9) can be applied to general knot sequences satisfying (1)–(3). Observe also that the mth order divided difference

$$[t_1, \ldots, t_{1+m}] \left\{ (a - \bullet)^0_+ g(\bullet) \right\}$$

is determined by the "partial sum" of the function and possible derivative values of g, depending on where the location of a is, as will be demonstrated in the following section.

§3. Explicit Formulas for Refinement Sequences

In this section, we establish the *explicit* expression of $\{p_{m,j,k;\ell}\}_{\ell \in \mathbb{Z}}$ in (7) for B-splines with nonuniform simple knot sequences that satisfy (4)–(5).

Theorem 3.1. *Let* $\mathbf{t}_j \subset \mathbf{t}_{j+1}$ *be nonuniform simple knot sequences satisfying* (1) (*with* $\leq$ *replaced by* $<$) *and* (2)–(3), *and generated via* (4)–(5). *Then the refinement sequences* $\{p_{m,j,k;\ell}\}_{\ell \in \mathbb{Z}}, k \in \mathbb{Z}$, *of* B-*splines* $M_{m,\mathbf{t}_j,k}$ *in* (7) *are given explicitly by*

$$p_{m,j,k;2k+\ell} = (t_{j+1,2k+2m} - t_{j+1,2k})$$

$$\times \sum_{r=0}^{\lfloor \ell/2 \rfloor} \frac{\displaystyle\prod_{n=1}^{m-1} (t_{j+1,2k+\ell+n} - t_{j+1,2k+2r})}{\displaystyle\prod_{\eta=0, \eta \neq r}^{m} (t_{j+1,2k+2\eta} - t_{j+1,2k+2r})}, \quad \ell = 0, \ldots, \left\lfloor \frac{m}{2} \right\rfloor, \quad (20)$$

$$p_{m,j,k;2k+m-\ell} = (t_{j+1,2k+2m} - t_{j+1,2k})$$

$$\times \sum_{r=0}^{\lfloor \ell/2 \rfloor} \frac{\displaystyle\prod_{n=1}^{m-1} (t_{j+1,2k+2m-2r} - t_{j+1,2k+m-\ell+n})}{\displaystyle\prod_{\eta=0, \eta \neq m-r}^{m} (t_{j+1,2k+2m-2r} - t_{j+1,2k+2\eta})},$$

$$\ell = 0, \ldots, \left\lfloor \frac{m-1}{2} \right\rfloor. \quad (21)$$

Remark 3.2. Observe that, by Theorem 3.1, the refinement sequence of the cardinal B-splines N_m can be recovered through the identity

$$\sum_{\xi=0}^{\lfloor \ell/2 \rfloor} (-1)^\xi \binom{m}{\xi} \binom{m + \ell - 2\xi - 1}{\ell - 2\xi} = \binom{m}{\ell}, \quad \ell = 0, \ldots, m.$$

Proof of Theorem 3.1. Notice first that, if $t_{j,k} < \cdots < t_{j,k+m}$, then

$$[t_{j,k}, \ldots, t_{j,k+m}] \left\{ (a - \bullet)^0_+ \, g(\bullet) \right\} = \sum_{r=0}^{m} \frac{(a - t_{j,k+r})^0_+ \, g(t_{j,k+r})}{\displaystyle\prod_{\eta=0, \eta \neq r}^{m} (t_{j,k+r} - t_{j,k+\eta})}. \qquad (22)$$

Hence, it is clear that $p_{m,j,k;\ell} = 0$, if $\ell < 2k$ or $\ell > 2k + m$. Next, for $\ell = 0, \ldots, m$, if we choose

$$\tau_{j+1,2k+\ell} = t_{j+1,2k+\ell+1}, \quad \ell = 0, \ldots, m,$$

then it follows from (22) that

$$[t_{j,k}, \ldots, t_{j,k+m}] \left\{ (\tau_{j+1,2k+\ell} - \bullet)^0_+ \prod_{n=1}^{m-1} (t_{j+1,2k+\ell+n} - \bullet) \right\}$$

$$= \sum_{r=0}^{m} \frac{(t_{j+1,2k+\ell+1} - t_{j,k+r})^0_+ \displaystyle\prod_{n=1}^{m-1} (t_{j+1,2k+\ell+n} - t_{j,k+r})}{\displaystyle\prod_{\eta=0, \eta \neq r}^{m} (t_{j,k+r} - t_{j,k+\eta})}$$

$$= \sum_{r=0}^{\lfloor \ell/2 \rfloor} \frac{\displaystyle\prod_{n=1}^{m-1} (t_{j+1,2k+\ell+n} - t_{j+1,2k+2r})}{\displaystyle\prod_{\eta=0, \eta \neq r}^{m} (t_{j+1,2k+2r} - t_{j+1,2k+2\eta})}$$

$$= (-1)^{m-1} \sum_{r=0}^{\lfloor \ell/2 \rfloor} \frac{\displaystyle\prod_{n=1}^{m-1} (t_{j+1,2k+\ell+n} - t_{j+1,2k+2r})}{\displaystyle\prod_{\eta=0, \eta \neq r}^{m} (t_{j+1,2k+2\eta} - t_{j+1,2k+2r})}.$$

Therefore, by applying (8) in Theorem 2.1 and using the symmetry of B-splines with respect to their knots, the formulas (20)–(21) of $p_{m,j,k;\ell}$, $\ell \in \mathbb{Z}$, are obtained. $\square$

As a simple example, for $m = 2$, it is clear from (20)–(21) that

$$M_{2,\mathbf{t}_j,k}(t) = \sum_{\ell=0}^{2} p_{2,j,k;2k+\ell} M_{2,\mathbf{t}_{j+1},2k+\ell}(t),$$

where

$$p_{2,j,k;2k} = \frac{t_{j+1,2k+1} - t_{j+1,2k}}{t_{j+1,2k+2} - t_{j+1,2k}},$$

$$p_{2,j,k;2k+1} = 1,$$

$$p_{2,j,k;2k+2} = \frac{t_{j+1,2k+4} - t_{j+1,2k+3}}{t_{j+1,2k+4} - t_{j+1,2k+2}}, \quad k \in \mathbb{Z}.$$

Similarly, for $m = 4$,

$$M_{4,\mathbf{t}_j,k}(t) = \sum_{\ell=0}^{4} p_{4,j,k;2k+\ell} M_{3,\mathbf{t}_{j+1},2k+\ell}(t),$$

where

$$p_{4,j,k;2k} = \frac{(t_{j+1,2k+1} - t_{j+1,2k})(t_{j+1,2k+3} - t_{j+1,2k})}{(t_{j+1,2k+4} - t_{j+1,2k})(t_{j+1,2k+6} - t_{j+1,2k})},$$

$$p_{4,j,k;2k+1} = \frac{t_{j+1,2k+3} - t_{j+1,2k}}{t_{j+1,2k+6} - t_{j+1,2k}},$$

$$\begin{aligned}
p_{4,j,k;2k+2} &= \frac{t_{j+1,2k+8} - t_{j+1,2k}}{t_{j+1,2k+2} - t_{j+1,2k}} \\
&\quad \times \left[\frac{(t_{j+1,2k+3} - t_{j+1,2k})(t_{j+1,2k+5} - t_{j+1,2k})}{(t_{j+1,2k+6} - t_{j+1,2k})(t_{j+1,2k+8} - t_{j+1,2k})} \right. \\
&\qquad\quad \left. - \frac{(t_{j+1,2k+3} - t_{j+1,2k+2})(t_{j+1,2k+5} - t_{j+1,2k+2})}{(t_{j+1,2k+6} - t_{j+1,2k+2})(t_{j+1,2k+8} - t_{j+1,2k+2})} \right] \\
&= \frac{t_{j+1,2k+8} - t_{j+1,2k}}{t_{j+1,2k+8} - t_{j+1,2k+6}} \\
&\quad \times \left[\frac{(t_{j+1,2k+8} - t_{j+1,2k+3})(t_{j+1,2k+8} - t_{j+1,2k+5})}{(t_{j+1,2k+8} - t_{j+1,2k})(t_{j+1,2k+8} - t_{j+1,2k+2})} \right. \\
&\qquad\quad \left. - \frac{(t_{j+1,2k+6} - t_{j+1,2k+3})(t_{j+1,2k+6} - t_{j+1,2k+5})}{(t_{j+1,2k+6} - t_{j+1,2k})(t_{j+1,2k+6} - t_{j+1,2k+2})} \right],
\end{aligned}$$

$$p_{4,j,k;2k+3} = \frac{t_{j+1,2k+8} - t_{j+1,2k+5}}{t_{j+1,2k+8} - t_{j+1,2k+2}},$$

$$p_{4,j,k;2k+4} = \frac{(t_{j+1,2k+8} - t_{j+1,2k+5})(t_{j+1,2k+8} - t_{j+1,2k+7})}{(t_{j+1,2k+8} - t_{j+1,2k+2})(t_{j+1,2k+8} - t_{j+1,2k+4})}.$$

§4. B-wavelets on $[a, b]$ as a Basis of Nonstationary Wavelets

In this section, we consider B-wavelets only with knot sequences on a finite interval $[a, b]$, so that all the spline and wavelet subspaces are finite dimensional. In addition, for each $j \in \mathbb{Z}_+$, the knot sequence $\mathbf{t}_j$ is given by

$$\mathbf{t}_j : \quad a = t_{j,1} = \cdots = t_{j,m} < t_{j,m+1} \leq t_{j,m+2} \leq \cdots$$
$$\leq t_{j,m+K_j} < t_{j,m+K_j+1} = \cdots = t_{j,K_j+2m} = b, \quad (23)$$

with

$$t_{j,k} < t_{j,k+m}, \qquad k = 1, \ldots, m + K_j. \quad (24)$$

A simple reason for studying finite knot sequences is clear from the following simple situation. Consider $\mathbf{t}_0$ as a bi-infinite nonuniform knot sequence, and $\mathbf{t}_1 = \mathbf{t}_0 \cup \{t_{1,\ell}\}$, where $t_{1,\ell} \in (t_{0,k}, t_{0,k+1})$ for some $k \in \mathbb{Z}$ and $t_{0,k} < t_{0,k+1}$. In other words, we only insert one simple knot to $\mathbf{t}_0$ to form $\mathbf{t}_1$. Certainly, the wavelet subspace W_0 of $\mathcal{S}_{2,1}$ relative to $\mathcal{S}_{2,0}$ is a one-dimensional space generated by a *single* function ψ. However, such a function ψ must be *infinitely supported* if all knots of $\mathbf{t}_0$ are simple. It could be compactly supported only if there are knots with multiplicity 2 to the left and right of $t_{1,\ell}$.

It follows from (23)–(24) that

$$\dim \mathcal{S}_{m,j} = m + K_j, \quad j \in \mathbb{Z}_+, \quad (25)$$

and, indeed,

$$\mathcal{S}_{m,j} = \mathrm{Clos}_{L^2} \mathrm{span} \{ M_{m,\mathbf{t}_j,k} : \quad k = 1, \ldots, m + K_j \}, \quad j \in \mathbb{Z}_+. \quad (26)$$

Certainly, it follows from $\mathbf{t}_0 \subset \mathbf{t}_1 \subset \mathbf{t}_2 \subset \cdots$ that $K_j \leq K_{j+1}$, $j \in \mathbb{Z}_+$. As usual, the L^2-orthogonal complementary subspace W_j of $\mathcal{S}_{m,j+1}$ relative to $\mathcal{S}_{m,j}$ is called the wavelet subspace, which has dimension $K_{j+1} - K_j$ and is defined by

$$\mathcal{S}_{m,j+1} = \mathcal{S}_{m,j} \oplus W_j, \quad j \in \mathbb{Z}_+. \quad (27)$$

For $K_j < K_{j+1}$, we investigate those $K_{j+1} - K_j$ minimally supported nontrivial spline functions, namely, B-wavelets, in $\mathcal{S}_{m,j+1}$ that span W_j. To begin with, we need additional notations. Denote by $m_\mathbf{x}(t)$ the multiplicity of t in a knot sequence $\mathbf{x}$, e.g., $m_{\mathbf{t}_j}(a) = m_{\mathbf{t}_j}(b) = m$, for each $j \in \mathbb{Z}_+$. Let

$$d_{j,k}^L = \max\{\eta \geq 1 : t_{j,k} = t_{j,k-\eta+1}\}, \quad (28)$$
$$d_{j,k}^R = \max\{\eta \geq 1 : t_{j,k} = t_{j,k+\eta-1}\}, \quad (29)$$

be the left- and right-multiplicities of $t_{j,k}$ in $\mathbf{t}_j$, respectively. In $\mathcal{S}_{m,j+1}$, all knots in $\mathbf{t}_j$ will be called old knots, and those in $\mathbf{t}_{j+1} \setminus \mathbf{t}_j$ will be named new knots. These new knots in $\mathbf{t}_{j+1}$ are inserted to $\mathbf{t}_j$ *from the right*. Hence, $t_{j+1,k}$ is a new knot if and only if $d_{j+1,k}^L > m_{\mathbf{t}_j}(t_{j+1,k})$. Define the index sets of all new and old knots by

$$I_j^{\text{new}} = \{k : m+1 \le k \le m + K_{j+1}, d_{j+1,k}^L > m_{\mathbf{t}_j}(t_{j+1,k})\}, \quad (30)$$

$$I_j^{\text{old}} = \{1, \cdots, K_{j+1} + 2m\} \setminus I_j^{\text{new}}. \quad (31)$$

Clearly, there are $K_{j+1} - K_j$ new knots in $\mathcal{S}_{m,j+1}$, so that

$$\operatorname{card} I_j^{\text{new}} = K_{j+1} - K_j.$$

If, for all $k \in I_j^{\text{new}}$, some linearly independent and minimally supported functions $\psi_{m,j,k}$, $k \in I_j^{\text{new}}$, are identified, W_j will be generated by these $K_{j+1} - K_j$ splines, to be called B-wavelets. To this end, we write

$$\psi_{m,j,k}(t) = \sum_{\xi=\ell_{j,k}}^{r_{j,k}-m} q_{m,j,k;\xi} M_{m,\mathbf{t}_{j+1},\xi}(t), \quad (32)$$

$$q_{m,j,k;\ell_{j,k}} \, q_{m,j,k;r_{j,k}-m} \ne 0, \qquad k \in I_j^{\text{new}}. \quad (33)$$

We call $\{q_{m,j,k;\xi}\}_{\xi \in \mathbf{Z}}$ the two-level sequence or inter-level sequence of $\psi_{m,j,k}$ from $\mathcal{S}_{m,j}$ to $\mathcal{S}_{m,j+1}$. Here, the values of $\ell_{j,k}$ and $r_{j,k}$ are determined to be the greatest and smallest integers so that there are $m - d_{j+1,\ell_{j,k}}^R$ and $m - d_{j+1,r_{j,k}}^L$ new knots on $(t_{j+1,\ell_{j,k}}, t_{j+1,k}]$ and $[t_{j+1,k}, t_{j+1,r_{j,k}})$, respectively, i.e.,

$$\ell_{j,k} = \max \Big\{ \eta : \ d_{j+1,\eta}^R$$

$$+ \operatorname{card}\{\xi \in I_j^{\text{new}} : \ \xi < k, \ t_{j+1,\eta} < t_{j+1,\xi} \le t_{j+1,k}\} = m \Big\}, \quad (34)$$

$$r_{j,k} = \min \Big\{ \eta : \ d_{j+1,\eta}^L$$

$$+ \operatorname{card}\{\xi \in I_j^{\text{new}} : \ \xi > k, \ t_{j+1,k} \le t_{j+1,\xi} < t_{j+1,\eta}\} = m \Big\}. \quad (35)$$

Observe that, since $d_{j+1,1}^R = d_{j+1,K_{j+1}+2m}^L = m$, it is always possible to find $\ell_{j,k}, r_{j,k} \in \{1, \ldots, K_{j+1} + 2m\}$, even if there is no new knot at all to the left (or to the right) of $t_{j+1,k}$. Observe also that, since $\ell_{j,k} \le k - m \le r_{j,k} - 2m$, there are $r_{j,k} - \ell_{j,k} - 2m$ interior old knots on $(t_{j+1,\ell_{j,k}}, t_{j+1,r_{j,k}})$, counting multiplicities. Assume that the first old knot is t_{j,k_1}, with k_1 determined from the value $d_{j,k_1}^L = 1$, then these $r_{j,k} - \ell_{j,k} - 2m$ old knots are $t_{j,k_1}, t_{j,k_1+1}, \ldots, t_{j,k_1+r_{j,k}-\ell_{j,k}-2m-1}$. By introducing a function of t from $N+1$ sufficiently continuous functions $u_0, \ldots, u_N$ and $\mu+1$ pairwise distinct real numbers $\tau_0, \ldots, \tau_\mu$, via the determinant of the collocation matrix, namely,

$$
D \left(\begin{array}{c} u_0(\bullet), u_1(\bullet), \cdots, u_N(\bullet) \\ t, \underbrace{\tau_0, \ldots, \tau_0}_{\ell_0}, \ldots, \underbrace{\tau_\mu, \ldots, \tau_\mu}_{\ell_\mu} \end{array} \right) ,
$$

where $N = \sum_{j=0}^{\mu} \ell_j$, we have the following.

Proposition 4.1. *Let* $\mathcal{S}_{m,j}$ *and* $\mathcal{S}_{m,j+1}$ *be the* m^{th} *order spline spaces with knot sequences* $\mathbf{t}_j$ *and* $\mathbf{t}_{j+1}$ *in* (23)–(24) *such that* $\mathbf{t}_j \subset \mathbf{t}_{j+1}$. *Let* W_j *be the* L^2-*orthogonal complementary subspace of* $\mathcal{S}_{m,j+1}$ *relative to* $\mathcal{S}_{m,j}$. *Then* W_j *can be generated by* $K_{j+1} - K_j$ *B-wavelets* $\psi_{m,j,k}, k \in I_j^{\text{new}}$, *which, up to constant multiples, are given by*

$$
\psi_{m,j,k}(t) =
$$
$$
\frac{d^m}{dt^m} D \left(\begin{array}{cccc} M_{2m,\mathbf{t}_{j+1},\ell_{j,k}}, & M_{2m,\mathbf{t}_{j+1},\ell_{j,k}+1}, & \cdots, & M_{2m,\mathbf{t}_{j+1},r_{j,k}-2m} \\ t, & t_{j,k_1}, & \cdots, & t_{j,k_1+r_{j,k}-\ell_{j,k}-2m-1} \end{array} \right), \quad (36)
$$

where $t_{j,k_1} \in (t_{j+1,\ell_{j,k}}, t_{j+1,r_{j,k}})$ *is the first old knot that satisfies* $d_{j,k_1}^L = 1$.

Remark 4.2. This successful construction of B-wavelets is based on the following fact. If $\psi \in W_j$ and $\psi = \Psi^{(m)}$, where $\Psi \in \mathcal{S}_{2m,j}$, then we have $\Psi \in C^{m-1}(\mathbb{R})$ and that, for $k \in \mathbb{Z}$,

$$
0 = \langle M_{m,\mathbf{t}_j,k}, \psi \rangle = \langle M_{m,\mathbf{t}_j,k}, \Psi^{(m)} \rangle \tag{37}
$$
$$
= \sum_{i=0}^{m-1} \int_{t_{j,k+i}}^{t_{j,k+i+1}} M_{m,\mathbf{t}_j,k}(t) \, \Psi^{(m)}(t) \, dt
$$
$$
= \sum_{i=0}^{m-1} \sum_{\ell=0}^{m-1} (-1)^\ell M_{m,\mathbf{t}_j,k}^{(\ell)}(t) \, \Psi^{(m-1-\ell)}(t) \Big|_{t_{j,k+i}}^{t_{j,k+i+1}}
$$
$$
= \sum_{\ell=0}^{m-1} (-1)^\ell \sum_{i=1}^{m-1} \left[M_{m,\mathbf{t}_j,k}^{(\ell)}(t_{j,k+i}+) - M_{m,\mathbf{t}_j,k}^{(\ell)}(t_{j,k+i}-) \right] \Psi^{(m-1-\ell)}(t_{j,k+i}),
$$

due to the facts that

$$
M_{m,\mathbf{t}_j,k}^{(\ell)}(t_{j,k}-) = 0, \quad M_{m,\mathbf{t}_j,k}^{(\ell)}(t_{j,k+m}+) = 0, \quad \ell = 0, \ldots, m-1,
$$

where $f^{(\ell)}(t+)$ and $f^{(\ell)}(t-)$ denote the ℓth order left and right side derivatives of f at t. Hence, by the definition or continuity conditions of $M_{m,\mathbf{t}_j,k}$,

$$
M_{m,\mathbf{t}_j,k}^{(\ell)}(t_{j,k+i}+) - M_{m,\mathbf{t}_j,k}^{(\ell)}(t_{j,k+i}-) = 0,
$$
$$
\ell = 0, \ldots, m-1 - m_{\mathbf{t}_j}(t_{j,k+i}), \quad i = 1, \ldots, m-1,
$$

the equalities in (37) are satisfied if $\Psi \in S_{2m,j}$ is interpolatory, namely,

$$\Psi^{(\ell)}(t_{j,k+i}) = 0, \quad \ell = 0, \ldots, m_{\mathbf{t}_j}(t_{j,k+i}) - 1, \quad i = 1, \ldots, m - 1.$$

Remark 4.3. The formulation (36) involves the mth order derivatives of $(2m)^{\text{th}}$ order B-splines. To get the two-level sequences $\{q_{m,j,k;\xi}\}_{\xi \in \mathbb{Z}}$ from $S_{m,j}$ to $S_{m,j+1}$, as in (32), recall that, with a knot sequence $\mathbf{u} = \{u_k\}_{k \in \mathbb{Z}}$, the B-spline $M_{m,\mathbf{u},k}$ has derivative

$$
\begin{aligned}
&M'_{m,\mathbf{u},k}(t) \\
&= \frac{m-1}{u_{k+m-1} - u_k} M_{m-1,\mathbf{u},k}(t) - \frac{m-1}{u_{k+m} - u_{k+1}} M_{m-1,\mathbf{u},k+1}(t), \quad (38)
\end{aligned}
$$

if $u_k < u_{k+m-1}$ and $u_{k+1} < u_{k+m}$. Since $\mathbf{t}_{j+1}$ satisfies (23)–(24), it is safe to apply (38) m times to get $M^{(m)}_{2m,\mathbf{t}_{j+1},k}$ in terms of $M_{m,\mathbf{t}_{j+1},k}, \ldots,$ $M_{m,\mathbf{t}_{j+1},k+m}$, namely,

$$M^{(m)}_{2m,\mathbf{t}_{j+1},k}(t) = \frac{(2m-1)!}{(m-1)!} \sum_{\ell=0}^{m} (-1)^{\ell} \alpha_{k,\ell} M_{m,\mathbf{t}_{j+1},k+\ell}(t), \quad (39)$$

where $\alpha_{k,\ell} > 0$, $\ell = 0, \ldots, m$, and, in particular, $\alpha_{k,0}$ and $\alpha_{k,m}$ are determined from

$$\frac{1}{\alpha_{k,0}} = \prod_{\ell=1}^{m} (t_{j+1,k+2m-\ell} - t_{j+1,k}), \quad \frac{1}{\alpha_{k,m}} = \prod_{\ell=1}^{m} (t_{j+1,k+2m} - t_{j+1,k+\ell}).$$

In general, the $m + 1$ coefficients $\alpha_{k,\ell}$ in (39) can be computed via the following simple algorithm.

Algorithm 4.4.

1) $\alpha(0,0) = 1$.
2) For $p = 1, \ldots, m$, evaluate

$$
\begin{aligned}
\alpha(p,0) &= \frac{\alpha(p-1,0)}{t_{j+1,k+2m-p} - t_{j+1,k}}, \\
\alpha(p,q) &= \frac{\alpha(p-1,q-1) + \alpha(p-1,q)}{t_{j+1,k+2m-p+q} - t_{j+1,k+q}}, \qquad q = 1, \ldots, p-1, \\
\alpha(p,p) &= \frac{\alpha(p-1,p-1)}{t_{j+1,k+2m} - t_{j+1,k+p}}.
\end{aligned}
$$

3) $\alpha_{k,\ell} = \alpha(m,\ell)$, $\ell = 0, \ldots, m$.

§5. Reconstruction and Decomposition Matrices

With the spline spaces $\mathcal{S}_{m,j}$ introduced in (26), and with general knot sequences that satisfy (23)–(24), all refinement equations can be rewritten into matrix forms, i.e.,

$$\mathbf{M}_j = P_j\,\mathbf{M}_{j+1}, \tag{40}$$

$$\mathbf{\Psi}_j = Q_j\,\mathbf{M}_{j+1}, \tag{41}$$

where, for each $j \in \mathbb{Z}_+$, $\mathbf{M}_j$ is the vector of B-splines that span $\mathcal{S}_{m,j}$, namely,

$$\mathbf{M}_j = \left(M_{m,\mathbf{t}_j,1}, \cdots, M_{m,\mathbf{t}_j,m+K_j} \right)^\top, \quad j \in \mathbb{Z}_+, \tag{42}$$

and P_j, Q_j are the $(m+K_j) \times (m+K_{j+1})$ and $(K_{j+1}-K_j) \times (m+K_{j+1})$ reconstruction matrices, respectively. The matrix P_j is composed of the refinement sequences $\{p_{m,j,k;\ell}\}_{\ell=1}^{m+K_{j+1}}$ of the B-splines $M_{m,\mathbf{t}_j,k}$, as in (7), i.e.,

$$P_j = \left(p_{m,j,k;\ell} \right)_{k=1,\ldots,m+K_j;\ \ell=1,\ldots,m+K_{j+1}}, \tag{43}$$

for which Theorem 2.1 can be applied. Analogously, the matrix Q_j is composed of the two-level sequences $\{q_{m,j,k;\ell}\}_{\ell=1}^{m+K_{j+1}}$ of the B-wavelets $\psi_{m,j,k}$, as in (32)–(33), i.e.,

$$Q_j = \left(q_{m,j,k;\ell} \right)_{k=1,\ldots,K_{j+1}-K_j;\ \ell=1,\ldots,m+K_{j+1}}. \tag{44}$$

On the other hand, $\mathbf{M}_{j+1}$ can be written as

$$\mathbf{M}_{j+1} = A_j\,\mathbf{M}_j + B_j\,\mathbf{\Psi}_j, \tag{45}$$

where A_j, B_j are the $(m+K_{j+1}) \times (m+K_j)$ and $(m+K_{j+1}) \times (K_{j+1}-K_j)$ decomposition matrices, respectively. Due to the linear independence of components of both $\mathbf{M}_j$ and $\mathbf{\Psi}_j$, it follows from (40)–(41) and (45) that

$$P_j\,A_j = I_{m+K_j}, \qquad P_j\,B_j = 0, \tag{46}$$

$$Q_j\,A_j = 0, \qquad Q_j\,B_j = I_{K_{j+1}-K_j}, \tag{47}$$

$$A_j\,P_j + B_j\,Q_j = I_{m+K_{j+1}}, \tag{48}$$

where I_k is the identity matrix of order k. Hence,

$$\begin{pmatrix} P_j \\ Q_j \end{pmatrix} \begin{pmatrix} A_j & B_j \end{pmatrix} = I_{m+K_{j+1}}, \tag{49}$$

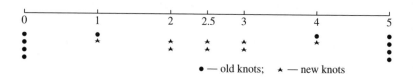

Fig. 1. Old and new knots distribution of $\mathcal{S}_{4,0}$ and $\mathcal{S}_{4,1}$.

which can be applied to finding A_j and B_j when P_j and Q_j are given. However, the formulation of the reconstruction matrix Q_j involves the Gramian matrix of $\mathbf{M}_{j+1}$. To this end, and to distinguish from the usual inner or *angular* product on $L^2[a,b]$, i.e., $\langle f, g \rangle = \int_a^b f(t)\overline{g}(t)dt$, we define the circular product of two vector-valued functions $\mathbf{f} = (f_1, \cdots, f_{r_1})^\top \in (L^2[a,b])^{r_1}$ and $\mathbf{g} = (g_1, \cdots, g_{r_2})^\top \in (L^2[a,b])^{r_2}$, as the following $r_1 \times r_2$ matrix

$$(\mathbf{f}, \mathbf{g}) := (\langle f_j, g_k \rangle)_{j=1,\ldots,r_1;\, k=1,\ldots,r_2}, \tag{50}$$

with a slight abuse of notation. Then, both the Gramian matrices of $\mathbf{M}_j$ and $\boldsymbol{\Psi}_j$, i.e., $(\mathbf{M}_j, \mathbf{M}_j)$ and $(\boldsymbol{\Psi}_j, \boldsymbol{\Psi}_j)$, are positive definite for each $j \in \mathbb{Z}_+$. In addition, $\mathcal{S}_{m,j} \perp W_j \Leftrightarrow (\mathbf{M}_j, \boldsymbol{\Psi}_j) = P_j (\mathbf{M}_{j+1}, \mathbf{M}_{j+1}) Q_j^\star = 0$, where $\star$ denotes the transpose of complex conjugation. Together with (40)–(41), we have the following identity

$$\begin{pmatrix} P_j \\ Q_j \end{pmatrix} (\mathbf{M}_{j+1}, \mathbf{M}_{j+1}) \begin{pmatrix} P_j^\star & Q_j^\star \end{pmatrix} = \begin{pmatrix} (\mathbf{M}_j, \mathbf{M}_j) & 0 \\ 0 & (\boldsymbol{\Psi}_j, \boldsymbol{\Psi}_j) \end{pmatrix}. \tag{51}$$

Without giving too many details, we will illustrate the formulation in the following section by a simple example.

§6. An Illustrative Example

To illustrate our formulation, we consider the following simple example. Let $m = 4$ and

$$\mathbf{t}_0 = \{t_{0,1}, \ldots, t_{0,10}\} = \{0, 0, 0, 0, 1, 4, 5, 5, 5, 5\},$$
$$\mathbf{t}_1 = \{t_{1,1}, \ldots, t_{1,18}\} = \{0, 0, 0, 0, 1, 1, 2, 2, 2.5, 2.5, 3, 3, 4, 4, 5, 5, 5, 5\},$$

be two general knot sequences. See Fig. 1. Then $K_0 = 2, K_1 = 10$, and $\mathcal{S}_{4,1} = \mathcal{S}_{4,0} \oplus W_0$ with dim $W_0 = 8$. First,

$$\mathbf{M}_0 = \left(M_{4,\mathbf{t}_0,1}, \cdots, M_{4,\mathbf{t}_0,6} \right)^\top, \quad \mathbf{M}_1 = \left(M_{4,\mathbf{t}_1,1}, \cdots, M_{4,\mathbf{t}_1,14} \right)^\top,$$

and the reconstruction matrices P_0, Q_0 are 6×14 and 8×14, respectively.

6.1. Reconstruction matrix P_0

It is clear from (43) that $P_0 = \left(p_{4,0,k;\ell}\right)_{k=1,\ldots,6;\,\ell=1,\ldots,14}$, with

$$p_{4,0,k;\ell} = (t_{0,k+4} - t_{0,k}) \left[t_{0,k}, \ldots, t_{0,k+4}\right] (\tau_{1,\ell} - \cdot)_+^0 \prod_{r=1}^{3} (t_{1,\ell+r} - \cdot),$$

and $\tau_{1,\ell} \in (t_{1,\ell}, t_{1,\ell+4})$. More explicitly,

$$p_{4,0,1;\ell} = \left[0,0,0,0,1\right] (\tau_{1,\ell} - \cdot)_+^0 \prod_{r=1}^{3} (t_{1,\ell+r} - \cdot),$$

$$p_{4,0,2;\ell} = 4\left[0,0,0,1,4\right] (\tau_{1,\ell} - \cdot)_+^0 \prod_{r=1}^{3} (t_{1,\ell+r} - \cdot),$$

$$p_{4,0,3;\ell} = 5\left[0,0,1,4,5\right] (\tau_{1,\ell} - \cdot)_+^0 \prod_{r=1}^{3} (t_{1,\ell+r} - \cdot),$$

$$p_{4,0,4;\ell} = 5\left[0,1,4,5,5\right] (\tau_{1,\ell} - \cdot)_+^0 \prod_{r=1}^{3} (t_{1,\ell+r} - \cdot),$$

$$p_{4,0,5;\ell} = 4\left[1,4,5,5,5\right] (\tau_{1,\ell} - \cdot)_+^0 \prod_{r=1}^{3} (t_{1,\ell+r} - \cdot),$$

$$p_{4,0,6;\ell} = \left[4,5,5,5,5\right] (\tau_{1,\ell} - \cdot)_+^0 \prod_{r=1}^{3} (t_{1,\ell+r} - \cdot),$$

with $\tau_{1,\ell} \in (t_{1,\ell}, t_{1,\ell+4})$, $\ell = 1, \ldots, 14$, which leads to

$$P_0 = \begin{pmatrix} 1 & 0 & 0 & 0 & 0 & 0 & 0 & 0 & 0 & 0 & 0 & 0 & 0 & 0 \\ 0 & 1 & \frac{3}{4} & \frac{3}{8} & \frac{1}{4} & \frac{1}{8} & \frac{3}{32} & \frac{3}{64} & \frac{1}{32} & 0 & 0 & 0 & 0 & 0 \\ 0 & 0 & \frac{1}{4} & \frac{21}{40} & \frac{11}{20} & \frac{1}{2} & \frac{15}{32} & \frac{25}{64} & \frac{11}{32} & \frac{1}{5} & \frac{1}{10} & 0 & 0 & 0 \\ 0 & 0 & 0 & \frac{1}{10} & \frac{1}{5} & \frac{11}{32} & \frac{25}{64} & \frac{15}{32} & \frac{1}{2} & \frac{11}{20} & \frac{21}{40} & \frac{1}{4} & 0 & 0 \\ 0 & 0 & 0 & 0 & 0 & \frac{1}{32} & \frac{3}{64} & \frac{3}{32} & \frac{1}{8} & \frac{1}{4} & \frac{3}{8} & \frac{3}{4} & 1 & 0 \\ 0 & 0 & 0 & 0 & 0 & 0 & 0 & 0 & 0 & 0 & 0 & 0 & 0 & 1 \end{pmatrix}.$$

6.2. Reconstruction matrix Q_0

Notice first that, intuitively,

$$I_0^{\text{new}} = \{6, 7, 8, 9, 10, 11, 12, 14\}. \tag{52}$$

Indeed, I_0^{new} can also be obtained from (30) and (28), since

$$\begin{aligned}
d_{1,6}^L &= 2 > m_{t_0}(t_{1,6}) = 0, & d_{1,10}^L &= 2 > m_{t_0}(t_{1,10}) = 0, \\
d_{1,7}^L &= 1 > m_{t_0}(t_{1,7}) = 0, & d_{1,11}^L &= 1 > m_{t_0}(t_{1,11}) = 0, \\
d_{1,8}^L &= 1 > m_{t_0}(t_{1,8}) = 0, & d_{1,12}^L &= 2 > m_{t_0}(t_{1,12}) = 0, \\
d_{1,9}^L &= 2 > m_{t_0}(t_{1,9}) = 0, & d_{1,14}^L &= 2 > m_{t_0}(t_{1,14}) = 0,
\end{aligned}$$

while $d_{1,5}^L = d_{1,13}^L = 1 = m_{t_0}(t_{1,5}) = m_{t_0}(t_{1,13})$. For each $j \in I_0^{\text{new}}$ in (52), we have, by applying (34)–(35), that

$$\begin{aligned}
\ell_{0,6} &= 1, & r_{0,6} &= 10; & \ell_{0,10} &= 6, & r_{0,10} &= 14; \\
\ell_{0,7} &= 2, & r_{0,7} &= 11; & \ell_{0,11} &= 7, & r_{0,11} &= 16; \\
\ell_{0,8} &= 3, & r_{0,8} &= 12; & \ell_{0,12} &= 8, & r_{0,12} &= 17; \\
\ell_{0,9} &= 5, & r_{0,9} &= 13; & \ell_{0,14} &= 9, & r_{0,14} &= 18.
\end{aligned}$$

Hence, if we denote $\psi_k = \psi_{4,0,k}$, $k \in I_0^{\text{new}}$ in (52), it follows from (36) that

$$\begin{aligned}
\psi_6(t) &= \frac{d^4}{dt^4} D \begin{pmatrix} M_{8,t_1,1} & M_{8,t_1,2} \\ t & t_{0,5} \end{pmatrix}, & \psi_{10}(t) &= \frac{d^4}{dt^4} D \begin{pmatrix} M_{8,t_1,6} \\ t \end{pmatrix}, \\
\psi_7(t) &= \frac{d^4}{dt^4} D \begin{pmatrix} M_{8,t_1,2} & M_{8,t_1,3} \\ t & t_{0,5} \end{pmatrix}, & \psi_{11}(t) &= \frac{d^4}{dt^4} D \begin{pmatrix} M_{8,t_1,7} & M_{8,t_1,8} \\ t & t_{0,6} \end{pmatrix}, \\
\psi_8(t) &= \frac{d^4}{dt^4} D \begin{pmatrix} M_{8,t_1,3} & M_{8,t_1,4} \\ t & t_{0,5} \end{pmatrix}, & \psi_{12}(t) &= \frac{d^4}{dt^4} D \begin{pmatrix} M_{8,t_1,8} & M_{8,t_1,9} \\ t & t_{0,6} \end{pmatrix}, \\
\psi_9(t) &= \frac{d^4}{dt^4} D \begin{pmatrix} M_{8,t_1,5} \\ t \end{pmatrix}, & \psi_{14}(t) &= \frac{d^4}{dt^4} D \begin{pmatrix} M_{8,t_1,9} & M_{8,t_1,10} \\ t & t_{0,6} \end{pmatrix}.
\end{aligned}$$

By applying (39) or Algorithm 4.4, with $j = 0$ and $m = 4$, we have

$$M_{8,t_1,k}^{(4)}(t) = 840 \sum_{\ell=0}^{4} (-1)^\ell \alpha_{k,\ell}\, M_{4,t_1,k+\ell}(t), \quad k = 1, \ldots, 10,$$

where $\alpha_{k,\ell}$, $\ell = 0, \ldots, 4$, could be obtained from Algorithm 4.4 as follows.

$$\alpha(1,0) = \frac{\alpha(0,0)}{t_{1,k+7} - t_{1,k}}, \qquad \alpha(3,2) = \frac{\alpha(2,1) + \alpha(2,2)}{t_{1,k+7} - t_{1,k+2}},$$

$$\alpha(1,1) = \frac{\alpha(0,0)}{t_{1,k+8} - t_{1,k+1}}; \qquad \alpha(3,3) = \frac{\alpha(2,2)}{t_{1,k+8} - t_{1,k+3}};$$

$$\alpha(2,0) = \frac{\alpha(1,0)}{t_{1,k+6} - t_{1,k}}, \qquad \alpha_{k,0} = \alpha(4,0) = \frac{\alpha(3,0)}{t_{1,k+4} - t_{1,k}},$$

$$\alpha(2,1) = \frac{\alpha(1,0) + \alpha(1,1)}{t_{1,k+7} - t_{1,k+1}}, \qquad \alpha_{k,1} = \alpha(4,1) = \frac{\alpha(3,0) + \alpha(3,1)}{t_{1,k+5} - t_{1,k+1}},$$

$$\alpha(2,2) = \frac{\alpha(1,1)}{t_{1,k+8} - t_{1,k+2}}; \qquad \alpha_{k,2} = \alpha(4,2) = \frac{\alpha(3,1) + \alpha(3,2)}{t_{1,k+6} - t_{1,k+2}},$$

$$\alpha(3,0) = \frac{\alpha(2,0)}{t_{1,k+5} - t_{1,k}}, \qquad \alpha_{k,3} = \alpha(4,3) = \frac{\alpha(3,2) + \alpha(3,3)}{t_{1,k+7} - t_{1,k+3}},$$

$$\alpha(3,1) = \frac{\alpha(2,0) + \alpha(2,1)}{t_{1,k+6} - t_{1,k+1}}, \qquad \alpha_{k,4} = \alpha(4,4) = \frac{\alpha(3,3)}{t_{1,k+8} - t_{1,k+4}}.$$

With

$$(M_{8,\mathbf{t}_1,1}(1), \ldots, M_{8,\mathbf{t}_1,4}(1)) = \left(\frac{763}{2000}, \frac{303}{1000}, \frac{43}{375}, \frac{1}{75} \right),$$

$$(M_{8,\mathbf{t}_1,7}(1), \ldots, M_{8,\mathbf{t}_1,10}(1)) = \left(\frac{1}{75}, \frac{43}{375}, \frac{303}{1000}, \frac{763}{2000} \right),$$

we have the reconstruction matrix Q_0 in (41) or (44) up to multiples of constant nonsingular diagonal matrices of order 8. For the convenience of typing in the matrix in a single page, we show $Q_0^\top$ instead of Q_0 in Fig. 2 below.

6.3. Decomposition matrices A_0 and B_0

To save space, we simply say that both decomposition matrices A_0 and B_0 could be obtained from (49) with $j = 0$, namely, $(A_0 \quad B_0) = \begin{pmatrix} P_0 \\ Q_0 \end{pmatrix}^{-1}$.

§7. Formulas for B-wavelets with Simple Interior Knots

In this section, we consider B-wavelets with nonuniform but simple knot sequences, with $\mathbf{t}_{j+1}$ being formed by inserting a new knot in $(t_{j,k}, t_{j,k+1})$ for each $k \in \{m, \ldots, m + K_j\}$, namely,

$$\mathbf{t}_j : \quad a = t_{j,1} = \cdots = t_{j,m} < t_{j,m+1} < t_{j,m+2} < \cdots$$

$$< t_{j,m+K_j} < t_{j,m+K_j+1} = \cdots = t_{j,K_j+2m} = b, \qquad (53)$$

$$t_{j+1,m+2\ell-1} \in (t_{j,m+\ell-1}, t_{j,m+\ell}), \quad \ell = 1, \ldots, K_j + 1, \qquad (54)$$

$$t_{j+1,m+2\ell} = t_{j,m+\ell}, \quad \ell = 1, \ldots, K_j, \qquad (55)$$

$$
\begin{pmatrix}
\frac{6363}{100} & 0 & 0 & 0 & 0 & 0 & 0 & 0 \\[2ex]
\frac{60333}{500} & \frac{1204}{125} & 0 & 0 & 0 & 0 & 0 & 0 \\[2ex]
\frac{10269}{400} & \frac{4473}{625} & \frac{56}{125} & 0 & 0 & 0 & 0 & 0 \\[2ex]
-\frac{1273251}{50000} & -\frac{35903}{3125} & -\frac{2072}{1875} & 0 & 0 & 0 & 0 & 0 \\[2ex]
-\frac{496384}{9375} & -\frac{1789466}{28125} & -\frac{43736}{1875} & \frac{280}{3} & 0 & 0 & 0 & 0 \\[2ex]
-\frac{42728}{1875} & -\frac{133483}{2250} & -\frac{15764}{375} & 280 & \frac{140}{3} & 0 & 0 & 0 \\[2ex]
0 & -\frac{2121}{100} & -\frac{4144}{75} & 770 & 560 & \frac{602}{25} & 0 & 0 \\[2ex]
0 & 0 & -\frac{602}{25} & 560 & 770 & \frac{4144}{75} & \frac{2121}{100} & 0 \\[2ex]
0 & 0 & 0 & \frac{140}{3} & 280 & \frac{15764}{375} & \frac{133483}{2250} & \frac{42728}{1875} \\[2ex]
0 & 0 & 0 & 0 & \frac{280}{3} & \frac{43736}{1875} & \frac{1789466}{28125} & \frac{496384}{9375} \\[2ex]
0 & 0 & 0 & 0 & 0 & \frac{2072}{1875} & \frac{35903}{3125} & \frac{1273251}{50000} \\[2ex]
0 & 0 & 0 & 0 & 0 & -\frac{56}{125} & -\frac{4473}{625} & -\frac{10269}{400} \\[2ex]
0 & 0 & 0 & 0 & 0 & 0 & -\frac{1204}{125} & -\frac{60333}{500} \\[2ex]
0 & 0 & 0 & 0 & 0 & 0 & 0 & -\frac{6363}{100}
\end{pmatrix}
$$

Fig. 2. The matrix $Q_0^\top$ for the example in Sect. 6.

for $j \geq 0$, where $K_0 \geq 0$. $K_{j+1} = 2K_j + 1$ follows from (53)–(55), so that

$$
\begin{aligned}
K_j &= 2^j(K_0 + 1) - 1, \\
\dim W_j &= K_j + 1 = 2^j(K_0 + 1), \\
I_j^{\text{new}} &= \{m + 1, m + 3, \ldots, m + 2(K_j + 1) - 1\}, \quad j \in \mathbb{Z}_+.
\end{aligned}
$$

It follows from Proposition 4.1 that, when K_j is sufficiently large, there are $m - 1$ left-boundary B-wavelets with a as the left endpoint of their supports, $K_j + 1 - 2(m - 1)$ interior B-wavelets, and $m - 1$ right-boundary B-wavelets with b as the right endpoint of their supports. However, when K_j is too small, there are $K_j + 1$ boundary B-wavelets only, with full support $[a, b]$, which we call 2-boundary B-wavelets. Precisely, we have the following.

Proposition 7.1. Let $\mathcal{S}_{m,j}$ and $\mathcal{S}_{m,j+1}$ be the mth order spline spaces with knot sequences $\mathbf{t}_j$ and $\mathbf{t}_{j+1}$ as in (53)–(55). Let W_j be the L^2-orthogonal complementary subspace of $\mathcal{S}_{m,j+1}$ relative to $\mathcal{S}_{m,j}$. Then the $K_j + 1$ B-wavelets $\psi_{m,j,m+2\ell-1}$ in W_j are given by the following.

1) When $K_j \leq m - 2$, all the B-wavelets are 2-boundary, given by

$$\psi_{m,j,m+2\ell-1}(t) =$$
$$= \frac{d^m}{dt^m} D \begin{pmatrix} M_{2m,\mathbf{t}_{j+1},\ell},\ M_{2m,\mathbf{t}_{j+1},\ell+1},\ \cdots,\ M_{2m,\mathbf{t}_{j+1},\ell+K_j} \\ t, \qquad t_{j,m+1}, \qquad \cdots, \qquad t_{j,m+K_j} \end{pmatrix},$$
$$\ell = 1, \ldots, K_j + 1.$$

2) When $m - 1 \leq K_j \leq 2m - 3$, there are $K_j + 1 - (m - 1), 2(m - 1) - (K_j + 1)$, and $K_j + 1 - (m - 1)$ left-, 2-, and right-boundary B-wavelets, respectively, given by

$$\psi_{m,j,m+2\ell-1}(t)$$
$$= \begin{cases} \dfrac{d^m}{dt^m} D \begin{pmatrix} M_{2m,\mathbf{t}_{j+1},\ell},\ M_{2m,\mathbf{t}_{j+1},\ell+1},\ \cdots,\ M_{2m,\mathbf{t}_{j+1},m+2\ell-2} \\ t, \qquad t_{j,m+1}, \qquad \cdots, \qquad t_{j,m+(m+\ell-2)} \end{pmatrix}, \\ \qquad\qquad\qquad\qquad \ell = 1, \ldots, K_j + 1 - (m - 1), \\[2em] \dfrac{d^m}{dt^m} D \begin{pmatrix} M_{2m,\mathbf{t}_{j+1},\ell},\ M_{2m,\mathbf{t}_{j+1},\ell+1},\ \cdots,\ M_{2m,\mathbf{t}_{j+1},\ell+K_j} \\ t, \qquad t_{j,m+1}, \qquad \cdots, \qquad t_{j,m+K_j} \end{pmatrix}, \\ \qquad\qquad\qquad\qquad \ell = K_j + 1 - (m - 2), \ldots, m - 1, \\[2em] \dfrac{d^m}{dt^m} D \begin{pmatrix} M_{2m,\mathbf{t}_{j+1},2\ell-m},\ M_{2m,\mathbf{t}_{j+1},2\ell-m+1},\ \cdots,\ M_{2m,\mathbf{t}_{j+1},\ell+K_j} \\ t, \qquad t_{j,\ell+1}, \qquad \cdots, \qquad t_{j,m+K_j} \end{pmatrix}, \\ \qquad\qquad\qquad\qquad \ell = m, \ldots, K_j + 1. \end{cases}$$

3) When $K_j \geq 2m - 2$, there are $m - 1, K_j + 1 - 2(m - 1)$, and $m - 1$ left-, interior-, and right-boundary B-wavelets, respectively, given by

$\psi_{m,j,m+2\ell-1}(t) =$

$$
= \begin{cases}
\dfrac{d^m}{dt^m} D \begin{pmatrix} M_{2m,\mathbf{t}_{j+1},\ell}, \ M_{2m,\mathbf{t}_{j+1},\ell+1}, \ \cdots, \ M_{2m,\mathbf{t}_{j+1},m+2\ell-2} \\[6pt] t, \qquad\qquad t_{j,m+1}, \qquad \cdots, \quad t_{j,m+(m+\ell-2)} \end{pmatrix}, \\[4pt]
\hphantom{xxxxxxxxxxxxxxxxxxxxxxxxxx} \ell = 1, \ldots, m-1, \\[10pt]
\dfrac{d^m}{dt^m} D \begin{pmatrix} M_{2m,\mathbf{t}_{j+1},2\ell-m}, \ M_{2m,\mathbf{t}_{j+1},2\ell-m+1}, \ \cdots, \ M_{2m,\mathbf{t}_{j+1},m+2\ell-2} \\[6pt] t, \qquad\qquad\quad t_{j,\ell+1}, \qquad\qquad \cdots, \qquad t_{j,\ell+2m-2} \end{pmatrix}, \\[4pt]
\hphantom{xxxxxxxxxxxxxxxxxxxxxx} \ell = m, \ldots, K_j + 1 - (m-1), \\[10pt]
\dfrac{d^m}{dt^m} D \begin{pmatrix} M_{2m,\mathbf{t}_{j+1},2\ell-m}, \ M_{2m,\mathbf{t}_{j+1},2\ell-m+1}, \ \cdots, \ M_{2m,\mathbf{t}_{j+1},\ell+K_j} \\[6pt] t, \qquad\qquad\quad t_{j,\ell+1}, \qquad\qquad \cdots, \qquad t_{j,m+K_j} \end{pmatrix}, \\[4pt]
\hphantom{xxxxxxxxxxxxxxxxx} \ell = K_j + 1 - (m-2), \ldots, K_j + 1.
\end{cases}
$$

The construction coincides with the classical cardinal B-wavelets when $t_{j,k} = k/2^j$, $j, k \in \mathbb{Z}$. Indeed, it follows from Proposition 7.1 that

$\psi_{m,0,2\ell+1}(t) =$

$$
\dfrac{d^m}{dt^m} D \begin{pmatrix} M_{2m,\mathbf{t}_1,2\ell-2(m-1)}, \ M_{2m,\mathbf{t}_1,2\ell-2(m-1)+1}, \ \cdots, \ M_{2m,\mathbf{t}_1,2\ell} \\[6pt] t, \qquad\qquad t_{0,\ell-(m-1)+1}, \qquad \cdots, \quad t_{0,\ell+m-1} \end{pmatrix}, \ \ell \in \mathbb{Z},
$$

so that, with $\ell = m-1$, the *classical* cardinal m^{th} order B-spline wavelets

$$
\psi_m(t) = \sum_{\ell=0}^{2(m-1)} (-1)^\ell N_{2m}(\ell+1) N_{2m}^{(m)}(2t - \ell)
$$

can be recovered from

$\psi_{m,0,2m-1}(t) =$

$$
\begin{aligned}
&= \dfrac{d^m}{dt^m} D \begin{pmatrix} N_{2m}(2\bullet), \ N_{2m}(2 \bullet -1), \ \cdots, \ N_{2m}(2 \bullet -2(m-1)) \\[6pt] t, \qquad\quad 1, \qquad \cdots, \qquad 2(m-1) \end{pmatrix} \\[6pt]
&= C_1 \sum_{\ell=0}^{2(m-1)} (-1)^\ell N_{2m}(\ell+1) N_{2m}^{(m)}(2t - \ell) = C_1 \, \psi_m(t), \qquad (56)
\end{aligned}
$$

where the constant C_1 in (56) is given by

$$C_1 = \det \left(N_{2m}(2j + 2 - k) \right)_{j,k=0,\ldots,2(m-2)}.$$

Here, in the second equality of (56), we have used the following two facts. First, the $(2m)^{\text{th}}$ order B-spline function

$$\Psi_m(t) := D \left(\begin{array}{cccc} N_{2m}(2\bullet), & N_{2m}(2 \bullet -1), & \cdots, & N_{2m}(2 \bullet -2(m-1)) \\ t, & 1, & \cdots, & 2(m-1) \end{array} \right)$$

is exactly the function Ψ in (37) with $\Psi_m(\ell) = 0, \ell = 1,\ldots,2(m-1)$. Secondly, the $2m - 1$ coefficients $\alpha_\ell := (-1)^\ell N_{2m}(\ell + 1), \ell = 0,\ldots,2(m-1)$, satisfy

$$\sum_{\xi=0}^{2(m-1)} \alpha_\xi N_{2m}(2t - \xi)\Big|_{t=\ell} = 0, \quad \ell = 1,\ldots,2(m-1).$$

§8. An Illustrative Example of B-wavelets

To illustrate the formulation of B-wavelets with nonuniform but simple knot sequences in Section 7, we end this paper by demonstrating the formulation by considering linear B-wavelets. It follows from Proposition 7.1 that when $K_j \geq 2$, there are one left- and one right-boundary B-wavelets, given by

$$\psi_{2,j,3}(t) = \frac{d^2}{dt^2} D \left(\begin{array}{cc} M_{4,\mathbf{t}_{j+1},1}, & M_{4,\mathbf{t}_{j+1},2} \\ t, & t_{j,3} \end{array} \right)$$

$$= M_{4,\mathbf{t}_{j+1},2}(t_{j+1,4}) M''_{4,\mathbf{t}_{j+1},1}(t) - M_{4,\mathbf{t}_{j+1},1}(t_{j+1,4}) M''_{4,\mathbf{t}_{j+1},2}(t),$$

$$\psi_{2,j,2K_j+3}(t) = \frac{d^2}{dt^2} D \left(\begin{array}{cc} M_{4,\mathbf{t}_{j+1},2K_j}, & M_{4,\mathbf{t}_{j+1},2K_j+1} \\ t & t_{j,2+K_j} \end{array} \right)$$

$$= M_{4,\mathbf{t}_{j+1},2K_j+1}(t_{j+1,2K_j+2}) M''_{4,\mathbf{t}_{j+1},2K_j}(t)$$

$$- M_{4,\mathbf{t}_{j+1},2K_j}(t_{j+1,2K_j+2}) M''_{4,\mathbf{t}_{j+1},2K_j+1}(t),$$

respectively. The $K_j - 1$ interior B-wavelets are given by

$$\psi_{2,j,2\ell+1}(t) = \frac{d^2}{dt^2} D \left(\begin{array}{ccc} M_{4,\mathbf{t}_{j+1},2\ell-2}, & M_{4,\mathbf{t}_{j+1},2\ell-1}, & M_{4,\mathbf{t}_{j+1},2\ell} \\ t, & t_{j,\ell+1}, & t_{j,\ell+2} \end{array} \right),$$

where $\ell = 2, \ldots, K_j$. To be specific, let $j = 0, K_0 = 2$, and

$$\mathbf{t}_0 = \{t_{0,1}, \ldots, t_{0,7}\} = \{0, 0, 1, 3, 4, 5, 5\},$$
$$\mathbf{t}_1 = \{t_{1,1}, \ldots, t_{1,11}\} = \{0, 0, 1/3, 1, 5/2, 3, 10/3, 4, 9/2, 5, 5\},$$

be two general knot sequences. Then $K_0 = 3, K_1 = 7, \mathcal{S}_{4,1} = \mathcal{S}_{4,0} \oplus W_0$. With $\dim W_0 = 4$, and

$$\mathbf{M}_0 = \left(M_{2,\mathbf{t}_0,1}, \cdots, M_{2,\mathbf{t}_0,5} \right)^\top, \quad \mathbf{M}_1 = \left(M_{2,\mathbf{t}_1,1}, \cdots, M_{2,\mathbf{t}_1,9} \right)^\top,$$

the reconstruction matrices P_0, Q_0 in (40)–(41), with $j = 0$, are 5×9 and 4×9, respectively.

Acknowledgments. This research was partially supported by NSF Grants CCR-9988289 and CCR-0098331 and ARO Grants DAAD 19-00-10512 and DAAD 19-01-10739.

References

1. Phillip, J. B., de Boor–Fix functionals and polar forms, Comput. Aided Geom. Design **7** (1990), 425–430.

2. de Boor, C., *A Practical Guide to Splines*, Springer Verlag (Berlin), 1978.

3. Chui, C. K. and J. M. De Villiers, Spline-wavelets with arbitrary knots on a bounded interval, orthogonal decomposition and computational algorithms, Comm. Appl. Anal. **2** (1998), 457–486.

4. Lyche, T. and K. Mørken, Making the Oslo algorithm more efficient, SIAM J. Numer. Anal. **23** (1986), 663–675.

5. Lyche, T. and K. Mørken, Spline-wavelets of minimal support, *Numerical Methods of Approximation Theory, Vol. 9*, D. Braess and L. L. Schumaker (eds.), Birkhäuser Verlag, Basel, 1992, 177–194.

6. Lyche, T., K. Mørken, and E. Quak, Theory and algorithms for nonuniform spline wavelets, *Multivariate Approximation and Applications*, N. Dyn, D. Leviatan, D. Levin, and A. Pinkus (eds.), Cambridge University Press, 2001.

7. Ramshaw, L., Blossoms are polar forms, Comput. Aided Geom. Design **6** (1989), 323–358.

8. Schumaker, L. L., *Spline Functions: Basic Theory*, John Wiley & Sons Publications (New York), 1981.

9. Seidel, H.-P., A new multiaffine approach to B-splines, Comput. Aided Geom. Design **6** (1989), 23–32.

Charles K. Chui
Dept. of Mathematics and Computer Science
Univ. of Missouri – St. Louis
St. Louis, MO 63121-4499

and

Dept. of Statistics
Stanford University
Stanford, CA 94305-4065
cchui@stat.stanford.edu

Jian-ao Lian
Dept. of Mathematics
Prairie View A&M University
Prairie View, TX 77446
unurbs@yahoo.com

Locally Stable Spline Bases
on Nested Triangulations

Oleg Davydov

Abstract. Given a nested sequence of triangulations $\triangle_0, \triangle_1, \ldots$ of a polygonal domain Ω, we construct for any $r \geq 1$, $d \geq 4r + 1$, locally stable bases for some spaces $\tilde{\mathcal{S}}_d^r(\triangle_0) \subset \tilde{\mathcal{S}}_d^r(\triangle_1) \subset \cdots \subset \tilde{\mathcal{S}}_d^r(\triangle_n) \subset \cdots$ of bivariate polynomial splines of smoothness r and degree d. In particular, the bases are stable and locally linearly independent simultaneously.

§1. Introduction

Given a triangulation $\triangle$ of a polygonal domain Ω in $\mathbb{R}^2$, let

$$\mathcal{S}_d^r(\triangle) = \{s \in C^r(\Omega) : s|_T \in \mathcal{P}_d \text{ for all triangles } T \in \triangle\},$$

be the space of **polynomial splines of degree d and smoothness r on** $\triangle$. Here $\mathcal{P}_d$ is the space of bivariate polynomials of total degree d.

Suppose the sequence of triangulations $\triangle_0, \triangle_1, \ldots, \triangle_n, \ldots$, of Ω obtained by consecutive refinements of an initial coarse triangulation $\triangle_0$ is shape regular, *i.e.*, the smallest angle of all triangles in each $\triangle_n$ is at least θ, for some $\theta > 0$ independent of n. Standard conforming smooth finite elements [4] with respect to $\triangle_n$ span the **superspline** subspaces

$$\mathcal{S}_d^{r,2r}(\triangle) = \{s \in \mathcal{S}_d^r(\triangle) : s \in C^{2r}(v) \text{ for all vertices } v \text{ of } \triangle\},$$

of $\mathcal{S}_d^r(\triangle)$, $d \geq 4r + 1$, [3,21], where $s \in C^{2r}(v)$ means that s is $2r$ times differentiable at v. Since these spaces are not nested, *i.e.*,

$$\mathcal{S}_d^{r,2r}(\triangle_n) \not\subset \mathcal{S}_d^{r,2r}(\triangle_{n+1}),$$

Approximation Theory X: Wavelets, Splines, and Applications
Charles K. Chui, Larry L. Schumaker, and Joachim Stöckler (eds.), pp. 231–240.

231

there are well known complications in using them for multiscale numerical methods, see *e.g.* [19]. As an alternative, it was suggested in [5,17] to use locally supported bases for the full spline spaces $\mathcal{S}_d^r(\triangle_n)$. However, the standard constructions [1,14,15,18] of locally supported bases for $\mathcal{S}_d^r(\triangle)$ in general lack the highly desirable property of stability, where a sequence of spline bases $(s_i^{[n]})_{i \in I_n} \subset \mathcal{S}_d^r(\triangle_n)$, $n = 0, 1, \ldots$, is called (L_p-)**stable** if for all choices of the coefficient vectors $c = (c_i)_{i \in I_n}$,

$$K_1 \|c\|_p \leq \left\| \sum_{i \in I_n} c_i s_i^{[n]} \right\|_{L_p(\Omega)} \leq K_2 \|c\|_p, \tag{1}$$

with constants K_1, K_2 depending only on r, d and θ. On the other hand, stable locally supported spline bases used in [2,16] span superspline subspaces of $\mathcal{S}_d^r(\triangle)$, $d \geq 3r + 2$, that are less restrictive than $\mathcal{S}_d^{r,2r}(\triangle)$ but also lack nestedness for nested triangulations.

Recently constructed [10,11] stable locally supported bases for $\mathcal{S}_d^r(\triangle)$, $d \geq 3r + 2$, solve this problem. Moreover, a construction of stable locally supported bases for the full spline spaces over regular triangulations of polyhedral domains in $n \geq 3$ variables is also available [7] if $d \geq r2^n + 1$. A certain drawback of these constructions is the fact that the dimension of $\mathcal{S}_d^r(\triangle)$ is *instable* since it depends on the geometry of the triangulation [18,20], thus allowing sudden changes in the number of basis functions as certain vertices are slightly perturbed. Computational aspects of this situation were discussed in [8] and numerical evidence was provided showing that despite the instability of the dimension the basis splines can be efficiently computed. However, it may be desirable to work with *nested subspaces of $\mathcal{S}_d^r(\triangle_n)$ with stable dimension*.

In this paper we present a construction of locally supported bases $(s_i^{[n]})_{i \in I_n}$ for certain nested subspaces $\tilde{\mathcal{S}}_d^r(\triangle_n)$ of $\mathcal{S}_d^r(\triangle_n)$ if $d \geq 4r + 1$. These subspaces satisfy

$$\mathcal{S}_d^{r,2r}(\triangle_n) \subset \tilde{\mathcal{S}}_d^r(\triangle_n) \subset \mathcal{S}_d^r(\triangle_n), \qquad n = 0, 1, \ldots, \tag{2}$$

and therefore include all polynomials in $\mathcal{P}_d$. The dimension of $\tilde{\mathcal{S}}_d^r(\triangle_n)$ is independent of the geometry of $\triangle_n$. Moreover, the sequence of bases is locally L_p-**stable** in the sense that for all choices of the coefficient vectors $c = (c_i)_{i \in I_n}$, and for each triangle $T \in \triangle_n$,

$$K_1 \left\| c|_{I_n(T)} \right\|_p \leq \left\| \sum_{i \in I_n(T)} c_i s_i^{[n]} \right\|_{L_p(T)} \leq K_2 \left\| c|_{I_n(T)} \right\|_p, \tag{3}$$

with constants K_1, K_2 depending only on r, d and θ, where

$$I_n(T) := \{ i \in I_n : T \subset \operatorname{supp} s_i^{[n]} \}.$$

Obviously, a locally L_p-stable basis is always locally linearly independent, *i.e.*, for each $T \in \triangle_n$, the functions in

$$\{s_i^{[n]}|_T : \; i \in I_n(T)\}$$

are linearly independent. Since $s_i^{[n]}|_T$ coincide with some polynomials in $\mathcal{P}_d$, it follows that $\#I_n(T) \le \binom{d+2}{2}$, and hence, a locally L_p-stable spline basis is also L_p-*stable* in the sense of (1).

We recall that local stability is a property of B-splines and finite-element bases, whereas it was shown in [10,11] that stability and local linear independence are not compatible in general for the bases of the full spline space $\mathcal{S}_d^r(\triangle)$ in two variables.

As a simple example to show that local linear independence and stability together do not necessarily imply local stability, consider the univariate spaces $S_n = \mathrm{span}\{s_1^{[n]}, s_2^{[n]}, s_3^{[n]}\}$, $n = 1, 2, \ldots$, where

$$s_1^{[n]} = \begin{cases} t, & t \in [0,1], \\ 0, & t \in [-1,0], \end{cases} \qquad s_2^{[n]} = \begin{cases} 0, & t \in [0,1], \\ -t, & t \in [-1,0], \end{cases}$$

$$s_3^{[n]} = \begin{cases} (1-t)/n, & t \in [0,1], \\ -(1+t)(t-1/n), & t \in [-1,0]. \end{cases}$$

(For more on locally linearly independent bases, see *e.g.* [6,12,13].)

§2. Spaces $\tilde{\mathcal{S}}_d^r(\triangle_n)$ and Construction of Bases

Let $\mathcal{V}_n$ and $\mathcal{E}_n$ be the sets of all vertices and all edges of the triangulation $\triangle_n$, respectively. Denote by E_n the union of all edges $e \in \mathcal{E}_n$. Since the triangulations $\triangle_n$, $n = 1, 2, \ldots$, are obtained by consecutive refinements of $\triangle_0$, we have

$$\mathcal{V}_0 \subset \mathcal{V}_1 \subset \cdots \subset \mathcal{V}_n \subset \cdots,$$
$$E_0 \subset E_1 \subset \cdots \subset E_n \subset \cdots.$$

Let

$$\tilde{\mathcal{V}}_0 = \emptyset, \qquad \tilde{\mathcal{V}}_n := \tilde{\mathcal{V}}_{n-1} \cup \left[(\mathcal{V}_n \setminus \mathcal{V}_{n-1}) \cap E_{n-1} \cap \mathrm{int}\,\Omega\right], \qquad n = 1, 2, \ldots.$$

For any $v \in \bigcup_{n \in \mathbf{N}} \tilde{\mathcal{V}}_n$, let $n_v := \min\{n : \; v \in \tilde{\mathcal{V}}_n\}$. Obviously, there is a unique edge e_v of $\triangle_{n_v-1}$, with adjacent triangles $T_v^+, T_v^- \in \triangle_{n_v-1}$, such that v lies in the interior of e_v. We set

$$\tilde{\mathcal{S}}_d^r(\triangle_n) = \{s \in \mathcal{S}_d^r(\triangle_n) : \; s \in C^{2r}(v) \; \text{for all } v \in \mathcal{V}_n \setminus \tilde{\mathcal{V}}_n, \text{ and}$$
$$s|_{T_v^+} \in C^{2r}(v), \; s|_{T_v^-} \in C^{2r}(v) \; \text{for all } v \in \tilde{\mathcal{V}}_n\}.$$

It is easy to see that (2) is satisfied and that the spaces are nested,

$$\tilde{\mathcal{S}}_d^r(\triangle_{n-1}) \subset \tilde{\mathcal{S}}_d^r(\triangle_n), \qquad n = 1, 2, \ldots. \tag{4}$$

To show the latter, we suppose that $s \in \tilde{\mathcal{S}}_d^r(\triangle_{n-1})$. Then $s \in \mathcal{S}_d^r(\triangle_{n-1}) \subset \mathcal{S}_d^r(\triangle_n)$. If $v \in \tilde{\mathcal{V}}_n$, then either $v \in \tilde{\mathcal{V}}_{n-1}$, or $n_v = n$, $T_v^+, T_v^- \in \triangle_{n-1}$ and v lies in the interior of the common edge e_v of these two triangles. Obviously, in both cases $s|_{T_v^+} \in C^{2r}(v)$, $s|_{T_v^-} \in C^{2r}(v)$. If, otherwise, $v \in \mathcal{V}_n \setminus \tilde{\mathcal{V}}_n$, then either $v \in \mathcal{V}_{n-1} \setminus \tilde{\mathcal{V}}_{n-1}$, or $v \in \Omega \setminus E_{n-1}$, or $v \in \partial\Omega \setminus \mathcal{V}_{n-1}$. In all three cases $s \in C^{2r}(v)$.

We first determine a basis for the dual space $(\tilde{\mathcal{S}}_d^r(\triangle_n))^*$ by using usual nodal functionals. Let D_e and $D_{e^\perp}$ denote the derivative in the direction parallel or perpendicular to an edge $e \in \mathcal{E}_n$, respectively. We will also need the same notation $D_{e_v}, D_{e_v^\perp}$ for the special edge e_v defined above for each $v \in \tilde{\mathcal{V}}_n$. The linear functional evaluating at $\xi \in \Omega$ any function f continuous at ξ will be denoted by δ_ξ.

Consider the set

$$\mathcal{N} = \left(\bigcup_{v \in \mathcal{V}_n} \mathcal{N}_v \right) \cup \left(\bigcup_{e \in \mathcal{E}_n} \mathcal{N}_e \right) \cup \left(\bigcup_{T \in \triangle_n} \mathcal{N}_T \right),$$

of nodal linear functionals on $\tilde{\mathcal{S}}_d^r(\triangle_n)$, where for each $T = \langle v_1, v_2, v_3 \rangle \in \triangle_n$,

$$\mathcal{N}_T = \{\delta_\xi : \xi \in \Xi_T\},$$
$$\Xi_T = \left\{ \xi = \frac{i_1 v_1 + i_2 v_2 + i_3 v_3}{d} : i_1 + i_2 + i_3 = d, \quad i_1, i_2, i_3 > r \right\},$$

for each edge $e = \langle v_1, v_2 \rangle \in \mathcal{E}_n$,

$$\mathcal{N}_e = \{\delta_\xi D_{e^\perp}^q : \xi \in \Xi_{e,q}, \quad q = 0, \ldots, r\},$$
$$\Xi_{e,q} = \left\{ \xi = \frac{i_1 v_1 + i_2 v_2}{d} : i_1 + i_2 = d, \quad i_1, i_2 > 2r - q \right\},$$

and for each vertex $v \in \mathcal{V}_n$ the set $\mathcal{N}_v$ is defined as follows. If $v \in \mathcal{V}_n \setminus \tilde{\mathcal{V}}_n$, then

$$\mathcal{N}_v = \{\delta_v D_x^\alpha D_y^\beta : \alpha + \beta \leq 2r\}.$$

If $v \in \tilde{\mathcal{V}}_n$, then

$$\mathcal{N}_v = \{\delta_v D_{e_v}^\alpha D_{e_v^\perp}^\beta : \alpha + \beta \leq 2r, \ \beta \leq r\}$$
$$\cup \{\delta_v^+ D_{e_v}^\alpha D_{e_v^\perp}^\beta : \alpha + \beta \leq 2r, \ \beta \geq r+1\}$$
$$\cup \{\delta_v^- D_{e_v}^\alpha D_{e_v^\perp}^\beta : \alpha + \beta \leq 2r, \ \beta \geq r+1\},$$

where $\delta_v^\pm f := \delta_v(f|_{T_v^\pm})$.

Theorem 1. *The set $\mathcal{N}$ is a basis for $(\tilde{\mathcal{S}}_d^r(\triangle_n))^*$.*

Proof: We first prove that $\mathcal{N}$ is a spanning set for $(\tilde{\mathcal{S}}_d^r(\triangle_n))^*$. This will follow if we show that $\eta s = 0$ for all $\eta \in \mathcal{N}$ implies $s = 0$ whenever $s \in \tilde{\mathcal{S}}_d^r(\triangle_n)$. Given a triangle $T \in \triangle_n$, with vertices v_1, v_2, v_3 and edges e_1, e_2, e_3, consider the set

$$\mathcal{N}(T) = \Big(\bigcup_{i=1}^{3} \mathcal{N}_{v_i}(T) \Big) \cup \Big(\bigcup_{i=1}^{3} \mathcal{N}_{e_i} \Big) \cup \mathcal{N}_T, \tag{5}$$

with $\mathcal{N}_v(T) := \mathcal{N}_v$ if $v \in \mathcal{V}_n \setminus \tilde{\mathcal{V}}_n$, and

$$\mathcal{N}_v(T) := \{\delta_v D_{e_v}^\alpha D_{e_v^\perp}^\beta : \alpha + \beta \le 2r, \ \beta \le r\}$$

$$\cup \begin{cases} \{\delta_v^+ D_{e_v}^\alpha D_{e_v^\perp}^\beta : \alpha + \beta \le 2r, \ \beta \ge r+1\}, & \text{if } T \subset T_v^+, \\ \{\delta_v^- D_{e_v}^\alpha D_{e_v^\perp}^\beta : \alpha + \beta \le 2r, \ \beta \ge r+1\}, & \text{if } T \subset T_v^-, \end{cases}$$

if $v \in \tilde{\mathcal{V}}_n$. Clearly, $\eta s = 0$ for all $\eta \in \mathcal{N}(T)$ implies that the polynomial $p_T = s|_T$ satisfies homogeneous interpolation conditions of the standard interpolation scheme used to define finite elements of the space $\mathcal{S}_d^{r,2r}(\triangle)$, see *e.g.* [21]. Therefore, $s|_T = 0$ for each T, and $s = 0$.

We now prove that $\mathcal{N}$ is linearly independent. To this end, we show that for any $a = (a_\eta)_{\eta \in \mathcal{N}} \in \mathbb{R}^{\#\mathcal{N}}$, there exists a spline $s \in \tilde{\mathcal{S}}_d^r(\triangle_n)$ such that

$$\eta s = a_\eta, \qquad \text{all } \eta \in \mathcal{N}. \tag{6}$$

We define the polynomial pieces p_T, $T \in \triangle_n$, using the finite-element interpolation scheme mentioned above, such that for each $T \in \triangle_n$,

$$\eta p_T = a_\eta, \qquad \text{all } \eta \in \mathcal{N}(T). \tag{7}$$

Setting $s|_T = p_T$ for all $T \in \triangle_n$, we have to show that the piecewise polynomial function s lies in $\tilde{\mathcal{S}}_d^r(\triangle_n)$.

Let $v \in \mathcal{V}_n \setminus \tilde{\mathcal{V}}_n$, and let T_1, T_2 are any two triangles attached to v. Since $\mathcal{N}_v \subset \mathcal{N}(T_1) \cap \mathcal{N}(T_2)$, we have by (7),

$$\delta_v D_x^\alpha D_y^\beta p_{T_1} = \delta_v D_x^\alpha D_y^\beta p_{T_2}, \qquad \text{all } \alpha + \beta \le 2r,$$

which ensures that $s \in C^{2r}(v)$. If $v \in \tilde{\mathcal{V}}_n$, then for any two triangles T_1, T_2 attached to v,

$$\delta_v D_{e_v}^\alpha D_{e_v^\perp}^\beta p_{T_1} = \delta_v D_{e_v}^\alpha D_{e_v^\perp}^\beta p_{T_2}, \qquad \text{all } \alpha + \beta \le 2r, \ \beta \le r.$$

Moreover, if either $T_1 \cup T_2 \subset T_v^+$ or $T_1 \cup T_2 \subset T_v^-$, then

$$\delta_v D_{e_v}^\alpha D_{e_v^\perp}^\beta p_{T_1} = \delta_v D_{e_v}^\alpha D_{e_v^\perp}^\beta p_{T_2}, \qquad \text{all } \alpha + \beta \le 2r,$$

which implies that $s|_{T_v^+} \in C^{2r}(v)$ and $s|_{T_v^-} \in C^{2r}(v)$.

Let $e = \langle v_1, v_2 \rangle$ be an interior edge of $\triangle_n$, with adjacent triangles T_1, T_2. From the above it follows that for all $\alpha + \beta \le 2r$, $\beta \le r$,

$$\delta_{v_1} D_e^\alpha D_{e\perp}^\beta p_{T_1} = \delta_{v_1} D_e^\alpha D_{e\perp}^\beta p_{T_2}, \qquad \delta_{v_2} D_e^\alpha D_{e\perp}^\beta p_{T_1} = \delta_{v_2} D_e^\alpha D_{e\perp}^\beta p_{T_2}.$$

Moreover, since $\mathcal{N}_e \subset \mathcal{N}(T_1) \cap \mathcal{N}(T_2)$, we have

$$\delta_\xi D_{e\perp}^q p_{T_1} = \delta_\xi D_{e\perp}^q p_{T_2}, \qquad \xi \in \Xi_{e,q}, \quad q = 0, \dots, r.$$

This implies that for each $q = 0, \dots, r$, the univariate polynomial

$$\tilde{p} := D_{e\perp}^q (p_{T_1} - p_{T_2})|_e$$

of degree $d - q$ satisfies homogeneous Hermite interpolation conditions

$$D_e^\mu \tilde{p}(v_1) = D_e^\mu \tilde{p}(v_2) = 0, \qquad \mu = 0, \dots, 2r - q,$$
$$\tilde{p}(\xi) = 0, \qquad \text{all } \xi \in \Xi_{e,q}.$$

Therefore, $\tilde{p} = 0$, *i.e.*,

$$D_{e\perp}^q p_{T_1}|_e = D_{e\perp}^q p_{T_2}|_e, \qquad q = 0, \dots, r,$$

which shows that s is r times continuously differentiable across e. Since $s \in \tilde{\mathcal{S}}_d^r(\triangle_n)$, (6) follows from (7), and the proof is complete. $\square$

By counting the number of functionals in each $\mathcal{N}_v$, $\mathcal{N}_e$, and $\mathcal{N}_T$, it is easy to check the following dimension formula,

$$\dim \tilde{\mathcal{S}}_d^r(\triangle_n) = \binom{d - 3r - 1}{2} \#\triangle_n + \frac{(r+1)(2d - 7r - 2)}{2} \#\mathcal{E}_n \\ + \binom{2r+2}{2} \#\mathcal{V}_n + \binom{r+1}{2} \#\tilde{\mathcal{V}}_n, \tag{8}$$

which shows that the dimension is independent of geometry.

The desired basis $(s_\eta^{[n]})_{\eta \in \mathcal{N}}$ for $\tilde{\mathcal{S}}_d^r(\triangle_n)$ can be determined by the duality condition

$$\mu s_\eta^{[n]} = \delta_{\mu,\eta}, \qquad \text{all } \mu, \eta \in \mathcal{N}. \tag{9}$$

Arguing similarly to the first part of the proof of Theorem 1, it is easy to show that

$$s_\eta^{[n]}|_T \ne 0 \quad \text{only if} \quad \eta \in \mathcal{N}(T), \tag{10}$$

which in particular implies that

$$\operatorname{supp} s_\eta^{[n]} \subset \begin{cases} T, & \text{if } \eta \in \mathcal{N}_T \text{ for some } T \in \mathcal{T}_n, \\ \text{star}(e), & \text{if } \eta \in \mathcal{N}_e \text{ for some } e \in \mathcal{E}_n, \\ \text{star}(v), & \text{if } \eta \in \mathcal{N}_v \text{ for some } v \in \mathcal{V}_n, \end{cases} \tag{11}$$

where star(e) and star(v), respectively, denote the union of all triangles in $\triangle_n$ attached to an edge e or vertex v, respectively.

§3. Local Stability

Theorem 2. *The above constructed sequence of bases* $(s_\eta^{[n]})_{\eta \in \mathcal{N}}$, $n = 0, 1, \ldots$, *is locally L_p-stable, $1 \le p \le \infty$, after a suitable normalization.*

Proof: By (10), we have $I_n(T) = \mathcal{N}(T)$, where

$$I_n(T) := \{\eta \in \mathcal{N} : T \subset \operatorname{supp} s_\eta^{[n]}\}.$$

Since $\#\mathcal{N}(T) = \binom{d+2}{2}$, we have

$$\left\| \sum_{\eta \in \mathcal{N}(T)} c_\eta s_\eta^{[n]} \right\|_{L_\infty(T)} \le \binom{d+2}{2} \max_{\eta \in \mathcal{N}(T)} \left\| c_\eta s_\eta^{[n]} \right\|_{L_\infty(T)}.$$

The functionals in $\mathcal{N}(T)$ constitute a standard finite-element interpolation scheme, see *e.g.* [21]. Therefore, the general estimates for the norms of the finite-element interpolation operators [4] imply that

$$\|s_\eta^{[n]}\|_{L_\infty(T)} \le K_1 h_T^{q(\eta)}, \tag{12}$$

where K_1 depends only on r, d and θ, h_T denotes the diameter of T, and $q(\eta)$ is the order of the derivative that defines η. On the other hand, by the Markov inequality, we have for each $\mu \in \mathcal{N}(T)$,

$$|c_\mu| = \mu\left(\sum_{\eta \in \mathcal{N}(T)} c_\eta s_\eta^{[n]} \right) \le K_2 h_T^{-q(\eta)} \left\| \sum_{\eta \in \mathcal{N}(T)} c_\eta s_\eta^{[n]} \right\|_{L_\infty(T)}, \tag{13}$$

where K_2 depends only on d and θ. Therefore,

$$K_2^{-1} \max_{\eta \in \mathcal{N}(T)} |c_\eta| \le \left\| \sum_{\eta \in \mathcal{N}(T)} c_\eta h_T^{-q(\eta)} s_\eta^{[n]} \right\|_{L_\infty(T)} \le K_1 \binom{d+2}{2} \max_{\eta \in \mathcal{N}(T)} |c_\eta|.$$

Since $T \subset \operatorname{supp} s_\eta^{[n]}$ and $\operatorname{supp} s_\eta^{[n]}$ is at most the star of a vertex, we have

$$h_T \le h_\eta \le K_3 h_T, \tag{14}$$

where h_η denotes the diameter of $\operatorname{supp} s_\eta^{[n]}$, and K_3 depends only on θ (see [16]). Therefore, we conclude that the sequence of bases

$$(h_\eta^{-q(\eta)} s_\eta^{[n]})_{\eta \in \mathcal{N}}, \qquad n = 0, 1, \ldots,$$

is locally L_∞-stable, which completes the proof for the case $p = \infty$.

Let $1 \leq p < \infty$. In this case we set

$$\hat{s}_\eta^{[n]} := A_\eta^{-1/p} h_\eta^{-q(\eta)} s_\eta^{[n]},$$

where A_η is the area of $\operatorname{supp} s_\eta^{[n]}$. Then

$$A_T \leq A_\eta \leq K_4 A_T, \tag{15}$$

where A_T is the area of T, and K_4 depends only on θ. We have

$$\left\| \sum_{\eta \in \mathcal{N}(T)} c_\eta s_\eta^{[n]} \right\|_{L_p(T)}^p = \int_T \left| \sum_{\eta \in \mathcal{N}(T)} c_\eta s_\eta^{[n]} \right|^p$$

$$\leq A_T \|s_\eta^{[n]}\|_{L_\infty(T)}^p (\#\mathcal{N}(T))^{p-1} \sum_{\eta \in \mathcal{N}(T)} |c_\eta|^p,$$

which by (12) implies that

$$\left\| \sum_{\eta \in \mathcal{N}(T)} c_\eta s_\eta^{[n]} \right\|_{L_p(T)} \leq K_5 A_T^{1/p} h_T^{q(\eta)} \| c|_{\mathcal{N}(T)} \|_p, \tag{16}$$

with a constant K_5 depending only on r, d and θ. On the other hand, since

$$\sum_{\eta \in \mathcal{N}(T)} c_\eta s_\eta^{[n]}|_T$$

is a polynomial of degree d, we have for any $\mu \in \mathcal{N}(T)$ by (13) and a Nikolskii-type inequality,

$$|c_\mu| \leq K_2 h_T^{-q(\eta)} \left\| \sum_{\eta \in \mathcal{N}(T)} c_\eta s_\eta^{[n]} \right\|_{L_\infty(T)}$$

$$\leq K_6 A_T^{-1/p} h_T^{-q(\eta)} \left\| \sum_{\eta \in \mathcal{N}(T)} c_\eta s_\eta^{[n]} \right\|_{L_p(T)},$$

where K_6 depends only on d and θ. Therefore,

$$\| c|_{\mathcal{N}(T)} \|_p = \left(\sum_{\mu \in \mathcal{N}(T)} |c_\mu|^p \right)^{1/p}$$

$$\leq (\#\mathcal{N}(T))^{1/p} K_6 A_T^{-1/p} h_T^{-q(\eta)} \left\| \sum_{\eta \in \mathcal{N}(T)} c_\eta s_\eta^{[n]} \right\|_{L_p(T)}.$$

Since $\#\mathcal{N}(T) = \binom{d+2}{2}$, this last inequality together with (14)–(16) imply the local L_p-stability of the sequence of bases

$$(\hat{s}_\eta^{[n]})_{\eta \in \mathcal{N}}, \qquad n = 0, 1, \ldots. \quad \square$$

Finally, we note that the stability of the bases and the fact that

$$\mathcal{P}_d \subset \tilde{\mathcal{S}}_d^r(\triangle_n), \qquad n = 0, 1, \ldots,$$

can be used in a standard way (see[16]) to show the optimal approximation power of the spaces $\tilde{\mathcal{S}}_d^r(\triangle_n)$.

References

1. Alfeld, P., B. Piper, and L. L. Schumaker, Minimally supported bases for spaces of bivariate piecewise polynomials of smoothness r and degree $d \geq 4r + 1$, Comput. Aided Geom. Design **4** (1987), 105–123.

2. Chui, C. K., D. Hong, and R.-Q. Jia, Stability of optimal order approximation by bivariate splines over arbitrary triangulations, Trans. Amer. Math. Soc. **347** (1995), 3301–3318.

3. Chui, C. K. and M.-J. Lai, Multivariate vertex splines and finite elements, J. Approx. Theory **60** (1990), 245–343.

4. Ciarlet, P. G., *The Finite Element Method for Elliptic Problems*, North-Holland, Netherlands, 1978.

5. Dahmen, W., P. Oswald, and X.-Q. Shi, C^1-hierarchical bases, J. Comput. Appl. Math. **51** (1994), 37–56.

6. Davydov, O., Locally linearly independent basis for C^1 bivariate splines, in *Mathematical Methods for Curves and Surfaces II,* Morten Dæhlen, Tom Lyche, Larry L. Schumaker (eds), Vanderbilt University Press, Nashville & London, 1998, 71–78.

7. Davydov, O., Stable local bases for multivariate spline spaces, J. Approx. Theory **111** (2001), 267–297.

8. Davydov, O., On the computation of stable local bases for bivariate polynomial splines, in *Trends in Approximation Theory*, Kirill Kopotun, Tom Lyche, Mike Neamtu (eds), Vanderbilt University Press, 2001, 85–94.

9. Davydov, O., G. Nürnberger, and F. Zeilfelder, Bivariate spline interpolation with optimal approximation order, Constr. Approx. **17** (2001), 181–208.

10. Davydov, O. and L. L. Schumaker, Stable local nodal bases for C^1 bivariate polynomial splines, in *Curve and Surface Fitting: Saint-Malo 99*, A. Cohen, C. Rabut, and L. L. Schumaker (eds), Vanderbilt University Press, Nashville TN, 2000, 171–180.

11. Davydov, O. and L. L. Schumaker, On stable local bases for bivariate polynomial spline spaces, Constr. Approx. **18** (2002), 87–116.

12. Davydov, O. and L. L. Schumaker, Locally linearly independent bases for bivariate polynomial splines, Advances in Comp. Math. **13** (2000), 355–373.

13. Davydov, O., M. Sommer, and H. Strauss, On almost interpolation and locally linearly independent bases, East J. Approx. **5** (1999), 67–88.

14. Hong, D., Spaces of bivariate spline functions over triangulation, Approx. Theory Appl. **7** (1991), 56–75.

15. Ibrahim, A. and L. L. Schumaker, Super spline spaces of smoothness r and degree $d \geq 3r + 2$, Constr. Approx. **7** (1991), 401–423.

16. Lai, M. J. and L. L. Schumaker, On the approximation power of bivariate splines, Advances in Comp. Math. **9** (1998), 251–279.

17. Le Méhauté, A., Nested sequences of triangular finite element spaces, in *Multivariate Approximation: Recent Trends and Results*, Werner Haussmann, K. Jetter and M. Reimer (eds), Akademie-Verlag, 1997, 133–145.

18. Morgan, J. and R. Scott, A nodal basis for C^1 piecewise polynomials of degree $n \geq 5$, Math. Comp. **29(131)** (1975), 736–740.

19. Oswald, P., *Multilevel Finite Element Approximation*, Teubner, Stuttgart, 1994.

20. Schumaker, L. L., On the dimension of spaces of piecewise polynomials in two variables, in *Multivariate Approximation Theory*, W. Schempp and K. Zeller (eds), Birkhäuser, Basel, 1979, 396–412.

21. Schumaker, L. L., On super splines and finite elements, SIAM J. Numer. Anal. **26** (1989), 997–1005.

Oleg Davydov
Mathematisches Institut
Justus-Liebig-Universität
D-35392 Giessen, Germany
oleg.davydov@math.uni-giessen.de

Numerical Optimization of Financial Portfolios

Uwe Depczynski

Abstract. In optimizing financial portfolios by using Value-at-Risk criteria, one is often faced with the problem of numerically solving quadratic or general non-linear optimization problems with non-linear constraints. We present a hybrid algorithm for multiperiod financial portfolio optimization, based on a two-step approach, by using an evolutionary algorithm to get starting values in the first stage and deterministic numerical methods for a refinement of the initial approximation in the second stage.

§1. Introduction

Consider the initial portfolio $p^0 = (p_1^0, \ldots, p_n^0) \in \mathbb{R}^n$, p_k^0 denoting the market value of investment in asset class $k = 1, \ldots, n$. Given $T \in \mathbb{N}$ periods, our aim is to find regroupings $\delta^t := (\delta_1^t, \ldots, \delta_n^t) \in \mathbb{R}^n$, $t = 0, \ldots, T$, such that the risk of losing money is minimized while at the same time the resulting final portfolio at the end of period T maximizes profit. Note that this is a *multiperiod* $(T > 1)$ portfolio optimization. For classical portfolio optimization in the single period case, see *e.g.* Markowitz [9], Schöneburg, Heinzmann, and Feddersen [10], and Steiner and Bruns [17].

Having $n \in \mathbb{N}$ different asset classes and $T \in \mathbb{N}$ periods, we denote for $t = 0, \ldots, T$ the portfolio at the beginning of period t by

$$p^t = (p_1^t, \ldots, p_n^t) \in \mathbb{R}^n,$$

with p_k^t the market value of the investment in asset class $k = 0, \ldots, n$. The portfolio after regrouping is written

$$\tilde{p}^t := p^t + \delta^t, \tag{1}$$

where $\delta^t := (\delta_1^t, \ldots, \delta_n^t)$ is the vector of regroupings δ_k^t in period t of investment in asset class k.

Approximation Theory X: Wavelets, Splines, and Applications   241
Charles K. Chui, Larry L. Schumaker, and Joachim Stöckler (eds.), pp. 241–260.
Copyright ⊕ 2002 by Vanderbilt University Press, Nashville, TN.
ISBN 0-8265-1416-2.

With yield estimates

$$r^t := (r_1^t, \ldots, r_n^t) \in \mathbb{R}^n, \quad t = 0, \ldots, T-1$$

the resulting portfolio at the end of period t or beginning of the next period $t+1$ is given by

$$p^{t+1} = (p_1^{t+1}, \ldots, p_n^{t+1}), \quad p_k^{t+1} := (1 + r_k^t) \cdot \tilde{p}_k^t. \tag{2}$$

§2. Optimization When Ignoring Risk

As a first step, we consider the case when optimizing the final portfolio $\tilde{p}^T$ ignoring risk. Because we want $\tilde{p}^T$ to have maximum performance, our objective function to optimize is the sum of all market values in the final portfolio after regrouping, *i.e.*,

$$|\tilde{p}^T| := \sum_{i=1}^n \tilde{p}_i^T \tag{3}$$

is to be maximized.

Using (1) and (2), we can formulate (3) in terms of the given initial portfolio p^0, regroupings δ^k and yield estimates r^k:

$$|\tilde{p}^T| = \sum_{i=1}^n \tilde{p}_i^T$$

$$= \sum_{i=1}^n \left(p_i^0 \cdot \prod_{t=0}^{T-1} (1 + r_i^t) \right) + \sum_{i=1}^n \sum_{k=0}^T \left(\delta_i^k \cdot \prod_{t=k}^{T-1} (1 + r_i^t) \right). \tag{4}$$

Obviously, for a fixed initial portfolio p^0 and fixed yield estimates r^k, (4) is a linear function in terms of the regroupings $\delta := (\delta^0, \ldots, \delta^T)$:

$$|\tilde{p}^T| = |\tilde{p}^T|(\delta) = c + \zeta^T \cdot \delta, \quad c \in \mathbb{R}, \quad \zeta, \delta \in \mathbb{R}^{n \cdot (T+1)}. \tag{5}$$

Although ignoring risk, we have to take some constraints into consideration, because we have to model different investment strategies and legal provisions. For each period, these are:

(1) constraints on absolute combinations of asset classes in a portfolio: $\sum_i a_i \cdot \tilde{p}_i \leq a$,

(2) constraints on relative combinations of asset classes: $\sum_i b_i \cdot \tilde{p}_i \leq b \cdot |\tilde{p}|$,

(3) constraints on absolute combinations of regroupings: $\sum_i c_i \cdot \delta_i \leq c$,

(4) constraints on relative combinations of regroupings: $\sum_i d_i \cdot \delta_i \leq d \cdot |\tilde{p}|$.

Using matrices A_t, B_t, C_t, D_t and vectors a_t, b_t, c_t, d_t, for $t = 0, \ldots, T$, these constraints can be written as

$$
\begin{aligned}
A_t \cdot \tilde{p}^t &\leq a_t, \\
B_t \cdot \tilde{p}^t &\leq |\tilde{p}^t| \cdot b_t, \\
C_t \cdot \delta^t &\leq c_t, \\
D_t \cdot \delta^t &\leq |\tilde{p}^t| \cdot d_t
\end{aligned}
\tag{6}
$$

(where here and in the following, "$\leq$" applied to vectors means componentwise "less or equal").

Because of (1) and (2), each $\tilde{p}_k^t$ can we written as an affine linear function of δ. Therefore, we can reformulate all constraints (6) in terms of the regroupings δ. Our final optimization problem when ignoring risk then reads

$$
\begin{aligned}
|\tilde{p}^T| &= c + \zeta^T \cdot \delta \to \max, \\
F \cdot \delta &\leq f, \quad F \in \mathbb{R}^{m \times (n \cdot (T+1))}, \quad f \in \mathbb{R}^m,
\end{aligned}
\tag{7}
$$

where $m \in \mathbb{N}$ depends on the number of constraints in (6). This is a standard linear optimization problem with linear constraints, and can easily be solved by the simplex algorithm.

§3. Optimization When Considering Risk

Now we come to the case of considering risk while optimizing the final portfolio. We will modify the objective function (3) and extend the constraints (6) in such a way that not only the final portfolio has maximum profit, but also the risk of losing money on the way to it will be minimized. To this end, we first introduce value-at-risk as a risk measure.

3.1 Value-at-Risk as risk measure

During the past few years, Value-at-Risk (VaR) has become very important in finance as an easy to understand measure for risk. Here, we will only give a short overview of the idea and will give some formulas, important for our optimization problem. For a more detailed introduction to the subject, see *e.g.* Dowd [4], Poddig, Dichtl, Petersmeier [10], or Steiner, Bruns [17].

Given a single asset class i and probability α, the corresponding value-at-risk $\mathrm{VaR}(\alpha, p_i^t)$ at time t is defined as

$$
P\left((p_i^{t+1} - p_i^t) \leq -\mathrm{VaR}(\alpha, p_i^t)\right) := \alpha.
\tag{8}
$$

This means that with probability α, you will lose more money between period t and $t+1$ than $\mathrm{VaR}(\alpha, p_i^t)$. In other words, with probability $1 - \alpha$, you will *not* lose more money than $\mathrm{VaR}(\alpha, p_i^t)$ between period t and $t+1$.

One way to calculate value-at-risk for a specific asset class is the so-called *analytical model* (other ways for VaR calculations are based on Monte Carlo simulations or time series analysis). The calculation is based on the assumption that the yields are normally distributed, *i.e.*,

$$\frac{p_i^{t+1} - p_i^t}{p_i^t} \sim N(\mu, \sigma^2). \tag{9}$$

Applying (9) to (8), we obtain

$$\text{VaR}(\alpha, p_i^t) = -(z_\alpha \cdot \sigma + \mu) \cdot p_i^t, \tag{10}$$

with z_α the α-quantile of the normal distribution (typical values for α are 0.01 or 0.05). In many applications, one also has the additional assumption that the yields have $\mu = 0$, so that VaR is approximated by

$$\text{VaR}(\alpha, p_i^t) \approx -z_\alpha \cdot \sigma \cdot p_i^t. \tag{11}$$

The last formula for the approximated value of risk of a single asset class can be extended to the value of risk of a whole portfolio:

$$\text{VaR}(\alpha, \tilde{p}^t) = -z_\alpha \cdot \sqrt{(\tilde{p}^t)^T \cdot \left(Cov(r_i^t, r_j^t)\right)_{i,j} \cdot \tilde{p}^t},$$

with $\left(Cov(r_i^t, r_j^t)\right)_{i,j}$ the positive semidefinite covariance matrix of the portfolio. In terms of the regroupings δ, the portfolio's value-at-risk can be written as

$$\text{VaR}(\alpha, \tilde{p}^t)^2 = \gamma^t + g_\alpha^t \cdot \delta + \delta^T \cdot G_\alpha^t \cdot \delta,$$

where $\gamma \in \mathbb{R}$, $g_\alpha^t \in \mathbb{R}^{n \cdot (T+1)}$ and $G_\alpha^t \in \mathbb{R}^{(n \cdot (T+1)) \times (n \cdot (T+1))}$.

3.2 Optimization under risk constraints

Now, we will extend our optimization problem in such a way, that we will not only maximize the profit of the final portfolio, but also minimize the risk of losing money.

First, we will modify the objective function (3) by introducing a risk parameter $\theta \in [0, 1]$, and define

$$|\tilde{p}^T| := \sum_{i=1}^{n} \left(p_i^0 \cdot \prod_{t=0}^{T-1} (1 + r_i^t) \right) + \theta \cdot \sum_{i=1}^{n} \sum_{k=0}^{T} \left(\delta_i^k \cdot \prod_{t=k}^{T-1} (1 + r_i^t) \right)$$

$$- (1 - \theta) \cdot \sum_{t=0}^{T} \text{VaR}^2(\alpha, \tilde{p}^t), \qquad \theta \in [0, 1] \tag{12}$$

$$= \tilde{\gamma} + \tilde{g}_{\theta,\alpha}^T \cdot \delta + \delta^T \cdot \tilde{G}_{\theta,\alpha} \cdot \delta,$$

where $\tilde{\gamma} \in \mathbb{R}$, $\tilde{g}_{\theta,\alpha} \in \mathbb{R}^{n \cdot (T+1)}$, $\tilde{G}_{\theta,\alpha} \in \mathbb{R}^{(n \cdot (T+1)) \times (n \cdot (T+1))}$.

With $\theta = 1$, the objective function (12) is equal to the original objective function (4) without risk consideration. Having $\theta = 0$ minimizes the risk only, and ignores profit maximization. Therefore, to maximize the value of the final portfolio and minimize risk at the same time, one has to select a suitable value for θ which is located between 0 and 1. Choose values near 1, if maximizing profit is most important. If avoiding risk is getting more important, select values of θ, which are near 0.

We also extend the constraints (6) by the additional requirement that in every period the value-at-risk of the portfolio is bounded by a given constant $v^t \in \mathbb{R}$, $t = 0, \ldots, T$. The resulting constraints read

$$F \cdot \delta \le f$$
$$\mathrm{VaR}(\alpha, \tilde{p}^t)^2 = \gamma^t + g_\alpha^t \cdot \delta + \delta^T \cdot G_\alpha^t \cdot \delta \le v^t, \qquad t = 0, \ldots, T. \tag{13}$$

From (12) and (13) it becomes clear that our multiperiod portfolio optimization problem considering additional value-at-risk constraints is a quadratic optimization problem with quadratic constraints.

Because the covariance matrix is positive semidefinite, this optimization problem is convex (strict convex, iff all eigenvalues of the covariance matrix are positive), and therefore every local maximum is a global maximum. We can therefore distinguish between the following three cases:

(1) The set of feasible points is empty: no solution exists.

(2) The set of feasible points is not empty and all eigenvalues of the covariance matrix are positive: a unique solution exists.

(3) The set of feasible points is not empty and the eigenvalues of the covariance matrix are nonnegative: There is a nonempty set of solutions to the optimization problem (all giving the same maximum value).

§4. Numerical Algorithm

To maximize the objective function (12) under risk constraints (13), we use a two-step approach. In the first step, we use an evolutionary algorithm (EA), to find appropriate approximations to the optimal solution. In case that the covariance matrix is positive definite, we skip this step, because then the solution is unique and we only have to provide one initial feasible point for the application of further iterative numerical methods. In the general case of a positive semidefinite covariance matrix we have more than one solution, and the final population found by the EA will give an approximation to the multidimensional space of all possible solutions. After that, we improve the approximations by deterministic numerical procedures of constrained optimization. This finally results in a set of representatives of the solution space (which contains at most one element, in the strict positive definite case).

4.1 Evolutionary algorithm

Given the initial portfolio p^0 and yield estimates r^t, $t = 0, \dots, T-1$, we start with a set $\Delta^{(0)} := \{\delta_i^{(0)}; i = 1, \dots, M\}$ of $M \in \mathbb{N}$ regrouping vectors $\delta_i^{(0)} \in \mathbb{R}^{n \cdot (T+1)}$, which only have to satisfy the constraints (13), *i.e.*, we do not suppose any kind of optimality at this initial stage. This is also called the initial population. Note that finding an initial set of feasible regroupings $\delta_i^{(0)}$ can be a very difficult task.

In the following, we use an evolutionary algorithm for an iterative improvement of the initial population $\Delta^{(0)}$. An introduction to these optimization algorithms can be found, *e.g.* in Fogel, Owens, Walsh [5], Holland [8], Rechenberg [13], Schöneburg, Heinzmann, Feddersen [14], Schwefel [15]. We describe the iteration from step k to $k + 1$:

As input we have

- regroupings $\Delta^{(k)} = \{\delta_1^{(k)}, \dots, \delta_M^{(k)}\}$, $\delta_i^{(k)} \in \mathbb{R}^{n \cdot (T+1)}$,
- variances for the evolution algorithm

$$\Sigma^{(k)} = \{\Sigma_1^{(k)}, \dots, \Sigma_M^{(k)}\}, \quad \Sigma_i^{(k)} \in \mathbb{R}^{n \cdot (T+1)}.$$

Using the variances, we can control how much the EA may change the already found regroupings. Usually, we allow large changes in the beginning of the algorithm. Later on, when the regroupings converge to approximations of optimal solutions, we will only allow smaller changes.

The idea behind EA is to apply small changes in the elements of already found solutions (mutate) and to check, if the resulting elements give better solutions. To do so, we first have to select the number of elements, we want to mutate:

1) Select a number $s \in \{0, \dots, M\}$ of elements to mutate;

Then select s elements (regroupings and variances) from the exisiting population:

2) select elements:

$$\tilde{\Delta}^{(k)} = \{\delta_{\pi(1)}^{(k)}, \dots, \delta_{\pi(s)}^{(k)}\},$$
$$\tilde{\Sigma}^{(k)} = \{\Sigma_{\pi(1)}^{(k)}, \dots, \Sigma_{\pi(s)}^{(k)}\},$$

with $\pi : \{0, \dots, s\} \to \{0, \dots, M\}$ injective;

Then modify (mutate) the selected regroupings and variances. We describe this procedure for one regrouping vector

$$\delta_{\pi(\ell)}^{(k)} = x^{(k)} = (x_1^{(k)}, \dots, x_N^{(k)}),$$

and the corresponding variance vector

$$\Sigma_{\pi(\ell)}^{(k)} = \sigma^{(k)} = (\sigma_1^{(k)}, \ldots, \sigma_N^{(k)}),$$

(*i.e.*, for one element of the regrouping population and one element of the variance population). In the algorithm, this procedure is applied to every regrouping and variance vector of the actual population. Note that this can be done in parallel.

3) Modify the variances $\sigma_i^{(k+1)} := \sigma_i^{(k)} \cdot e^\nu$, with random variable $\nu \sim N(0, \sigma)$, $\sigma \geq 0$.

4) Modify the regroupings $x_i^{(k+1)} := x_i^{(k)} + \epsilon_i^{(k)}$, $\epsilon_i^{(k)} \sim N(0, \sigma_i^{(k+1)})$.

In case the modified regroupings $x_i^{(k+1)}$ do not satisfy the given constraints, reject them and keep the older ones, *i.e.*,, let $x_i^{(k+1)} := x_i^k$. Additionally, reduce the variances $\sigma_i^{(k+1)} := \sigma_i^{(k)}/2$.

Applying steps 3) and 4) to all elements of the population, we obtain the

5) modified elements:

$$\hat{\Delta}^{(k)} = \{\delta_{\pi(1)}^{(k+1)}, \ldots, \delta_{\pi(s)}^{(k+1)}\},$$
$$\hat{\Sigma}^{(k)} = \{\Sigma_{\pi(1)}^{(k+1)}, \ldots, \Sigma_{\pi(s)}^{(k+1)}\}.$$

The next generation is now created by selecting the best elements from the old population and the modified elements of step 5). Here, "best" is understood in terms of the objective function (12), *i.e.*, element δ_1 is better that δ_2 iff $|\tilde{p}^T|(\delta_1) > |\tilde{p}^T|(\delta_2)$ (in case of maximization).

4.2 Sequential quadratic programming

After P iterations, we obtain a set of approximate solutions $\Delta^{(P)}$ to our portfolio optimization problem. To improve those solutions, we use every single element $\delta_i^{(P)} \in \Delta^{(P)}$ as the initial value for further optimization by sequential quadratic programming SQP (see *e.g.* Fletcher [6], Gill, Murray, and M. H. Wright [7], and Stoer [18]).

The basic idea behind SQP applied to the non-linear optimization problem

$$f(x) = \max ,$$
$$g_i(x) \leq 0 , \; i = 1, \ldots, N, \tag{14}$$

is to reduce it to a sequence of quadratic programs, which can then be solved using standard methods. This reduction is done by quadratic approximation of the Lagrangian function

$$L(x, \lambda) = f(x) + \sum_{i=1}^{N} \lambda_i \cdot g_i(x)$$

of (14) and by linearization of the constraints g_i. Doing so results in the quadratic program

$$f'(x_k)^T \cdot x + \frac{1}{2} x^T \cdot f''(x_k) \cdot x = \max,$$

$$g_i(x_k) + g_i'(x_k)^T \cdot x \le 0, \qquad i = 1, \ldots, N. \tag{15}$$

The solution $x_{\min}$ of (15) is used to calculate the next approximate solution x_{k+1} of (14) by performing a line search along $x_{\min}$ to find some suitable $\alpha_k \in \mathbb{R}$ with $x_{k+1} = x_k + \alpha_k \cdot x_{\min}$.

In our computations, we use the function `fmincon` of the Matlab Optimization Toolbox [1] to solve the nonlinear maximization problem (14) by SQP. Because `fmincon` solves minimization problems only, we have to provide $(-f)$ as objective function (instead of f) to calculate the solution of (14).

Finally, after having applied SQP using the approximate solutions from $\Delta^{(P)}$ as initial values, we obtain a set $\tilde{\Delta}^{(P)}$ of (not necessarily different) improved regroupings $\tilde{\delta}_1^{(P)}, \ldots, \tilde{\delta}_M^{(P)}$. These can be understood as different ways or strategies to optimize the final portfolio. Further selections can be done using application specific rating functions (e.g. net present value). A unique solution is obtained iff the covariance matrix is positive definite.

§5. Examples

This section describes the application of our algorithm to two examples. In the first example, we consider the convex case, which is typical in portfolio optimization, as can be seen from (12) and (13) and from the fact, that the covariance matrix $\left(Cov(r_i^t, r_j^t) \right)_{i,j}$ has only nonnegative eigenvalues.

As a second example, we study the application to the more difficult case of non-convex optimization, on the basis of the kissing number problem. Although this is more than needed in our presently given portfolio optimization model, we like to demonstrate the applicability of the algorithm also in this case. It shows that the computational procedures presented in this paper can be used, even if the portfolio optimization model will be extended using nonconvex elements in the future.

5.1 Convex example

For $n, p \in \mathbb{N}$, $p \le n$, let the set $\mathcal{D}_{n,p}$ of feasible points be given by

$$\mathcal{D}_{n,p} := \{ x = (x_1, \ldots, x_n) \in \mathbb{R}^n \ ; \ x_i \ge 0, i = 1, \ldots, n, \text{ and } \|x\|_2 \le 1 \} \tag{16}$$

and define the objective function

$$f_p(x) := \sum_{i=1}^{p} (x_i - 1)^2 \, , \, x \in \mathcal{D}_{n,p} \, . \tag{17}$$

Obviously, on the set $\mathcal{D}_{n,p}$ the function f_p has global maximum p, which is attained exactly by the elements of the set

$$\mathcal{S}_{n,p} := \{x = (x_1, \ldots, x_n) \in \mathcal{D}_{n,p} \, ; \, x_1 = \cdots = x_p = 0\} \, . \tag{18}$$

By using matrices the objective function (17) can be rewritten as

$$f_p(x) = x' \cdot G_{n,p} \cdot x - 2 \, g_{n,p}' \cdot x + p,$$

with matrix $G_{n,p} = \text{diag}(1, \ldots, 1, 0, \ldots, 0) \in \mathbb{R}^{n \times n}$ (first p elements on the diagonal are 1) and vector $g_{n,p} = \sum_{i=1}^{p} e_i \in \mathbb{R}^n$.

The constraints in the feasible set (16) read

$$h_i(x) := x' \cdot A_i \cdot x + b_i' \cdot x + c_i \le 0, \qquad i = 0, 1, \ldots, n,$$

where $A_0 = I \in \mathbb{R}^{n \times n}$, $b_0 = 0 \in \mathbb{R}^n$, $c = 0$, and for $i = 1, \ldots, n$ we have $A_i = 0 \in \mathbb{R}^{n \times n}$, $b_i = -e_i \in \mathbb{R}^n$, and $c_i = 0$. Here and in the following, e_i denotes the i-th unit vector in $\mathbb{R}^n$. As can easily be seen, f_p and all h_i are convex functions and f_p is strict convex, iff $n = p$.

We demonstrate the algorithm in the two-dimensional case $n = 2$ and $p = 1$. Here, the set of maximizing points is given by

$$\mathcal{S}_{2,1} = \{(x_1, x_2) \, ; \, x_1 = 0, \, x_2 \in [0, 1]\}.$$

The evolutionary algorithm is parameterized to compute 6 generations with population size 30. The course of the computation is plotted in Fig. 1. It begins in the upper left corner, where the randomly generated initial population is given. After that, all six generations are plotted.

From the pictures it becomes clear, that the evolutionary process converges very fast to an approximation of the solution set $\mathcal{S}_{2,1}$. The population of the final generation is given at the lower left corner. To improve this approximation, every element of that generation is used as an initial point for maximizing the objective function f_1 by SQP. The final solution found by the algorithm is plotted at the lower right corner of Fig. 1. As expected, we obtain a set of different points, all contained in the solution space $\mathcal{S}_{2,1}$.

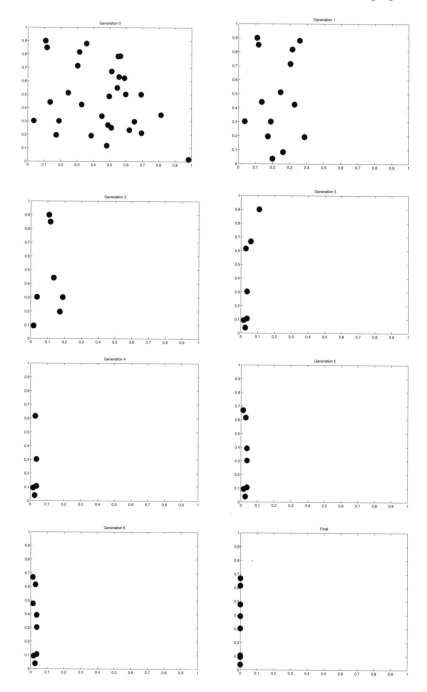

Fig. 1. Optimizing in the convex case for $n = 2$ and $p = 1$.

5.2 The kissing number problem

We apply our algorithm to the problem of arranging m unit spheres around the fixed unit sphere located at the origin in $\mathbb{R}^n$, such that the sum of the distances to these unit spheres is as small as possible. To avoid trivial cases, we always require $m \geq n$. This problem belongs to the family of sphere packing problems and is associated to the kissing number problem (see *e.g.* Conway and Sloane [3]). It is a common test problem in nonlinear programming and part of CUTE, the *Constrained and Unconstrained Testing Environment* [2] (problem KISSING2). Note that this is a nonconvex optimization problem.

For given $n, m \in \mathbb{N}$, define the objective function

$$f_{n,m}(p_1, \ldots, p_m) := \sum_{i=1}^{m} p_i^T \cdot p_i \,, \qquad p_1, \ldots, p_m \in \mathbb{R}^n. \tag{19}$$

If p_i denotes the center of the i-th sphere, we must have $\|p_i\|_2 \geq 2$, because otherwise sphere i would intersect the fixed unit sphere. Also, because two distinct spheres may not intersect, we require $\|p_i - p_j\|_2 \geq 2$ for all $i < j$. Together, this gives $N_c := m \cdot (m+1)/2$ constraints. The resulting optimization problem reads

$$\begin{aligned}
f_{n,m}(p_1, \ldots, p_m) &= \min, \\
(p_i - p_j)^T \cdot (p_i - p_j) &\geq 4, \quad i = 1, \ldots, m-1, \; j = i+1, \ldots, m \,, \\
p_i^T \cdot p_i &\geq 4, \quad i = 1, \ldots, m.
\end{aligned} \tag{20}$$

In case $\|p_i\|_2 = 2$ for all i, that is, the m spheres can be arranged in such a way that they all touch the fixed sphere, m is called a kissing number for dimension n. For such an arrangement, the objective function $f_{n,m}$ has absolute minimum $4m$. A table of the highest kissing numbers presently known can be found at N.J.A. Sloane's webpage [16]. We list the first five kissing numbers in Tab. 1 below.

Before applying our algorithm to problem (20), we reduce the problem size by fixing components and therefore avoid simple modifications by rotations. To do this we require

$$\text{span} \langle p_1, \ldots, p_k \rangle \perp \text{span} \langle e_{k+1}, \ldots, e_n \rangle, \quad k = 1, \ldots, n-1, \tag{21}$$

which is equivalent to

$$p_i = \sum_{\ell=1}^{i} x_i^\ell e_\ell \,, \; i = 1, \ldots, n-1, \tag{22}$$

dim n	kissing number m	problem size N_v	constraints N_c
1	2	2	3
2	6	11	21
3	12	33	78
4	24	90	300
5	40	190	820

Tab. 1. The first five kissing numbers.

with numbers $x_i^\ell \in \mathbb{R}$. Considering (21), problem (20) has

$$N_v := \frac{n \cdot (n+1)}{2} + (m - n) \cdot n$$

variables (for $m \geq n$).

For numbers $x_i^\ell \in \mathbb{R}$, $i = 1, \ldots, m$, $\ell = 1, \ldots, \min\{i, n\}$, let

$$x := (x_1^1, x_2^1, x_2^2, \ldots, x_i^1, \ldots, x_i^{\min\{i,n\}}, \ldots, x_m^1, \ldots x_m^n) \in \mathbb{R}^{N_v},$$

such that $p_i = \sum_{\ell=1}^{\min\{i,n\}} x_i^\ell \, e_\ell$. Using this, the objective function $f_{n,m}$ from (19) can be expressed as a function of x:

$$f_{n,m}(x) = \sum_{i=1}^{m} \sum_{\ell=1}^{\min\{i,n\}} \left(x_i^\ell\right)^2 = x^T \cdot x. \tag{23}$$

To describe the constraints in (20) using x, we define matrices $S_{i,j} \in \mathbb{R}^{N_v \times N_v}$ by

$$x^T \cdot S_{i,j} \cdot x := p_i^T \cdot p_j. \tag{24}$$

Note that $S_{i,j}$ is sparse, because it has only $\min\{i, j, n\}$ nonzero entries.

Using (23) and (24), our optimization problem (20) can be written in terms of x as

$$f_{n,m}(x) = x^T \cdot x = \min$$
$$x^T \cdot (-S_{i,i} + 2S_{i,j} - S_{j,j}) + 4 \leq 0, \qquad i < j, \tag{25}$$
$$-x^T \cdot S_{i,i} \cdot x + 4 \leq 0, \qquad i = 1, \ldots, m \,.$$

We apply our algorithm to (25) using $m = 6, 9, 18$ and $n = 2$. The corresponding problem dimensions and number of constraints are listed in Tab. 2.

n	m	problem size N_v	constraints N_c
2	6	11	21
2	9	17	45
2	18	35	171

Tab. 2. Numerical examples.

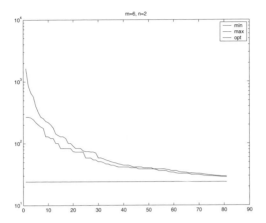

Fig. 2. EA convergence minimizing $f_{6,2}$.

In all examples, we start the evolutionary algorithm using initial population size 100 and compute 80 generations. The initial population is randomly selected from the set $\{p \in \mathbb{R}^n \; ; \; 2 \le \|p\| \le 20\}$, such that all constraints are satisfied.

The first example, $n = 2$ and $m = 6$, has the exact solution 24, because $m = 6$ is a kissing number in dimension 2. In Fig. 2, we plotted the minimum and maximum value of $f_{6,2}$ for every generation of the evolutionary algorithm. Also, the value of the optimal solution is plotted (as a horizontal line). As one can see, the evolutionary algorithm already yields a good approximation to the exact solution. In Fig. 3 we plotted the best configurations of spheres during the EA run for generation $10, 20, \ldots, 80$. The best configuration, found after applying SQP to the approximate solutions of the last generation, is given in Fig. 4.

In our second example, we increased the number of spheres around the fixed unit sphere from $m = 6$ to $m = 9$, thus increasing the problem dimension by 50% and doubling the number of quadratic constraints. As becomes clear from Fig. 5, also in this case the evolutionary algorithm will produce already a very good approximation to the final solution. Again, we plotted the best sphere configurations, as found by the EA in generation $10, 20, \ldots, 80$, in Fig. 6.

Finally, we set the number of spheres to $m = 18$. In this case, Fig. 8 shows us, that for a population size of 100 configurations, 80 generations are not enough for a reasonable approximation of the minimum. Better results can be obtained by calculating more generations (Fig. 10) and additionally choosing a larger population size (see Fig. 11). Nevertheless, the approximate solutions found, are good enough for SQP to converge. The optimal sphere configuration found after applying SQP is plotted in Fig. 9.

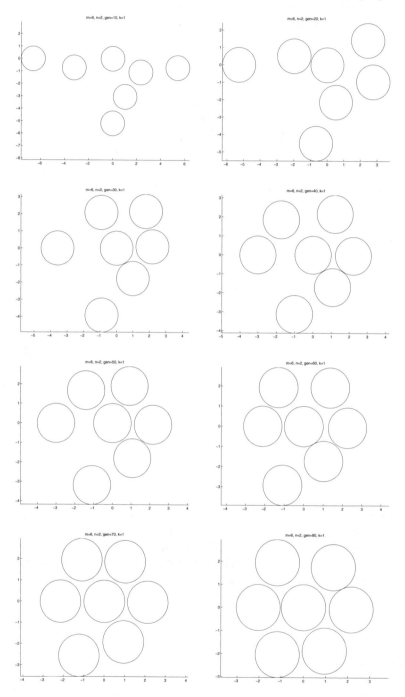

Fig. 3. EA optimizing $f_{6,2}$.

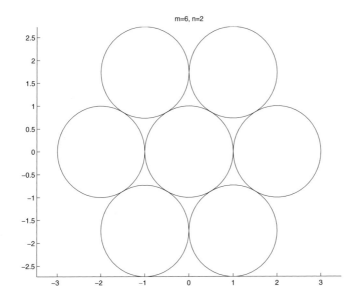

Fig. 4. Minimum of $f_{6,2}$ after SQP.

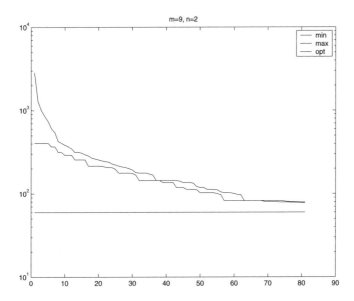

Fig. 5. EA convergence minimizing $f_{9,2}$.

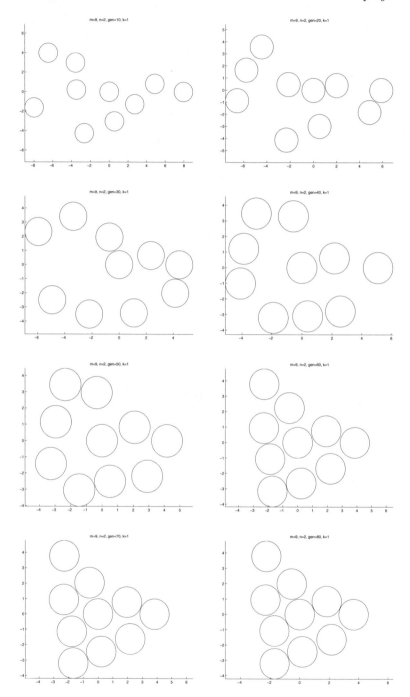

Fig. 6. EA optimizing $f_{9,2}$.

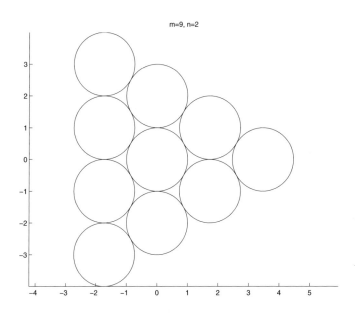

Fig. 7. Minimum of $f_{9,2}$ after SQP.

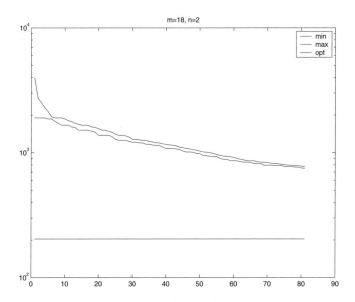

Fig. 8. EA convergence minimizing $f_{18,2}$.

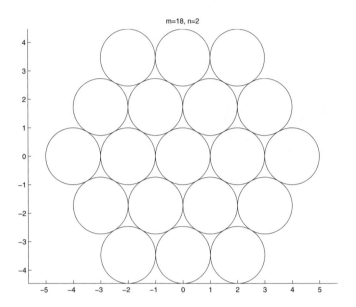

Fig. 9. Minimum of $f_{18,2}$ after SQP.

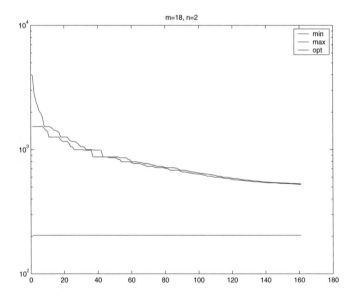

Fig. 10. EA convergence minimizing $f_{18,2}$, 160 generations.

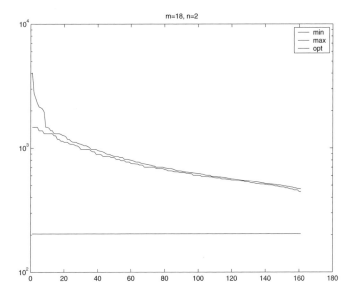

Fig. 11. EA convergence minimizing $f_{18,2}$, 160 generations, population size 200.

References

1. Coleman, Th., M.A. Branch, and A. Grace, *Optimization Toolbox User's Guide*, The Math Works, Inc., 1999.

2. Conn, A. R., I. Bongartz, N. I. M. Gould, and Ph. L. Toint, CUTE: Constrained and Unconstrained Testing Environment, technical report TR/PA 93/10, CERFACS, Toulouse, France, 1993.

3. Conway, J. H., and N. J. A. Sloane, *Sphere Packings, Lattices and Groups*, Springer-Verlag, NY, 1988.

4. Dowd, K., *Beyond Value at Risk*, John Wiley & Sons, 1998.

5. Fogel, L. J., A. J. Owens, and M. J. Walsh, *Artificial Intelligence through Simulated Evolution*, Wiley, New York, 1966.

6. Fletcher, R., *Practical Methods of Optimization*, Vol. 2, John Wiley and Sons., 1980.

7. Gill, P. E., W. Murray, and M. H. Wright, *Practical Optimization*, Academic Press, London, 1981.

8. Holland, J., *Adaption in Natural and Artificial Systems*, MIT Press, Cambridge, Mass., 1992.

9. Markowitz, H. M., Portfolio selection, The Journal of Finance, Vol. VII, 1952, 77–91.

10. Poddig, T., H. Dichtl, and K. Petersmeier, *Statistik, Ökonometrie, Optimierung*, Uhlenbruch Verlag, 2000.

11. Powell, M. J. D., A fast algorithm for nonlineary constrained optimization calculations, in *Numerical Analysis*, G. A. Watson (ed.), Lecture Notes in Mathematics, Springer Verlag, Vol. 630, 1978.

12. Powell, M. J. D., The convergence of variable metric methods for nonlinearly constrained optimization calculations, Nonlinear Programming 3, O. L. Mangasarian, R. R. Meyer, and S. M. Robinson (eds.), Academic Press, 1978.

13. Rechenberg, I., *Evolutionsstrategie*, Stuttgart, 1973.

14. Schöneburg, E., F. Heinzmann, and S. Feddersen, *Genetische Algorithmen und Evolutionsstrategien*, Addison-Wesley, 1994.

15. Schwefel, H.-P., *Numerische Optimierung von Computer-Modellen mittels der Evolutionsstrategie*, Basel, 1977.

16. Sloane, N. J. A., A table of the highest kissing numbers presently known, see `http://www.research.att.com/~njas/lattices/kiss.html`, 1999.

17. Steiner, M., and Ch. Bruns, *Wertpapiermanagement*, Stuttgart, 1998.

18. Stoer, J., Principles of sequential quadratic programming methods for solving nonlinear programs, in *Computational Mathematical Programming*, K. Schittkowski (ed.), Springer, 1985, 165–207.

Uwe Depczynski
FACT Unternehmensberatung GmbH
Goetheplatz 5
D-60313 Franfurt/Main, Germany
`u.depczynski@fact.de`

Quadratic Convergence of Newton's Method for Constrained Interpolation

Asen L. Dontchev, Houduo Qi, and Liqun Qi

Abstract. We give a review of our recent results on convergence of nonsmooth Newton-type methods for convex interpolation, convex smoothing, and shape-preserving interpolation.

§1. Convex Interpolation

Given nodes $a = t_1 < t_2 < \cdots < t_{N+2} = b$ and values $y_i = f(t_i)$, $i = 1, \ldots, N + 2$, $N \geq 3$, of an unknown function $f : [a, b] \to \mathbb{R}$, the standard interpolation problem consists in finding a function s from a given set S of interpolants such that $s(t_i) = y_i$, $i = 1, \ldots, N+2$. When S is the set of twice continuously differentiable piecewise cubic polynomials across t_i, we deal with cubic spline interpolation. The problem of cubic spline interpolation can be viewed in various ways; the closest to this paper is the classical Holladay variational characterization, according to which the natural cubic interpolating spline can be defined as the unique solution of the following optimization problem:

$$\text{minimize } \|f''\|_2 \quad \text{subject to} \quad f(t_i) = y_i, \ i = 1, \ldots, N + 2, \quad (1)$$

where $\| \cdot \|_2$ denotes the norm of $L^2[a, b]$. With a simple transformation, this problem can be written as a nearest point problem in $L^2[a, b]$: find the projection of the origin on the intersection of the hyperplanes

$$\left\{ u \in L^2[a, b] : \int_a^b u(t) B_i(t) dt = d_i, \quad i = 1, \ldots, N \right\},$$

Approximation Theory X: Wavelets, Splines, and Applications   261
Charles K. Chui, Larry L. Schumaker, and Joachim Stöckler (eds.), pp. 261–270.
Copyright © 2002 by Vanderbilt University Press, Nashville, TN.
ISBN 0-8265-1416-2.

where B_i are the piecewise linear normalized B-splines with support $[t_i, t_{i+2}]$ and d_i are the second divided differences of (t_i, y_i).

In this paper we consider problems of the type (1) with additional constraints on the "shape" of the interpolant. For example, if we add to the problem (1) the constraint $f'' \geq 0$ for a.e. $t \in [a, b]$, or equivalently, f should be convex, we obtain a convex interpolation problem the solution of which "preserves the shape" of the data provided that the data are "convex". If we add the constraint $f' \geq 0$, we obtain a monotone interpolation problem. Finally, the constraint $f \geq 0$ added to (1) results in the positive interpolation problem.

The area of constrained interpolation has a long history; its "modern" period starts with the path-breaking paper of Micchelli et al. [8] which linked, somewhat implicitly, these type of problems to duality in variational analysis. A subsequent paper by Irvine, Marin and Smith [7] laid the ground for the numerical analysis of constrained interpolation problems. In particular, they proposed a Newton-type method for shape-preserving interpolation and, based on numerical experience, conjectured its fast (quadratic) theoretical convergence. In a recent series of three papers, we studied their question. In our first paper [4] we proved that Newton's method of [7], applied to the convex interpolation problem, is locally superlinearly convergent. Later, in [5] we proved local quadratic convergence of the method for the problems of convex interpolation and convex smoothing. And finally, in [6] we showed local quadratic convergence for the general shape-preserving interpolation problem. In the present paper we review the general framework of these results; the interested reader may find full proofs and discussion of related works in [4-6].

Compared with the earlier work on constrained interpolation originated with Micchelli et al. [8], for a survey see [2], we approach this class of problems in a new way, by using recent advances in nonsmooth analysis. We will demonstrate the main idea on the convex interpolation problem of the form (1) with the additional constraint

$$f \text{ convex on } [a, b]. \tag{2}$$

If we assume that the second divided differences d_i are all positive, then we obtain existence of a solution of problem (1)–(2), and automatically uniqueness. The solution of (1)–(2) is a cubic spline whose second derivative is the positive part of a piecewise linear function across the grid $\{t_i\}$. Moreover, from the viewpoint of duality, the requirement $d_i > 0$ is a constraint qualification condition and then the problem is equivalent to its Lagrange dual which is an unconstrained finite-dimensional optimization problem of the form

$$\max_{\lambda \in \mathbf{R}^N} \left[-\frac{1}{2} \int_a^b \left(\sum_{i=1}^N \lambda_i B_i(t) \right)_+^2 dt + \sum_{i=1}^N \lambda_i d_i \right], \tag{3}$$

where $\lambda_1, \ldots, \lambda_N$ are the (unique) Lagrange multipliers associated with the interpolation conditions and $a_+ := \max\{0, a\}$. The objective function in (3) is concave and continuously differentiable, hence the problem (3) is equivalent to the standard optimality condition "the derivative equals zero," thus it reduces to the equation

$$F(\lambda) = d, \tag{4}$$

where $d = (d_1, \ldots, d_N)^T$ and the ith component of F is

$$F_i(\lambda) = \int_a^b \left(\sum_{l=1}^{N} \lambda_l B_l(t) \right)_+ B_i(t) dt. \tag{5}$$

It turns out, see [5] for details, that the function F is Lipschitz continuous but not (Frechét) differentiable everywhere in $\mathbb{R}^N$. Moreover, it is not even piecewise continuously differentiable, in the sense that it is not a continuous selection of a set-valued mapping whose graph is the union of the graphs of finitely many functions that are continuously differentiable.

To study the equation (4) we use nonsmooth analysis. Specifically, we first show that (4) is a strongly semismooth equation and then consider the associated version of Newton's method. For a locally Lipschitz continuous function $G : \mathbb{R}^n \to \mathbb{R}^n$, the generalized Jacobian $\partial G(x)$ of G at x in the sense of Clarke is the convex hull of all limits obtained along sequences on which G is differentiable:

$$\partial G(x) = \mathrm{co} \left\{ \lim_{x^j \to x} \nabla G(x^j) : G \text{ is differentiable at } x^j \in \mathbb{R}^n \right\}.$$

A function $G : \mathbb{R}^n \to \mathbb{R}^m$ is **strongly semismooth** at x if it is locally Lipschitz and directionally differentiable at x, and for all $h \to 0$ and $V \in \partial G(x+h)$ one has $G(x+h) - G(x) - Vh = O(\|h\|^2)$. The iteration of the generalized (semismooth) Newton method for the equation $G(x) = 0$ has the following form:

$$V_k(x^{k+1} - x^k) = -G(x^k), \quad V_k \in \partial G(x^k). \tag{6}$$

The following fundamental result is a generalization of the classical theorem of quadratic convergence of the Newton method.

Theorem 1. [11, Theorem 3.2] *Let $G : \mathbb{R}^n \to \mathbb{R}^n$ be strongly semismooth at x^* and let $G(x^*) = 0$. Assume that all elements V of the generalized Jacobian $\partial G(x^*)$ are nonsingular matrices. Then every sequence generated by the method (6) is quadratically convergent to x^* provided that the starting point x^0 is sufficiently close to x^*.*

In our paper [5] we proved that if the second divided differences d_i of (1) are positive, then the function F in (4) is strongly semismooth and

that all elements of the generalized Jacobian $\partial F(\lambda^*)$ at the solution λ^* are
nonsingular. Thus, from Theorem 1 above, for any choice of the elements
V_k, the generalized Newton method (6) has local quadratic convergence.
Furthermore, we showed in [5] that the Newton-type method proposed by
Irvine et al. [7] is a particular case of (6). Specifically, they chose

$$(M(\lambda))_{ij} = \int_a^b \left(\sum_{l=1}^{N} \lambda_l B_l(t) \right)_+^0 B_i(t) B_j(t) dt \tag{7}$$

with

$$(\tau)_+^0 = \begin{cases} 1, & \text{if } \tau > 0 \\ 0, & \text{if } \tau \le 0 \end{cases}$$

and define the Newton iterate for (4) as follows:

$$M(\lambda^k)(\lambda^{k+1} - \lambda^k) = -F(\lambda^k) + d. \tag{8}$$

The key observation in [5] is that $M(\lambda^k)$ is an element of the generalized
Jacobian of F; hence (8) is a particular case of (6) and Theorem 1 applies.

In the following section we review the problem of convex smoothing.
In Section 3 we deal with shape-preserving interpolation.

§2. Convex Smoothing

The convex smoothing problem is a minimization problem of the form

$$\text{minimize} \ \ \|f''\|_2 + \sum_{i,j=1}^{N+2} q_{ij}(f(t_i) - y_i)(f(t_j) - y_j) \tag{9}$$

$$\text{subject to} \quad f \text{ convex on } [a, b], \quad f \in W^{2,2}[a, b],$$

where, as before, $a = t_1 < t_2 < \cdots < t_{N+2} = b$ and y_i, $i = 1, \ldots, N+2$ are
given data. Here $W^{2,2}[a, b]$ denotes the Sobolev space of functions with
absolutely continuous first derivatives and second derivatives in $L^2[a, b]$.
In the problem (9) the second term represents least squares with weights
q_{ij} while the first one is a regularizing term representing the "energy"
of the solution. Without the convexity constraint, the solution to this
problem is the well-known smoothing spline. The convex smoothing problem was considered by Irvine, Marine and Smith in [7]; they showed that
its (unique) solution is still a spline whose second derivative is the positive part of a piecewise linear function across the grid points t_i. For
far-reaching generalizations of this result see Micchelli and Utreras [9].
Elfving and Andersson [1] proposed a Newton-type method for solving
(9) analogous to the method (8) for convex interpolation. They proved

quadratic convergence of this method on an assumption that the zeros of $\sum_{i=1}^{N} \lambda_i^* B_i$ are simple, where λ^* is the (unique) solution of (9). This turns out to be equivalent to the condition that the dual functional is twice continuously differentiable. Here we show that the local quadratic convergence of the Newton method easily follows from the analysis outlined in Section 1 without this assumption.

Analogously to the convex interpolation problem (1)–(2), problem (9) can be reformulated as a nonsmooth equation. Given data $((t_1, z_1), \ldots, (t_{N+2}, z_{N+2}))$, we denote the vector of the second divided differences by $d = (d_1, \ldots, d_N)$; then d satisfies $d = Kz$. Here K is the three-diagonal matrix whose elements of the i-th row are the coefficients of the second divided difference for t_i, t_{i+1}, t_{i+2}. As is well known, the matrix K has full rank. Define the vector function $B(t) = (B_1(t), \ldots, B_N(t))^T$. Integration by parts gives us

$$\int_a^b f''(t) B(t) dt = Kz, \quad z_i = f(t_i), \quad i = 1, 2, \ldots, N + 2.$$

Denote by Q the square matrix (q_{ij}) which we assume symmetric and positive definite, and let $y = (y_1, y_2, \ldots, y_{N+2})$. Then the problem (9) can be written in the following way:

$$\text{minimize} \quad \frac{1}{2}(\|u\|_2^2 + \langle z - y, Q(z - y) \rangle) \tag{10}$$

$$\text{subject to} \quad \int_a^b u(t) B(t) dt = Kz, \ u \geq 0 \ \text{a.e.,} \ u \in L^2[a, b], \ z \in \mathbb{R}^{N+2}.$$

This is a convex optimization problem which has a unique solution. The first-order optimality condition associated with the Lagrange multiplier rule for this problem is of the form

$$u(t) = \left(\sum_{i=1}^{N} \lambda_i B_i(t) \right)_+, \quad Q(z - y) + K^T \lambda = 0,$$

where λ is the vector of Lagrange multipliers associated with the equality constraint. The Lagrange dual problem to (10) becomes

$$\max_{\lambda \in \mathbb{R}^N} \left[-\frac{1}{2} \int_a^b \left(\sum_{i=1}^{N} \lambda_i B_i(t) \right)_+^2 dt - \frac{1}{2} \langle \lambda, KQ^{-1}K^T \lambda \rangle + \langle \lambda, Ky \rangle \right]. \tag{11}$$

This is a finite-dimensional convex optimization problem whose objective function is strongly concave and continuously differentiable. Therefore it is equivalent to the equation

$$S(\lambda) := F(\lambda) + KQ^{-1}K^T \lambda - Ky = 0, \tag{12}$$

where the components $F_i(\lambda)$ are given by (5).

The Newton-type method proposed by Andersson and Elfving [1] has the form

$$(M(\lambda^k) + KQ^{-1}K^T)(\lambda^{k+1} - \lambda^k) = -S(\lambda^k), \qquad (13)$$

where $M(\lambda)$ is given in (7); that is, as we already know, it is an element of the generalized Jacobian of F. Moreover, from the results mentioned at the end of the preceding section, the function S in the equation (12) is strongly semismooth since it is the sum of the strongly semismooth function F and a linear function. Also, the generalized Jacobian of S is nonsingular everywhere since each of its elements is the sum of the positive definite matrix $KQ^{-1}K^T$ and a nonnegative definite matrix which is an element of the generalized Jacobian of F (because F is monotone – note that we do not assume in this section that the second divided differences are positive). Thus, all conditions of Theorem 1.1 are satisfied and hence we automatically obtain local quadratic convergence of the method (13). Moreover, any method of the form (13), where $M(\lambda^k)$ is arbitrarily chosen from the generalized Jacobian of F, has local quadratic convergence.

§3. Shape-preserving Interpolation

The general problem of shape-preserving interpolation can be treated in an analogous way to the convex interpolation; while the general idea remains the same, the specific steps of the proof require more delicate analysis reflecting the more complicated structure of the constraints.

Let $\{(t_i, y_i)\}_1^{N+2}$ be given interpolation data and let $d_i, i = 1, 2, \ldots, N$ be the associated second divided differences. We assume that $d_i \neq 0$ for all $i = 1, \ldots, N$; we will discuss this assumption later. Define the following subsets $\Omega_i, i = 1, 2, 3$, of $[a, b]$:

$$\Omega_1 = \{[t_i, t_{i+1}] : d_{i-1} > 0 \text{ and } d_i > 0\},$$
$$\Omega_2 = \{[t_i, t_{i+1}] : d_{i-1} < 0 \text{ and } d_i < 0\},$$
$$\Omega_3 = \{[t_i, t_{i+1}] : d_{i-1}d_i < 0\}.$$

Also, let

$$[t_1, t_2] \subset \begin{cases} \Omega_1, & \text{if } d_1 > 0 \\ \Omega_2, & \text{if } d_1 < 0 \end{cases}, \qquad [t_{N+1}, t_{N+2}] \subset \begin{cases} \Omega_1, & \text{if } d_N > 0 \\ \Omega_2, & \text{if } d_N < 0. \end{cases}$$

The problem of shape-preserving interpolation stated by Micchelli et al. [8] is as follows:

$$\text{minimize } \|f''\|_2 \qquad (14)$$

$$\text{subject to } \begin{cases} f(t_i) = y_i, & i = 1, 2, \ldots, N + 2, \\ f''(t) \geq 0, & t \in \Omega_1, \\ f''(t) \leq 0, & t \in \Omega_2, \\ f \in W^{2,2}[a, b]. \end{cases}$$

In [8, Theorem 4.3] they showed that the solution of the problem (14) exists, is unique, and its second derivative has the following form:

$$f''(t) = \left(\sum_{i=1}^{N} \lambda_i B_i(t) \right)_+ \mathcal{X}_{\Omega_1}(t) - \left(\sum_{i=1}^{N} \lambda_i B_i(t) \right)_- \mathcal{X}_{\Omega_2}(t)$$

$$+ \left(\sum_{i=1}^{N} \lambda_i B_i(t) \right) \mathcal{X}_{\Omega_3}(t), \tag{15}$$

where $\lambda = (\lambda_1, \ldots, \lambda_N)^T \in \mathbb{R}^N$, $(a)_- = (-a)_+$, and $\mathcal{X}_\Omega$ is the characteristic function of the set Ω. As for the convex interpolation, the first-order optimality condition of the problem dual to (14) has the form of the nonlinear equation

$$F(\lambda) = d, \tag{16}$$

where $d = (d_1, \ldots, d_N)^T$ is the vector of the second divided differences and the vector function $F : \mathbb{R}^N \to \mathbb{R}^N$ has components

$$F_i(\lambda) = \int_{[t_i, t_{i+2}] \cap \Omega_1} \left(\sum_{l=1}^{N} \lambda_l B_l(t) \right)_+ B_i(t) dt$$

$$- \int_{[t_i, t_{i+2}] \cap \Omega_2} \left(\sum_{l=1}^{N} \lambda_l B_l(t) \right)_- B_i(t) dt$$

$$+ \int_{[t_i, t_{i+2}] \cap \Omega_3} \left(\sum_{l=1}^{N} \lambda_l B_l(t) \right) B_i(t) dt, \quad i = 1, 2, \ldots, N.$$

Irvine, Marine and Smith [7] proposed the following method for solving the equation (16): Given $\lambda^0 \in \mathbb{R}^N$, λ^{k+1} is a solution of the linear system

$$M(\lambda^k)(\lambda^{k+1} - \lambda^k) = -F(\lambda^k) + d. \tag{17}$$

Here $M(\lambda) \in \mathbb{R}^{N \times N}$ is the tridiagonal symmetric matrix with components

$$(M(\lambda))_{ij} = \int_a^b P(\lambda, t) B_i(t) B_j(t) dt,$$

where

$$P(\lambda, t) := \left(\sum_{l=1}^{N} \lambda_l B_l(t) \right)_+^0 \mathcal{X}_{\Omega_1}(t) + \left(\sum_{l=1}^{N} \lambda_l B_l(t) \right)_-^0 \mathcal{X}_{\Omega_2}(t) + \mathcal{X}_{\Omega_3}(t).$$

Following the pattern for the convex interpolation, in our paper [6] we proved that the method (17) is a particular case of the general nonsmooth Newton method (6). Moreover, the general method (6) applied to the problem (14) is locally quadratically convergent. In the proof, we first show that the function F in (16) is strongly semismooth and, if the second divided differences are all nonzero, then the generalized Jacobian of F at the solution contains positive definite matrices only. Compared to the convex interpolation, the argument here is more involved and needs special care of the regions where the constraints change from nonactive to active and back.

Local convergence does not imply global convergence of the Newton method even when the function minimized is convex, coercive and sufficiently smooth. It is, however, possible to modify the standard method to obtain global convergence. In [6] we proposed a damped version of the Newton method by using the following merit function:

$$L(\lambda) = \frac{1}{2} \int_a^b \left(\sum_{l=1}^{N} \lambda_l B_l(t) \right)^2 \mathcal{X}_{\Omega_1}(t)dt + \frac{1}{2} \int_a^b \left(\sum_{l=1}^{N} \lambda_l B_l(t) \right)^2 \mathcal{X}_{\Omega_2}(t)dt$$

$$+ \frac{1}{2} \int_a^b \left(\sum_{l=1}^{N} \lambda_l B_l(t) \right)^2 \mathcal{X}_{\Omega_3}(t)dt - \sum_{l=1}^{N} \lambda_l d_l.$$

This function is convex and continuously differentiable, with $\nabla L(\lambda) = F(\lambda) - d$. In [6] we showed that it is also coercive, that is, $|L(\lambda)| \to \infty$ as $\|\lambda\| \to \infty$. Then, any minimizing sequence will be bounded and therefore, since the solution is unique, will converge to the solution. Such problems are called well-posed in the sense of Tikhonov. Since $L(\lambda)$ is convex and coercive, and $\nabla L(\lambda) = F(\lambda) - d$, finding a solution of (16) is equivalent to solving the unconstrained optimization problem

$$\text{minimize } L(\lambda) \text{ subject to } \lambda \in \mathbb{R}^N. \tag{18}$$

We apply the following damped Newton method to the problem (18), which uses the Newton direction (17).

Algorithm 1.

0) Choose $\lambda^0 \in \mathbb{R}^N$, $\rho \in (0,1), \sigma \in (0,1/2)$, and tolerance tol > 0. $k := 0$.

1) (Termination criterion) If $\epsilon_k = \|F(\lambda^k) - d\| \leq$ tol, then stop; else go to 2.

2) (Direction generation) Choose s^k as a solution of the linear system

$$(M(\lambda^k) + \epsilon_k I)s = -\nabla L(\lambda^k).$$

3) (Line search) Choose m_k as the smallest nonnegative integer m satisfying

$$L(\lambda^k + \rho^m s^k) - L(\lambda^k) \leq \sigma \rho^m \nabla L(\lambda^k)^T s^k.$$

4) (Update) Set $\lambda^{k+1} = \lambda^k + \rho^{m_k} s^k$, $k := k + 1$, and return to step 1.

In [6] we proved that, for an arbitrary $\lambda^0 \in \mathbb{R}^N$, there is a unique sequence $\{\lambda^k\}$ generated by Algorithm 1 and this sequence converges quadratically to the solution. We also reported the results of extensive numerical experiments with the method. Overall, the method is fast and robust, achieving accuracy of order 10^{-13} in 10-12 iterations for a mid-size problem.

§4. Discussion

The main contribution of our research outlined in the present paper is to link constrained approximation with nonsmooth analysis and to demonstrate that the latter can be successfully used to solve a problem in the former that has been left open for quite a while. From the point of view of a nonsmooth analyst, the results are by far not unexpected. It is now well understood that, in general, traditional analytic methods may not work for optimization problems with constraints; however, such problems can be reformulated as nonsmooth problems. Here, the nonsmoothness is the "trade-off" for the constraints and is quite often easier to handle. The corresponding theory of nonsmooth analysis emerged already in the 70s, and is now becoming a standard tool for solving both theoretical and practical problems.

The present paper opens wide avenues going "deep" and "wide", to explore other constrained approximation problems or to study the convergence of other numerical methods for solving such problems. A challenging problem in this area is the problem of best approximation in a strip, which has the form of (1) with the constraint (2) replaced by

$$l(t) \leq f(t) \leq u(t) \quad \text{for a.e. } t \in [a, b],$$

where l and u are given functions. Some attempts at numerical treatment of such problems have been made e.g. in [3,10], but, in our opinion, no satisfactory understanding has yet been achieved as to how to approach their solution.

Regarding optimization methods, the attractive features of the trust region methods and the proximal point methods for solving nonsmooth problems make them prime candidates for applications to constrained approximation problems. There are a number of issues here that should be addressed, such as conditioning of the problems considered and complexity of the algorithms. To the best of our knowledge, conditioning of nonsmooth equations is a completely unexplored area which seems to be quite attractive for future research.

An attractive and potentially very important, but apparently very difficult, direction of future research would be to use ideas from constrained approximation of functions of one variable to build constrained curves and surfaces in higher dimensions.

References

1. Elfving, T. and L.-E. Andersson, An algorithm for computing constrained smoothing spline functions, Numer. Math. **52** (1988), 583–595.

2. Deutsch, F., Constrained best approximation. Approximation theory VI, Vol. I (College Station, TX, 1989), 165–173, Academic Press, Boston, MA, 1989.

3. Dontchev, A. L. and I. Kolmanovsky, Best interpolation in a strip. II. Reduction to unconstrained convex optimization, Comput. Optim. Appl. **5** (1996), 233–251.

4. Dontchev, A. L., H.-D. Qi and L. Qi, Convergence of Newton's method for convex best interpolation, Numer. Math. **87** (2001), 435–456.

5. Dontchev, A. L., H.-D. Qi and L. Qi, Quadratic convergence of Newton's method for convex interpolation and smoothing, Technical Report, School of Mathematics, the University of New South Wales, Sydney 2052, Australia, October 2000.

6. Dontchev, A. L., H.-D. Qi, L. Qi and H. Yin, A Newton method for shape-preserving interpolation, SIAM J. Optim., to appear.

7. Irvine, L. D., S. P. Marin and P. W. Smith, Constrained interpolation and smoothing, Constr. Approx. **2** (1986), 129–151.

8. Micchelli, C. A., P. W. Smith, J. Swetits and J. D. Ward, Constrained L_p approximation, Constr. Approx. **1** (1985), 93–102.

9. Micchelli, C. A. and F. I. Utreras, Smoothing and interpolation in a convex subset of a Hilbert space, SIAM J. Sci. Statist. Comput. **9** (1988), 728–747.

10. Oberle, H. J. and G. Opfer, Nonnegative splines, in particular of degree five, Numer. Math. **79** (1998), 427–450.

11. Qi, L. and J. Sun, A nonsmooth version of Newton's method, Math. Programming **58** (1993), 353–367.

Asen L. Dontchev
Mathematical Review
Ann Arbor, MI 48107, USA
ald@ams.org

Houduo Qi and Liqun Qi
Department of Applied Mathematics
The Hong Kong Polytechnic University
Hung Hom, Kowloon
Hong Kong
{mahdqi, maqilq}@polyu.edu.hk

Matrix-free Multilevel
Moving Least-Squares Methods

Gregory E. Fasshauer

Abstract. We investigate matrix-free formulations for polynomial-based moving least-squares approximation. The well-known Shepard's method is one such formulation that leads to $\mathcal{O}(h)$ approximation order. We are interested in methods with higher approximation orders. Several possible approaches are identified, and one of them – based on the analytic solution of small linear systems – is presented here. Numerical experiments with a multilevel residual updating algorithm are also presented.

§1. Introduction

Moving least-squares (MLS) methods have emerged as the basis of numerous meshless approximation methods that are being suggested by engineers as an alternative to the traditional finite element method (see, e.g., [1] and the references therein). The general MLS method seems to be attributable to Lancaster and Šalkauskas [6] with special cases going back to McLain [8] and Shepard [10]. In the mathematics literature this method has so far not attracted too much attention. Besides the original paper [6] there are only a few more papers by Šalkauskas as well as Farwig (see, e.g., [3]). One of the reasons MLS has not been seen as a viable approximation method for multivariate scattered data may be the much cited paper [5] by Franke in which a large number of scattered data approximation methods were evaluated, and the MLS variants did not get very high marks. Very recently, however, mathematicians have shown renewed interest in the MLS method, see e.g. [7,9,11].

In this paper we intend to make a few suggestions which may lead to more efficient computation of the MLS approximation. The paper is organized as follows. In the following two sections we review two alternate

Approximation Theory X: Wavelets, Splines, and Applications   271
Charles K. Chui, Larry L. Schumaker, and Joachim Stöckler (eds.), pp. 271–281.
Copyright ℗ 2002 by Vanderbilt University Press, Nashville, TN.
ISBN 0-8265-1416-2.

approaches to MLS approximation that show that evaluation of the MLS approximation at any given point involves the solution of a (small) system of linear equations. In Sect. 4 we summarize some of the main properties of MLS methods, and in Sect. 5 we present one possible way of obtaining a matrix-free MLS method. Then we investigate the use of MLS approximation within a multilevel approximation algorithm as studied previously in [4]. This investigation is supported by numerical experiments, as well as comments on connections to previous theoretical work. We conclude the paper with some remarks.

§2. MLS via Discrete Least Squares

We consider the following approximation problem. Given data $\{(\boldsymbol{x}_i, f(\boldsymbol{x}_i))\}_{i=1}^n \subset \mathbb{R}^d \times \mathbb{R}$ with distinct data sites $\boldsymbol{x}_i$ and f some (smooth) function, as well as an approximation space $\mathcal{U} = \mathrm{span}\{u_1, \ldots, u_m\}$ (with $m < n$), along with an inner product

$$\langle u, v \rangle_{\phi(\boldsymbol{x})} = \sum_{i=1}^n u(\boldsymbol{x}_i)v(\boldsymbol{x}_i)\phi_i(\boldsymbol{x}), \qquad \boldsymbol{x} \in \mathbb{R}^d \text{ fixed}, \tag{1}$$

where the positive weight functions ϕ_i, $i = 1, \ldots, n$, depend on the evaluation point $\boldsymbol{x}$, we wish to find the best approximation from $\mathcal{U}$ to f at the point $\boldsymbol{x}$ with respect to the norm induced by (1).

This means we will obtain the approximation (at the point $\boldsymbol{x}$) as

$$s(\boldsymbol{x}) = \sum_{j=1}^m c_j(\boldsymbol{x})u_j(\boldsymbol{x}), \tag{2}$$

where the coefficients $c_j(\boldsymbol{x})$ are such that

$$\sum_{i=1}^n \left[s(\boldsymbol{x}_i) - f(\boldsymbol{x}_i)\right]^2 \phi_i(\boldsymbol{x}) \to \min.$$

Due to the definition of the inner product (1) whose weight function "moves" with the evaluation point $\boldsymbol{x}$, the coefficients c_j in (2) depend also on $\boldsymbol{x}$. As a consequence, one has to solve the normal equations

$$\sum_{j=1}^m c_j \langle u_j, u_k \rangle_{\phi(\boldsymbol{x})} = \langle f, u_k \rangle_{\phi(\boldsymbol{x})}, \qquad k = 1, \ldots, m, \tag{3}$$

anew each time the evaluation point is changed. In matrix notation (3) becomes

$$G\boldsymbol{c} = \boldsymbol{f}, \tag{4}$$

with the positive definite Gram matrix $G = \left(\langle u_j, u_k \rangle_{\phi(\boldsymbol{x})}\right)_{j,k=1}^m$, coefficient vector $\boldsymbol{c}$ and right-hand side vector $\boldsymbol{f}$ as in (3) all depending on $\boldsymbol{x}$.

In this paper we will focus on the case when $\mathcal{U} = \mathcal{P}_Q^d$, the space of d-variate polynomials of total degree at most Q.

Example: Shepard's Method. If we take $\mathcal{U} = \mathcal{P}_0^d = \text{span}\{1\}$, then there is only a single normal equation

$$c_1(\boldsymbol{x})\langle 1, 1\rangle_{\phi(\boldsymbol{x})} = \langle f, 1\rangle_{\phi(\boldsymbol{x})},$$

which implies that

$$c_1(\boldsymbol{x}) = \frac{\sum_{i=1}^n f(\boldsymbol{x}_i)\phi_i(\boldsymbol{x})}{\sum_{i=1}^n \phi_i(\boldsymbol{x})},$$

and therefore (from (2))

$$s(\boldsymbol{x}) = \sum_{i=1}^n f(\boldsymbol{x}_i)\frac{\phi_i(\boldsymbol{x})}{\sum_{\ell=1}^n \phi_\ell(\boldsymbol{x})}. \tag{5}$$

This method is known as Shepard's method [10]. Note that Shepard's method (5) is now an explicit formula (valid for any $\boldsymbol{x} \in \mathbb{R}^d$). It no longer requires solution of a linear system.

It is our goal to derive explicit formulas similar to (5), but based on higher-degree polynomial spaces. As a first observation we note that we can simplify (2) by using shifted monomials, i.e., $\mathcal{U} = \mathcal{P}_Q^d = \text{span}\{(\cdot - \boldsymbol{x})^{\boldsymbol{\alpha}} : |\boldsymbol{\alpha}| \leq Q\}$. Here $\boldsymbol{\alpha} = (\alpha_1, \ldots, \alpha_d) \in \mathbb{N}^d$ is a multi-index with length $|\boldsymbol{\alpha}| = \sum_{i=1}^d |\alpha_i|$. For this choice of local basis we have $s(\boldsymbol{x}) = c_1(\boldsymbol{x})$ for any degree Q.

A second observation is mostly of theoretical interest. If one were to use an orthonormal basis (with respect to $\langle\ ,\ \rangle_{\phi(\boldsymbol{x})}$), then (since $G = I$) (3) implies

$$c_k(\boldsymbol{x}) = \langle f, u_k\rangle_{\phi(\boldsymbol{x})}, \qquad k = 1, \ldots, m,$$

and (from (2))

$$s(\boldsymbol{x}) = \sum_{j=1}^m \langle f, u_j\rangle_{\phi(\boldsymbol{x})} u_j(\boldsymbol{x}). \tag{6}$$

In this way MLS approximation can be viewed as a generalization of approximation by (truncated) Fourier series. Again, no linear system needs to be solved. However, discrete multivariate orthogonal polynomials over scattered centers are very difficult to construct.

§3. MLS via Constrained Optimization

The following connection between MLS approximation and constrained least squares minimization was first described by Bos and Šalkauskas [2]. It is referred to as the Backus-Gilbert formulation. The Backus-Gilbert formulation was also the approach used in [7] and [11] to obtain convergence order estimates for MLS approximation.

We now start with a quasi-interpolant of the form

$$\hat{f}(\boldsymbol{x}) = \sum_{i=1}^{n} f(\boldsymbol{x}_i)\gamma_i(\boldsymbol{x}), \qquad (7)$$

where the data $\{(\boldsymbol{x}_i, f(\boldsymbol{x}_i))\}$ are as before. The goal is to determine the functions γ_i in such a way that (7) is easy to compute and has high approximation order. To this end we enforce polynomial reproduction, i.e.,

$$\sum_{i=1}^{n} p(\boldsymbol{x}_i)\gamma_i(\boldsymbol{x}) = p(\boldsymbol{x}), \qquad \text{for all } p \in \mathcal{P}_Q^d.$$

In addition, a weighted norm of the γ functions is minimized according to

$$\frac{1}{2}\sum_{i=1}^{n}\gamma_i^2(\boldsymbol{x})w_i(\boldsymbol{x}) \to \min. \qquad (8)$$

Here the w_i are positive weight functions whose relation to the weights ϕ_i used earlier we point out below. The norm minimization (8) is motivated physically in the Backus-Gilbert theory, since this minimizes the "spread" of γ (see [2]).

Using Lagrange multipliers $\lambda_j(\boldsymbol{x})$ (which will again depend on the evaluation point $\boldsymbol{x}$) we can compute

$$\gamma_i(\boldsymbol{x}) = \frac{1}{w_i(\boldsymbol{x})}\sum_{j=1}^{m}\lambda_j(\boldsymbol{x})p_j(\boldsymbol{x}_i), \qquad i = 1, \dots, n, \qquad (9)$$

where the Lagrange multipliers $\lambda_j(\boldsymbol{x})$ are obtained as the solution of

$$G\boldsymbol{\lambda} = \boldsymbol{q}. \qquad (10)$$

Here $\boldsymbol{q} = [p_1(\boldsymbol{x}), \dots, p_m(\boldsymbol{x})]^T$, and the matrix G is identical with the Gram matrix G in (4) provided we identify $\dfrac{1}{w_i(\boldsymbol{x})} = \phi_i(\boldsymbol{x})$.

If we choose shifted monomials as basis of the polynomial reproduction space, then $\boldsymbol{q} = [1, 0, \dots, 0]^T$ and the formulas we obtain for the λ_j below simplify. Moreover, the numerical computations become more stable since all terms are expanded about the evaluation point $\boldsymbol{x}$.

One can also see the connection to Fourier series approximation using the Backus-Gilbert formulation, but it is not quite as obvious. In this case we get (using (10) with $G = I$)

$$\lambda_k(\boldsymbol{x}) = p_k(\boldsymbol{x}), \qquad k = 1, \dots, m,$$

and (from (9) with $\frac{1}{w_i(\boldsymbol{x})} = \phi_i(\boldsymbol{x})$)

$$\gamma_i(\boldsymbol{x}) = \phi_i(\boldsymbol{x}) \sum_{j=1}^{m} p_j(\boldsymbol{x}) p_j(\boldsymbol{x}_i). \tag{11}$$

Via (7), (11), and (1) this leads to

$$\hat{f}(\boldsymbol{x}) = \sum_{i=1}^{n} f(\boldsymbol{x}_i) \gamma_i(\boldsymbol{x}) = \sum_{i=1}^{n} f(\boldsymbol{x}_i) \phi_i(\boldsymbol{x}) \sum_{j=1}^{m} p_j(\boldsymbol{x}) p_j(\boldsymbol{x}_i)$$

$$= \sum_{j=1}^{m} \left[\sum_{i=1}^{n} f(\boldsymbol{x}_i) p_j(\boldsymbol{x}_i) \phi_i(\boldsymbol{x}) \right] p_j(\boldsymbol{x}) = \sum_{j=1}^{m} \langle f, p_j \rangle_{\phi(\boldsymbol{x})} p_j(\boldsymbol{x}).$$

§4. Properties of MLS Approximation

We briefly summarize some of the main features of MLS approximation. First, an approximation (computed either via (2) or via (7)) is valid at one point, $\boldsymbol{x}$, only, and a new linear system needs to be solved when the evaluation point is changed. If one considers rendering of approximation surfaces, then thousands of evaluations are required, and this may be one of the main reasons for the reluctance among mathematicians to accept MLS approximation as a viable scattered data approximation method.

If the weight functions ϕ_i have local support then all summations are local, i.e., instead of summing from $i = 1$ to n we need only sum over all points $\boldsymbol{x}_i$ in a neighborhood $N_{\boldsymbol{x}}$ of $\boldsymbol{x}$ referred to below with the index set $I_{\boldsymbol{x}}$. Coupled with the direct formulas proposed here, this leads to very fast approximation methods. This shows that earlier objections concerning the computational cost of MLS methods (see [5]) were not justified.

MLS methods are also very accurate. In fact, Levin [7] and Wendland [11] (and to some extent Farwig [3]) showed that if $\mathcal{U} = \mathcal{P}_Q^d$, the support of the weight function is local, and proportional to some "meshsize" h then one has approximation order $\mathcal{O}(h^{Q+1})$ provided the data come from a smooth enough function f. According to the cited references, the choice of weight function has no effect on the approximation order. The weight function only determines the smoothness of the approximation. If higher-order polynomial reproduction is enforced, then undesirable features such as "flat spots" will be avoided (another objection in [5]).

By construction, functions in $\mathcal{U}$ are reproduced exactly by the MLS method, and it was shown in earlier papers (e.g., [6]) that the choice of singular weight functions leads to *interpolation* at the data points.

§5. Matrix-free Formulations

The first possibility to obtain a matrix-free MLS formulation – outlined above – would be the use of generalized Fourier series (6) (with compactly supported weight functions). Due to the difficulty associated with finding discrete multivariate orthogonal polynomials, this approach seems to be impractical.

Another approach may be the construction of weight functions which already have certain approximation properties built in. This would then lead to an approximation analogous (in structure) to Shepard's method (5). We reserve this approach for a future paper.

In the present work we suggest solving small (e.g., $m \leq 4$) systems for λ_j in the Backus-Gilbert approach (10) analytically. This means that the resulting formulas can be directly programmed into the computer code, and no linear systems need to be solved at run-time.

For the following examples we define the moments

$$\mu_{\boldsymbol{\alpha}} = \sum_{i \in I_{\boldsymbol{x}}} (\boldsymbol{x}_i - \boldsymbol{x})^{\boldsymbol{\alpha}} \phi_i(\boldsymbol{x}), \qquad \boldsymbol{x} \in \mathbb{R}^d, \qquad |\boldsymbol{\alpha}| \leq 2Q,$$

with $\boldsymbol{\alpha}$ a multi-index, and $I_{\boldsymbol{x}}$ the index set mentioned above.

Example 1. $d = 1$, $Q = 1$, $m = 2$. We choose the polynomial basis $\mathcal{U} = \mathrm{span}\{p_1(s) = 1, \ p_2(s) = s - x\}$. Then

$$\lambda_1(x) = \frac{\mu_2}{\mu_0\mu_2 - \mu_1^2}, \qquad \lambda_2(x) = \frac{\mu_1}{\mu_1^2 - \mu_0\mu_2},$$

and according to (9) the quasi-interpolant (7) is formed with

$$\gamma_i(x) = [\lambda_1(x) + \lambda_2(x)(x_i - x)] \phi_i(x), \qquad i = 1, \ldots, n,$$

where the ϕ_i are arbitrary (positive) weight functions.

Example 2. $d = 1$, $Q = 2$, $m = 3$. We choose $\mathcal{U} = \mathrm{span}\{p_1(s) = 1, \ p_2(s) = s - x, \ p_3(s) = (s - x)^2\}$. Then

$$\lambda_1(x) = \frac{\mu_2\mu_4 - \mu_3^2}{D}, \qquad \lambda_2(x) = \frac{\mu_2\mu_3 - \mu_1\mu_4}{D}, \qquad \lambda_3(x) = \frac{\mu_1\mu_3 - \mu_2^2}{D},$$

where $D = 2\mu_1\mu_2\mu_3 - \mu_0\mu_3^2 - \mu_2^3 - \mu_1^2\mu_4 + \mu_0\mu_2\mu_4$. Now the quasi-interpolant is computed with

$$\gamma_i(x) = \left[\lambda_1(x) + \lambda_2(x)(x_i - x) + \lambda_3(x)(x_i - x)^2\right] \phi_i(x), \qquad i = 1, \ldots, n.$$

Example 3. $d = 2$, $Q = 1$, $m = 3$. If $\mathcal{U} = \mathrm{span}\{p_1(s,t) = 1,\ p_2(s,t) = s - x,\ p_3(s,t) = (t - y)\}$ then

$$\lambda_1(x,y) = \frac{1}{D}\left[\mu_{11}^2 - \mu_{20}\mu_{02}\right], \qquad \lambda_2(x,y) = \frac{1}{D}\left[\mu_{10}\mu_{02} - \mu_{01}\mu_{11}\right],$$

$$\lambda_3(x,y) = \frac{1}{D}\left[\mu_{20}\mu_{01} - \mu_{10}\mu_{11}\right],$$

with $D = \mu_{10}^2\mu_{02} + \mu_{20}\mu_{01}^2 - \mu_{00}\mu_{20}\mu_{02} - 2\mu_{10}\mu_{01}\mu_{11} + \mu_{00}\mu_{11}^2$. Here we set $\gamma_i(x,y) = [\lambda_1(x,y) + \lambda_2(x,y)(x_i - x) + \lambda_3(x,y)(y_i - y)]\,\phi_i(x,y)$.

§6. Experiments with a Multilevel Approximation Algorithm

In [4] we studied an iterative multilevel Newton algorithm with postconditioning based on convolution smoothing. In its simplest form (used here) the algorithm can be described as follows.

Algorithm:

1) Create nested point sets $\mathcal{X}_1 \subset \cdots \subset \mathcal{X}_K \subset \mathbb{R}^d$, define a sequence of convolution kernels Φ_k, and let $s(\boldsymbol{x}) = 0$.

2) For $k = 1, 2, \ldots, K$:

   (a) Solve $g(\boldsymbol{x}) = f(\boldsymbol{x}) - s(\boldsymbol{x})$ on $\mathcal{X}_k$.

   (b) Smooth the Newton update $\tilde{g}(\boldsymbol{x}) = (\Phi_k * g)(\boldsymbol{x})$.

   (c) Update $s(\boldsymbol{x}) = s(\boldsymbol{x}) + \tilde{g}(\boldsymbol{x})$.

In this context the MLS quasi-interpolant (7) can be interpreted as a discrete convolution representing step 2(b) in the above algorithm. The solution step 2(a) in this case is trivial. We just sample the residual $f - s$ on the computational grid $\mathcal{X}_k$.

We present two sets of experiments below. In 1D we use MLS quasi-interpolants based on the formulas given in Examples 1 and 2, and the test function

$$f(x) = \frac{1}{2}e^{-(9x-2)^2/4} + \frac{3}{4}e^{-(9x+1)^2/49} + \frac{1}{2}e^{-(9x-7)^2/4} - \frac{1}{5}e^{-(9x-4)^2}.$$

In 2D we use Shepard's method and the formulas of Example 3 which reproduce linear polynomials locally. The test function for those experiments is Franke's function

$$f(x,y) = \frac{3}{4}e^{-\left((9x-2)^2+(9y-2)^2\right)/4} + \frac{3}{4}e^{-(9x+1)^2/49-(9y+1)^2/10}$$
$$+ \frac{1}{2}e^{-\left((9x-7)^2+(9y-3)^2\right)/4} - \frac{1}{5}e^{-(9x-4)^2-(9y-7)^2}.$$

The weight functions we use for both sets of experiments are the compactly supported radial basis functions (listed together with their smoothness)

$$\varphi(r) = (1 - r)_+^4 (4r + 1), \qquad\qquad (C^2)$$

$$\varphi(r) = (1 - r)_+^6 (\frac{35}{3}r^2 + 6r + 1), \qquad\qquad (C^4)$$

$$\varphi(r) = (1 - r)_+^8 (32r^3 + 25r^2 + 8r + 1), \qquad (C^6)$$

where $r = \| \cdot -\boldsymbol{x}\|$, and $(\cdot)_+^\beta$ is the truncated power function. We then have $\phi_i(\boldsymbol{x}) = \varphi(\|\boldsymbol{x}_i - \boldsymbol{x}\|)$. For the multilevel algorithm we need scaled versions of φ, i.e., $\varphi_\rho(r) = \varphi(r/\rho)$, where ρ is the radius of support of φ.

The results of our 1D experiments are presented in Tables 1 and 2 as well as in the left part of Figure 1. The computational grids $\mathcal{X}_k$ are given by $2^k + 1$ equally spaced points in the interval $[0, 1]$. According to the theory (see Sect. 4) we need to scale the support of the weight function proportionally to the meshsize. We do this, and take the initial support radius equal to 1 for Table 1 and $\rho = 2$ for Table 2 (since local reproduction of quadratic polynomials requires more points within the support for the method to be well-posed). The ℓ_∞ errors are computed on a grid of equally spaced points one level finer than the finest level used in the algorithm, i.e., 524289 points for Table 1, and 65537 points for Table 2. The reason we terminated the algorithm earlier in Table 2 is not runtime (execution time is only a few seconds per level), but rather the size of the error which is approaching machine accuracy.

According to the convergence theory for MLS approximation (see Sect. 4) we should expect $\mathcal{O}(h^{Q+1})$ convergence (without a multilevel algorithm). The rate-columns contain the approximate convergence exponent ($\approx Q+1$), i.e., $\ln(e_{k-1}/e_k)/\ln 2$, for the multilevel algorithm. These rates can also be observed as the slopes of the curves in Figure 1. The tables also contain a line labeled "direct" in which we list the error obtained by computing the MLS approximation directly on the finest level. From the data in the tables, one can infer that neither the multilevel algorithm nor a change of weight function have a significant effect on the accuracy of the MLS method. This needs to be contrasted with radial basis function interpolation (using the same weight functions we use) where a multilevel algorithm is essential to obtaining an accurate approximation in a computationally efficient way (see, e.g., [9]).

For our 2D experiments (summarized in Tables 3 and 4) we use $(2^k + 1)^2$ equally spaced points in the unit square, and the errors are computed on a grid of 263169 equally spaced points. The initial support size of the radial weights is $\rho = 1$, and it is halved in each subsequent iteration.

		C^2		C^4		C^6	
k	# pts	ℓ_∞ error	rate	ℓ_∞ error	rate	ℓ_∞ error	rate
2	5	1.512(-1)		1.234(-1)		1.137(-1)	
3	9	6.543(-2)	1.21	5.696(-2)	1.11	5.977(-2)	0.93
4	17	1.168(-2)	2.49	9.961(-3)	2.52	1.162(-2)	2.36
5	33	1.210(-3)	3.27	1.476(-3)	2.75	2.185(-3)	2.41
6	65	2.277(-4)	2.41	2.394(-4)	2.62	5.409(-4)	2.01
7	129	5.926(-5)	1.94	6.155(-5)	1.96	1.400(-4)	1.95
$\vdots$	$\vdots$	$\vdots$	$\vdots$	$\vdots$	$\vdots$	$\vdots$	$\vdots$
17	131073	5.762(-11)	2.00	6.048(-11)	2.00	6.122(-11)	2.00
18	262145	1.441(-11)	2.00	1.513(-11)	2.00	1.531(-11)	2.00
	direct	1.036(-10)		9.036(-11)		8.754(-11)	

Tab. 1. 1D-multilevel approximation with linear polynomial reproduction.

		C^2		C^4		C^6	
k	# pts	ℓ_∞ error	rate	ℓ_∞ error	rate	ℓ_∞ error	rate
2	5	2.273(-1)		2.058(-1)		1.795(-1)	
3	9	8.714(-2)	1.38	6.780(-2)	1.60	5.366(-2)	1.74
4	17	1.259(-2)	2.79	8.476(-3)	3.00	5.926(-3)	3.18
5	33	5.933(-4)	4.41	3.194(-4)	4.73	2.347(-4)	4.66
6	65	4.266(-5)	3.80	1.227(-5)	4.70	2.079(-5)	3.50
7	129	4.610(-6)	3.21	1.139(-6)	3.43	3.253(-6)	2.68
$\vdots$	$\vdots$	$\vdots$	$\vdots$	$\vdots$	$\vdots$	$\vdots$	$\vdots$
14	16385	3.680(-12)	1.88	3.082(-13)	2.96	3.568(-12)	2.83
15	32769	6.303(-13)	2.55	4.208(-14)	2.87	5.039(-13)	2.82
	direct	2.863(-13)		1.976(-13)		1.626(-13)	

Tab. 2. 1D-multilevel approximation with quadratic polynomial reproduction.

§7. Concluding Remarks

The multilevel algorithm has virtually no effect on the convergence behavior for (scattered) data approximation. In some of the examples the error using the multilevel method was slightly smaller than for the direct solution (but also more expensive to obtain). However, initial experiments on a simple two point boundary value problem indicate that a multilevel approach will probably be essential for the solution of differential equations via collocation. The effect of changing the weight function on the accuracy is minimal. The error for the direct solution improves slightly for smoother weights.

k	# pts	C^2 ℓ_∞ error	rate	C^4 ℓ_∞ error	rate	C^6 ℓ_∞ error	rate
1	9	6.318(-1)		6.140(-1)		6.017(-1)	
2	25	2.548(-1)	1.31	1.894(-1)	1.70	1.648(-1)	1.87
3	81	7.517(-2)	1.76	6.433(-2)	1.56	5.552(-2)	1.57
4	289	2.047(-2)	1.88	1.634(-2)	1.98	1.693(-2)	1.71
5	1089	3.055(-3)	2.74	4.689(-3)	1.80	6.932(-3)	1.29
6	4225	1.136(-3)	1.43	1.822(-3)	1.36	3.008(-3)	1.20
7	16641	4.202(-4)	1.43	7.332(-4)	1.31	1.251(-3)	1.27
8	66049	1.604(-4)	1.39	2.985(-4)	1.30	5.417(-4)	1.21
	direct	1.474(-3)		8.508(-4)		4.920(-4)	

Tab. 3. 2D-multilevel approximation with reproduction of constants.

k	# pts	C^2 ℓ_∞ error	rate	C^4 ℓ_∞ error	rate	C^6 ℓ_∞ error	rate
1	9	6.163(-1)		6.174(-1)		6.178(-1)	
2	25	2.741(-1)	1.17	2.297(-1)	1.43	2.201(-1)	1.49
3	81	7.580(-2)	1.85	6.999(-2)	1.71	7.192(-2)	1.61
4	289	2.082(-2)	1.86	1.731(-2)	2.02	1.748(-2)	2.04
5	1089	2.378(-3)	3.13	2.426(-3)	2.83	3.607(-3)	2.28
6	4225	4.271(-4)	2.48	4.146(-4)	2.55	8.767(-4)	2.04
7	16641	1.085(-4)	1.98	1.043(-4)	1.99	2.141(-4)	2.03
8	66049	2.783(-5)	1.96	2.624(-5)	1.99	4.743(-5)	2.17
	direct	1.433(-4)		1.283(-4)		1.256(-4)	

Tab. 4. 2D-multilevel approximation with linear polynomial reproduction.

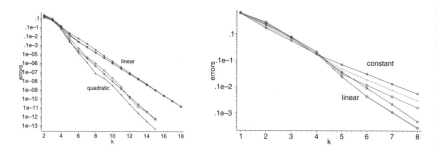

Fig. 1. Logplot of errors for 1D (left) and 2D approximation (right).

References

1. Belytschko, T., Y. Krongauz, D. Organ, M. Fleming, and P. Krysl, Meshless methods: an overview and recent developments, Computer Methods in Applied Mechanics and Engineering **139** (1996), 3–47.

2. Bos, L. P. and K. Šalkauskas, Moving least-squares are Backus-Gilbert optimal, J. Approx. Theory **59** (1989), 267–275.

3. Farwig, R., Multivariate interpolation of scattered data by moving least squares methods, in *Algorithms for the Approximation of Functions and Data*, J. C. Mason and M. G. Cox (eds), Oxford Univ. Press, Oxford, 1987, 193–211.

4. Fasshauer, G. E. and J. W. Jerome, Multistep approximation algorithms: Improved convergence rates through postconditioning with smoothing kernels, Advances in Comp. Math. **10** (1999), 1–27.

5. Franke, R., Scattered data interpolation: tests of some methods, Math. Comp. **48** (1982), 181–200.

6. Lancaster, P. and K. Šalkauskas, Surfaces generated by moving least squares methods, Math. Comp. **37** (1981), 141–158.

7. Levin, D., The approximation power of moving least-squares, Math. Comp. **67** (1998), 1517–1531.

8. McLain, D. H., Drawing contours from arbitrary data points, Comput. J. **17** (1974), 318–324.

9. Schaback, R., Remarks on meshless local construction of surfaces, *The Mathematics of Surfaces*, IX (Cambridge, 2000), Springer, London, 34–58.

10. Shepard, D., A two dimensional interpolation function for irregularly spaced data, Proc. 23rd Nat. Conf. ACM, 1968, 517–523.

11. Wendland, H., Local polynomial reproduction and moving least squares approximation, IMA J. Numer. Anal. **21** (2001), 285–300.

Greg Fasshauer
Department of Applied Mathematics
Illinois Institute of Technology
Chicago, IL 60616
fass@amadeus.math.iit.edu
http://amadeus.math.iit.edu/~fass

Shape-preserving, Multiscale Fitting of Bivariate Data by Cubic L_1 Smoothing Splines

David E. Gilsinn and John E. Lavery

Abstract. Bivariate cubic L_1 smoothing splines are introduced. The coefficients of a cubic L_1 smoothing spline are calculated by minimizing the weighted sum of the L_1 norms of second derivatives of the spline and the ℓ_1 norm of the residuals of the data-fitting equations. Cubic L_1 smoothing splines are compared with conventional cubic smoothing splines based on the L_2 and ℓ_2 norms. Computational results for fitting a challenging data set consisting of discontinuously connected flat and quadratic areas by C^1-smooth Sibson-element splines on a tensor-product grid are presented. In these computational results, the cubic L_1 smoothing splines preserve the shape of the data while cubic L_2 smoothing splines do not.

§1. Introduction

Among the current options for approximating bivariate data are tensor-product, polynomial and thin-plate smoothing splines [1,4,7,8,13,16,17,18], multiquadrics [2] and wavelets [3,5]. Tensor-product, polynomial and thin-plate smoothing splines often have extraneous, "nonphysical" oscillation, especially near multiscale phenomena, that is, near regions where the magnitude of the data or the sizes of the cells in the grid change abruptly. The oscillation in these smoothing splines can be mitigated or eliminated by shifting the positions of nodes, adjusting the number of nodes, adding various constraints or penalties and a posteriori filtering, often with significant amounts of human interaction. At additional computational expense, multiquadrics and wavelets can avoid nonphysical oscillation. Development of

Approximation Theory X: Wavelets, Splines, and Applications
Charles K. Chui, Larry L. Schumaker, and Joachim Stöckler (eds.), pp. 283–293.

computationally inexpensive smoothing splines that preserve shape without requiring human interaction would be of great benefit in modeling objects with multiscale phenomena such as urban and natural terrain, mechanical objects and images.

In [10,11,12], new classes of univariate L_1 interpolating splines, univariate L_1 smoothing splines and bi- and multivariate L_1 interpolating splines were proposed. These splines preserve shape for smooth data as well as for data with abrupt changes in magnitude and spacing and for smooth sets of spline nodes as well as for those with abrupt changes in spacing. In the present paper, we extend the results of [11,12] by creating a new class of bivariate cubic L_1 smoothing splines. Our focus here is mainly on bivariate C^1-smooth cubic L_1 smoothing splines on tensor-product grids.

The objective of this paper is to present two case studies of approximation of simulated urban structures by bivariate L_1 smoothing splines. The urban structures were simulated so that data sets would be devoid of signal noise and image contamination due to preprocessing. Although it is crucial to be able to deal with signal noise and image contamination, the primary focus here is to study the performance of L_1 smoothing splines on "clean" data sets.

§2. Bivariate Data, Grids and Sibson Shape Elements

We consider fitting the data $(\hat{x}_m, \hat{y}_m, \hat{z}_m)$, $m = 1, 2, \ldots, M$. The weight of the mth data point is a positive real number $\hat{w}_m$. We will create bivariate cubic L_1 smoothing splines on tensor-product grids with nodes x_i, $i = 0, 1, \ldots, I$ and y_j, $j = 0, 1, \ldots, J$ that are strictly monotonic partitions of the finite real intervals $[x_0, x_I]$ and $[y_0, y_J]$, respectively. The domain of the spline will be $D = [x_0, x_I] \times [y_0, y_J]$.

Sibson elements [6,9,12] will be used for the computational results in the present paper. To create a Sibson element on a rectangle $(x_i, x_{i+1}) \times (y_j, y_{j+1})$, one first divides the rectangle into four triangles by drawing the two diagonals of the rectangle. The Sibson element is a shape function $z(x, y)$ that is cubic in each triangle, is C^1 at the lines separating the four triangles, is C^1 with the Sibson elements in the adjacent rectangles, has derivative $\partial z / \partial x$ that is linear along the edges $x = x_i$ and $x = x_{i+1}$ of the rectangle and has derivative $\partial z / \partial y$ that is linear along the edges $y = y_j$ and $y = y_{j+1}$. The Sibson element z in a given rectangle depends only on the values of z, $\partial z / \partial x$ and $\partial z / \partial y$ at the corners of that rectangle (12 parameters per rectangle) as described in [6] and in Sec. 2 of [12]. The values of z, $\partial z / \partial x$ and $\partial z / \partial y$ at node (x_i, y_j) will be denoted by z_{ij}, z_{ij}^x and z_{ij}^y, respectively. The vectors of the values of the z_{ij}, z_{ij}^x and z_{ij}^y, $i = 0, 1, \ldots, I$, $j = 0, 1, \ldots, J$, will be denoted by z, z^x and z^y, respectively.

§3. Minimization Principle for Cubic L_1 Smoothing Splines

A cubic L_1 smoothing spline is a function that for a given α, $0 < \alpha < 1$, minimizes the following expression:

$$E = \alpha \sum_{m=1}^{M} \hat{w}_m \left| z(\hat{x}_m, \hat{y}_m) - \hat{z}_m \right|$$
$$+ (1 - \alpha) \iint_D \left[\left| \frac{\partial^2 z}{\partial x^2} \right| + 2 \left| \frac{\partial^2 z}{\partial x \partial y} \right| + \left| \frac{\partial^2 z}{\partial y^2} \right| \right] dx \, dy \tag{1}$$

over all surfaces z of a given class. The balance parameter α determines the trade-off between the closeness with which the data are fitted, represented by the sum in (1), and the tendency of the spline to be close to a piecewise planar surface, represented by the double integral in (1). The double integral in (1) is the functional that defines a cubic L_1 interpolating spline [12]. This double integral could be replaced by a double integral with different weights on the three terms in the integrand, as in expression (4) of [12], or by a double integral with different terms, as in expression (6) of [12].

Cubic L_1 smoothing splines based on Sibson elements exist because, as a function of the coefficients z, z^x and z^y, functional (1) is continuous, bounded below by 0 and convex and tends to ∞ uniformly as the Euclidean norm of the coefficients tends to ∞ (cf. Theorem 1 of [12], which states this result for interpolating splines). However, cubic L_1 smoothing splines under this definition need not be unique because functional (1) is not necessarily strictly convex. When there are several candidates for an L_1 smoothing spline, the candidate with (in some metric) the smallest absolute values of the first derivatives, that is, the flattest surface, is typically the choice of most users. For this reason, we add to E a "regularization" term:

$$E + \sum_{i=0}^{I} \sum_{j=0}^{J} \left[\varepsilon_{1ij} |z_{ij}^x| + \varepsilon_{2ij} |z_{ij}^y| \right], \tag{2}$$

where the regularization parameters ε_{1ij} and ε_{1ij} are small nonnegative numbers. Functional (2) is still not necessarily strictly convex and can therefore achieve its minimum for more than one set of coefficients z, z^x, z^y. However, standard interior-point algorithms, including the primal affine method used for the computational results presented in this paper, yield a unique set of coefficients that minimize (a discretization of) functional (2).

For comparison with cubic L_1 smoothing splines, we will calculate "cubic L_2 smoothing splines" by minimizing the functional

$$\alpha^2 \sum_{m=1}^{M} \left[\hat{w}_m \left(z(\hat{x}_m, \hat{y}_m) - \hat{z}_m \right) \right]^2$$

$$+ (1-\alpha)^2 \iint_D \left[\left(\frac{\partial^2 z}{\partial x^2} \right)^2 + 4 \left(\frac{\partial^2 z}{\partial x \partial y} \right)^2 + \left(\frac{\partial^2 z}{\partial y^2} \right)^2 \right] dx\,dy \qquad (3)$$

$$+ \sum_{i=0}^{I} \sum_{j=0}^{J} \left[\left(\varepsilon_{1ij} z_{ij}^x \right)^2 + \left(\varepsilon_{2ij} z_{ij}^y \right)^2 \right].$$

Functional (3) is the same as the functional (2) except that the squares of the ℓ_2 and L_2 norms have replaced the ℓ_1 and L_1 norms. The integral in the functional for cubic L_2 smoothing splines is not the same as the integral for thin-plate splines because the coefficient of the middle term in the integrand is 4, not 2. L_2 smoothing splines could, of course, be calculated with a thin-plate-spline functional replacing the double integral in (3). However, L_2 smoothing splines based on minimization principle (3) were chosen for comparison with L_1 smoothing splines because the main goal of this paper is to demonstrate that the fundamental solution of the shape-preservation problem is a proper choice of the function spaces. The most relevant comparisons are therefore those in which only the function spaces (and not, for example, the coefficients in the integrals) differ. Comparison of L_1 smoothing splines of many different types (including types A_1 and B described in [12]) with L_2 smoothing splines of many different types (including thin-plate smoothing splines and smoothing splines of types A_1 and B) is an important issue for future investigation.

§4. Algorithm and Computational Results

We calculate the coefficients of a Sibson-element L_1 smoothing spline by minimizing the functional (2) in which the integral is discretized by the scheme used in [12], which can be summarized as follows. Let $N \geq 2$ be an integer. Divide each rectangle $(x_i, x_{i+1}) \times (y_j, y_{j+1})$ into N^2 subrectangles. Approximate the double integral over the rectangle by $1/[2N(N-1)]$ times the sum of the $2N(N-1)$ values of the integrand at the midpoints of the sides of the subrectangles that are in the interior of the rectangle. This discretization was chosen because it uses values of the integrand only in the interiors (and not on the boundaries) of the 4 triangles that make up each rectangle.

    Minimization of (2) with the integral discretized in this manner was carried out by the primal affine method of Vanderbei, Meketon and Freedman [12,14,15] coded by the authors of this paper. The primal affine algorithm is known to converge globally to a unique solution no matter whether

Tab. 1. Data Set 1: Simulated high-rise hotel complex.

$\hat{z}$	Begin $\hat{\imath}$	End $\hat{\imath}$	Begin $\hat{\jmath}$	End $\hat{\jmath}$
150	71	135	16	35
25	101	125	51	75
10	16	135	100	120
10	16	35	121	155
100	36	115	121	155
10	116	135	121	155
10	16	135	156	175
0	Otherwise	Otherwise	Otherwise	Otherwise

functional (2) is strictly convex or only (non-strictly) convex. Further information on the convergence of the primal affine method can be found in the second paragraph of Sec. 3 of [11] and the second-to-last paragraph of Sec. 4 of [12], which cite the original results in [14,15] and elsewhere.

For comparison with cubic L_1 smoothing splines, cubic L_2 smoothing splines with the integral in (3) discretized in the same manner as the integral in (2) were calculated. For all of the computational results presented below, $N = 5$. The authors have computational experience with the data sets of Figs. 1 and with other data sets not only with $N = 5$ but also with $N = 3$. No significant differences have been noted. Investigation of how L_1 and L_2 smoothing splines vary with N may be of interest. However, before commencing such an investigation, it may be useful to determine whether functional (2) can be minimized without discretizing the integral. Research by the optimization community may answer this question within the next few years.

The weights $\hat{w}_m$ were chosen to be 1 for all m. The regularization parameters ε_{1ij} and ε_{2ij} were chosen to be 10^{-4} for all i and j.

Cubic L_1 and L_2 smoothing splines were computed for two data sets. Data set 1, shown on the left in Fig. 1, is a set of discontinuously connected flat areas that represents a high-rise hotel complex. This data set consists of 201×201 data points at equal, 1-unit spacing. Let $\hat{\imath} = 0, 1, \ldots, 200$ and $\hat{\jmath} = 0, 1, \ldots, 200$ be the coordinates of the data locations. The data points $\hat{z}$ of data set 1 are given in Table 1.

Data set 2, shown on the right in Fig. 1, is a simulated sports stadium consisting of a quadratic surface discontinuously embedded in a flat area. This data set consists of 128×128 data points at equal, 1-unit spacing. Letting $\hat{\imath} = 0, 1, \ldots, 128$ and $\hat{\jmath} = 0, 1, \ldots, 128$ be the coordinates of the

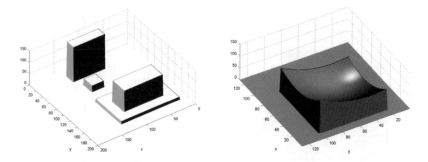

Fig. 1. Simulated high-rise hotel complex (left, data set 1) and sports stadium (right, data set 2).

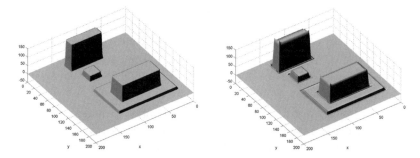

Fig. 2. L_1 smoothing spline (left) and L_2 smoothing spline (right) with $\alpha = 0.10$ for data set 1.

data locations, the data points $\hat{z}$ of data set 2 are given by

$$\hat{z}(\hat{i}, \hat{j}) = \begin{cases} 20 + 0.024 * (\hat{i} - 62)^2 + 0.016 * (\hat{j} - 42)^2, & \text{if } \begin{cases} \hat{i} = 12 : 92 \\ \hat{j} = 22 : 102 \end{cases} \\ 0, & \text{otherwise.} \end{cases}$$

(4)

For data set 1, cubic L_1 and L_2 smoothing splines were computed on spline grids consisting of 100×100 equal cells, each of size 2 units by 2 units, with $\alpha = 0.8$. For these smoothing splines, the "raw compression ratio," that is, the number of floating-point storage locations for the original data $\hat{z}(\hat{i}, \hat{j})$ divided by the number of floating-point storage locations for the smoothing spline parameters $z_{i,j}$, $z_{i,j}^x$, $z_{i,j}^y$, $i = 0, 1, \ldots, 100$, $j = 0, 1, \ldots, 100$ is $201^2/(3 * 101^2) = 1.32$. This case was chosen because it shows the different capabilities of L_1 and L_2 splines at a low compression ratio. The L_1 and L_2 smoothing spline approximations are shown in Fig. 2. Note the sharp and accurate approximations of edges and corners in the L_1 smoothing spline. In contrast, the L_2 spline has over- and undershoot at the edges of the buildings and has oscillation near the edges.

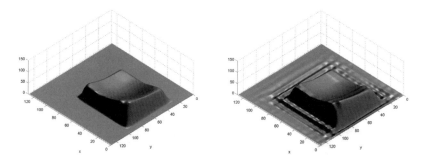

Fig. 3. L_1 smoothing spline (left) and L_2 smoothing spline (right) with $\alpha = 0.8$ for data set 2.

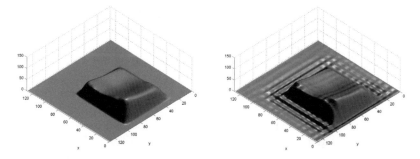

Fig. 4. L_1 smoothing spline (left) and L_2 smoothing spline (right) with $\alpha = 0.75$ for data set 2.

For data set 2, cubic L_1 and L_2 smoothing splines were computed on spline grids consisting of 16×16 equal cells, each of size 8 units by 8 units, with $\alpha = 0.99, 0.9, 0.85, 0.8, 0.75, 0.65, 0.6, 0.5, 0.4$ and 0.3. For these smoothing splines, the raw compression ratio is $129^2/(3*17^2) = 19.19$. This case was chosen because it shows the capabilities of L_1 and L_2 splines at a medium compression ratio. The L_1 and L_2 splines with $\alpha = 0.8$ and 0.75 are shown in Figs. 3 and 4, resp. In both of these figures, the L_1 smoothing splines fit the data well and have very little over/undershoot and extraneous oscillation. In contrast, the L_2 smoothing splines have considerable "nonphysical" oscillation.

In the above paragraph, the measure for the performance of the smoothing splines is visual inspection because it is not yet known how to measure shape preservation quantitatively. Nevertheless, it is appropriate to give some quantitative information about the performance of the smoothing splines. To do so, we will use the following norms: 1) the (normalized) ℓ_1 norm $|| \cdot ||_{\ell_1}$ (sum of the absolute values of the 129^2 values divided by 129^2), 2) the (normalized) ℓ_2 norm $|| \cdot ||_{\ell_2}$, also known as the RMS or root-mean-square norm (square root of the quotient that consists

Tab. 2. Norms of errors of smoothing splines for data set 2.

$i =$	1		2	
weight $\alpha =$	0.8	0.75	0.8	0.75
$\|z_{[L_i,\alpha]} - \hat{z}\|_{\ell_1} =$	8.373	9.517	8.637	9.729
$\|z_{[L_i,\alpha]} - \hat{z}\|_{\ell_2} =$	22.80	23.81	22.18	23.23
$\|z_{[L_i,\alpha]} - \hat{z}\|_{\ell_\infty} =$	137.6	137.6	130.7	132.4

of the sum of the squares of the 129^2 values divided by 129^2) and 3) the ℓ_∞ norm $\| \cdot \|_{\ell_\infty}$ (maximum of the 129^2 absolute values). In Table 2, we present the ℓ_1, ℓ_2 and ℓ_∞ norms of the error between the values z of the L_1 and L_2 smoothing splines and the $\hat{z}$ of data set 2. In this table, we denote L_1 smoothing splines by $z_{[L_1,\alpha]}$ and L_2 smoothing splines by $z_{[L_2,\alpha]}$.

From the information in Table 2, one could not determine whether L_1 smoothing splines are better or worse than L_2 smoothing splines. Since L_1 and L_2 smoothing splines fit data in the spaces ℓ_1 and ℓ_2, respectively, it is perhaps not surprising that L_1 smoothing splines perform better in the ℓ_1 norm and that L_2 smoothing splines perform better in the ℓ_2 (and ℓ_∞) norm. However, most observers interested in geometric modeling agree that the L_1 smoothing splines represent the original data better than the L_2 smoothing splines. The error norms in Table 2 confirm what is well known in the image processing community, namely, that the ℓ_1, ℓ_2 and ℓ_∞ norms are not good measures of shape preservation.

§5. Convergence Issues

The primal affine method, which has so far been the method of choice for calculating L_1 smoothing splines, typically performs well for small data sets. However, for the large data sets of interest in this paper, the primal affine method converged slowly (100-600 iterations) for some α, converged incompletely (difference between iterations decreased until a certain point and then increased) for other α and diverged for yet other α. The computational results shown above were, of course, for cases of complete convergence. Alternative strategies for calculating L_1 smoothing splines are under investigation. One of these strategies is domain decomposition, in which the global domain is broken up into many slightly overlapping subdomains and the global L_1 smoothing spline is patched together from the local L_1 smoothing splines. Also, linear and nonlinear programming algorithms to replace the primal affine method are under investigation. The primal affine method in its current global implementation has been sufficient to "prove the principle" (in the present paper and in [10,11,12]) that L_1 interpolating and smoothing splines are able to preserve the shape of data much better than conventional L_2 splines. In the future, it is likely that other algorithms for calculating L_1 splines will be of great interest and use.

§6. Conclusion

In this paper, we have focused on providing evidence that the choice of the function spaces in smoothing spline minimization principles has far greater influence than previously expected. The seemingly minor change of the function spaces from the conventional choices ℓ_2 and L_2 to the unconventional choices ℓ_1 and L_1 results in a vast improvement in the shape-preserving, multiscale capabilities of cubic smoothing splines. The contribution of this article is not to prove that L_1 smoothing splines preserve shape better than L_2 smoothing splines but merely to observe that that is so for a limited set of test cases. A full proof that L_1 smoothing splines preserve shape better than L_2 smoothing splines requires quantitative understanding of shape preservation, something that does not yet exist.

The smoothing splines presented here were calculated with Sibson elements and with no adaptivity in the spline grids. Investigation of bivariate L_1 smoothing splines using various Sibson and non-Sibson elements on quadrangulations and triangulations in nonadaptive and adaptive settings would be of large interest. Comparison with other widely used methods for modeling irregular, multiscale data (TINs, wavelets, JPEG, etc.) needs be carried out.

One could, of course, choose to fit the data of Figs. 1 and 2 by L_2 smoothing splines on subdomains that do not cross the lines of discontinuity and therefore have much less extraneous oscillation. However, if one does so, one must identify the lines of discontinuity, introduce topology into the fitting procedure and handle issues of matching of splines at the boundaries of the subdomains. The cost of doing this has to be balanced against the advantages of having one "terrain skin," the L_1 smoothing spline, that requires none of this. Different users will make different choices that fit their needs. L_1 smoothing splines do not replace other options but do add a new, attractive option to the set of options from which the user can choose.

Cubic L_1 smoothing splines are a promising new technique for geometric modeling, especially modeling of objects with multiscale phenomena such as urban and natural terrain, mechanical objects and images. The preliminary results in the present paper indicate that further investigation of L_1 smoothing splines may have high payoff.

References

1. Anthony, G. T., and M. G. Cox, The fitting of extremely large data sets by bivariate splines, in Mason, J.C., and M.G. Cox, eds., *Algorithms for Approximation*, Clarendon Press, Oxford, 1987, 5–20.

2. Carlson, R. E., and B. K. Natarajan, Sparse approximate multi-quadric interpolation, Comput. Math. Appl. $\tilde{2}7$ (1994), 99–108.

3. Cohen, A., Nonlinear wavelet approximation and image compression, in Chui, C.K., and L.L. Schumaker, eds., *Wavelets and Multilevel Approximation, Approximation Theory VIII*, vol. 2, World Scientific, Singapore, 1995, 17–38.

4. Dierckx, P., *Curve and Surface Fitting with Splines*, Clarendon Press, New York, 1993.

5. Dubuc, S., and G. Deslauriers, eds., *Spline Functions and the Theory of Wavelets*, Proceedings and Lectures Notes, vol. 18, American Mathematical Society, Providence, Rhode Island, 1999.

6. Gilsinn, D. E., Constructing Sibson Elements for a Rectangular Mesh, NISTIR 6718, National Institute of Standards and Technology, Gaithersburg, Maryland, 20899.

7. Gmelig Meyling, R. H. J., Approximation by cubic C^1 splines on arbitrary triangulations, Numer. Math. **51** (1987), 65–85.

8. Golitschek, M. von, and L. L. Schumaker, Data fitting by penalized least squares, in Mason, J.C., and M.G. Cox, eds., *Algorithms for Approximation II*, Chapman & Hall, London, 1990, 210–227.

9. Han, L., and L. L. Schumaker, Fitting monotone surfaces to scattered data using C^1 piecewise cubics, SIAM J. Numer. Anal. **34** (1997), 569–585.

10. Lavery, J. E., Univariate cubic L_p splines and shape-preserving, multiscale interpolation by univariate cubic L_1 splines, Comput. Aided Geom. Design **17** (2000), 319–336.

11. Lavery, J. E., Shape-preserving, multiscale fitting of univariate data by cubic L_1 smoothing splines, Comput. Aided Geom. Design **17** (2000), 715–727.

12. Lavery, J. E., Shape-preserving, multiscale interpolation by bi- and multivariate cubic L_1 splines, Comput. Aided Geom. Design **18** (2001), 321–343.

13. McMahon, J., R. John, and R. Franke, Knot selection for least squares thin plate splines, SIAM J. Sci. Stat. Comp. **13** (1992), 484–498.

14. Vanderbei, R. J., Affine-scaling for linear programs with free variables, Mathematical Programming **43** (1989), 31–44.

15. Vanderbei, R. J., M. J. Meketon, and B.A. Freedman, A modification of Karmarkar's linear programming algorithm, Algorithmica **1** (1986), 395–407.

16. Wahba, G., Multivariate thin plate spline smoothing with positivity and other linear inequality constraints, in Wegman, E.J., and D.J. dePriest, eds., *Statistical Image Processing and Graphics*, Marcel Dekker, New York, 1985, 275–290.

17. Wahba, G., *Spline Models for Observational Data*, Society for Industrial and Applied Mathematics, Philadelphia, Pennsylvania, 1990.

18. Willemans, K., and P. Dierckx, Surface fitting using convex Powell-Sabin splines, J. Comput. Appl. Math. **56** (1994), 263–282.

David E. Gilsinn
National Institute of Standards and Technology
100 Bureau Drive, Stop 8910
Gaithersburg, MD 20899-8910
dgilsinn@nist.gov

John E. Lavery
Computing and Information Sciences Division
Army Research Office
Army Research Laboratory
P.O. Box 12211
Research Triangle Park, NC 27709-2211
lavery@arl.aro.army.mil

Orthogonal Spline Projections
for Linear Boundary Value Problems

Manfred v. Golitschek

Abstract. Results by C. de Boor, A. Shadrin and others on the L_∞-norm of the L_2- and ℓ_2-projections onto spline spaces $\mathcal{S}_r(\triangle)$ of order $r \geq 2$ are applied for the orthogonal spline projections which arise when the *least-squares method* is applied for the solution of the linear boundary value problem $Ly := y'' + cy' + dy = q$, $y(a) = y(b) = 0$, on $[a, b]$. We compare these least-squares projections with the spline projections of the Galerkin method.

§1. Introduction

Let $X \subseteq C[a, b]$ be a linear space of continuous real-valued functions, and $\langle \cdot, \cdot \rangle$ be a semi-definite inner product on X. We denote the corresponding semi-norm on X by $\| \cdot \| := \langle \cdot, \cdot \rangle^{1/2}$. For $r \geq 2$, let $\mathcal{S}_r(\triangle) \subset C^{r-2}[a, b]$ be the linear space of polynomial splines of order r with knots

$$\triangle : a = x_0 < x_1 < \cdots < x_n = b, \tag{1.1}$$

and let $\mathcal{S} \subseteq \mathcal{S}_r(\triangle)$ be a linear subspace of $\mathcal{S}_r(\triangle)$. We assume throughout this paper that the restriction of $\| \cdot \|$ onto $\mathcal{S}$ is a norm on $\mathcal{S}$. For each $f \in X$ there exists a unique $u_f \in \mathcal{S}$ such that

$$\|f - u_f\| = \min\{\|f - u\| : u \in \mathcal{S}\}, \tag{1.2}$$

and u_f is characterized by the orthogonality relations

$$\langle f - u_f, u \rangle = 0, \quad u \in \mathcal{S}. \tag{1.3}$$

The mapping $\mathbf{P} : X \rightarrow \mathcal{S}$ defined by $\mathbf{P}f := u_f$ is linear and satisfies $\mathbf{P}u = u$ for $u \in \mathcal{S}$. $\mathbf{P}$ is called the **orthogonal projection** of X onto $\mathcal{S}$ with respect to $\langle \cdot, \cdot \rangle$.

Approximation Theory X: Wavelets, Splines, and Applications   295
Charles K. Chui, Larry L. Schumaker, and Joachim Stöckler (eds.), pp. 295–304.
Copyright ⊘ 2002 by Vanderbilt University Press, Nashville, TN.
ISBN 0-8265-1416-2.

For numerical purposes it is important that the L_∞-norm of $\mathbf{P}$ is small. It is defined by

$$\|\mathbf{P}\|_\infty := \sup\left\{\frac{\|\mathbf{P}f\|_\infty}{\|f\|_\infty} : f \in X, f \neq 0\right\}$$
$$= \sup\{\|\mathbf{P}f\|_\infty : f \in X, \|f\|_\infty = 1\},$$

where $\|f\|_\infty := \max\{|f(x)| : x \in [a,b]\}$. The norm $\|\mathbf{P}\|_\infty$ describes how changes of the data function f can influence the values $(\mathbf{P}f)(x)$. Indeed,

$$\|\mathbf{P}[f - g]\|_\infty \leq \|\mathbf{P}\|_\infty \|f - g\|_\infty, \qquad (1.4)$$

and the number $\|\mathbf{P}\|_\infty$ in (1.4) is best possible (with regard to all $f, g \in X$). Moreover, since

$$\|f - \mathbf{P}f\|_\infty \leq (1 + \|\mathbf{P}\|_\infty)\text{dist}(f, \mathcal{S})_\infty,$$

small norms $\|\mathbf{P}\|_\infty$ imply that the uniform error $\|f - \mathbf{P}f\|_\infty$ is not much larger than the *distance*

$$\text{dist}(f, \mathcal{S})_\infty := \min_{u \in \mathcal{S}} \|f - u\|_\infty.$$

In the last decades, many people have studied properties of $\mathbf{P}$ for $X := C[a,b]$, $\mathcal{S} := \mathcal{S}_r(\triangle)$, and the L_2 inner product $\langle f, g \rangle := \int_a^b fg$. Only recently, A. Shadrin [6] succeeded in proving the famous conjecture by C. de Boor [1,2] that for each $r \geq 2$, there exists an absolute constant C_r only depending on r so that with respect to the L_2-inner product on $[a,b]$,

$$\|\mathbf{P}\|_\infty \leq C_r$$

for all partitions $\triangle$ of the form (1.1).

In numerical applications, if the function f is only known on a finite data set $T \subset [a,b]$, the ℓ_2-projection $\mathbf{P}$ with respect to

$$\langle f, g \rangle_T := \sum_{t \in T} f(t)g(t), \quad f, g \in C[a,b],$$

serves to reconstruct f. Some properties of the ℓ_2-projection $\mathbf{P}$ have been described by the author [4]. Corresponding results for functions f in two variables and piecewise polynomial spaces $\mathcal{S}$ on triangulations are described in v. Golitschek and Schumaker [5].

In the present paper we consider the linear boundary value problem

$$y''(x) + c(x)y'(x) + d(x)y(x) = q(x), \quad a \leq x \leq b, \quad y(a) = y(b) = 0, \quad (1.5)$$

where $c, d \in C[a, b]$. The least-squares method can be described by an orthogonal projection $\mathbf{P}^*$ as follows: Let $X = \{f \in C^2[a, b] : f(a) = f(b) = 0\}$, $\mathcal{S} := \{u \in \mathcal{S}_r(\triangle) : u(a) = u(b) = 0\}$, and $L : X \to C[a, b]$ be the linear differential operator of the form

$$Ly := y'' + cy' + dy, \quad y \in X.$$

Then

$$\langle f, g \rangle := \int_a^b (Lf)(Lg), \quad f, g \in X, \tag{1.6}$$

is a semi-definite inner product on X. We assume that $Lu = 0$ implies $u = 0$ for $u \in \mathcal{S}$. Then (1.6) is an inner product on $\mathcal{S}$, and the orthogonal projection $u_f =: \mathbf{P}^* f$ of f is defined by

$$\int_a^b \left(L[f - u_f] \right)^2 = \min_{u \in \mathcal{S}} \int_a^b \left(L[f - u] \right)^2. \tag{1.7}$$

Therefore, if $y \in X$ is the solution of the linear boundary value problem (1.5), then its least squares approximation $\mathbf{P}^* y$ is defined by

$$\int_a^b \left(q - L[\mathbf{P}^* y] \right)^2 = \min_{u \in \mathcal{S}} \int_a^b \left(q - Lu \right)^2, \tag{1.8}$$

and is characterized by

$$\int_a^b L[\mathbf{P}^* y] \ Lu = \int_a^b q \ Lu, \quad \text{for all } u \in \mathcal{S}. \tag{1.9}$$

The paper is organized as follows: Section 2 contains a short review of the L_2-projection onto $\mathcal{S}_r(\triangle)$. Section 3 yields the main theoretical result about the orthogonal spline projection $\mathbf{P}^*$ of the least-squares method. After a short review of the Galerkin method in Section 4, we compare in Section 5 the least-squares and Galerkin methods by numerical examples.

§2. The L_∞-norm of the L_2-projection

In this section we want to review some properties of the L_2-projection which will be used in Section 3.

Let $X = C[a, b]$, $\langle f, g \rangle = \int_a^b fg$, and let $\mathbf{P}$ be the L_2-projection of X onto the spline space $\mathcal{S} := \mathcal{S}_r(\triangle)$ of order $r \geq 2$. The dimension of $\mathcal{S}$ is $M := n + r - 1$. Extending $\triangle$ to $\triangle^* : x_{-r+1} < \cdots < x_0 = a < \cdots < x_n = b < \cdots < x_{n+r-1}$ let $\mathcal{B}_r = \{N_{k,r}\}_{k=1}^M$ be the B-spline basis of $\mathcal{S}$ for $\triangle^*$ such that $\sum_{k=1}^M N_{j,r}(x) = 1$, $x \in [a, b]$ and $\text{supp}(N_{k,r}) = [x_{k-r}, x_k]$, $k = 1, \ldots, M$.

For $f \in C[a,b]$ and $\mathbf{P}f =: \sum_{k=1}^{M} \alpha_k N_{k,r}$, the orthogonality relations (1.3) are equivalent to the system of M linear equations

$$\sum_{k=1}^{M} \alpha_k \int_a^b N_{k,r} N_{j,r} = \int_a^b f N_{j,r}, \quad j = 1, \dots, M. \tag{2.1}$$

We write

$$a_{j,k} := \frac{1}{x_j - x_{j-r}} \int_a^b N_{k,r} N_{j,r} \quad , \quad b_j := \frac{1}{x_j - x_{j-r}} \int_a^b f N_{j,r},$$

$A := (a_{j,k})_{j,k=1}^{M}$ and denote the inverse matrix of A by $A^{-1} =: (\beta_{j,k})_{j,k=1}^{M}$. Its row-sum norm is

$$\mathrm{lub}_\infty(A^{-1}) := \max_{1 \le j \le M} \sum_{k=1}^{M} |\beta_{j,k}|. \tag{2.2}$$

Since $\|N_{j,r}\|_\infty \le 1$, it follows that $|b_j| \le \|f\|_\infty$ and thus that

$$\max_{1 \le j \le M} |\alpha_j| \le \mathrm{lub}_\infty(A^{-1})\|f\|_\infty.$$

Since $N_{j,r} \ge 0$ and $\sum_{k=1}^{M} N_{j,r}(x) = 1$ on $[a,b]$, it follows that

$$\|\mathbf{P}\|_\infty \le \mathrm{lub}_\infty(A^{-1}). \tag{2.3}$$

It is well known (see [1,2]) that $\mathrm{lub}_\infty(A^{-1}) \le C_r$ for some $C_r > 0$ which depends on r and the ratio

$$\kappa := \frac{\max\{x_j - x_{j-1} : j = 1, \dots, n\}}{\min\{x_j - x_{j-1} : j = 1, \dots, n\}}.$$

A. Shadrin [6] proved recently that C_r depends only on r.

§3. Least-Squares Method for Linear Boundary Value Problems

Let $r \ge 4$. Let $X = \{f \in C^2[0,1] : f(a) = f(b) = 0\}$ and $\mathcal{S} := \{u \in \mathcal{S}_r(\triangle) : u(a) = u(b) = 0\}$. Let $L : X \to C[a,b]$ be a linear differential operator of the form

$$Ly := y'' + cy' + dy, \quad y \in X, \tag{3.1}$$

where $c, d \in W_\infty^1[a,b]$. In the proof of the next theorem we will need that there exists a positive constant δ such that

$$\int_a^b (Lf)^2 \ge \delta \int_a^b (f'')^2, \quad \text{for all } f \in X. \tag{3.2}$$

The assumptions on L are strong enough to guarantee that the differential operator $L : X \to C[a,b]$ is injective.

Theorem 3.1. Let $\mathbf{P}^* : X \to \mathcal{S}$ be the linear projection onto $\mathcal{S}$ of the least-squares method defined by (1.7). Under assumption (3.2), $\mathbf{P}^*$ satisfies

$$\|(\mathbf{P}^* f)''\|_\infty \le C \ \delta^{-1/2} \mathrm{lub}_\infty(A^{-1}) \ \|f''\|_\infty, \quad f \in C^2[a,b], \qquad (3.3)$$

where A is the $(n + r - 3) \times (n + r - 3)$-matrix

$$A := (\alpha_{j,k})_{j,k=1}^{n+r-3} := \left\{ \frac{1}{x_j - x_{j-r+2}} \int_a^b N_{k,r-2} N_{j,r-2} \right\}_{j,k=1}^{n+r-3}, \qquad (3.4)$$

and where C depends only on $\|c\|_\infty$, $\|d\|_\infty$, $\|c'\|_\infty$, $\|d'\|_\infty$.

Proof: For convenience, we may assume that $[a, b] = [0, 1]$. The minimum problem (1.7) can be formulated in the setting of §1. Indeed, assumption (3.2) guarantees that

$$\langle f, g \rangle := \int_0^1 (Lf)(Lg), \quad f, g \in X,$$

is an inner product on X. The orthogonality relations for (1.7) are

$$\int_0^1 L[f - \mathbf{P}^* f] \cdot Lu = 0, \quad u \in \mathcal{S}. \qquad (3.5)$$

Let $M := n + r - 3$ and let $\mathcal{B}_{r-2} = \{N_{k,r-2}\}_{k=1}^M$ be the B-spline basis for $\mathcal{S}_{r-2}(\Delta)$. We define the functions $u_k \in \mathcal{S}$, $k = 1, \ldots, M$ by

$$u_k \in \mathcal{S} \quad , \quad u_k'' = N_{k,r-2}, \quad k = 1, \ldots, M.$$

Trivially, $\mathcal{B}_r^* := \{u_k\}_{k=1}^M$ is a basis of $\mathcal{S}$. By writing $w := \mathbf{P}^* f$, the orthogonality relations (3.5) are equivalent to

$$\int_0^1 (f'' - w'')u_j'' = -\int_0^1 (f'' - w'')(cu_j' + du_j)$$
$$-\int_0^1 \{(c(f' - w') + d(f - w)\}u_j''$$
$$-\int_0^1 \{(c(f' - w') + d(f - w)\}(cu_j' + du_j).$$

Since $u_j(0) = u_j(1) = 0$, there exists $\xi_j \in (0, 1)$ so that $u_j'(\xi_j) = 0$. Hence

$$|u_j'(x)| \le \int_0^1 |u_j''(t)| \le x_j - x_{j-r+2}, \quad 0 \le x \le 1,$$

and
$$|u_j(x)| \le x_j - x_{j-r+2}, \quad 0 \le x \le 1,$$
since $u_j'' = N_{j,r-2}$, $j = 1, \ldots, M$. Moreover, since $f(0) - w(0) = 0$ and $f(1) - w(1) = 0$,
$$\|f - w\|_\infty \le \int_0^1 |f' - w'| \le \int_0^1 |f'' - w''|.$$

It follows by simple computations that each of the integrals
$$\left| \int_0^1 (f'' - w'')(cu_j' + du_j) \right|,$$
$$\left| \int_0^1 \{(c(f' - w') + d(f - w)\}u_j'' \right|,$$
$$\left| \int_0^1 \{(c(f' - w') + d(f - w)\}(cu_j' + du_j) \right|,$$

are bounded from above by
$$\gamma_1(x_j - x_{j-r+2}) \int_0^1 |f'' - w''|,$$

where the positive number γ_1 (and later γ_2, γ_3) depend only on $\|c\|_\infty$, $\|d\|_\infty$, $\|c'\|_\infty$, $\|d'\|_\infty$. Therefore,
$$\left| \int_0^1 (f'' - w'')u_j'' \right| \le \gamma_2(x_j - x_{j-r+2}) \int_0^1 |f'' - w''|.$$

This implies that
$$\left| \int_0^1 w'' u_j'' \right| \le \left| \int_0^1 f'' u_j'' \right| + \gamma_2 \int_0^1 |f'' - w''|$$
$$\le (x_j - x_{j-r+2})\left(\|f''\|_\infty + \gamma_2 \int_0^1 |f'' - w''| \right).$$

Next we apply
$$\int_0^1 |f'' - w''| \le \left(\int_0^1 |f'' - w''|^2 \right)^{1/2} \le \left(\frac{1}{\delta} \int_0^1 L[f - w]^2 \right)^{1/2}$$
$$\le \left(\frac{1}{\delta} \int_0^1 (Lf)^2 \right)^{1/2} \le \gamma_3 \delta^{-1/2} \|f''\|_\infty$$

to get
$$\frac{1}{x_j - x_{j-r+2}} \left| \int_0^1 w'' u_j'' \right| \le C \, \delta^{-1/2} \|f''\|_\infty, \quad j = 1, \ldots, M. \qquad (3.6)$$

We can now repeat the arguments of §2 replacing r by $r-2$. Indeed, let $w = \mathbf{P}^* f = \sum_{k=1}^{M} c_k u_k$ so that $w'' = \sum_{k=1}^{M} c_k u_k''$. We define $b = (b_1, \ldots, b_M)^T$ by $b_j := \sum_{k=1}^{M} \alpha_{j,k} c_k$. Then (3.6) is equivalent to

$$|b_j| \leq C \; \delta^{-1/2} \|f''\|_\infty, \quad j = 1, \ldots, M.$$

It follows that $c = (c_1, \ldots, c_m)^T = A^{-1}b$ satisfies

$$\max_{1 \leq j \leq M} |c_j| \leq C \; \delta^{-1/2} \mathrm{lub}_\infty(A^{-1}) \; \|f''\|_\infty.$$

Since $u_k''(x) \geq 0$ and $\sum_{k=1}^{M} u_k''(x) = 1$ on $[0,1]$, we obtain (3.3). $\square$

Corollary 3.2. *Under the assumptions of Theorem 3.1, for each $y \in X$,*

$$\|y'' - (\mathbf{P}^* y)''\|_\infty \leq \left(1 + C \; \delta^{-1/2} C_{r-2}\right) \mathrm{dist}(y'', \mathcal{S}_{r-2}(\triangle))_\infty, \qquad (3.7)$$

where C is the constant in Theorem 3.1, and where C_{r-2} is defined in Section 2. Moreover, for $y \in C^r[a,b] \cap X$,

$$\|y'' - (\mathbf{P}^* y)''\|_\infty \leq K_r \left(1 + C \; \delta^{-1/2} C_{r-2}\right) h^{r-2} \|y^{(r)}\|_\infty, \qquad (3.8)$$

where $h := \max\{x_j - x_{j-1} : j = 1, \ldots, n\}$ and where K_r depends only on r.

Proof: For each $v \in \mathcal{S}_{r-2}(\triangle)$ there exists a unique $u \in \mathcal{S}$ so that $u'' = v$. Since $u = \mathbf{P}^* u$, it follows that

$$\begin{aligned} \|y'' - (\mathbf{P}^* y)''\|_\infty &\leq \|y'' - u''\|_\infty + \|u'' - (\mathbf{P}^* y)''\|_\infty \\ &= \|y'' - v\|_\infty + \|(\mathbf{P}^*[u-y])''\|_\infty. \end{aligned}$$

Applying the last theorem to $u - y$, it follows from $\mathrm{lub}_\infty(A^{-1}) \leq C_{r-2}$ and (3.3) that

$$\|y'' - (\mathbf{P}^* y)''\|_\infty \leq \|y'' - v\|_\infty + C \; \delta^{-1/2} C_{r-2} \; \|u'' - y''\|_\infty.$$

Since $u'' = v$ and since $v \in \mathcal{S}_{r-2}(\triangle)$ is arbitrary, we obtain (3.7). The error estimate (3.8) follows from well-known estimates for $\mathrm{dist}(y'', \mathcal{S}_{r-2}(\triangle))_\infty$. $\square$

Remark 1. It is not known if Theorem 3.1 and Corollary 3.2 are true under weaker conditions than (3.2).

Remark 2. It follows from (3.8) that

$$\|y - (\mathbf{P}^* y)\|_\infty = \mathcal{O}(h^{r-2}) \|y^{(r)}\|_\infty, \quad y \in C^r[a,b] \cap X. \qquad (3.9)$$

It is not known under which conditions the factor h^{r-2} in (3.9) can be replaced by h^r.

§4. The Galerkin Method

Let X, $\mathcal{S}$ and the linear differential operator L be as in Section 3. For $f \in X$, the Galerkin approximation $G = G_f \in \mathcal{S}$ is characterized by

$$\int_a^b (L[f - G_f])u(x) = 0, \quad \text{for all } u \in \mathcal{S}. \tag{4.1}$$

We assume that $G_f \in \mathcal{S}$ is uniquely defined by (4.1) for all $f \in X$. Setting $\mathbf{G}f := G_f$ (4.1) yields a linear projection $\mathbf{G} : X \to \mathcal{S}$ from X onto $\mathcal{S}$. Let

$$h := \max\{h_j : j = 0, \ldots, n - 1\}, \quad h_j := x_{j+1} - x_j,$$

and let $\kappa := \max_{j,k} h_j h_k^{-1}$. Douglas, Dupont, Wahlbin [3, Theorem 1.1] claim that under very mild assumptions on the functions c and d of L, the Galerkin solution $\mathbf{G}f$ satisfies

$$\|f - \mathbf{G}f\|_\infty \leq Ch^r\|f^{(r)}\|_\infty, \quad f \in C^r[a, b], \tag{4.2}$$

where the constant $C > 0$ in (4.2) depends only on r, κ, $\|c\|_\infty$, $\|d\|_\infty$, $\|c'\|_\infty$, $\|d'\|_\infty$.

§5. Numerical Examples

In this section we compare the least-squares approximations $\mathbf{P}^*y$ and the Galerkin approximation $\mathbf{G}y$ for some linear differential operators $Ly := y'' + cy' + dy$ on $[a, b]$. For the computations we do not use the basis of the proof of Theorem 3.1. Instead we take the modified B-spline basis

$$u_1 := N_1 - \frac{N_1(a)}{N_2(a)} N_2, \qquad u_2 := N_1 - \frac{N_1(a)}{N_3(a)} N_3$$

$$u_k := N_{k+1}, \quad k = 3, \ldots, n - 1$$

$$u_n := N_{n+3} - \frac{N_{n+3}(b)}{N_{n+1}(b)} N_{n+1}, \qquad u_{n+1} := N_{n+3} - \frac{N_{n+3}(b)}{N_{n+2}(b)} N_{n+2}.$$

In all examples below, we take cubic spline spaces $\mathcal{S}$ with equidistant knots

$$\triangle : a = x_0 < \cdots < x_k := a + kh < \cdots < x_n = b, \quad h := (b - a)/n,$$

and we write $S := \mathbf{P}^*y$ for the least-squares approximation, and $G := \mathbf{G}y$ for the Galerkin approximation. The examples indicate that the Galerkin approximations $G = \mathbf{G}y$ are often superior to the least-squares approximation $S = \mathbf{P}^*y$. In particular in Example 2, when $\lambda \to \pi$ approaches the case of non-uniqueness of y, the Galerkin approximation G is much better than S. In the tables below we write $|\cdot|$ for the L_∞-norm on $[-1, 1]$ (instead of $\|\cdot\|_\infty$), and $\|\cdot\|_2$ for the L_2-norm on $[-1, 1]$.

Example 1. Let $Ly := y'' - 2.0 \cos(5x)y' + 1.1 \cos(9x)y = q$ for $-1 \le x \le 1$, where q is chosen so that $y(x) := (2\pi)^{-4} \sin(2\pi x)$ is the solution of $Ly = q$, $y(-1) = y(1) = 0$. Note that $\|y^{(4)}\|_\infty = 1$.

n	$|y - S|$	$|y' - S'|$	$|y'' - S''|$	$\|q - LS\|_2$
5	$5.3 * 10^{-4}$	$2.0 * 10^{-3}$	$1.5 * 10^{-2}$	$5.7 * 10^{-3}$
10	$3.9 * 10^{-5}$	$1.6 * 10^{-4}$	$3.4 * 10^{-3}$	$1.2 * 10^{-3}$
20	$1.9 * 10^{-6}$	$1.2 * 10^{-5}$	$8.1 * 10^{-4}$	$2.8 * 10^{-4}$
40	$1.1 * 10^{-7}$	$1.2 * 10^{-6}$	$2.1 * 10^{-4}$	$6.7 * 10^{-5}$
80	$6.8 * 10^{-9}$	$1.4 * 10^{-7}$	$5.2 * 10^{-5}$	$1.7 * 10^{-5}$

n	$|y - G|$	$|y' - G'|$	$|y'' - G''|$	$\|q - LG\|_2$
5	$9.3 * 10^{-5}$	$1.2 * 10^{-3}$	$2.6 * 10^{-2}$	$6.9 * 10^{-3}$
10	$3.7 * 10^{-6}$	$8.6 * 10^{-5}$	$3.4 * 10^{-3}$	$1.2 * 10^{-3}$
20	$1.5 * 10^{-7}$	$8.5 * 10^{-6}$	$8.2 * 10^{-4}$	$2.8 * 10^{-4}$
40	$8.9 * 10^{-9}$	$1.0 * 10^{-6}$	$2.1 * 10^{-4}$	$6.7 * 10^{-5}$
80	$5.5 * 10^{-10}$	$1.3 * 10^{-7}$	$5.2 * 10^{-5}$	$1.7 * 10^{-5}$

Example 2. Let $Ly := y'' + \lambda^2 y = (\lambda^2 - \pi^2)\sin(\pi x)$ for $-1 \le x \le 1$. Then $y(x) := \sin(\pi x)$ solves $Ly = q$, $y(-1) = y(1) = 0$, for $q(x) := (\lambda^2 - \pi^2)\sin(\pi x)$. We always take $n = 20$ and vary λ. Notice that for $\lambda = \pi$, one has $q = 0$ so that $Y(x) = \alpha \sin \pi x$, for all $\alpha \in \mathbb{R}$, are solutions of $LY = 0$, $Y(-1) = Y(1) = 0$.

λ	$|y - S|$	$|y' - S'|$	$|y'' - S''|$	$\|q - LS\|_2$
1.0	$2.9 * 10^{-5}$	$7.9 * 10^{-4}$	$8.1 * 10^{-2}$	0.0260
2.0	$5.1 * 10^{-5}$	$8.0 * 10^{-4}$	$8.1 * 10^{-2}$	0.0260
3.0	$1.8 * 10^{-3}$	$5.7 * 10^{-3}$	$6.4 * 10^{-2}$	0.0259
3.1	$1.9 * 10^{-2}$	$6.1 * 10^{-2}$	0.23	0.0257
3.14	0.93	2.92	9.18	0.0068
3.15	0.32	1.02	3.20	0.0213

λ	$|y - G|$	$|y' - G'|$	$|y'' - G''|$	$\|q - LG\|_2$
1.0	$1.4 * 10^{-5}$	$7.9 * 10^{-4}$	$8.2 * 10^{-2}$	0.0260
2.0	$1.4 * 10^{-5}$	$7.9 * 10^{-4}$	$8.2 * 10^{-2}$	0.0260
3.0	$1.3 * 10^{-5}$	$7.9 * 10^{-4}$	$8.2 * 10^{-2}$	0.0259
3.1	$1.3 * 10^{-5}$	$7.9 * 10^{-4}$	$8.2 * 10^{-2}$	0.0259
3.14	$4.4 * 10^{-5}$	$8.0 * 10^{-4}$	$8.1 * 10^{-2}$	0.0259
3.15	$2.0 * 10^{-5}$	$7.9 * 10^{-4}$	$8.2 * 10^{-2}$	0.0259

Remark 3. In Example 1, $\|y - S\|_\infty$ and $\|y - G\|_\infty$ behave like $\mathcal{O}(h^4)$, $\|y' - S'\|_\infty$ and $\|y' - G'\|_\infty$ like $\mathcal{O}(h^3)$, and $\|y'' - S''\|_\infty = \mathcal{O}(h^2)$, $\|y'' - G''\|_\infty = \mathcal{O}(h^2)$, $\|q - LS\|_\infty = \mathcal{O}(h^2)$, $\|q - LG\|_\infty = \mathcal{O}(h^2)$. It is worth mentioning that $\|y - G\|_\infty$ is always smaller than $\|y - S\|_\infty$. Moreover, the Galerkin error functions $y(x) - G(x)$, $-1 \le x \le 1$, have many zeros in $[-1, 1]$, but $y(x) - S(x)$ has only a few.

Remark 4. It is also worth noting that in Example 2 for the almost "singular case" $\lambda = 3.1415$, the Galerkin approximation G for $n = 20$ approximates y very well, namely

$$\|y - G\|_\infty = 5.7 * 10^{-4},$$

$$\|y' - G'\|_\infty = 2.0 * 10^{-3},$$

$$\|y'' - G''\|_\infty = 7.6 * 10^{-2},$$

$$\|q - LG\|_2 = 2.6 * 10^{-2}.$$

References

1. de Boor, C., A bound on the L_∞-norm of the L_2-approximation by splines in terms of a global mesh ratio, Math. Comp. **30** (1976), 765-771.

2. de Boor, C., On a max-norm bound for the least-square spline approximant, in *Approximation and Function Spaces*, Z. Ciesielski (ed.), Proceedings of the International Conference (Gdansk, August 27-31, 1979), PWN, Warzawa, 1981, 163–175.

3. Douglas, J., T. Dupont, and L. Wahlbin, Optimal L_∞ error estimates for Galerkin approximations to solutions of two-point boundary value problems, Mathematics of Computation **29** (1975), 475–483.

4. v. Golitschek, M., The L_∞-norm of orthogonal and interpolation projections, in *Trends in Approximation Theory*, K. Kopotun, T. Lyche, M. Neamtu (eds.), Vanderbilt University Press, Nashville, 2001, 123–141.

5. v. Golitschek, M., and L. L. Schumaker, Bounds on projections onto bivariate polynomial spline spaces with stable bases, Constructive Approximation, to appear.

6. Shadrin, A. Yu., The L_∞-norm of the L_2-projector is bounded independently of the knot sequence: A proof of de Boor's conjecture, Acta Math. **187** (2001), 59–137.

Manfred v. Golitschek
Institut für Angewandte Mathematik und Statistik
Universität Würzburg
97074 Würzburg, Germany
goli@mathematik.uni-wuerzburg.de

Stable Refinable Generators of Shift Invariant Spaces with Certain Regularities and Vanishing Moments

Tian-Xiao He

Abstract. In this paper, we discuss the stable refinable functions that generate shift invariant (SI) spaces and possess the largest possible regularities and required vanishing moments. The stability of the corresponding complementary spaces is also discussed.

§1. Introduction

We start by setting some notation. We define a low-pass filter as

$$m_0(\xi) = 2^{-1} \sum_n h_n e^{-in\xi}. \tag{1}$$

Here, we assume that only finitely many h_n are nonzero. However, some of our results can be extended to infinite sequences that have sufficient decay for $|n| \to \infty$. Next, we define ϕ by

$$\hat{\phi}(\xi) = \prod_{j=1}^{\infty} m_0(2^{-j}\xi). \tag{2}$$

This infinite product converges only if $m_0(0) = 1$, i.e., if $\sum_n h_n = 2$. In this case, the infinite products in (2) converge uniformly and absolutely on compact sets, so that $\hat{\phi}$ is a well-defined C^∞ function. Obviously, $\hat{\phi}(\xi) = m_0(\xi/2)\hat{\phi}(\xi/2)$, or, equivalently, $\phi(t) = \sum_n h_n \phi(2t - n)$ at least in the sense of distributions. From Lemma 3.1 in [9], ϕ has compact support.

We now consider the simplest possible masks $m_0(\xi)$ with the following form.

Approximation Theory X: Wavelets, Splines, and Applications   305
Charles K. Chui, Larry L. Schumaker, and Joachim Stöckler (eds.), pp. 305–322.
Copyright ⊘ 2002 by Vanderbilt University Press, Nashville, TN.
ISBN 0-8265-1416-2.

Definition 1. *Denote by Φ the set of all functions $\phi(t)$ that have Fourier transform $\hat{\phi}(\xi) = m_0(\xi/2)\hat{\phi}(\xi/2)$. Here the filter $m_0(\xi) = 2^{-1}\sum_n h_n e^{-in\xi}$ is in the set M that contains all filters of the form*

$$m_0^N(\xi) = \left(\frac{1+e^{-i\xi}}{2}\right)^N F(\xi), \tag{3}$$

where

$$F(\xi) = e^{-ik'\xi}\sum_{j=0}^{k} a_j e^{-ij\xi}. \tag{4}$$

Here, all coefficients of $F(\xi)$ are real, $F(0) = 1$; N and k are positive integers; and $k' \in \mathbb{Z}$. Hence, the corresponding ϕ can be written as

$$\hat{\phi}(\xi) = \left(\frac{1+e^{-i\xi/2}}{2}\right)^N F(\xi/2)\hat{\phi}(\xi/2). \tag{5}$$

Clearly, ϕ is a B-spline of order N if $F(\xi) = 1$. The vanishing moments of ϕ are completely controlled by the exponents of its "spline factor," $\left(\frac{1+e^{-i\xi}}{2}\right)^N$. In addition, the regularity of ϕ is justified by the factors $F(\xi)$, and are independent of their vanishing moments.

A shift invariant (SI) space is a closed subspace of $L_2(\mathbb{R})$ that is invariant under the operator $S_k(f) := f(\cdot - k)$ $(k \in \mathbb{Z})$. For $\phi \in L_2(\mathbb{R})$, we say that $V = S(\phi) := \overline{span}\{\phi(\cdot - k) : k \in \mathbb{Z}\}$ is generated by ϕ. In addition, if ϕ is refinable, then ϕ is said to be a **refinable generator** of $S(\phi)$, and $S(\phi)$ is called a **refinable SI space**. Each element $\phi \in \Phi$ is a refinable generator of the corresponding SI space $S(\phi)$. A refinable generator is said to be a **pseudo-scaling (refinable) generator** if it satisfies $\hat{\phi}(\xi) = m_0(\xi/2)\hat{\phi}(\xi/2)$ and $|m_0(\xi)|^2 + |m_0(\xi + \pi)|^2 = 1$.

In [1,2,11], the following concepts were introduced that are important in our discussion.

Definition 2. *The bracket operator $[,] : L_2(\mathbb{R}) \times L_2(\mathbb{R}) \to L_1(\mathbb{T})$, $\mathbb{T} = [0, 2\pi)$, is defined by*

$$[f,g] = \sum_{k\in\mathbb{Z}} f(\xi + 2\pi k)\overline{g(\xi + 2\pi k)}. \tag{6}$$

*For $f \in L_2(\mathbb{R})$ the function $[f, f] \in L_1(\mathbb{T})$ is called the **auto-correlation** of f.*

If f, g are compactly supported, then $[\hat{f}, \hat{g}]$ is a trigonometric polynomial and has the Fourier expansion

$$[\hat{f},\hat{g}](\xi) = \sum_{k\in\mathbb{Z}} \langle f(\cdot), g(\cdot + k)\rangle e^{ik\xi}. \tag{7}$$

Definition 3. *Let $S(\phi)$ be a shift invariant space that is generated by ϕ. $\{\phi(\cdot - k)\}_{k \in \mathbb{Z}}$ is called a* **stable basis** *of $S(\phi)$ if there exist constants $0 < A \leq B < \infty$ such that for every $\mathbf{c} = \{c_k\}_{k \in \mathbb{Z}} \in \ell_2(\mathbb{Z})$*

$$A \left\| \mathbf{c} \right\|_{\ell_2(\mathbb{Z})}^2 \leq \left\| \sum_{k \in \mathbb{Z}} c_k \phi(\cdot - k) \right\|_{L_2(\mathbb{R})}^2 \leq B \left\| \mathbf{c} \right\|_{\ell_2(\mathbb{Z})}^2 . \tag{8}$$

Obviously, a stable basis of $S(\phi)$ is a basis of $S(\phi)$.

Theorem 4. *[16] Let $\phi \in L_2(\mathbb{R})$ and let $0 < A \leq B < \infty$. Then (8) and*

$$A \leq \left[\hat{\phi}, \hat{\phi} \right] \leq B, \ a.e.$$

are equivalent.

In Section 2 we will discuss the conditions for the coefficients $\{a_j\}$ such that the corresponding function ϕ is in $L_2(\mathbb{R})$ and is stable. The stability of the corresponding complementary spaces will be also discussed. Section 3 will give applications of the refinable generators in the construction of the biorthogonal and orthogonal scaling functions (the original generators) and wavelets (the generators of the corresponding complementary spaces) that possess the largest possible regularities and required vanishing moments. Since the biorthogonality and orthogonality imply the stability of the integer translates of the generators, we obtain simpler and sufficient conditions of the stability of a refinable generator with the largest possible regularities and required vanishing moments.

§2. Stable Generators of Shift Invariant Spaces

In [13] we have the following result. For the reader's convenience, we include a simpler alternative proof here.

Lemma 5. *Let $\phi \in \Phi$ be defined as in Definition 1; i.e.,*

$$\hat{\phi} = \prod_{j=1}^{\infty} m_0^N (2^{-j} \xi),$$

where $m_0^N(\xi) \in M$ is defined by (3) and (4):

$$m_0^N(\xi) = \left(\frac{1 + e^{-i\xi}}{2} \right)^N F(\xi)$$

and $F(\xi) = e^{-ik'\xi} \sum_{j=0}^{k} a_j e^{-ij\xi}$, $N, k \in \mathbb{Z}_+$ and $k' \in \mathbb{Z}$, where $F(0) = 1$. If $F(\pi) \neq -1$ and the coefficients of $F(\xi)$ satisfy

$$(k+1) \sum_{j=0}^{k} a_j^2 < 2^{2N-1}, \tag{9}$$

then ϕ is in $L_2(\mathbb{R})$. In addition, (9) can be replaced by the weaker condition

$$C(\{a_j\}, k) < 2^{2N-1}, \qquad (9)'$$

where $C(\{a_j\}, k)$ equals $k \sum_{j=0}^{k} a_j^2$ if $k \geq 1$ and equals a_0^2 if $k = 0$.

Proof: We first prove that $\phi \in L_2(\mathbb{R})$; i.e.,

$$C(\{a_j\}, k) < 2^{2N-1},$$

implies $\hat{\phi} = \prod_{j=1}^{\infty} m_0^N(2^{-j}\xi)$ is in $L^2(\mathbb{R})$. It is sufficient to prove the boundedness of the following integral

$$\int_{|\xi| \geq \pi} |\hat{\phi}(\xi)|^2 d\xi$$

$$= \sum_{\ell=1}^{\infty} \int_{2^{\ell-1}\pi \leq |\xi| \leq 2^{\ell}\pi} \prod_{j=1}^{\infty} \left| \left(\frac{1 + e^{-i2^{-j}\xi}}{2} \right)^N F(2^{-j}\xi) \right|^2 d\xi$$

$$= \sum_{\ell=1}^{\infty} \int_{2^{\ell-1}\pi \leq |\xi| \leq 2^{\ell}\pi} \left| \frac{1 - e^{-i\xi}}{i\xi} \right|^{2N} \prod_{j=1}^{\infty} |F(2^{-j}\xi)|^2 d\xi$$

$$\leq C \sum_{\ell=1}^{\infty} \int_{2^{\ell-1}\pi \leq |\xi| \leq 2^{\ell}\pi} \frac{1}{|\xi|^{2N}} \prod_{j=1}^{\infty} |F(2^{-j}\xi)|^2 d\xi$$

$$\leq C \sum_{\ell=1}^{\infty} \frac{1}{2^{2\ell N}} \int_{2^{\ell-1}\pi \leq |\xi| \leq 2^{\ell}\pi} \prod_{j=1}^{\ell} |F(2^{-j}\xi)|^2 \prod_{j=1}^{\infty} |F(2^{-\ell-j}\xi)|^2 d\xi$$

$$\leq C \sum_{\ell=1}^{\infty} 4^{-\ell N} \int_{2^{\ell-1}\pi \leq |\xi| \leq 2^{\ell}\pi} \prod_{j=1}^{\ell} |F(2^{-j}\xi)|^2 d\xi. \qquad (10)$$

We now prove the boundedness of the last integral in (10). Denote

$$Tf(\xi) = \left| F\left(\frac{\xi}{2}\right) \right|^2 f\left(\frac{\xi}{2}\right) + \left| F\left(\frac{\xi}{2} + \pi\right) \right|^2 f\left(\frac{\xi}{2} + \pi\right). \qquad (11)$$

Hence, for any 2π-periodic continuous function f, we have

$$\int_{2^{\ell-1}\pi \leq |\xi| \leq 2^{\ell}\pi} f(2^{-\ell}\xi) \prod_{j=1}^{\ell} \left| F(2^{-j}\xi) \right|^2 d\xi$$

$$= \int_{-\pi}^{\pi} T^{\ell} f(\xi) d\xi \leq \sqrt{2\pi} \|T^{\ell} f\|_{L^2} \leq \sqrt{2\pi} \|f\|_{L^2} \|T^{\ell}\|. \qquad (12)$$

Let $\rho(T)$ be the spectral radius of the operator T. Since $F(0) = 1$ and $F(\pi) \neq -1$, it can be shown as follows that $\rho(T) > 0$ (see also [6]). Considering the Fourier expansion

$$|F(\xi)|^2 = \sum_{\ell=-k}^{k} b_\ell e^{i\ell\xi},$$

where k is a positive integer and $b_t = \sum_{j=0}^{k-|t|} a_{k-|t|-j} a_{k-j}$ $(t = -k, \cdots, k)$, we find that the matrix of T restricted to

$$E_k = \{ \sum_{\ell=-k}^{k} c_\ell e^{i\ell\xi} : (c_{-k}, \ldots, c_k) \in \mathbf{C}^{2k+1} \}$$

is

$$M_T = (2b_{i-2j})_{i,j=-k,\ldots,k} = 2 \begin{bmatrix} b_k & 0 & 0 & \cdots & 0 \\ b_{k-2} & b_{k-1} & b_k & \cdots & 0 \\ \vdots & \vdots & \vdots & \cdots & \vdots \\ b_{-k} & b_{-k+1} & b_{-k+2} & \cdots & b_k \\ 0 & 0 & b_{-k} & \cdots & b_{k-2} \\ \vdots & \vdots & \vdots & \cdots & \vdots \\ 0 & 0 & 0 & \cdots & b_{-k} \end{bmatrix}. \quad (13)$$

Noting that $|F(0)|^2 = \sum_{\ell=-k}^{k} b_\ell = 1$ and $|F(\pi)|^2 = \sum_{\ell=-k}^{k} (-1)^\ell b_\ell = \alpha \neq -1$, it follows that

$$\sum_{\ell} b_{2\ell} = \sum_{\ell} b_{2\ell+1} = (\alpha + 1)/2,$$

and for the vector $u = (1, \ldots, 1) \in \mathbf{C}^{2k+1}$,

$$Tu = uM = (\alpha + 1)u.$$

Thus, T has at least one eigenvalue $\alpha + 1 \neq 0$.

For every $\epsilon > 0$, there is an integer $\ell(\epsilon)$ such that

$$\|T^\ell\| \leq (\rho(T) + \epsilon)^\ell, \qquad \ell > \ell(\epsilon).$$

It follows from (10) that

$$\int_{|\xi| \geq \pi} |\hat{\phi}(\xi)|^2 d\xi \leq C \sum_{\ell=1}^{\ell(\epsilon)} 4^{-N\ell} \|T^\ell\| + C \sum_{\ell=\ell(\epsilon)+1}^{\infty} 4^{-N\ell} (\rho(T) + \epsilon)^\ell,$$

so $\rho(T)$ must be estimated if we are to choose an $\epsilon > 0$ small enough for the series to converge. Regardless of how small an $\epsilon > 0$ is chosen, the contribution

$$C \sum_{\ell=1}^{\ell(\epsilon)} 4^{-N\ell} \|T^\ell\| \leq C \sum_{\ell=1}^{\ell(\epsilon)} 4^{-N\ell} \|T\|^\ell$$

is finite, although possibly very large.

To evaluate $\rho(T)$, we consider the matrix of T, M_T, which was given in (13). It is clear that $\rho(T) = \rho(M_T)$. We write $M_T = 2H$, where

$$H = (b_{i-2j})_{i,j=-k,\cdots,k}.$$

Obviously, b_β can be written as

$$b_\beta = \sum_{j=0}^{k-|\beta|} a_{k-|\beta|-j} a_{k-j}, \quad \beta = -k, \cdots, k.$$

Hence, $b_\beta = b_{-\beta}$, for all $\beta = -k, \cdots, k$. It is also clear that b_k is an eigenvalue of H with multiplicity 2. To estimate bounds of the eigenvalues of H, we establish

$$|b_\beta| \leq b_0, \quad \beta = -k, \cdots, k.$$

In fact,

$$|b_\beta| \leq \sum_{j=0}^{k-|\beta|} |a_{k-|\beta|-j} a_{k-j}|$$

$$\leq \sum_{j=0}^{k-|\beta|} [\frac{1}{2} a_{k-|\beta|-j}^2 + \frac{1}{2} a_{k-j}^2]$$

$$= \frac{1}{2} \sum_{j=\beta}^{k} a_{k-j}^2 + \frac{1}{2} \sum_{j=0}^{k-\beta} a_{k-j}^2$$

$$\leq \sum_{j=0}^{k} a_{k-j}^2 = b_0.$$

It is obvious that the spectral radius of H is b_0 if $k = 0$. For $k \geq 1$, the characteristic polynomial of H is $(b_k - \lambda)(b_{-k} - \lambda)$ multiplied by the characteristic polynomial of the core matrix, H_c, which consists of all rows and columns of H except its first and last rows and columns. Hence, the spectral radius of H is

$$\rho(H) = \max\{b_k, \rho(H_c)\} \leq \max\{b_k, \|H_c\|_1\}$$

$$= \max\{b_k, \sum_{i=-k+1}^{k-1} |b_{i-2j}| : j = -k+1, \cdots, k-1\} \leq kb_0.$$

Therefore, $\rho(T) = 2\rho(H) \leq 2C(\{a_j\}, k)$. Here, $C(\{a_j\}, k)$ equals $k \sum_{j=0}^{k} a_j^2$ if $k \geq 1$ and equals a_0^2 if $k = 0$. If $C(\{a_j\}, k) < 2^{2N-1}$, then $\rho(T) < 2^{2N}$, so we choose

$$\epsilon = \frac{1}{2} \left(2^{2N} - \rho(T) \right).$$

Thus

$$\rho(T) + \epsilon < 2^{2N},$$

and we obtain the estimate

$$\int_{|\xi| \geq \pi} \left| \hat{\phi}(\xi) \right|^2 d\xi \leq C \sum_{\ell=1}^{\ell(\epsilon)} 4^{-N\ell} \|T\|^\ell + \sum_{\ell=\ell(\epsilon)+1}^{\infty} \left(\frac{\rho(T) + \epsilon}{4^N} \right)^\ell.$$

The tail of the series is a convergent geometric series, thus completing the proof of $\phi \in L_2(\mathbb{R})$ if condition (9)', $C(\{a_j\}, k) < 2^{2N-1}$, holds. Since $C(\{a_j\}, k) \leq (k+1) \sum_{j=0}^{k} a_j^2$, the proof is complete. $\square$

We now discuss the stability of ϕ. From [15], we obtain a necessary and sufficient condition for a refinable function defined as in Definition 1 to be stable.

Lemma 6. *The function ϕ defined in Definition 1 is stable if and only if $F(\xi)$ satisfies the following two conditions.*

(i) *$F(\xi)$ does not have any symmetric zeroes on $\mathbb{T} = [0, 2\pi)$;*

(ii) *For any odd integer $m > 1$ and a primitive mth root $\omega = e^{-i2n\pi/m}$ of unity (i.e., n is an integer relatively prime to m), there exists an integer d, $0 \leq d < ord_m 2$, such that $F\left(-2^{d+1}n\pi/m\right) \neq 0$, where $p = ord_m 2$ is the smallest positive integer with $2^p \equiv 1 (mod \ m)$.*

In addition, if ϕ is stable, then $|\hat{\phi}(\xi)|^2$ and

$$\bar{F}(\xi) = \prod_{j=1}^{\infty} \left| F\left(2^{-j}(\xi + 2\pi u)\right) \right|^2 / \prod_{j=1}^{\infty} \left| F\left(2^{-j}\xi\right) \right|^2$$

$$= \frac{\prod_{j=1}^{\infty} \left| \sum_{\ell=0}^{k} a_\ell e^{-i\ell 2^{-j}(\xi + 2\pi u)} \right|^2}{\prod_{j=1}^{\infty} \left| \sum_{\ell=0}^{k} a_\ell e^{-i\ell 2^{-j}\xi} \right|^2} \tag{14}$$

have no roots in $\mathbb{T}$.

Proof: By using Lemma 6.6 of [5] and Theorem 1 of [15], we obtain that ϕ is stable. In addition, since

$$\left| \hat{\phi}(\xi) \right|^2 = \left| \frac{1 - e^{-i\xi}}{i\xi} \right|^{2N} \prod_{j=1}^{\infty} \left| F\left(2^{-j}\xi\right) \right|^2, \tag{15}$$

we have

$$\left|\hat{\phi}(\xi + 2\pi u)\right|^2 = \bar{F}(\xi) \left|\hat{\phi}(\xi)\right|^2 \frac{\xi^{2N}}{(\xi + 2\pi u)^{2N}},$$

where $\bar{F}(\xi)$ is defined as (14). Therefore,

$$\left[\hat{\phi}, \hat{\phi}\right] = \sum_{u \in \mathbb{Z}} \left|\hat{\phi}(\xi + 2\pi u)\right|^2$$

$$= \left|\hat{\phi}(\xi)\right|^2 \bar{F}(\xi)(\xi)^{2N} \sum_{u \in \mathbb{Z}} \frac{1}{(\xi + 2\pi u)^{2N}}.$$

By using formula (4.2.7) in [6], we can rewrite the last expression as

$$\left[\hat{\phi}, \hat{\phi}\right] = \left|\hat{\phi}(\xi)\right|^2 \frac{\bar{F}(\xi)(\xi/2)^{2N}}{\sin^{2N}(\xi/2)} \sum_{u=-\infty}^{\infty} \left|\hat{B}_N(\xi + 2\pi u)\right|^2 \qquad (16)$$

where $\hat{B}_N(\xi)$ is the Fourier transform of the B-spline of order N. In addition, the sum on the right-hand side of (16) can be evaluated by using formula (4.2.10) in [6]:

$$\sum_{u=-\infty}^{\infty} \left|\hat{B}_N(2\xi + 2\pi u)\right|^2 = \frac{-\sin^{2N}\xi}{(2N-1)!} \frac{d^{2N-1}}{d\xi^{2N-1}} \cot \xi.$$

Therefore, noting that ϕ is stable and applying Theorem 4 to the above $\left[\hat{\phi}, \hat{\phi}\right]$, we immediately know that $|\hat{\phi}(\xi)|^2, \bar{F}(\xi) \neq 0$ for all $\xi \in \mathbb{T}$. This completes the proof. $\square$

Combining Lemmas 5 and 6, we obtain the following result.

Theorem 7. *Let $\phi \in \Phi$ be defined as in Definition 1. If $F(\xi)$ satisfies $F(\pi) \neq -1$, conditions (i) and (ii) in Lemma 6, and (9) or (9)', then the corresponding ϕ is in $L_2(\mathbb{R})$ and is stable.*

Remark. Obviously, the stability condition described in Lemma 6 is not easy to check. We will give a simpler sufficient condition in the next section.

Let $V_j := \overline{span}\{\phi(2^j t - k) : k \in \mathbb{Z}\}$. Following [11], for any $\phi \in L_2(\mathbb{R})$, we define the (natural) dual $\tilde{\phi}$ by its Fourier transform

$$\hat{\tilde{\phi}} := \frac{\hat{\phi}}{\left[\hat{\phi}, \hat{\phi}\right]},$$

where we interpret $0/0 = 0$. Thus, for ϕ given in Theorem 7, from (16), the dual function's Fourier transform is

$$\widehat{\tilde{\phi}} = \frac{\hat{\phi}}{\left[\hat{\phi}, \hat{\phi}\right]} = \frac{\sin^{2N}(\xi/2)}{(\xi/2)^{2N}\overline{\hat{\phi}}\bar{F}(\xi)\sum_{u=-\infty}^{\infty}\left|\hat{B}_N(\xi + 2\pi u)\right|^2}.$$

The properties of the dual function $\tilde{\phi}$ can be found in [6,12].

It is clear that $S(\phi) \subset V_1 = \overline{span}\{\phi(2 \cdot -k) : k \in \mathbb{Z}\}$. We now consider the complementary space of $S(\phi)$ in V_1, which is generated by a function $\psi \in V_1$. We say that a function $f \in L_2(\mathbb{T})$ is in **W**, the Wiener Algebra, if its Fourier series $\sum_{k\in\mathbb{Z}} f_k e^{-ik\xi}$ satisfies $\{f_k\} \in \ell_1(\mathbb{Z})$. From [11], we can establish the following theorem.

Theorem 8. *Let $\phi \in \Phi$ satisfy all conditions of Theorem 7, where Φ is defined in Definition 1, and let $\psi \in V_1 := \overline{span}\{\phi(2 \cdot -k) : k \in \mathbb{Z}\}$ have the symbol $m_1(\xi) \in$ **W**, the Wiener Algebra, such that*

$$|m_1(\xi)|^2 + |m_1(\xi + \pi)|^2 > 0, \quad \xi \in \mathbb{T}. \tag{17}$$

Then ψ is stable (i.e., a stable generator for $S(\psi)$).

Proof: Since $\psi \in V_1$, using the two-scale relation of ψ yields

$$\left[\hat{\psi}, \hat{\psi}\right](\xi) = \sum_{i\in\mathbb{Z}} \left|m_1\left(\frac{\xi}{2} + \pi i\right)\right|^2 \left|\hat{\phi}\left(\frac{\xi}{2} + \pi i\right)\right|^2$$

$$= \sum_{i\in\mathbb{Z}} \left|m_1\left(\frac{\xi}{2} + 2\pi i\right)\right|^2 \left|\hat{\phi}\left(\frac{\xi}{2} + 2\pi i\right)\right|^2$$

$$+ \sum_{i\in\mathbb{Z}} \left|m_1\left(\frac{\xi}{2} + \pi + 2\pi i\right)\right|^2 \left|\hat{\phi}\left(\frac{\xi}{2} + \pi + 2\pi i\right)\right|^2$$

$$= \left|m_1\left(\frac{\xi}{2}\right)\right|^2 \left[\hat{\phi}\left(\frac{\xi}{2}\right), \hat{\phi}\left(\frac{\xi}{2}\right)\right] + \left|m_1\left(\frac{\xi}{2} + \pi\right)\right|^2 \times$$

$$\left[\hat{\phi}\left(\frac{\xi}{2} + \pi\right), \hat{\phi}\left(\frac{\xi}{2} + \pi\right)\right].$$

From Theorem 7, ϕ is in $L_2(\mathbb{R})$ and is stable. Thus, Theorem 4 shows there exist $0 < A \leq B < \infty$ such that

$$A \leq \left[\hat{\phi}, \hat{\phi}\right] \leq B \quad a.e.$$

Thus we can bound the auto-correlation of ψ by

$$AM_1(\xi) \leq \left[\hat{\psi}, \hat{\psi}\right](\xi) \leq BM_1(\xi),$$

where

$$M_1(\xi) := \left| m_1 \left(\frac{\xi}{2} \right) \right|^2 + \left| m_1 \left(\frac{\xi}{2} + \pi \right) \right|^2.$$

Since $m_1(\xi) \in (\mathbb{T})$, from condition (17) we have

$$\bar{A} := \min_{\xi \in \mathbb{T}} M_1(\xi) > 0.$$

On the other hand, $\bar{B} := \| m_1(\xi) \|_{C(\mathbb{T})} < \infty$. Therefore,

$$0 < \bar{A} A \leq \left[\hat{\psi}, \hat{\psi} \right] (\xi) \leq \bar{B} B < \infty, \quad a.e.$$

By using Theorem 4, we have proved that ψ is a stable generator for $S(\psi)$.
□

§3. Applications of Stable Refinable Generators

Obviously, the stability condition given in Lemma 6 is not easy to apply. Recalling Cohen's following result (see [7]), we can give a simpler and sufficient condition for the stability of the integer translates of a refinable generator. In [7], Cohen showed that a refinable generator ϕ with $\hat{\phi}(0) = 1$ has orthonormal integer translates (i.e., $\{\phi(\cdot - k)\}_{k \in \mathbb{Z}}$ is an orthonormal set) if and only if ϕ is a stable and pseudo-scaling generator. Hence, by extension of the results given in the author's paper [13] on the biorthogonal refinable generators, we can find stable refinable generators that possess the largest possible regularities and required vanishing moments.

Biorthogonal refinable generators defined by Definition 1 were discussed in [13]. Similarly, from Cohen and Daubechies' result in [8], the biorthogonality of the integer translates implies the stability of the translate system. The biorthogonal system associated with the integer translates of $\phi \in \Phi$ is the set of the integer translates of another refinable generator $\tilde{\phi} \in \Phi$. Here, $\tilde{\phi}(t) = \sum_n \tilde{h}_n \tilde{\phi}(2t - n)$ or equivalently, $\hat{\tilde{\phi}}(\xi) = \tilde{m}_0(\xi/2) \hat{\tilde{\phi}}(\xi/2)$ with $\tilde{m}_0(\xi) = 2^{-1} \sum_n \tilde{h}_n e^{-in\xi} \in M$, which is defined in Definition 1. Therefore, we can write

$$\tilde{m}_0(\xi) = \tilde{m}_0^{\tilde{N}}(\xi) = \left(\frac{1 + e^{-i\xi}}{2} \right)^{\tilde{N}} \tilde{F}(\xi), \tag{18}$$

where

$$\tilde{F}(\xi) = e^{-i\tilde{k}'\xi} \sum_{j=0}^{\tilde{k}} \tilde{a}_j e^{-ij\xi}.$$

Here, all coefficients of $\tilde{F}(\xi)$ are real, $\tilde{F}(0) = 1$; $\tilde{N}$ and $\tilde{k}$ are positive integers; and $\tilde{k}' \in \mathbb{Z}$. Hence, the corresponding ϕ and $\tilde{\phi}$ can be written as

$$\hat{\phi}(\xi) = \left(\frac{1 + e^{-i\xi/2}}{2}\right)^N F(\xi/2)\hat{\phi}(\xi/2),$$

$$\hat{\tilde{\phi}}(\xi) = \left(\frac{1 + e^{-i\xi/2}}{2}\right)^{\tilde{N}} \tilde{F}(\xi/2)\hat{\tilde{\phi}}(\xi/2). \tag{19}$$

We also define the corresponding ψ and $\tilde{\psi}$ by

$$\hat{\psi}(\xi) = e^{i\xi/2}\overline{\tilde{m}_0(\xi/2 + \pi)}\hat{\phi}(\xi/2) \quad \hat{\tilde{\psi}}(\xi) = e^{i\xi/2}\overline{m_0(\xi/2 + \pi)}\hat{\tilde{\phi}}(\xi/2), \tag{20}$$

or, equivalently,

$$\psi(x) = \sum_n (-1)^{n-1}\tilde{h}_{-n-1}\phi(2x - n)$$

$$\tilde{\psi}(x) = \sum_n (-1)^{n-1}h_{-n-1}\phi(2x - n). \tag{21}$$

Similar to ϕ, $\tilde{\phi}$ is also a B-spline of order $\tilde{N}$ if $\tilde{F}(\xi) = 1$. Since vanishing moment conditions $\int x^\ell \psi(x)dx = 0$, $\ell = 0, 1, \cdots, L$, are equivalent to $\frac{d^\ell}{d\xi^\ell}\hat{\psi}|_{\xi=0} = 0$, $\ell = 0, 1, \cdots, L$, we immediately know that the maximum orders of vanishing moment for ψ and $\tilde{\psi}$ are $N - 1$ and $\tilde{N} - 1$, respectively. Therefore, the vanishing moments of ϕ and $\tilde{\phi}$ are completely controlled by the exponents of their respective "spline factors," $\left(\frac{1+e^{-i\xi}}{2}\right)^N$ and $\left(\frac{1+e^{-i\xi}}{2}\right)^{\tilde{N}}$. In addition, as we pointed out at the beginning of this paper, the regularities of ϕ and $\tilde{\phi}$ are justified by the factors $F(\xi)$ and $\tilde{F}(\xi)$, respectively, and are independent of their vanishing moments.

From Lemma 5, if $\tilde{\phi} \in \Phi$ and the coefficients of the corresponding $\tilde{F}(\xi)$ satisfy

$$(\tilde{k} + 1)\sum_{j=0}^{\tilde{k}} \tilde{a}_j^2 < 2^{2\tilde{N}-1} \tag{22}$$

and $\tilde{F}(\pi) \neq -1$, then $\tilde{\phi} \in L_2(\mathbb{R})$.

In addition, from [8], the stability of ϕ, $\tilde{\phi}$, ψ, and $\tilde{\psi}$ are implied by their biorthogonality. In fact, [8] gave the following results.

Lemma 9. [8] *If $\phi, \tilde{\phi} \in L_2(\mathbb{R})$ satisfy $\langle \phi(t), \tilde{\phi}(t - n)\rangle = \delta_{n0}$, then $\{\phi(t - k)\}_{k\in\mathbb{Z}}$ and $\{\tilde{\phi}(t - k)\}_{k\in\mathbb{Z}}$ are stable; i.e., they are stable bases (Riesz bases) in the subspace that they generate. In addition, the corresponding biorthogonal wavelet functions ψ and $\tilde{\psi}$ are also stable; i.e., they are stable bases (Riesz bases) in the subspace that they generate.*

Hence, from the following result established in [13], we can construct biorthogonal (stable) refinable generators with the largest possible regularities and required vanishing moments by using the algorithm shown in [13].

Theorem 10. [13] *Let* $\phi, \tilde{\phi} \in \Phi$ *be defined in Definition 1; that is* $\phi = \prod_{j=1}^{\infty} m_0^N(2^{-j}\xi)$ *and* $\tilde{\phi} = \prod_{j=1}^{\infty} \tilde{m}_0^{\tilde{N}}(2^{-j}\xi)$, *where* $m_0^N(\xi)$ *and* $\tilde{m}_0^{\tilde{N}}(\xi)$ *are as in (3) and (18). Suppose* $F(\pi) \neq -1$, $\tilde{F}(\pi) \neq -1$; *the coefficients of* $F(\xi)$ *and* $\tilde{F}(\xi)$ *satisfy respectively conditions (9) and (22); and*

$$\sum_{j=\mu}^{\nu} \sum_{\ell=0}^{k} \sum_{\tilde{\ell}=0}^{\tilde{k}} \binom{\tilde{N}}{j - \tilde{\ell} - \tilde{k}'} \binom{N}{j + 2n - \ell - k'} \tilde{a}_{\tilde{\ell}} a_\ell = 2^{N+\tilde{N}-1} \delta_{n0}, \quad (23)$$

where $\mu = \min\{k', \tilde{k}'\}$; $\nu = \max\{N + k + k', \tilde{N} + \tilde{k} + \tilde{k}'\}$; δ_{n0} *is the Kronecker symbol; and* $n = 0, \pm 1, \pm 2, \cdots$; *then* $\phi, \tilde{\phi} \in L_2(\mathbb{R})$ *are stable and* $\langle \phi(t), \tilde{\phi}(t - i) \rangle = \delta_{i,0}$ *for all* $i \in \mathbb{Z}$. *The corresponding biorthogonal wavelets* ψ *and* $\tilde{\psi}$ *defined as (20) or (21) are in* C^α *and* $C^{\tilde{\alpha}}$, *respectively. Here* α *and* $\tilde{\alpha}$ *satisfy*

$$\alpha > N - \frac{1}{2} \log_2 \left((k+1) \sum_{j=0}^{k} a_j^2 \right),$$

and

$$\tilde{\alpha} > \tilde{N} - \frac{1}{2} \log_2 \left((\tilde{k}+1) \sum_{j=0}^{\tilde{k}} \tilde{a}_j^2 \right), \quad (24)$$

respectively. Equivalently, the Sobolev exponents

$$\alpha := \sup\{s \geq 0 : \int_R (1 + |\xi|^2)^s \left| \hat{\phi}(\xi) \right|^2 d\xi < \infty\}$$

and

$$\tilde{\alpha} := \sup\{s \geq 0 : \int_R (1 + |\xi|^2)^s \left| \hat{\tilde{\phi}}(\xi) \right|^2 d\xi < \infty\}$$

satisfy (24).

If we consider the case of $\tilde{\phi} = \phi$ in Theorem 10, then we obtain the following result.

Theorem 11. *Let* $\phi \in \Phi$ *be defined by Definition 1. If* F *also satisfies* $F(\pi) \neq -1$ *and its coefficients satisfy condition (9) and*

$$\sum_{j=k'}^{N+k+k'} \sum_{\ell=0}^{k} \sum_{\tilde{\ell}=0}^{k} \binom{N}{j - \tilde{\ell} - k'} \binom{N}{j + 2n - \ell - k'} a_{\tilde{\ell}} a_\ell = 2^{2N-1} \delta_{n0}, \quad (25)$$

where δ_{n0} is the Kronecker symbol and $n = 0, \pm 1, \pm 2, \cdots$, then $\phi \in L_2(\mathbb{R})$ and is a stable pseudo-scaling generator. In addition, The corresponding ψ defined as (20) or (21) is an orthogonal wavelet in C^α. Here α is more than

$$N - \frac{1}{2} \log_2 \left((k+1) \sum_{j=0}^{k} a_j^2 \right).$$

The regularity of ϕ (i.e., the Sobolev exponent of ϕ),

$$\alpha := sup\{s \geq 0 : \int_R (1 + |\xi|^2)^s \left| \hat{\phi}(\xi) \right|^2 d\xi < \infty\},$$

satisfies

$$\alpha > N - \frac{1}{2} \log_2 \left((k+1) \sum_{j=0}^{k} a_j^2 \right).$$

Here, the condition (9) can be replaced by the following weaker condition:

$$C(\{a_j\}, k) < 2^{2N-1}, \tag{9$'$}$$

where $C(\{a_j\}, k)$ equals $k \sum_{j=0}^{k} a_j^2$ if $k \geq 1$ and equals a_0^2 if $k = 0$. Hence, the corresponding regularities of ψ is determined by $\psi \in C^{\alpha'}$, where α' is more than $N - \frac{1}{2} \log_2 (2C(\{a_j\}, k))$. And the regularity of ϕ is $\alpha > N - \frac{1}{2} \log_2 (2C(\{a_j\}, k))$.

Proof: Let $\phi \in \Phi$ be the function defined by Definition 1 that satisfies $F(\pi) \neq -1$ and (9). Then, from Lemma 5, ϕ is in $L_2(\mathbb{R})$. Noting Theorem 10, we have that (9) and (25) imply

$$\langle \phi(t), \phi(t - n) \rangle = \delta_{n0},$$

which is equivalent to ϕ being a stable refinable pseudo-scaling generator (see [7]). To prove that the generator ψ of the complementary space of $S(\phi)$ is an orthogonal wavelet, it is sufficient (see [14]) to prove that it satisfies

$$\sum_{j \in \mathbb{Z}} \left| \hat{\psi} \left(2^j \xi \right) \right|^2 = 1 \qquad a.e. \tag{26}$$

and

$$\sum_{j \geq 0} \hat{\psi} \left(2^j \xi \right) \overline{\hat{\psi} \left(2^j (\xi + 2q\pi) \right)} = 0 \quad a.e., \tag{27}$$

for all $q \in 2\mathbb{Z} + 1$; i.e., for all odd integers, q. First,

$$\sum_{j \in \mathbb{Z}} \left| \hat{\psi} \left(2^j \xi \right) \right|^2 = \sum_{j \in \mathbb{Z}} \left| m_0 \left(2^{j-1} \xi + \pi \right) \right|^2 \left| \hat{\phi} \left(2^{j-1} \xi \right) \right|^2$$

$$= \lim_{n \to \infty} \sum_{j=-n}^{n} \left| m_0 \left(2^{j-1} \xi + \pi \right) \right|^2 \left| \hat{\phi} \left(2^{j-1} \xi \right) \right|^2$$

$$= \lim_{n \to \infty} \sum_{j=-n}^{n} \left[1 - \left| m_0 \left(2^{j-1} \xi \right) \right|^2 \right] \left| \hat{\phi} \left(2^{j-1} \xi \right) \right|^2$$

$$= \lim_{n \to \infty} \left[\left| \hat{\phi} \left(2^{-n-1} \xi \right) \right|^2 - \left| \hat{\phi} \left(2^n \xi \right) \right|^2 \right].$$

Since $\phi \in L_2(\mathbb{R})$, $\lim\limits_{n \to \infty} \left| \hat{\phi} \left(2^n \xi \right) \right|^2 = 0$ for *a.e.* ξ. From the condition $F(0) = 1$, we have $\lim\limits_{n \to \infty} \left| m_0 \left(2^{-n} \xi \right) \right| = 1$. Hence, taking limit $n \to \infty$ on both sides of equation $\hat{\phi}(\xi) = m_0(\xi/2)\hat{\phi}(\xi/2)$ and noting that

$$\lim_{n \to \infty} \left| \hat{\phi}(\xi) \right| = \lim_{n \to \infty} \prod_{j=1}^{\infty} \left| m_0(2^{-j}\xi) \right| \neq 0,$$

we obtain

$$\lim_{n \to \infty} \left| \hat{\phi} \left(2^{-n} \xi \right) \right| = 1,$$

which shows $\sum_{j \in \mathbb{Z}} \left| \hat{\psi} \left(2^j \xi \right) \right|^2 = 1$. Therefore, (26) holds for our ψ. Secondly, for any odd integer q,

$$\sum_{j \geq 0} \hat{\psi} \left(2^j \xi \right) \overline{\hat{\psi} \left(2^j \left(\xi + 2q\pi \right) \right)}$$

$$= \sum_{j > 0} e^{i 2^{j-1} \xi} \overline{m_0 \left(2^{j-1} \xi + \pi \right)} \hat{\phi} \left(2^{j-1} \xi \right) \times$$

$$\overline{e^{i 2^{j-1} \xi} \overline{m_0 \left(2^{j-1} \xi + \pi \right)} \hat{\phi} \left(2^{j-1} \xi + 2^j q\pi \right)}$$

$$+ e^{i 2^{-1} \xi} \overline{m_0 \left(2^{-1} \xi + \pi \right)} \hat{\phi} \left(2^{-1} \xi \right) \overline{e^{i 2^{-1} \xi + \pi} \overline{m_0 \left(2^{-1} \xi \right)} \hat{\phi} \left(2^{-1} \xi + q\pi \right)}$$

$$= \sum_{j > 0} \left| m_0 \left(2^{j-1} \xi + \pi \right) \right|^2 \hat{\phi} \left(2^{j-1} \xi \right) \overline{\hat{\phi} \left(2^{j-1} \xi + 2^j \pi \right)}$$

$$- \overline{m_0 \left(2^{-1} \xi + \pi \right)} \hat{\phi} \left(2^{-1} \xi \right) m_0 \left(2^{-1} \xi \right) \overline{\hat{\phi} \left(2^{-1} \xi + \pi \right)}$$

$$= \sum_{j > 0} \left[1 - \left| m_0 \left(2^{j-1} \xi \right) \right|^2 \right] \hat{\phi} \left(2^{j-1} \xi \right) \overline{\hat{\phi} \left(2^{j-1} \xi + 2^j \pi \right)} - \hat{\phi}(\xi) \overline{\hat{\phi}(\xi + 2\pi)}$$

$$= \sum_{j>0} \left[\hat{\phi}\left(2^{j-1}\xi\right) \overline{\hat{\phi}\left(2^{j-1}\xi + 2^j\pi\right)} - \hat{\phi}\left(2^j\xi\right) \overline{\hat{\phi}\left(2^j\xi + 2^{j+1}\pi\right)} \right]$$

$$- \hat{\phi}(\xi)\overline{\hat{\phi}(\xi + 2\pi)}$$

$$= - \lim_{n\to\infty} \hat{\phi}\left(2^n\xi\right) \overline{\hat{\phi}\left(2^n\xi + 2^{n+1}\pi\right)} = 0,$$

where $\hat{\phi}\left(2^{j-1}\xi + 2^j q\pi\right) = \hat{\phi}\left(2^{j-1}\xi + 2^j\pi\right)$ for all $q \in 2\mathbb{Z} + 1$ because $m_0(\xi)$ is $2\pi-$ periodic. Therefore, (27) also holds, and the proof of Theorem 11 is complete. $\square$

Remark. From Theorem 11, we immediately know that a refinable generator $\phi \in \Phi$ defined by Definition 1 is a stable pseudo-scaling generator if its corresponding $F(\xi)$ satisfies $F(\pi) \neq -1$ and equations (9) and (25).

We now give a general algorithm to construct the pseudo-scaling generator ϕ such that it possesses the largest possible regularity and the required vanishing moments. In fact, this method can be described as an optimization problem of finding suitable $F(\xi)$, or, equivalently, a suitable coefficient set $\mathbf{a} = \{a_0, \cdots, a_k\}$, of $F(\xi)$, such that $\sum_{j=0}^{k} a_j^2$ is the minimum under all conditions shown in Theorem 11. Thus, the above optimization problem can be written as follows:

$$\min_{\mathbf{a}} \quad \sum_{j=0}^{k} a_j^2, \tag{28}$$

$$\text{subject to} \quad \left(\sum_{j=0}^{k}(-1)^j a_j + 1\right)^2 > 0, \tag{29}$$

$$\sum_{j=0}^{k} a_j^2 < 2^{2N-1}/(k+1), \tag{30}$$

$$\sum_{j=k'}^{N+k+k'} \left(\sum_{\ell=0}^{k} \binom{N}{j + 2n - \ell - k'} a_\ell\right) \left(\sum_{\tilde{\ell}=0}^{k} \binom{N}{j - \tilde{\ell} - k'} a_{\tilde{\ell}}\right)$$

$$= 2^{2N-1}\delta_{n0},$$

$$n = 0, \pm1, \pm2, \cdots, \tag{31}$$

where the objective function (28) gives the largest possible regularity, condition (29) is from the definition of $F(\pi) \neq -1$, and conditions (30) and (31) come from conditions (9) and (25) of Theorem 11.

Problem (28)–(31) can be written in a form without the inequality conditions by defining parameters $s, t \neq 0$ as $s^2 = 2^{2N-1}/(k+1) - \sum_{j=0}^{k} a_j^2$

and $t^2 = \left(\sum_{j=0}^{k}(-1)^j a_j + 1\right)^2$, respectively. Hence, the optimization problem becomes

$$\min_{\mathbf{a}} \quad \sum_{j=0}^{k} a_j^2 + \frac{1}{s^2} + \frac{1}{t^2},$$

$$\text{subject to } t^2 = \left(\sum_{j=0}^{k}(-1)^j a_j + 1\right)^2$$

$$s^2 + \sum_{j=0}^{k} a_j^2 = 2^{2N-1}/(k+1)$$

$$\sum_{j=k'}^{N+k+k'} \left(\sum_{\ell=0}^{k} \binom{N}{j+2n-\ell-k'} a_\ell\right) \left(\sum_{\tilde{\ell}=0}^{k} \binom{N}{j-\tilde{\ell}-k'} a_{\tilde{\ell}}\right)$$

$$= 2^{2N-1} \delta_{n0},$$

$$n = 0, \pm 1, \pm 2, \cdots,$$

As examples, we choose $N = 1$ and $k = k' = 0$. Then the solution of problem (28)–(31) is $a_0 = 1$, and we obtain the Haar function. If we choose $N = 2$, $k = 1$, and $k' = 0$, then the solutions of the problem are $a_0 = \frac{1 \pm \sqrt{3}}{2}$ and $a_1 = \frac{1 \mp \sqrt{3}}{2}$. Hence, the corresponding ϕ is defined by $\hat{\phi}(\xi) = \prod_{j=1}^{\infty} \tilde{m}_0(2^{-j}\xi)$, where

$$m_0(\xi) = \frac{1 \pm \sqrt{3}}{2} \left(1 - e^{-i\xi}\right) \left(\frac{1 + e^{-i\xi}}{2}\right)^2.$$

In addition, the regularity of ϕ is more than $2 - \log_2(4)/2 = 1$ and $\psi \in C^1$.

Finally, we discuss the error of L_2 approximation from $S(\phi)$. Denote

$$E(f, S(\phi))_{L_2} := \inf_{g \in S(\phi)} \|f - g\|_{L_2}$$

where ϕ is a stable function. Since ϕ defined in Theorem 7 is compactly supported and has N vanishing moments (i.e., accuracy N), it has approximation order N. In addition, from [4], the approximation coefficient is

$$C_\phi^N = \frac{1}{m!} \sqrt{\sum_{u \neq 0} \left|\hat{\phi}^{(m)}(2\pi u)\right|^2}.$$

Therefore, for any function $f \in W_2^{N+1}(\mathbb{R})$, the Sobolev space,

$$E\left(f, S(\phi)^h\right)_2 = C_\phi^N \|f\|_{W_2^N(\mathbb{R})} + O\left(h^{N+1}\right),$$

where $S(\phi)^h := \{f(\cdot/h)|f \in S(\phi)\}$.

Acknowledgments. I would like to thank the referees and the editor for their comments that helped improve the presentation of this paper.

References

1. C. de Boor, R. DeVore, and A. Ron, Approximation from shift invariant subspaces of $L_2(\mathbb{R}^d)$, Trans. Amer. Math. Soc. **341** (1994), 787–806.

2. C. de Boor, R. DeVore, and A. Ron, The structure of finitely generated shift-invariant spaces in $L_2(\mathbb{R}^d)$, J. of Functional Analysis **119** (1994), 37–78.

3. C. de Boor, R. DeVore, and A. Ron, Approximation orders of FSI spaces in $L_2(\mathbb{R}^d)$, Constr. Approx. **14** (1998), 631–652.

4. T. Blu, and M. Unser, Approximation error for quasi-interpolators and (multi-) wavelet expansions, Appl. Comp. Harmonic Anal., **6** (1999), 219–251.

5. A. S. Cavaretta, W. Dahmen, and C. A. Micchelli, *Stationary Subdivision*, Mem. Amer. Math. Soc., 453, Amer. Math. Soc., Providence, RI, 1991.

6. C. K. Chui, *An Introduction to Wavelets*, Academic Press, Boston, 1992.

7. A. Cohen, *Ondelettes, analyses multirésolutions et traitement numerique du signal*, Ph.D. Thesis, Université de Paris IX (Dauphine), France, 1990.

8. A. Cohen and I. Daubechies, A stability criterion for biorthogonal wavelet bases and their related subband coding scheme, Duke Math. J. **68** (1992), 313–335.

9. A. Cohen, and I. Daubechies, and J. C. Feauveau, J.C., Biorthogonal bases of compactly supported wavelets, Comm. Pure Appl. Math. **45** (1992), 485–560.

10. I. Daubechies, *Ten Lectures on Wavelets*, CBMS-NSF Series in Appl. Math., SIAM Publ., Philadelphia, 1992.

11. S. Dekel and D. Leviatan, Wavelet decompositions of non-refinable shift invariant spaces, Appl. Comp. Harmonic Anal., to appear.

12. T. X. He, Short time Fourier transform, integral wavelet transform, and wavelet functions associated with splines, J. Math. Anal. & Appl., **224** (1998), 182–200.

13. T. X. He, Biorthogonal wavelets with certain regularities, Appl. Comp. Harmonic Anal. **11** (2001), 227–242.

14. E. Hernández and G. Weiss, *A First Course on Wavelets*, CRC Press, New York, 1996.

15. R. Q. Jia and J. Wang, Stability and linear independence associated with wavelet decompositions, Proc. Amer. Math. Soc. **117** (1993), 1115–1124.

16. Y. Meyer, *Wavelets and Operators*, Cambridge, 1992.

17. G. Strang and G. Fix, A Fourier analysis of the finite element variational method, in *Constructive Aspects of Functional Analysis*, G. Geymonat (ed.) C.I.M.E., II Ciclo 1971, 793–840.

Tian-Xiao He
Department of Mathematics and Computer Science
Illinois Wesleyan University
Bloomington, IL 61702-2900
the@sun.iwu.edu

A Short-support Dual Mask to the Piecewise Linears on a Uniform Triangulation

Bruce Kessler

Abstract. Much work has been done in finding duals and wavelets for a single scaling function on a triangulation (see [2,4,6] among others). The CBC algorithm allows the construction of duals with arbitrary approximation order at the expense of a larger support. This is an independent approach to the construction of a dual mask to the piecewise linear "hat" function on a uniform triangulation with a focus on maintaining the same hexagonal support. The associated wavelet masks are also constructed with only slightly larger supports.

§1. Introduction

Let $\triangle$ be the type-1 triangulation generated by $\epsilon_1 = (1,0)$ and $\epsilon_2 = (\frac{1}{2}, \frac{\sqrt{3}}{2})$. Define h as the piecewise linear "hat" function on $\triangle$, where $h(\mathbf{0}) = 1$ and $h = 0$ at all other vertices of $\triangle$. Define the translation function $t_{m,n} := \cdot - m\epsilon_1 - n\epsilon_2$ and the dilation function $d_{m,n} := N \cdot -m\epsilon_1 - n\epsilon_2$ for fixed $N \in \mathbb{Z}$, $N > 1$. Define the rotation function r^n as the $60n°$ clockwise rotation about $\mathbf{0}$ for $n = 0, \ldots, 5$. Let

$$S(\phi) = \mathrm{clos}_{L^2}\mathrm{span}\{\phi \circ t_{m,n} : m, n \in \mathbb{Z}\}.$$

Definition 1. *A distribution ϕ on $\mathbb{R}^2$ is* refinable *at dilation $N \in \mathbb{Z}$, $N > 1$, if it satisfies*

$$\phi = \sum_{m,n \in \mathbb{Z}} c(m,n)\phi \circ d_{m,n}, \tag{1}$$

for some finite sequence $c(m,n)$, called the mask *of ϕ.*

Approximation Theory X: Wavelets, Splines, and Applications   323
Charles K. Chui, Larry L. Schumaker, and Joachim Stöckler (eds.), pp. 323–332.
Copyright © 2002 by Vanderbilt University Press, Nashville, TN.
ISBN 0-8265-1416-2.

Definition 2. *If* $c(m,n)$ *satisfies*

$$\sum_{\beta \in \mathbb{Z}^2} c(\beta) = N^2, \tag{2}$$

then there exists a unique compactly supported distribution ϕ *that satisfies (1) (see [1]) such that* $\hat{\phi}(0) = 1$, *called the* normalized solution *of (1).*

Definition 3. *A pair of distributions* ϕ *and* $\tilde{\phi}$ *on* $\mathbb{R}^2$ *are said to be* biorthogonal *if*

$$\langle \phi, \tilde{\phi} \circ t_{m,n} \rangle = \delta_{0,m}\delta_{0,n}, \, m, n \in \mathbb{Z}. \tag{3}$$

The distribution $\tilde{\phi}$ *is said to be a* dual *to* ϕ.

Definition 4. *A finite sequence* $\tilde{c}(m,n)$ *satisfying the equation*

$$\sum_{\beta \in \mathbb{Z}^2} \overline{c(N\alpha + \beta)}\tilde{c}(\beta) = N^2\delta_{\alpha,\mathbf{0}}, \qquad \text{for all } \alpha \in \mathbb{Z}^2 \tag{4}$$

is called a dual mask *of* $c(m,n)$. *If a mask* $c(m,n)$ *has a dual mask, then* $c(m,n)$ *is called a* primal mask.

Two refinable distributions ϕ and $\tilde{\phi}$ satisfying (3) will necessarily have masks satisfying (4), but (4) is not sufficient to guarantee (3). Each distribution also needs to be stable. For a larger discussion on stability, see [3,5,8,9,11].

Definition 5. *For a positive integer* k, *the sequence* $c(m,n)$ *for* $m, n \in \mathbb{Z}$ *satisfies the* sum rules of order k *if*

$$\sum_{\alpha \in \mathbb{Z}^2} c(N\alpha + \beta)p \circ d_\beta(\alpha) = \sum_{\alpha \in \mathbb{Z}^2} c(N\alpha)p \circ d_\mathbf{0}(\alpha) \text{ for all } \beta \in \mathbb{Z}^2, \, p \in \Pi_{k-1},$$
$$\tag{5}$$

where Π_{k-1} *is the set of polynomials of degree* $k - 1$ *or less. If* $c(m,n)$ *also satisfies (1), then* $S(\phi)$ *has approximation order* k *(see [7]).*

§2. Dual Mask Construction

Let $\phi = h$ and set dilation $N = 3$. Then $\phi = \sum c(m,n)\phi \circ d_{m,n}$ as illustrated in Figure 1.

Theorem 1. *The mask* $c(m,n)$ *for* $\phi = h$ *has a family of duals* $\tilde{c}(m,n)$ *so that the normalized solution of* $\tilde{\phi} = \sum_{m,n \in \mathbb{Z}} \tilde{c}(m,n)\tilde{\phi} \circ d_{m,n}$ *has the same support as* ϕ *and* $\tilde{\phi} \circ r^1 = \tilde{\phi}$.

Proof: Let $\tilde{\phi} = \sum \tilde{c}(m,n)\tilde{\phi} \circ d_{m,n}$ as in Figure 1 so that supp(ϕ) = supp($\tilde{\phi}$) and $\tilde{\phi} \circ r^1 = \tilde{\phi}$. The biorthogonality condition in (4) with $\alpha = \mathbf{0}$ gives the equation

$$1 + 6\left(\frac{2}{3}\right)s + 6\left(\frac{1}{3}\right)t + 6\left(\frac{1}{3}\right)u = 9,$$

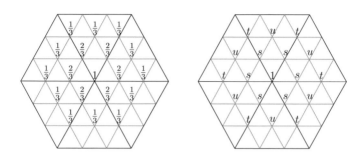

Fig. 1. The masks for ϕ and $\tilde{\phi}$.

and so,

$$2s + t + u = 4. \tag{6}$$

The condition in (4) with $\alpha = (1,0)$ gives the equation

$$\frac{1}{3}s + \frac{2}{3}t + 2\left(\frac{1}{3}\right)u = 0,$$

and so

$$s + 2t + 2u = 0. \tag{7}$$

Equations (6) and (7) have the solution $s = \frac{8}{3}$ and $u = -\left(t + \frac{4}{3}\right)$, which satisfies equation (2). □

Note that $N = 3$ is the smallest integer dilation factor that will allow (4) and the conditions in Theorem 1 to be met. The only choice of u and t that satisfy the sum rules of order 1, a basic prerequisite for stability, is $t = -\frac{5}{3}$ and $u = \frac{1}{3}$. However, the distribution $\tilde{\phi}$ with $t = -\frac{5}{3}$ is not L^2-stable, since the transition operator does not satisfy Condition E (see [9]). Still, we may construct perfect reconstruction filters from the masks of ϕ, $\tilde{\phi}$, and their associated wavelet masks.

§3. Wavelet Mask Construction

Due to the lack of stability of the $\tilde{\phi}$, we focus only on constructing the $3^2 - 1$ biorthogonal wavelet masks $g^k(m,n)$ and $\tilde{g}^k(m,n)$, $k = 1, \ldots, 8$, that satisfy

$$\sum_{\beta \in \mathbb{Z}^2} \overline{g^k(3\alpha + \beta)} \tilde{c}(\beta) = 0 \text{ and } \sum_{\beta \in \mathbb{Z}^2} \overline{\tilde{g}^k(3\alpha + \beta)} c(\beta) = 0 \tag{8}$$

for all $\alpha \in \mathbb{Z}^2$ and $k \in \{1, \ldots, 8\}$, and

$$\sum_{\beta \in \mathbb{Z}^2} \overline{g^k(3\alpha + \beta)} \tilde{g}^j(\beta) = 9\delta_{\alpha,\mathbf{0}}\delta_{k,j} \tag{9}$$

for $k, j \in \{1, \ldots, 8\}$.

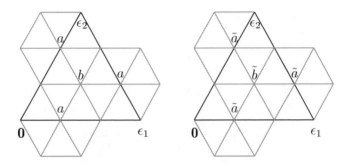

Fig. 2. The masks for ψ^1 and $\tilde{\psi}^1$.

Two wavelet masks with very short support

Suppose $\psi^1 = \sum g^1(m,n)\phi \circ d_{m,n}$ and $\tilde{\psi}^1 = \sum \tilde{g}^1(m,n)\tilde{\phi} \circ d_{m,n}$ as in Figure 2. The conditions (8) and (9) applied to the masks of ψ^1 and $\tilde{\psi}^1$ give a system of equations with the solution

$$
\begin{aligned}
&a \neq 0, &&\tilde{a} = -\tfrac{3}{4a}(3t+4),\\
&b = \tfrac{3t+8}{3t+4}a, &\text{and} \qquad &\tilde{b} = \tfrac{9}{4a}(3t+4),
\end{aligned}
$$

with $t \neq -\tfrac{4}{3}$. Choose a nonzero a and let b, $\tilde{a}$, and $\tilde{b}$ be defined as above. Let $\psi^2 = \phi^1 \circ r^1$ and $\tilde{\psi}^2 = \tilde{\phi}^1 \circ r^1$. Then the masks $g^2(m,n)$ and $\tilde{g}^2(m,n)$ automatically satisfy (8) and (9).

The remaining wavelets

In order to explicitly construct the masks g^k and $\tilde{g}^k$, $k = 3,\ldots,8$, we first build some sequences that satisfy some of the biorthogonality conditions that form a bases for the remaining wavelet mask spaces and then extract the biorthogonal masks.

Suppose $\mu = \sum f(m,n)\phi \circ d_{m,n}$ and $\tilde{\mu} = \sum \tilde{f}(m,n)\tilde{\phi} \circ d_{m,n}$ as in Figure 3. The conditions

$$
\sum_{\beta \in \mathbb{Z}^2} \overline{f(3\alpha+\beta)}\tilde{c}(\beta) = 0 \text{ and } \sum_{\beta \in \mathbb{Z}^2} \overline{\tilde{f}(3\alpha+\beta)}c(\beta) = 0, \ \alpha \in \mathbb{Z}^2 \qquad (10)
$$

and

$$
\sum_{\beta \in \mathbb{Z}^2} \overline{f(3\alpha+\beta)}\tilde{g}^k(\beta) = 0 \text{ and } \sum_{\beta \in \mathbb{Z}^2} \overline{\tilde{f}(3\alpha+\beta)}g^k(\beta) = 0, \ \alpha \in \mathbb{Z}^2 \qquad (11),
$$

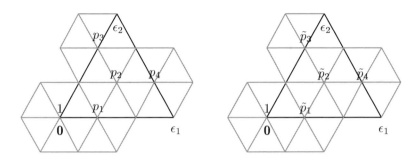

Fig. 3. The masks for μ and $\tilde{\mu}$.

applied to the masks of μ and $\tilde{\mu}$ for $k = 1, 2$ give a system of equations with the solution

$$
\begin{aligned}
p_1 &= -\frac{9t^2 - 36t + 128}{4(9t^2 - 24t + 64)}, & \tilde{p}_1 &= -\frac{1}{12}(3t + 20), \\
p_2 &= -\frac{1}{4}, & \tilde{p}_2 &= \frac{1}{4}(3t + 4), \\
p_3 &= -\frac{9t^2 + 32}{4(9t^2 - 24t + 64)}, & \tilde{p}_3 &= -\frac{1}{12}(3t + 8), \\
p_4 &= -\frac{9t^2 - 36t + 32}{4(9t^2 - 24t + 64)}, & \tilde{p}_4 &= \frac{1}{12}(4 - 3t).
\end{aligned}
$$

Set p_k and $\tilde{p}_k$, $k = 1, \ldots, 4$, as above. Notice that the masks for $\mu \circ r^n$ and $\tilde{\mu} \circ r^n$ also satisfy (10) and (11) for $n = 1, \ldots, 5$.

Now, suppose $\nu = \sum f(m, n) \phi \circ d_{m,n}$ and $\tilde{\nu} = \sum \tilde{f}(m, n) \tilde{\phi} \circ d_{m,n}$ as in Figure 4. The conditions in (10) and (11) applied to the masks of ν and $\tilde{\nu}$ give a system of equations with the solution

$$
\begin{aligned}
q_1 &= -\frac{3t + 4}{4(3t - 8)}, & \tilde{q}_1 &= \frac{1}{4}(3t + 4), \\
q_2 &= -\frac{3t - 20}{4(3t - 8)}, & \tilde{q}_2 &= -\frac{3}{4}(t + 4), \\
q_3 &= -\frac{3t}{4(3t - 8)}, & \tilde{q}_3 &= -\frac{1}{4}(3t + 4), \\
q_4 &= -\frac{3t - 2}{2(3t - 8)}, & \tilde{q}_4 &= 1, \\
q_5 &= \frac{2}{3t - 8}, & \tilde{q}_5 &= \frac{1}{2}(3t + 4), \\
q_6 &= \frac{3t + 4}{4(3t - 8)}, & \tilde{q}_6 &= -\frac{1}{4}(3t + 4),
\end{aligned}
$$

for $t \neq \frac{8}{3}$.

Finally, let $\gamma = v_1\mu + v_2\mu \circ r^1 + \nu$ and $\tilde{\gamma} = \tilde{v}_1\tilde{\mu} + \tilde{v}_2\tilde{\mu} \circ r^1 + \tilde{\nu}$ and let f^k and $\tilde{f}^k$ be the masks of $\gamma \circ r^k$ and $\tilde{\gamma} \circ r^k$, respectively, $k = 0, \ldots, 5$. The condition

$$
\sum_{\beta \in \mathbb{Z}^2} \overline{f^k(3\alpha + \beta)} \tilde{f}^j(\beta) = 0, \qquad \alpha \in \mathbb{Z}^2,\ \alpha \neq \mathbf{0},\ i \neq j,
$$

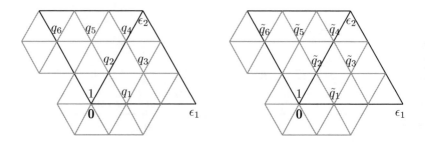

Fig. 4. The masks for ν and $\tilde{\nu}$.

applied to the masks of γ and $\tilde{\gamma}$ for $k = 1, \ldots, 6$ gives a system of equations with the solutions

$$v_1 = -\frac{9t^2 - 18t + 80 \pm (3t - 8)\sqrt{9t^2 - 24t + 52}}{12(3t - 8)},$$

$$v_2 = \frac{9t^2 - 18t + 80 \pm (3t - 8)\sqrt{9t^2 - 24t + 52}}{12(3t - 8)},$$

$$\tilde{v}_1 = \frac{1}{12}(3t + 14 \mp \sqrt{9t^2 - 24t + 52}),$$

$$\tilde{v}_2 = -\frac{1}{12}(3t + 14 \mp \sqrt{9t^2 - 24t + 52}),$$

for $t \neq \frac{8}{3}$. Choose one of the two solutions and set v_k and $\tilde{v}_k$, $k = 1, \ldots, 4$, as above. The resulting masks $f^k(m, n)$ and $\tilde{f}^k(m, n)$ do not completely satisfy equation (9), but sequences $g^k(m, n)$ and $\tilde{g}^k(m, n)$, $k = 3, \ldots, 8$ satisfying (9) can be extracted using a biorthogonal version of the Gram-Schmidt process or from the following lemma.

Lemma 2. *The sequences* $g^k(m, n)$ *and* $\tilde{g}^k(m, n)$, *where*

$$
\begin{array}{ll}
h^3 = \sum_{j=1}^{6} f^j, & \tilde{h}^3 = \sum_{j=1}^{6} \tilde{f}^j, \\
h^4 = \sum_{j=1}^{6} (-1)^{j-1} f^j, & \tilde{h}^4 = \sum_{j=1}^{6} (-1)^{j-1} \tilde{f}^j, \\
h^5 = 2f^1 - f^3 - f^5, & \tilde{h}^5 = 2\tilde{f}^1 - \tilde{f}^3 - \tilde{f}^5, \\
h^6 = 2f^2 - f^4 - f^6, & \tilde{h}^6 = 2\tilde{f}^2 - \tilde{f}^4 - \tilde{f}^6, \\
h^7 = f^3 - f^5, & \tilde{h}^7 = \tilde{f}^3 - \tilde{f}^5, \\
h^8 = f^4 - f^6, & \tilde{h}^8 = \tilde{f}^4 - \tilde{f}^6,
\end{array}
$$

and

$$g^k = \frac{3h^k}{\langle h^k, \tilde{h}^k \rangle}, \qquad \tilde{g}^k = \frac{3\tilde{h}^k}{\langle h^k, \tilde{h}^k \rangle},$$

for $k = 3, \ldots, 8$, *satisfy the biorthogonality conditions in* (8) *and* (9) *for* $t \neq -\frac{4}{3}, \frac{8}{3}$.

Fig. 5. The distribution $\tilde{\phi}$ for $t = -\frac{5}{3}$.

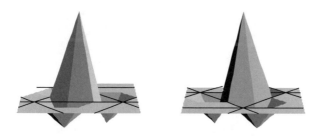

Fig. 6. Wavelets ψ^1 and ψ^2.

Fig. 7. Wavelets $\tilde{\psi}^1$ and $\tilde{\psi}^2$.

Proof: As linear combinations of the previously constructed f^k, respectively $\tilde{f}^k$, that satisfied equation (8), the g^k, respectively $\tilde{g}^k$, will also. The second condition in (9) can be quickly verified for the g^k and $\tilde{g}^k$ using symbolic manipulation software, like *Mathematica*. $\square$

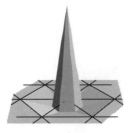

Fig. 8. Wavelets ψ^3 and ψ^4.

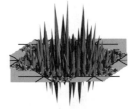

Fig. 9. Wavelets $\tilde{\psi}^3$ and $\tilde{\psi}^4$.

Fig. 10. Wavelets ψ^5 and ψ^6.

Fig. 11. Wavelets $\tilde{\psi}^5$ and $\tilde{\psi}^6$.

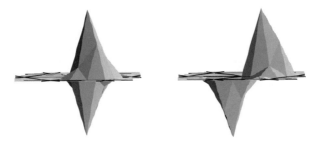

Fig. 12. Wavelets ψ^7 and ψ^8.

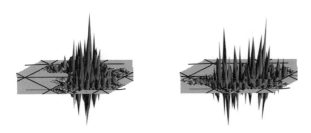

Fig. 13. Wavelets $\tilde{\psi}^7$ and $\tilde{\psi}^8$.

Theorem 3. *The sequences $g^k(m,n)$ and $\tilde{g}^k(m,n)$, $k = 1,\ldots,8$ form dual wavelet mask bases to $\phi = h$ and $\tilde{\phi}$ from Theorem 1 for $t \neq -\frac{4}{3}, \frac{8}{3}$.*

Proof: By construction, the sequences satisfy the biorthogonality conditions (8) and (9). At issue is whether the dilated functions $\phi \circ d_{m,n}$, respectively $\tilde{\phi} \circ d_{m,n}$, $m, n \in \{0, 1, 2\}$, can be built as linear combinations of translates of ϕ, respectively $\tilde{\phi}$, and translates of the ψ^k, respectively $\tilde{\psi}^k$. This is quickly verified using symbolic manipulation software. $\square$

§4. Illustrations

Figures 5–13 show the normalized solution $\tilde{\phi}$ for the mask $\tilde{c}(m,n)$ that satisfies the sum rules of order 1 (Figure 5) and the distributions ψ^k and $\tilde{\psi}^k$ associated with the wavelet masks $g^k(m,n)$ and $\tilde{g}^k(m,n)$ as in Lemma 2. The distributions shown in Figures 7, 9, 11, and 13 are approximations.

Acknowledgments. I would like to thank Qingtang Jiang and Doug Hardin for their helpful comments on stability and smoothness of distributions based upon the mask.

References

1. Cavaretta, A. S., W. Dahmen, and C. A. Micchelli, Stationary subdivision, Memoirs of Amer. Math. Soc. **93** (1991), 1–186.

2. Chen, D. R., B. Han, and S. D. Riemenschneider, Construction of multivariate biorthogonal wavelets with arbitrary vanishing moments, Advances in Comp. Math. **13** (2000), 131–165.

3. Han, B., and R. Q. Jia, Multivariate refinement equations and convergence of subdivision schemes, SIAM J. Math. Anal. **29** (1998), 1177–1199.

4. Han, B., and S. D. Riemenschneider, Interpolatory biorthogonal wavelets and CBC algorithm, Proceedings of the Conference on Wavelet Analysis and Applications (2000), 117–136.

5. Hogan, T., Stability and independence of the shifts of a multivariate refinable function, *Approximation Theory VIII, Vol. 2: Wavelets*, Charles K. Chui and Larry L. Schumaker (eds.), World Scientific Publishing Co., Inc., Singapore, 1995, 159–166.

6. Ji, H., S. D. Riemenschneider, and Z. Shen, Multivariate compactly supported fundamental refinable functions, duals and biorthogonal wavelets, Studies in Appl. Math. **102** (1999), 173–204.

7. Jia, R. Q., Approximation properties of multivariate wavelets, Math. Comp. **67** (1998), 647–665.

8. Jia, R. Q., Characterization of smoothness of multivariate refinable functions in Sobolev spaces, Trans. Amer. Math. Soc. **351** (1999), 4089–4112.

9. Jiang, Q., and P. Oswald, On the analysis of $\sqrt{3}$-subdivision schemes, preprint (2001).

10. Riemenschneider, S. D., and Z. Shen, Multidimensional interpolatory subdivision schemes, SIAM J. Numer. Anal. **34** (1997), 2357–2381.

11. Shen, Z., Refinable function vectors, SIAM J. Math. Anal. **29** (1998), 235–250.

Bruce Kessler
1 Big Red Way
Department of Mathematics
Western Kentucky University
Bowling Green, KY 42101
`bruce.kessler@wku.edu`

Wavelet Analysis of Cardinal L-splines and Construction of Multivariate Prewavelets

Ognyan Kounchev and Hermann Render

Abstract. In the fundamental paper [1] non-stationary multiresolution analysis was developed and applied to the case of cardinal L−splines. In the first part of this paper we review wavelet analysis of cardinal L−splines in more detail by exploiting special properties of the Euler-Frobenius polynomial for cardinal L−splines. For example, the Riesz bounds of the mother and father wavelets of cardinal L−splines are determined in a constructive way. The main result states that the system of all translations of the wavelets ψ_j over all levels $j \in \mathbb{Z}$ induces a stable basis of $L^2(\mathbb{R})$. In the second part we use these results in order to discuss multivariate prewavelets based on polysplines.

§1. Introduction

Wavelet analysis of polysplines has been recently introduced by the first author in [7]. This approach is based on the definition of a *cardinal polyspline* (for definition see Section 5) — a concept which was introduced by the first author in [6] and which was successfully exploited in the papers [8,9,10]. In this paper it is not assumed that the reader is familiar with the results in the monograph [7]: we will present here parts of the theory in a rather compact form (taking advantage of rather general results in [1]) and we shall bring some of the results in a new perspective. However, our main result in Section 4 is an interesting improvement of a stability result in [7].

The strength of the above-mentioned approach depends on the *computational character of the concept of polysplines. Indeed, the study of cardinal polysplines can be reduced to the study of a sequence of one-dimensional objects.* This fact depends on the orthogonal decomposition of

Approximation Theory X: Wavelets, Splines, and Applications   333
Charles K. Chui, Larry L. Schumaker, and Joachim Stöckler (eds.), pp. 333–353.

a function $f \in L^2(\mathbb{R}^n)$ via spherical harmonics where $L^2(\mathbb{R}^n)$ denotes the set of all square-integrable functions $f \colon \mathbb{R}^n \to \mathbb{C}$ with respect to Lebesgue measure. Let $Y_{k,l}(\theta)$, $k = 0, 1, 2, ..., l = 1, ..., a_k$, be a standard basis of the set of all spherical harmonics where k denotes the degree of the spherical harmonic (for definition see Section 5) and S^{n-1} denotes the unit sphere $\{x \in \mathbb{R}^n : |x| = 1\}$. For the reader who is not familiar with spherical harmonics, it might be useful just to consider the two-dimensional case: identify S^1 with $[0, 2\pi]$ and choose $Y_0 = \frac{1}{\sqrt{2\pi}}$, and for $k \in \mathbb{N}$

$$Y_{k,1}(t) = \frac{1}{\sqrt{\pi}} \cos kt \quad \text{and} \quad Y_{k,2}(t) = \frac{1}{\sqrt{\pi}} \sin kt \tag{1}$$

which gives a standard basis of spherical harmonics. Roughly speaking, polysplines S of order p can be characterized as functions of the form (where $x = r\theta, \theta \in S^{n-1}$ and $r > 0$)

$$S(r\theta) = \sum_{k=0}^{\infty} \sum_{l=1}^{a_k} S_{k,l}(\log|x|) Y_{k,l}(\theta), \tag{2}$$

with the property that for each $k \in \mathbb{N}_0$ the function $v \mapsto S_{k,l}(v)$ is a cardinal $L-$spline with respect to a certain linear differential operator with constant coefficients of order $2p$ which depends on the degree k. In order to describe wavelet analysis of polysplines, it is necessary to have a good picture of the theory of cardinal $L-$splines. For this reason we shall continue a detailed discussion of wavelet analysis of polysplines in Section 5. The reader who is already familiar with wavelet analysis of cardinal $L-$splines therefore may skip the next sections.

We begin with a discussion of wavelet analysis of cardinal $L-$splines. Important results can already be found in [1], and a detailed account is given in [7]. Further in [11] a discussion of $L-$spline wavelets with arbitrary knots is given. Let us recall the definition of a cardinal $L-$spline: let L be a linear differential operator of order $N + 1$ with constant coefficients. A function $u \colon \mathbb{R} \to \mathbb{R}$ is called a **cardinal $L-$spline** on the mesh $2^{-j}\mathbb{Z}$ if u is $(N - 1)-$times continuously differentiable and if for each $k, j \in \mathbb{Z}$ there exists $f_{k,j} \in U_L := \{f \in C^{\infty}(\mathbb{R}) : Lf(x) = 0 \text{ in } \mathbb{R}\}$ such that $u(t) = f_{k,j}(t)$ for all $t \in (2^{-j}k, 2^{-j}(k+1))$. Throughout the paper we assume that the linear differential operator L is given by

$$L = M_\Lambda := \prod_{j=1}^{N+1} \left(\frac{d}{dx} - \lambda_j \right), \tag{3}$$

where the parameter vector $\Lambda = (\lambda_1, ..., \lambda_{N+1})$ consists of *real numbers* and at least some λ_j is non-negative. The space of cardinal $L-$splines on the mesh $2^{-j}\mathbb{Z}$ is denoted by $\mathcal{S}_{2^{-j}\mathbb{Z}}(\Lambda)$.

To a large extent wavelet analysis of cardinal L–splines can be developed analogously to the polynomial case which corresponds to $\Lambda = (0, ..., 0)$, or equivalently, $L = \left(\frac{d}{dx}\right)^{N+1}$. Scaling spaces are defined by

$$V_j(\Lambda) = L^2(\mathbb{R})\text{-closure } \left(L^2(\mathbb{R}) \cap \mathcal{S}_{2^{-j}\mathbb{Z}}(\Lambda)\right). \tag{4}$$

It is clear that the spaces $V_j(\Lambda)$ are $2^{-j}\mathbb{Z}$-invariant, and obviously

(i) $V_j(\Lambda) \subset V_{j+1}(\Lambda)$ for all $j \in \mathbb{Z}$.

Theorem 4.3 in [1] in combination with (5) and (6) shows that

(ii) $\bigcup_{j\in\mathbb{Z}} V_j(\Lambda)$ is dense in $L^2(\mathbb{R})$.

Finally it can be proved that

(iii) $\bigcap_{j\in\mathbb{Z}} V_j = \{0\}$ provided that there exists $\lambda_j \geq 0$ in Λ, cf. [1] and [8].

The **wavelet spaces** are defined by $W_j(\Lambda) := V_{j+1}(\Lambda) \ominus V_j(\Lambda)$. Properties (ii) and (iii) show that $L^2(\mathbb{R})$ is equal to the orthogonal sum of the wavelet spaces $W_j(\Lambda)$, $j \in \mathbb{Z}$. Hence, according to [1], the sequence $(V_j(\Lambda))_{j\in\mathbb{Z}}$ is a **multiresolution** provided some λ_j is non-negative.

An important tool in L–spline analysis is the existence of a basic spline which will denoted by Q_Λ: it has the property that $Q_\Lambda(x) > 0$ for all $x \in (0, N+1)$ and $Q_\Lambda(x) = 0$ for all $x \in \mathbb{R} \setminus (0, N+1)$, and it is uniquely determined up to a positive constant through this property, see [13]. The Fourier transform of Q_Λ is given by

$$\widehat{Q_\Lambda}(\xi) = \frac{\prod_{j=1}^{N+1}\left(e^{-\lambda_j} - e^{-i\xi}\right)}{\prod_{j=1}^{N+1}(i\xi - \lambda_j)}. \tag{5}$$

Note that in [1] the basic spline is denoted by N_Λ and it differs by a constant, namely $Q_\Lambda = e^{-c_\Lambda} N_\Lambda$, where $c_\Lambda = \lambda_1 + \cdots + \lambda_{N+1}$.

It is well known that the space $V_0(\Lambda)$ is the $L^2(\mathbb{R})$-closure of the linear hull of the translates $Q_\Lambda(\cdot - l)$, $l \in \mathbb{Z}$. More generally, we have that

$$V_j(\Lambda) = L^2(\mathbb{R})\text{-closure of linear span}\left(Q_{2^{-j}\Lambda}\left(2^j x - k\right), \quad k \in \mathbb{Z}\right). \tag{6}$$

It is important to note that V_j is not the 2^j-dilate of V_0. Indeed, the function $Q_\Lambda(2x)$ is *not* in V_1 since in general it is not a L–spline on $\frac{1}{2}\mathbb{Z}$ with respect to Λ. Thus, we are in the case of non-stationary multiresolution analysis where the scaling function φ_j is not just the 2^j-dilate of φ_0. As already noted in [1], the discussion of stability (for definition see Section 4) in the non-stationary case might be much harder.

The paper is organized as follows. In Section 2 we use the fundamental results in [1, p. 152] to describe wavelets and Riesz bounds of cardinal

L−splines. It was proved in [1] that the wavelet space $W_j(\Lambda)$ is generated by $2^{-j}\mathbb{Z}$-shifts of a single function $\psi_{j,0}$. Further the system $\psi_{j,l}, l \in \mathbb{Z}$, (cf. formula (17)) is stable: Roughly speaking, the Riesz bounds for the mother and father wavelet can be expressed by the bounds of the function

$$S_{Q_\Lambda}(\xi) := \sum_{j=-\infty}^{\infty} \left| \widehat{Q_\Lambda}(\xi + 2\pi j) \right|^2. \tag{7}$$

In Section 3 the Riesz bounds are described by the *Euler-Frobenius polynomial* $\Pi_{\widetilde{\Lambda}}(0, e^{i\xi})$ (see Section 3) of the symmetrized vector

$$\widetilde{\Lambda} := (\lambda_1, ..., \lambda_{N+1}, -\lambda_1, ..., -\lambda_{N+1}) =: (\Lambda, -\Lambda). \tag{8}$$

As a consequence the following *constructive bounds* for S_{Q_Λ} are exact: using (25) and (21) (the latter applied to the symmetrized vector $\widetilde{\Lambda}$) we obtain for all $\xi \in \mathbb{R}$

$$e^{c_\Lambda} \left| \sum_{k=0}^{2N+1} (-1)^{N-k} Q_{\widetilde{\Lambda}}(k) \right| \le |S_{Q_\Lambda}(\xi)| \le e^{c_\Lambda} \sum_{k=0}^{2N+1} Q_{\widetilde{\Lambda}}(k), \tag{9}$$

where $c_\lambda := \lambda_1 + \cdots + \lambda_{N+1}$.

It follows from (25) and a well-known criterion for stability (Theorem 3.24 in [3, p. 76]) that the family $Q_\Lambda(\cdot - l), l \in \mathbb{Z}$, is stable, and the Riesz bounds are described in (9). Similarly the stability of the wavelet system $\psi_j(x - l), l \in \mathbb{Z}$, for fixed level $j \in \mathbb{Z}$ can be proved as done in [1]. However, as already pointed out in [1], this does not imply that the full system $\psi_{j,l} := N_j \psi_j \left(2^j \cdot -l \right), j, l \in \mathbb{Z}$, provides a stable basis for $L^2(\mathbb{R})$ for suitable constants N_j. Our main result, given in Section 4, says that N_j can be chosen in such a way that $N_j \psi_{j,l}, j, l \in \mathbb{Z}$, is indeed a stable basis. The proof requires a deeper analysis of the Euler-Frobenius polynomial for cardinal L−splines. In Section 5 a description of wavelet analysis for polysplines is given.

§2. Basic Constructions

Let φ be a function in $L^2(\mathbb{R})$ and let $\widehat{\varphi}(\xi)$ be the Fourier transform of φ. Then the function

$$S_\varphi(\xi) := \sum_{j=-\infty}^{\infty} |\widehat{\varphi}(\xi + 2\pi j)|^2 \tag{10}$$

is of fundamental importance in wavelet analysis. In the notation of [1], we have $S_\varphi = [\widehat{\varphi}, \widehat{\varphi}]$, where the bracket product of two functions $f, g \in L^2(\mathbb{R})$ is defined by $[f, g] := \sum_{j=-\infty}^{\infty} f(\cdot + 2\pi j) \overline{g(\cdot + 2\pi j)}$.

Let us recall the construction of wavelets in [1, Theorem 5.5]: Let φ and η be functions in $L^2(\mathbb{R})$ such that the support of $\widehat{\varphi}$ and $\widehat{\eta}$ is equal to $\mathbb{R}$. Assume that V_0 is generated by integer shifts of φ and V_1 is generated by $\frac{1}{2}\mathbb{Z}$-shifts of the function η. Let A be a 4π-periodic function such that $\widehat{\varphi} = A\widehat{\eta}$. Define a function $\psi \in L^2(\mathbb{R})$ by its Fourier transform

$$\widehat{\psi}(\xi) = 2e^{-\frac{1}{2}i\xi}\overline{A(\xi + 2\pi)}\left(\widetilde{\widehat{\eta}}(\xi + 2\pi)\right)^2 \widehat{\eta}(\xi), \qquad (11)$$

where the function $\widetilde{\widehat{\eta}}$ is defined by $\widetilde{\widehat{\eta}}(x) = \left(\sum_{k=-\infty}^{\infty}|\widehat{\eta}(x + 4\pi k)|^2\right)^{\frac{1}{2}}$. Then the wavelet space $W_0 = V_1 \ominus V_0$ is the L^2-closure of the linear span of the translates $\psi(\cdot - l), l \in \mathbb{Z}$. The Riesz bounds of this family can be computed by the bounds of the function S_ψ for which the following identity holds:

$$S_\psi(\xi) = 4\left(\widetilde{\widehat{\eta}}(\xi + 2\pi)\right)^2 \left(\widetilde{\widehat{\eta}}(\xi)\right)^2 S_\varphi(\xi). \qquad (12)$$

Let us apply this construction to cardinal $L-$splines. Let Q_Λ be the basic spline. Let us define $\varphi(x) = Q_\Lambda(x)$ and $\eta(x) = Q_{\frac{1}{2}\Lambda}(2x)$. As in [1, p. 151] define a 4π-trigonometric polynomial by

$$A_\Lambda(\xi) = 2^{-N}\prod_{j=1}^{N+1}\left(e^{\lambda_j - i\frac{\xi}{2}} + 1\right). \qquad (13)$$

A short computation (using (5) and (13) and $\widehat{\eta}(\xi) = \frac{1}{2}\widehat{Q_{\frac{1}{2}\Lambda}}\left(\frac{\xi}{2}\right)$) shows that

$$\widehat{\varphi}(\xi) = \widehat{Q_\Lambda}(\xi) = A_{\frac{1}{2}\Lambda}(\xi)e^{-\frac{1}{2}(\lambda_1 + \ldots \lambda_{N+1})}\widehat{\eta}(\xi). \qquad (14)$$

Therefore we define $A(\xi) := A_{\frac{1}{2}\Lambda}(\xi)e^{-\frac{1}{2}(\lambda_1 + \ldots \lambda_{N+1})}$. Moreover, we have

$$\left(\widetilde{\widehat{\eta}}(\xi + 2\pi)\right)^2 = \sum_{k=-\infty}^{\infty}\frac{1}{4}\left|\widehat{Q_{\frac{1}{2}\Lambda}}\left(\frac{\xi}{2} + \pi + 2\pi k\right)\right|^2 = \frac{1}{4}S_{Q_{\frac{1}{2}\Lambda}}\left(\frac{\xi}{2} + \pi\right).$$

Hence we *define for given* Λ *a wavelet* $\psi_\Lambda \in W_0(\Lambda)$ through the formula

$$\widehat{\psi_\Lambda}(\xi) = \frac{1}{4}e^{-\frac{1}{2}(\lambda_1 + \ldots + \lambda_{N+1})}e^{-\frac{1}{2}i\xi}\overline{A_{\frac{1}{2}\Lambda}(\xi + 2\pi)}S_{Q_{\frac{1}{2}\Lambda}}\left(\frac{\xi}{2} + \pi\right)\widehat{Q_{\frac{1}{2}\Lambda}}\left(\frac{\xi}{2}\right). \qquad (15)$$

Moreover, (12) shows that

$$S_{\psi_\Lambda}(\xi) = \frac{1}{4}S_{Q_{\frac{1}{2}\Lambda}}\left(\frac{\xi}{2} + \pi\right)S_{Q_{\frac{1}{2}\Lambda}}\left(\frac{\xi}{2}\right)S_{Q_\Lambda}(\xi). \qquad (16)$$

Let us now consider the case of arbitrary scaling.

Proposition 1. Let Λ be fixed, $j \in \mathbb{Z}$, and let $\psi_{2^{-j}\Lambda}$ be defined by (15). Then $W_j(\Lambda)$ is the closure of the linear hull of

$$\psi_{j,l}(x) := \psi_{2^{-j}\Lambda}(2^j x - l) \quad \text{for } l \in \mathbb{Z}. \tag{17}$$

Proof: Clearly $\psi_{2^{-j}\Lambda}$ is in $W_0(2^{-j}\Lambda) = V_1(2^{-j}\Lambda) \ominus V_0(2^{-j}\Lambda)$: thus it is a cardinal $L-$spline with respect to the differential operator associated to $2^{-j}\Lambda$ defined on mesh $\mathbb{Z}$, and the orthogonality relations

$$\int \psi_{2^{-j}\Lambda}(y)\, \overline{r(y)} dy = 0 \tag{18}$$

hold for all $r \in V_0(2^{-j}\Lambda)$. Define now $\psi_j(x) := \psi_{2^{-j}\Lambda}(2^j x)$. Then ψ_j is a cardinal $L-$spline with respect to Λ defined on the mesh $2^{-j}\mathbb{Z}$, hence $\psi_j \in V_{j+1}(\Lambda)$. By making the substitution $y = 2^j x$ in (18) we see that $\psi_j \in W_j(\Lambda) = V_{j+1}(\Lambda) \ominus V_j(\Lambda)$. $\square$

§3. Riesz Bounds and the Euler-Frobenius Polynomial

In Section 2 we have seen that the Riesz bounds of the translates of the mother or father wavelet can be characterized by the bounds of the function S_{Q_Λ}. We now want to relate S_{Q_Λ} to the Euler-Frobenius polynomial for cardinal $L-$splines. For this reason let us recall some basic facts and definitions: Define $e^\Lambda = \{e^{\lambda_j} : j = 1, ..., N+1\}$. The function $A_\Lambda : \mathbb{R} \times (\mathbb{C} \setminus e^\Lambda) \to \mathbb{C}$ (cf. [13], p. 223) is defined by

$$A_\Lambda(x, \lambda) = \frac{1}{2\pi i} \int_\Gamma \frac{1}{q_\Lambda(z)} \frac{e^{xz}}{e^z - \lambda} dz, \quad \text{where } q_\Lambda(z) = \prod_{j=1}^{N+1}(z - \lambda_j). \tag{19}$$

Here Γ is a closed simple curve in the complex plane surrounding all λ_j, $j = 1, ..., N+1$, and having the zeros of the function $e^z - \lambda$ in the exterior of Γ. The Euler-Frobenius function is defined by

$$\Pi_\Lambda(x, \lambda) := A_\Lambda(x, \lambda) \cdot r_\Lambda(\lambda), \quad \text{where } r_\Lambda(\lambda) = \prod_{j=1}^{N+1}(e^{\lambda_j} - \lambda). \tag{20}$$

The following fundamental formula relates the Euler-Frobenius function with the basic spline (cf. [13, p. 221 and p. 222]):

$$\sum_{j=0}^{N} \lambda^{N-j} Q_\Lambda(x + j) = \frac{(-1)^N}{e^{(\lambda_1 + ... + \lambda_{N+1})}} \cdot \Pi_\Lambda(x, \lambda). \tag{21}$$

It follows that the function $\Pi_\Lambda(x, \lambda)$ is a polynomial of degree $\leq N$ in the variable λ. For $x = 0$ it is called the **Euler-Frobenius polynomial** which is of degree $\leq N - 1$ since $Q_\Lambda(0) = 0$. Next we consider

$$\Phi_\Lambda(x, \lambda) := \sum_{j \in \mathbb{Z}} \lambda^j Q_\Lambda(x - j). \qquad (22)$$

For fixed λ the function $x \rightarrow \Phi_\Lambda(x, \lambda)$ is a cardinal $L-$spline and it satisfies the "exponential equation"

$$\Phi_\Lambda(x + 1, \lambda) = \lambda \Phi_\Lambda(x, \lambda), \qquad (23)$$

which reminds of the exponential equation $\lambda^{x+1} = \lambda \lambda^x$. Since Q_Λ has support in $[0, N + 1]$, we have for $0 \leq x < 1$

$$\sum_{j=0}^{N} \lambda^{N-j} Q_\Lambda(x + j) = \lambda^N \sum_{j=0}^{N} \lambda^{-j} Q_\Lambda(x + j) = \lambda^N \sum_{j=-\infty}^{\infty} \lambda^{-j} Q_\Lambda(x + j).$$

This formula and (21) yields the following identity valid for all $0 \leq x < 1$ and for all λ:

$$\Phi_\Lambda(x, \lambda) = \frac{(-1)^N}{e^{(\lambda_1 + \cdots + \lambda_{N+1})} \lambda^N} \cdot \Pi_\Lambda(x, \lambda). \qquad (24)$$

Theorem 1. *Suppose that $\Lambda \in \mathbb{R}^{N+1}$. Then the following identity holds for all $\xi \in \mathbb{R}$:*

$$S_{Q_\Lambda}(\xi) = (-1) e^{-(\lambda_1 + \cdots + \lambda_{N+1})} e^{-Ni\xi} \cdot \Pi_{\widetilde{\Lambda}}(0, e^{i\xi}) \neq 0, \qquad (25)$$

where $\Pi_{\widetilde{\Lambda}}(0, \lambda)$ is the Euler–Frobenius polynomial with respect to the symmetrized vector $\widetilde{\Lambda} = (\Lambda, -\Lambda)$. Furthermore we have for all $\xi \in \mathbb{R}$

$$|S_{Q_\Lambda}(-\pi)| \leq |S_\Lambda(\xi)| \leq |S_{Q_\Lambda}(0)|. \qquad (26)$$

Proof: Note that the Fourier transform of Q_Λ can be estimated by $\left|\widehat{Q_\Lambda}(\xi)\right| \leq C \frac{1}{|\xi|^{N+1}}$. By Theorem 2.28 in [3], we conclude that (note that Q_Λ is real-valued)

$$S_{Q_\Lambda}(\xi) = \sum_{j=-\infty}^{\infty} \left\{ \int_{-\infty}^{\infty} Q_\Lambda(y + j) Q_\Lambda(y) \, dy \right\} e^{-ij\xi}. \qquad (27)$$

It is easy to see that $S_{Q_\Lambda}(-\xi) = S_{Q_\Lambda}(\xi)$, since

$$\int_{-\infty}^{\infty} Q_\Lambda(y - j) Q_\Lambda(y) \, dy = \int_{-\infty}^{\infty} Q_\Lambda(x) Q_\Lambda(x + j) \, dy.$$

An application of the Hermite-Genocchi formula (cf. Proposition 3 in the appendix) shows that the coefficients of the above Fourier series can be computed by

$$\int_{-\infty}^{\infty} Q_\Lambda \left(y + j \right) Q_\Lambda \left(y \right) dy = e^{-(\lambda_1 + \cdots + \lambda_{N+1})} Q_{(\Lambda, -\Lambda)} \left(N + 1 + j \right).$$

This fact and (27) and definition (22) imply that

$$S_{Q_\Lambda} \left(\xi \right) = e^{-(\lambda_1 + \cdots + \lambda_{N+1})} \Phi_{\widetilde{\Lambda}} \left(N + 1, e^{-i\xi} \right). \tag{28}$$

By (23) we have $\Phi_{\widetilde{\Lambda}} \left(N + 1, \lambda \right) = \lambda^{N+1} \Phi_{\widetilde{\Lambda}} \left(0, \lambda \right)$. Hence (24) applied to the symmetrized vector $\widetilde{\Lambda}$ (which has length $2N + 2$) shows that

$$\Phi_{\widetilde{\Lambda}} \left(N + 1, \lambda \right) = \lambda^{N+1} \frac{(-1)^{2N+1}}{\lambda^{2N+1}} \cdot \Pi_{\widetilde{\Lambda}} \left(0, \lambda \right). \tag{29}$$

Hence the identity (21) is proven if we take $\lambda = e^{-ix}$ recalling that $S_{Q_\Lambda} \left(-\xi \right) = S_{Q_\Lambda} \left(\xi \right)$.

It is well known that the polynomial $\lambda \longmapsto \Pi_{\widetilde{\Lambda}} \left(0, \lambda \right)$ has $2N$ zeros (since $\widetilde{\Lambda}$ has length $2N + 2$) which are simple, real and negative, cf. [13]. If v is a zero then, by the symmetry of $\widetilde{\Lambda}$, v^{-1} is a zero as well. From this it follows that $\lambda = -1$ is not a zero of $\Pi_{\widetilde{\Lambda}} \left(0, \lambda \right)$ since otherwise $\lambda = -1$ is not a simple zero. It follows that $\Pi_{\widetilde{\Lambda}} \left(0, e^{i\xi} \right) \neq 0$ for all $\xi \in \mathbb{R}$.

By the above we can write

$$\Pi_{\widetilde{\Lambda}} \left(0, \lambda \right) = D \prod_{j=1}^{N} \left(\lambda - v_j \right) \left(\lambda - v_j^{-1} \right) \tag{30}$$

for a suitable constant D. A simple computation yields

$$\left| \Pi_{\widetilde{\Lambda}} \left(0, e^{i\xi} \right) \right| = D \prod_{j=1}^{N} \frac{\left| e^{i\xi} - v_j \right|^2}{|v_j|} = D \prod_{j=1}^{N} \frac{1 - 2v_j \cos \xi + v_j^2}{|v_j|}.$$

Since $v_j < 0$ we obtain that $\left| \Pi_{\widetilde{\Lambda}} \left(0, e^{i\xi} \right) \right| \leq \left| \Pi_{\widetilde{\Lambda}} \left(0, 1 \right) \right|$ and similarly we have $\left| \Pi_{\widetilde{\Lambda}} \left(0, e^{i\xi} \right) \right| \geq \left| \Pi_{\widetilde{\Lambda}} \left(0, -1 \right) \right|$ for all $\xi \in \mathbb{R}$. $\square$

§4. Proof of the Main Result

Let I be a countable (index) set. Then $l^2(I)$ is the Hilbert space of all sequences $c = \{c_j\}_{j \in I}$ for which the l^2-norm $\|c\|_{\ell_2(I)} = \left(\sum_{j \in I} |c_j|^2\right)^{1/2}$ is finite. Recall that a family of functions $f_j \in L^2(\mathbb{R})$, $j \in I$, is **stable** or satisfies the **Riesz condition** if there exist two constants $0 < A \le B < \infty$ such that

$$A \|c\|_{\ell_2(I)}^2 \le \left\| \sum_{j=-\infty}^{\infty} c_j f_j(x) \right\|_{L_2(\mathbb{R})}^2 \le B \|c\|_{\ell_2(I)}^2. \tag{31}$$

The optimal bounds A, B in (31) are called **Riesz bounds**. Now we formulate our main result.

Theorem 2. *Assume that* $\Lambda \in \mathbb{R}^{N+1}$. *Then there exist positive constants* N_j *such that the system of functions* $N_j \psi_{2^{-j}\Lambda}(2^j x - l)$ $j, l \in \mathbb{Z}$, *is stable.*

Proof: Define $f_{j,l}(x) := \psi_{2^{-j}\Lambda}(2^j x - l)$. Let N_j be numbers which will be specified later, and let $(c_{j,l})_{j,l \in \mathbb{Z}}$ be in $l^2(\mathbb{Z} \times \mathbb{Z})$. Then by the orthogonality of the different levels j

$$M := \left\| \sum_{j,l \in \mathbb{Z}} c_{j,l} N_j f_{j,l} \right\|^2 = \sum_{j \in \mathbb{Z}} N_j^2 \left\| \sum_{l \in \mathbb{Z}} c_{j,l} f_{j,l} \right\|^2. \tag{32}$$

Now let $0 < A_j \le B_j$ be Riesz bounds of the system of functions $f_{j,l}, l \in \mathbb{Z}$. It is not difficult to see that the system $N_j f_{j,l}$, $j, l \in \mathbb{Z}$, is stable if and only if B_j/A_j, $j \in \mathbb{Z}$, is bounded. In that case N_j can be choosen as $N_j = \frac{1}{\sqrt{A_j}}$.

Let us now consider the Riesz bounds $A_j \le B_j$ of the system $f_{j,l}(x) = \psi_{2^{-j}\Lambda}(2^j x - l)$, $l \in \mathbb{Z}$. By a transformation of variables we see that $A_j = 2^{-j}\widetilde{A_j}$ where $\widetilde{A_j}$ is the lower Riesz bound of the system $\psi_{2^{-j}\Lambda}(\cdot - l), l \in \mathbb{Z}$. It is well known that a family $\varphi(x-l)$, $l \in \mathbb{Z}$, is stable (see Theorem 3.24 in [3, p. 76]) with bounds $A \le B$ in (31) if and only if $A \le |S_\varphi(\xi)| \le B$ for almost every $\xi \in \mathbb{R}$. The function $S_{\psi_{2^{-j}\Lambda}}$ can be estimated according to (16) and Theorem 1 by

$$\left(S_{Q_{\frac{1}{2}2^{-j}\Lambda}}(-\pi)\right)^2 S_{Q_{2^{-j}\Lambda}}(-\pi) \le 4|S_{\psi_{2^{-j}\Lambda}}(\xi)|$$
$$\le \left(S_{Q_{\frac{1}{2}2^{-j}\Lambda}}(0)\right)^2 S_{Q_{2^{-j}\Lambda}}(0).$$

Hence it suffices to show that

$$\frac{S_{Q_{2^{-j}\Lambda}}(0)}{S_{Q_{2^{-j}\Lambda}}(-\pi)} \tag{33}$$

is bounded for $j \in \mathbb{Z}$. Let us put $w_j = 2^j$ for $j \in \mathbb{Z}$. Theorem 1 shows that

$$\left| S_{Q_{w_j \Lambda}} (\xi) \right| = e^{-w_j(\lambda_1 + \cdots + \lambda_{N+1})} \left| \Pi_{\widetilde{w_j \Lambda}} \left(0, e^{i\xi} \right) \right|. \tag{34}$$

Note that $\widetilde{w_j \Lambda} = w_j \widetilde{\Lambda}$. In the first case we assume that $j \to -\infty$, hence $w_j \to 0$. It follows that $w_j \widetilde{\Lambda}$ converges to the zero vector 0. Lemma 1 (see below, applied to the symmetrized vector $\widetilde{\Lambda}$) shows that $\Pi_{w_j \widetilde{\Lambda}} (0, \lambda)$ converges to $\Pi_{0\widetilde{\Lambda}} (0, \lambda)$ for $w_j \to 0$. Theorem 1 tells us that $\Pi_{0\widetilde{\Lambda}} (0, \pm 1) \neq 0$, hence the quotient $\Pi_{w_j \widetilde{\Lambda}} (0, 1) / \Pi_{w_j \widetilde{\Lambda}} (0, -1)$ is bounded for $w_j \to 0$.

In the second case we assume that $j \to \infty$, hence $w_j \to \infty$. Recall that $\Pi_{w\widetilde{\Lambda}} (0, \lambda) = r_{w\widetilde{\Lambda}} (\lambda) A_{w\widetilde{\Lambda}} (0, \lambda)$, cf. (20). By (34) and (33) it suffices to show that

$$\frac{A_{w\widetilde{\Lambda}} (0, 1)}{A_{w\widetilde{\Lambda}} (0, -1)} \tag{35}$$

is bounded for all $w \geq 1$, since clearly

$$\frac{r_{w\widetilde{\Lambda}} (1)}{r_{w\widetilde{\Lambda}} (-1)} = \prod_{k=1}^{2N+2} \frac{\left(e^{w_j \lambda_k} - 1 \right)}{\left(e^{w_j \lambda_k} + 1 \right)}$$

is bounded: each factor is of the type $(x - 1)/(x + 1)$ with $x \in (0, \infty)$.

In the first subcase we assume that 0 does not occur in Λ. Then Lemma 2 applied to $\widetilde{\Lambda}$ shows that for $w \to \infty$

$$w^{2N+1} A_{w\widetilde{\Lambda}} (0, \lambda) \to \frac{-1}{\lambda} \frac{1}{2\pi i} \int_{\Gamma_1} \frac{1}{q_{\widetilde{\Lambda}} (z)} dz,$$

and Lemma 3 shows that the last number is non-zero. Hence the quotient (35) is bounded for $w \to \infty$.

In the second subcase, assume that the multiplicity of 0 is strictly positive. Then the symmetrized vector has multiplicity $m > 1$. Then Lemma 2 shows that $w^{2N+1-(m-1)} A_{w\widetilde{\Lambda}} (0, \lambda)$ converges to the constant $D_{\widetilde{\Lambda}} (\lambda)$ defined in (45), and it is easy to see that this constant is non-zero for $\lambda = -1$. Hence we have proved that the quotient (35) is bounded for all $j \in \mathbb{Z}$ and the theorem is proved. $\square$

Lemma 1. *Suppose that $\Lambda_j \in \mathbb{R}^{N+1}$, $j \in \mathbb{N}$, converges to Λ in $\mathbb{R}^{N+1}$. Then $\Pi_{\Lambda_j} (0, \lambda)$ converges to $\Pi_\Lambda (0, \lambda)$ for $j \to \infty$ uniformly on compact sets.*

Proof: Recall that $\Pi_{\Lambda_j} (0, \lambda) = r_{\Lambda_j} (\lambda) A_{\Lambda_j} (0, \lambda)$ for all $\lambda \notin e^{\Lambda_j}$. Clearly $r_{\Lambda_j} (\lambda)$ converges to $r_\Lambda (\lambda)$ uniformly on compact sets. Now let $\lambda \neq e^{\lambda_j}$, where $\Lambda = (\lambda_1, \ldots, \lambda_{N+1})$, and let K be a compact neighborhood of λ

such that $K \cap e^\Lambda$ is empty. Then for large j we know that $K \cap e^{\Lambda_j}$ is empty, and we can consider

$$A_{\Lambda_j}(0, \lambda) = \frac{1}{2\pi i} \int_\Gamma \frac{1}{q_{\Lambda_j}(z)} \frac{1}{e^z - \lambda} dz \tag{36}$$

for all $\lambda \in K$, where Γ is a cycle surrounding all λ's in Λ_j for large j and avoiding all zeros of $z \longmapsto e^z - \lambda$. Since $q_{\Lambda_j}(z)$ converges to $q_\Lambda(z)$ on Γ we conclude that $A_{\Lambda_j}(0, \lambda)$ converges to $A_\Lambda(0, \lambda)$ on K. Hence $\Pi_{\Lambda_j}(0, \lambda)$ converges to $\Pi_\Lambda(0, \lambda)$ on K which is disjoint to e^Λ. Since $\Pi_{\Lambda_j}(0, \lambda)$ and $\Pi_\Lambda(0, \lambda)$ are polynomials, the maximum principle shows that the convergence holds for all points $\lambda \in \mathbf{C}$. The proof is complete. $\square$

Lemma 2. *Let* $\Lambda = (\lambda_1, \ldots, \lambda_{N+1})$ *and* $w > 0$. *Let* Γ_1 *be a path only surrounding the negative values of* Λ. *Then*

1) *Suppose that* $\lambda_j \neq 0$ *for all* $j = 1, \ldots, N+1$. *Then for* $w \to \infty$,

$$w^N A_{w\Lambda}(0, \lambda) \to \frac{-1}{\lambda} \frac{1}{2\pi i} \int_{\Gamma_1} \frac{1}{q_\Lambda(z)} dz. \tag{37}$$

2) *If* Λ *contains* 0 *with multiplicity* $m > 1$ *(clearly* $m \leq N+1$*), then for* $w \to \infty$

$$w^{N-(m-1)} A_{w\Lambda}(0, \lambda) \to D_\Lambda(\lambda), \tag{38}$$

where $D_\Lambda(\lambda)$ *is defined in (45) below.*

Proof: 1) Since $q_{w\Lambda}(z) = w^{N+1} q_\Lambda\left(\frac{z}{w}\right)$, a short computation shows that

$$A_{w\Lambda}(0, \lambda) = \frac{1}{w^N} \frac{1}{2\pi i} \int_\Gamma \frac{1}{q_\Lambda(z)} \frac{1}{e^{wz} - \lambda} dz. \tag{39}$$

Let us split the integral in (39) into three terms: let Γ_1 be a suitable path contained in the left half plane such that it surrounds exactly the negative λ's in Λ. Further Γ_2 should be a path in the right half plane which surrounds the positive λ's in Λ, and finally let Γ_3 be a path surrounding zero. Define

$$R_j(w, \lambda) := \frac{1}{2\pi i} \int_{\Gamma_j} \frac{1}{q_\Lambda(z)} \frac{1}{e^{wz} - \lambda} dz. \tag{40}$$

Hence we have

$$w^N A_{w\Lambda}(0, \lambda) = R_1(w, \lambda) + R_2(w, \lambda) + R_3(w, \lambda). \tag{41}$$

2) The path Γ_1 is contained in the left half plane. Hence $1/(e^{wz} - \lambda)$ converges to $-1/\lambda$ for $w \to \infty$ uniformly for all $z \in \Gamma_1$ and for $w \to \infty$. This shows that

$$R_1(w, \lambda) \to -\frac{1}{\lambda} \frac{1}{2\pi i} \int_{\Gamma_1} \frac{1}{q_\Lambda(z)} dz. \tag{42}$$

The path Γ_2 is contained in the right half plane, hence $\frac{1}{e^{wz}-\lambda} \to 0$ uniformly for all $z \in \Gamma_3$ and for $w \to \infty$. Hence $R_2(w, \lambda) \to 0$.

3) If $m = 0$ then by definition $R_3(w, \lambda)$ is equal to zero and the first claim is obvious.

4) Assume now that the multiplicity $m > 0$. For $j = 3$ we make the transformation $y = wz$ and obtain

$$R_3(w, \lambda) = \frac{1}{2\pi i} \int_{\Gamma_3} \frac{1}{q_\Lambda\left(\frac{y}{w}\right)} \frac{1}{e^y - \lambda} \frac{dy}{w}. \tag{43}$$

Let us write (if $m = N + 1$ then put $f_\Lambda = 1$)

$$q_\Lambda\left(\frac{y}{w}\right) = \frac{y^m}{w^m} f_\Lambda(y), \quad \text{where } f_\Lambda(y) = \prod_{\lambda_j \neq 0}\left(\frac{y}{w} - \lambda_j\right). \tag{44}$$

It follows that for $w \to \infty$

$$w^{-(m-1)} R_3(w, \lambda) \to \frac{(-1)^{N+1-m}}{\prod_{\lambda_j \neq 0} \lambda_j} \frac{1}{2\pi i} \int_{\Gamma_3} \frac{1}{y^m} \frac{1}{e^y - \lambda} dy =: D_\Lambda(\lambda). \tag{45}$$

Finally we see that $w^{N-(m-1)} A_{w\Lambda}(0, \lambda)$, which is equal to

$$w^{-(m-1)} R_1(w, \lambda) + w^{-(m-1)} R_2(w, \lambda) + w^{-(m-1)} R_3(w, \lambda),$$

converges for $w \to \infty$ to the constant $D(\lambda)$ defined in (45) provided that $m > 1$ since then ($w^{-(m-1)} \to 0$). $\square$

Lemma 3. *Suppose* $\Lambda = (\lambda_1, \ldots, \lambda_{N+1}) \in \mathbb{R}^{N+1}$ *is given where* $N \geq 0$ *and* $0 \neq \lambda_j$ *for all* $j = 1, \ldots, N+1$. *Let* Γ_1 *be a path only surrounding the negative values of the symmetrized vector* $\widetilde{\Lambda}$. *Then the following identity holds:*

$$\frac{1}{2\pi i} \int_{\Gamma_1} \frac{1}{q_{\widetilde{\Lambda}}(z)} dz = \frac{(-1)^{N+1}}{2\pi} \int_{-\infty}^{\infty} \frac{1}{\prod_{j=1}^{N+1}(x^2 + \lambda_j^2)} dx \neq 0. \tag{46}$$

Proof: We can assume that Γ_1 consists of the path γ_R from $-iR$ to iR defined by $\gamma_R(t) = it$ and the path ρ_R defined by $\rho_R(t) = R \cdot e^{it}$ with $t \in \left[\frac{\pi}{2}, \frac{3\pi}{2}\right]$, where $R > 0$ is taken to be so large such that the path contains all negative values of $\widetilde{\Lambda}$. Since $q_{\widetilde{\Lambda}}(z)$ is a polynomial of degree ≥ 2, it is clear that

$$\lim_{R \to \infty} \frac{1}{2\pi i} \int_{\rho_R} \frac{1}{q_{\widetilde{\Lambda}}(z)} dz = 0. \tag{47}$$

For the second integral we obtain

$$\lim_{R \to \infty} \frac{1}{2\pi i} \int_{\gamma_R} \frac{1}{q_{\widetilde{\Lambda}}(z)} dz = \frac{1}{2\pi} \lim_{R \to \infty} \int_{-R}^{R} \frac{1}{\prod_{j=1}^{N+1}(ix + \lambda_j)(ix - \lambda_j)} dx.$$

The proof is complete. $\square$

§5. Wavelet Analysis of Polysplines

A function $S : \mathbb{R}^n \setminus \{0\} \to \mathbb{C}$ is called a cardinal polyspline (on spheres) of order p if S is $(2p - 2)$-times continuously differentiable and the restriction of S to each open annulus

$$A_{0,l} := \left\{ x \in \mathbb{R}^n : e^l < |x| < e^{l+1} \right\} \tag{48}$$

is a polyharmonic function of order p. Recall that a function f defined on an open set U in Euclidean space $\mathbb{R}^n$ is polyharmonic of order p if f is $2p$–times continuously differentiable and $\Delta^p f (x) = 0$ for all $x \in U$, where $\Delta = \frac{\partial^2}{\partial x_1^2} + \cdots + \frac{\partial^2}{\partial x_n^2}$ is the Laplace operator and Δ^p its p–th iterate.

In analogy to the case of cardinal splines (cf. the definition of the scaling spaces $V_j(\Lambda)$ as the $L^2(\mathbb{R})$-closure of $\mathcal{S}_{2^{-j}\mathbb{Z}}(\Lambda) \cap L^2(\mathbb{R})$), we define P_j to be the set of all functions $S : \mathbb{R}^n \setminus \{0\} \to \mathbb{C}$ which are $(2p - 2)$-times continuously differentiable and whose restriction to each open annulus

$$A_{j,l} := \left\{ x \in \mathbb{R}^n : e^{2^{-j} l} < |x| < e^{2^{-j}(l+1)} \right\} \tag{49}$$

is a polyharmonic function of order p. Then we define for $j \in \mathbb{Z}$

$$PV_j := L^2 - \text{closure of } P_j \cap L^2 (\mathbb{R}^n) . \tag{50}$$

Our goal in this section is to prove the following result.

Theorem 3. *The sequence $(PV_j)_{j \in \mathbb{Z}}$ satisfies the following conditions: (i) $PV_j \subset PV_{j+1}$ for all $j \in \mathbb{Z}$, (ii) the set $\bigcup_{j \in \mathbb{Z}} PV_j$ is dense in $L_2 (\mathbb{R}^n)$ and (iii) $\bigcap_{j \in \mathbb{Z}} PV_j = \{0\}$.*

In the terminology of [1], the sequence $(PV_j)_{j \in \mathbb{Z}}$, forms a multiresolution. In this general framework it is possible to define wavelet spaces $PW_j, \; j \in \mathbb{Z}$ as the orthogonal complement of PV_j in PV_{j+1}, i.e. that $PW_j := PV_{j+1} \ominus PV_j$. By property (ii) and (iii) we obtain the orthogonal decomposition

$$L_2 (\mathbb{R}^n) = \bigoplus_{j \in \mathbb{Z}} PW_j. \tag{51}$$

We mention that PV_j is invariant under *rotations* while in the classical approaches, based on tensor products, box splines or radial basis function methods, the scaling spaces are shift-invariant under the discrete action $2^{-j}\mathbb{Z}^n$.

In the following we need some facts about polyharmonic functions: Each $x \in \mathbb{R}^n$ will be written in spherical coordinates $x = r\theta$ with $r \geq 0$ and $\theta \in S^{n-1}$. Recall that a function $Y : S^{n-1} \to \mathbb{C}$ is a spherical harmonic of degree $k \in \mathbb{N}_0$ if there exists a homogeneous harmonic polynomial $P(x)$

of degree k such that $P(\theta) = Y(\theta)$ for all $\theta \in S^{n-1}$. The set $\mathcal{H}_k$ of all spherical harmonics of degree exactly k is a linear space of dimension

$$a_k := \dim \mathcal{H}_k = \binom{n+k-1}{k} - \binom{n+k-3}{k-2}. \tag{52}$$

We denote by $Y_{k,l}$ with $l = 1, 2, \ldots, a_k$ an orthonormal basis of $\mathcal{H}_k$ with respect to the usual inner product

$$\int_{S^{n-1}} f(\theta)\overline{g(\theta)}d\theta.$$

For a detailed account we refer to [15].

Let $u : (R_1, R_2) \to \mathbb{C}$ be infinitely differentiable and $Y_k \in \mathcal{H}_k$. Then it is well known that

$$\Delta(u(r) \cdot Y_k(\theta)) = Y_k(\theta) \cdot L_{(k)}u(r), \tag{53}$$

where we have put

$$L_{(k)} = \frac{d^2}{dr^2} + \frac{n-1}{r}\frac{d}{dr} - \frac{k(k+n-2)}{r^2}. \tag{54}$$

By iteration we have $\Delta^p(u(r)Y_k(\theta)) = Y_k(\theta) \cdot [L_{(k)}]^p u(r)$. Let us put for convenience

$$\Lambda_+(k,p) = (k, k+2, \ldots, k+2p-2) \tag{55}$$

$$\Lambda_-(k,p) = (-k-n+2, -k-n+4, \ldots, -k-n+2p) \tag{56}$$

The following result describes the solutions of the differential operator $[L_{(k)}]^p$ explicitly:

Proposition 2. *The space of solutions of the equation $L_{(k)}^p f(r) = 0$ which are C^∞ for $r > 0$ is generated by a simple basis subdivided into the following parts:*

$$r^j \quad \text{for all } j \text{ in } \Lambda_+(k,p) \text{ or } \Lambda_-(k,p),$$

$$r^j \log r \quad \text{for all } j \text{ in } \Lambda_+(k,p) \text{ and } \Lambda_-(k,p).$$

It will be convenient to make a transformation of variables from r to $v = \log r$. Then a solution of the form r^j as in Proposition 2 will be transformed to e^{jv}. A solution of the form $r^j \log r$ with j in $\Lambda_+(k,p)$ and $\Lambda_-(k,p)$ is transformed to ve^{jv}. We see immediately that all solutions to the equation $L_{(k)}^p f(r) = 0$ are transformed to solutions of the equation $M_{\Lambda(k)}g(v) = 0$, where $M_{\Lambda(k)}$ is defined by (3) with respect to the vector

$$\Lambda_k := (k, k+2, \ldots, k+2(p-1), -(k+n)+2, \ldots, -(k+n)+2p).$$

The dependence on the parameter p and n will be suppressed. Now we are able to formulate the following result.

Theorem 4. *Let $S : \mathbb{R}^n \setminus \{0\} \to \mathbb{R}$ be a polyspline of order p. Then the function $S_{k,l} : \mathbb{R} \to \mathbb{R}$ defined by*

$$S_{k,l}(v) := \int_{S^{n-1}} S(e^v \theta) Y_{k,l}(\theta) \, d\theta \tag{57}$$

is a cardinal L-spline with respect to the linear differential operator $M_{\Lambda(k)}$.

Proof: Since S is $(2p - 2)$-times continuously differentiable in $\mathbb{R}^n \setminus \{0\}$, it is clear that $S_{k,l}$ is $(2p - 2)$-times continuously differentiable in $\mathbb{R}$ with respect to the variable v. Let $j \in \mathbb{Z}$. By definition S is polyharmonic on the open annulus $A_{0,l}$. Then (see [16] or [7]) there exist infinitely differentiable functions $\psi_{k_1,l_1} : (e^j, e^{j+1}) \to \mathbb{C}$, such that

$$S(r\theta) = \sum_{k_1=0}^{\infty} \sum_{l_1=1}^{a_{k_1}} \psi_{k_1,l_1}(r) Y_{k_1,l_1}(\theta) \tag{58}$$

with respect to convergence on compact subsets of the annulus $A_{0,l}$ defined in (48) and ψ_{k_1,l_1} are solutions of the linear differential equation $L_{(k_1)}^p \psi_{k_1,l_1}(r) = 0$ for $e^j < r < e^{j+1}$. Inserting (58) in (57) and interchanging integration and summation, we obtain for $v \in (j, j+1)$

$$S_{k,l}(v) = \sum_{k_1=0}^{\infty} \sum_{l_1=1}^{a_k} \psi_{k_1,l_1}(e^v) \int_{S^{n-1}} Y_{k_1,l_1}(\theta) Y_{k,l}(\theta) \, d\theta = \psi_{k,l}(e^v),$$

where for the last equality we have used the orthogonality relations for spherical harmonics. This shows that $\psi_{k,l}(r) = S_{k,l}(\log r)$ for all $r \in (e^j, e^{j+1})$, $j \in \mathbb{Z}$. Since $L_{(k)}^p \psi_{k,l}(r) = 0$ on the annulus, the function $S_{k,l}(v)$ is a solution of the equation $M_{\Lambda(k)}\left(\frac{d}{dv}\right) = 0$ for all $v \in (j, j+1)$ with $j \in \mathbb{Z}$. Hence $S_{k,l}$ is a cardinal L-spline with respect to $M_{\Lambda(k)}$. $\square$

Recall that P_j is the set of all cardinal polysplines on the mesh defined in (49). By a straightforward modification of the proof of Theorem 4, it follows that for $S \in P_j$ the function $S_{k,l}$ defined in (57) is a cardinal L-spline on the mesh $2^{-j}\mathbb{Z}$ with respect to Λ_k.

We now want to characterize the $L^2(\mathbb{R}^n)$-closure of $P_j \cap L^2(\mathbb{R}^n)$ which we have denoted by PV_j. It is a temptation to assume that for $S \in PV_j$ the Fourier-Laplace coefficient defined through formula (57) will be in $V_j(\Lambda_k)$, i.e. in the closure of $\mathcal{S}_{2^{-j}\mathbb{Z}}(\Lambda_k) \cap L^2(\mathbb{R})$. This is not true since the transformation rule will give us an additional weight in the following formula which is valid for all $f \in L_2(\mathbb{R}^n)$:

$$\int_{\mathbb{R}^n} |f(x)|^2 \, dx = \int_0^\infty \int_{S^{n-1}} |f(r\theta)|^2 \, r^{n-1} d\theta dr. \tag{59}$$

Fortunately, this problem can be easily remedied: let us define

$$\overline{\Lambda_k} = \left(\frac{n}{2}, \ldots, \frac{n}{2}\right) + \Lambda_k. \tag{60}$$

Theorem 5. *For each $k \in \mathbb{N}_0, l = 1, \ldots, a_k$, the map $S \longmapsto \overline{S}_{k,l}$, defined on the domain $P_j \cap L^2(\mathbb{R}^n)$ by*

$$\overline{S}_{k,l}(v) := e^{\frac{n}{2}v} \int_{S^{n-1}} S(e^v \theta) Y_{k,l}(\theta) \, d\theta, \tag{61}$$

maps into $\mathcal{S}_{2^{-j}\mathbb{Z}}\left(\overline{\Lambda_k}\right) \cap L^2(\mathbb{R}, dv)$ and is continuous with respect to the L^2-norms on $\mathbb{R}^n$ and $\mathbb{R}$, respectively. Hence there exists a continuous extension which maps

$$PV_j \to V_j\left(\overline{\Lambda_k}\right).$$

Proof: Recall that the set of spherical harmonics $Y_{k,l}(\theta)$, for $k \in \mathbb{N}_0$, $l = 1, \ldots, a_k$, is an orthonormal basis of the space $\mathcal{H}_k$ of all spherical harmonics with respect to $d\theta$. Let $S \in P_j \cap L^2(\mathbb{R}^n)$. For all $k \in \mathbb{N}_0$, and $l = 1, \ldots, a_k$ the Fourier–Laplace coefficients of the integrable function $\theta \longmapsto S(r\theta)$ are defined by

$$S_{k,l}(r) := \int_{S^{n-1}} S(r\theta) Y_{k,l}(\theta) \, d\theta. \tag{62}$$

Formula (59) and the orthonormality of the system $Y_{k,l}$ show that

$$\int_{\mathbb{R}^n} |S(x)|^2 \, dx = \sum_{k=0}^{\infty} \sum_{l=1}^{a_k} \int_0^{\infty} |S_{k,l}(r)|^2 \, r^{n-1} dr. \tag{63}$$

By the substitution $v = \log r$, we obtain the formula

$$\int_{\mathbb{R}^n} |S(x)|^2 \, dx = \sum_{k=0}^{\infty} \sum_{l=1}^{a_k} \int_{-\infty}^{\infty} \left|e^{\frac{n}{2}} S_{k,l}(e^v)\right|^2 \, dv. \tag{64}$$

By the remark after Theorem 4, the function $S_{k,l}$ is in $\mathcal{S}_{2^{-j}\mathbb{Z}}(\Lambda_k)$. It is easy to see that $\overline{S}_{k,l}$ defined by $\overline{S}_{k,l}(v) = e^{\frac{n}{2}} S_{k,l}(e^v)$ is in $\mathcal{S}_{2^{-j}\mathbb{Z}}\left(\overline{\Lambda_k}\right)$, since it is clearly $2(p-1)$-times continuously differentiable and a solution of the differential operator $M_{\overline{\Lambda_k}}$ on the open interval $\left(2^{-j}l, 2^{-j}(l+1)\right)$ (we know already that $v \longmapsto S_{k,l}(e^v)$ is a solution of the differential operator M_{Λ_k}), hence it is in $\mathcal{S}_{2^{-j}\mathbb{Z}}\left(\overline{\Lambda_k}\right)$. Formula (64) shows that

$$\int_{\mathbb{R}^n} |S(x)|^2 \, dx = \sum_{k=0}^{\infty} \sum_{l=1}^{a_k} \int_{-\infty}^{\infty} \left|\overline{S}_{k,l}(v)\right|^2 \, dv. \tag{65}$$

Hence $\overline{S}_{k,l}$ is square-integrable and clearly the map $S \longmapsto \overline{S}_{k,l}$ is continuous and linear. Hence it can be extended to PV_j. $\square$

Theorem 6. *Let* $V_j\left(\overline{\Lambda_k}\right), j \in \mathbb{Z}$, *be the scaling spaces of the cardinal* L*-splines with respect to* $M_{\overline{\Lambda_k}}$ *and let* $W_j\left(\overline{\Lambda_k}\right), j \in \mathbb{Z}$ *be the associated wavelet spaces. Then the scaling spaces* PV_j *of polysplines of order* p *are isomorphic to*

$$V_j := \bigoplus_{k \in \mathbb{N}_0, l=1,\ldots,a_k} V_j\left(\overline{\Lambda_k}\right), \tag{66}$$

and PW_j *is isomorphic to*

$$W_j := \bigoplus_{k \in \mathbb{N}_0, l=1,\ldots,a_k} W_j\left(\overline{\Lambda_k}\right). \tag{67}$$

Proof: The claim is obvious from the above. $\square$

The following characterization shows that the scaling spaces PV_j could have been introduced without using the term "polyharmonic", using only terms like spherical harmonics and cardinal L-splines (with respect to vectors $\Lambda_k \in \mathbb{R}^{2p}$).

Theorem 7. *Let* $C_c\left(\mathbb{R}^n\right)$ *be the set of all continuous functions with compact support in* $\mathbb{R}^n$. *For each* $j \in \mathbb{Z}$ *the set* PV_j *is equal to the* $L^2\left(\mathbb{R}^n\right)$*-closure of the following subspaces (where* $\theta = \frac{x}{|x|}$*):*

$$A_j = \left\{ \sum_{k=0}^{N} \sum_{l=1}^{a_k} S_{k,l}\left(\log|x|\right) Y_{k,l}\left(\theta\right) : N \in \mathbb{N}_0, \right.$$

$$\left. S_{k,l} \in \mathcal{S}_{2^{-j}\mathbb{Z}}\left(\Lambda_k\right) \cap C_c\left(\mathbb{R}\right) \right\},$$

$$B_j = \left\{ \sum_{k=0}^{N} \sum_{l=1}^{a_k} S_{k,l}\left(\log|x|\right) Y_{k,l}\left(\theta\right) : N \in \mathbb{N}_0, \right.$$

$$\left. S_{k,l} \in \mathcal{S}_{2^{-j}\mathbb{Z}}\left(\Lambda_k\right) \cap L^2\left(\mathbb{R}\right) \right\}.$$

Proof: The inclusion $A_j \subset B_j$ is trivial. Since each member $S \in B_j$ is clearly $2\left(p-1\right)$-times continuously differentiable, and since S is polyharmonic on annuli with radii $e^{l2^{-j}}$ and $e^{(l+1)2^{-j}}, l \in \mathbb{Z}$, it follows that B_j is contained in $P_j \cap L^2\left(\mathbb{R}^n\right)$, cf. the remarks after formula (53). Since PV_j is defined as the closure of the latter space, we conlcude that $\overline{A_j} \subset \overline{B_j} \subset PV_j$. Hence the proof is complete if we can show that each $S \in P_j \cap L^2\left(\mathbb{R}^n\right)$ can be approximated by some elements of A_j.

Let $S \in P_j \cap L^2(\mathbb{R}^n)$. Recall that $\overline{S}_{k,l}(v) = e^{\frac{n}{2}} S_{k,l}(e^v)$. By formula (65) we can find for each $\varepsilon > 0$ a natural number K such that

$$I_\varepsilon := \left| \int_{\mathbb{R}^n} |S(x)|^2 \, dx - \sum_{k=0}^{K} \sum_{l=1}^{a_k} \int_{-\infty}^{\infty} \left| \overline{S}_{k,l}(v) \right|^2 dv \right| \leq \frac{1}{2}\varepsilon. \qquad (68)$$

Define $M = \sum_{k=0}^{K} \sum_{l=1}^{a_k} 1$. By well known properties of the basic cardinal L−spline, there exists for each $\overline{S}_{k,l}$ an L−spline $T_{k,l} \in \mathcal{S}_{2^{-j}\mathbb{Z}}(\overline{\Lambda_k}) \cap C_c(\mathbb{R})$ such that

$$\left\| \overline{S}_{k,l} - T_{k,l} \right\|_{L^2(\mathbb{R})} \leq \frac{1}{2M}\varepsilon. \qquad (69)$$

Define $T_K(x) := \sum_{k=0}^{K} \sum_{l=1}^{a_k} e^{-\frac{n}{2}} T_{k,l}(\log |x|) Y_{k,l}(\theta)$. As before, it is easy to see that $v \mapsto e^{-\frac{n}{2}} T_{k,l}(v)$ is in $\mathcal{S}_{2^{-j}\mathbb{Z}}(\Lambda_k)$, hence T_k is in A_j. Define $S_K(x) = \sum_{k=0}^{K} \sum_{l=1}^{a_k} S_{k,l}(\log |x|) Y_{k,l}(\theta)$. Then

$$A := \|S - T_K\|_{L^2(\mathbf{R}^N)} \leq \|S_K - T_K\|_{L^2(\mathbf{R}^N)} + \|S_K - T_K\|_{L^2(\mathbf{R}^N)}.$$

The transformation rule now shows that the first summand is less than or equal to $I_\varepsilon \leq \frac{1}{2}\varepsilon$. Similarly, by (69) and the definition of M, the second summand is less than or equal to $\frac{1}{2}\varepsilon$. □

Finally we are able to give

Proof of Theorem 3:. This says that the sequence of scaling spaces $(PV_j)_{j \in \mathbb{Z}}$ forms a multiresolution. The first statement $PV_j \subset PV_{j+1}$ follows directly from the definition. Assume now that $S \in PV_j$ for all $j \in \mathbb{Z}$. Then $\overline{S}_{k,l} \in V_j(\overline{\Lambda_k})$ for all $j \in \mathbb{Z}$. Since we already know that $V_j(\overline{\Lambda_k}), j \in \mathbb{Z}$, forms a multiresolution we conclude that $\overline{S}_{k,l} = 0$. It follows that $S = 0$. Finally we have to show that the union of all PV_j is dense in $L^2(\mathbb{R}^n)$. But it is easy to see that already the union of all subspaces $A_j, j \in \mathbb{Z}$, is dense. The proof is complete. □

Theorem 6 shows that the wavelet space PW_j of the scaling spaces PV_j can be decomposed into a direct sum of wavelet spaces $W_j(\overline{\Lambda_k})$, where $k = 0, 1, 2, ..$ and $l = 1, \ldots, a_k$. The space $W_j(\overline{\Lambda_k})$ is generated by the stable system $\psi_{j,m}, m \in \mathbb{Z}$, as we have seen in Section 3. But it is a fact that the Riesz bounds $0 < A_{j,k} \leq B_{j,k}$ depend on k and $j \in \mathbb{Z}$. An important result in [7] states that for each $j \in \mathbb{Z}$ the sequence $B_{j,k}/A_{j,k}$ is bounded (for the variable $k \in \mathbb{Z}$). This fact allows one to carry over the decomposition and reconstruction algorithm of the component spaces $W_j(\overline{\Lambda_k}), j \in \mathbb{Z}$, to the spaces PW_j for $j \in \mathbb{Z}$. For further details we refer to [7].

§6. Appendix: The Hermite-Genocchi Formula

The Hermite Genocchi formula for cardinal $L-$splines was already derived in [5]. For our purposes we need a variant of this formula, and for convenience of the reader we give a quick proof. The formula for $\widehat{Q_\Lambda}$ in (5) shows that $\widehat{Q_{\{\lambda_1\}}}(\xi)$ is equal to

$$\frac{e^{-\lambda_1} - e^{-i\xi}}{i\xi - \lambda_1} = e^{-\lambda_1}\frac{e^{\lambda_1 - i\xi} - 1}{\lambda_1 - i\xi} = e^{-\lambda_1}\int_0^1 e^{\lambda_1 x}e^{-i\xi x}dx. \tag{70}$$

Let $\Lambda_2 = (\mu_1, \ldots, \mu_{M+1})$ and set $(\Lambda_1, \Lambda_2) := (\lambda_1, \ldots, \lambda_{N+1}, \mu_1, \ldots, \mu_{M+1})$. By formula (5) we obtain

$$\widehat{Q_{(\Lambda_1,\Lambda_2)}}(\xi) = \widehat{Q_{\Lambda_1}}(\xi) \cdot \widehat{Q_{\Lambda_2}}(\xi) = \widehat{Q_{\Lambda_1} * Q_{\Lambda_2}}(\xi). \tag{71}$$

The inverse Fourier transform yields $Q_{(\Lambda_1,\Lambda_2)}(x) = Q_{\Lambda_1} * Q_{\Lambda_2}(x)$. This implies the recursive formula

$$Q_{\Lambda \cup \{\mu\}}(x) = e^{-\mu}\int_0^1 Q_\Lambda(x - y)e^{\mu y}dy. \tag{72}$$

Theorem 8. *Let* $f \in L^1(\mathbb{R}) \cap L^2(\mathbb{R})$ *and* Q_Λ *be the basic spline with respect to* $\Lambda = (\lambda_1, \ldots, \lambda_{N+1})$. *Define* $C_\Lambda = e^{-(\lambda_1 + \cdots + \lambda_{N+1})}$. *Then*

$$\int_{-\infty}^{\infty} f(x)Q_\Lambda(x)\,dx$$
$$= C_\Lambda \int_0^1 \cdots \int_0^1 f(x_1 + \cdots + x_{N+1}) \prod_{j=1}^{N+1} e^{\lambda_j x_j}\,dx_1 \cdots dx_{N+1}.$$

Proof: Recall that Q_Λ is real-valued. The Parseval identity implies that

$$I := \int_{-\infty}^{\infty} f(x)\,Q_\Lambda(x)\,dx = \frac{1}{2\pi}\int_{-\infty}^{\infty} \widehat{f}(\xi) \cdot \overline{\widehat{Q_\Lambda}(\xi)}d\xi \tag{73}$$

Formula (70) and (71) show that

$$\overline{\widehat{Q_\Lambda}(\xi)} = \prod_{j=1}^{N+1} e^{-\lambda_j} \cdot \prod_{j=1}^{N+1}\int_0^1 e^{\lambda_j x}e^{i\xi x}dx. \tag{74}$$

It follows that I is equal to

$$\frac{C_\Lambda}{2\pi}\int_0^1 \cdots \int_0^1 \int_{-\infty}^{\infty} \widehat{f}(\xi) \cdot e^{i\xi(x_1 + \cdots + x_{N+1})} \prod_{j=1}^{N+1} e^{\lambda_j x_j}\,d\xi dx_1 \cdots dx_{N+1}.$$

The proof is accomplished by using the inverse Fourier transform formula for the function f. $\square$

Proposition 3. *Let* $\Lambda = (\lambda_1, \ldots, \lambda_{N+1})$. *Then the following identity holds:*

$$\int_{-\infty}^{\infty} Q_\Lambda(x+j) Q_\Lambda(x) \, dx = e^{-(\lambda_1 + \cdots + \lambda_{N+1})} Q_{(\Lambda, -\Lambda)}(N+1+j) \quad (75)$$

Proof: Apply Theorem 8 to $f(x) = Q_\Lambda(x+j)$. Then the right-hand side of the Hermite-Genocchi formula is equal to

$$e^{-(\lambda_1 + \cdots + \lambda_{N+1})} \int_{[0,1]^{N+1}} \prod_{j=1}^{N+1} e^{\lambda_j x_j} \cdot Q_\Lambda(x_1 + \cdots + x_{N+1} + j) \, dx_1 \cdots dx_{N+1}.$$

By substituting $x_j = 1 - y_j$ for $j = 1, \ldots, N+1$, we obtain the expression

$$\int_0^1 \cdots \int_0^1 \prod_{j=1}^{N+1} e^{-\lambda_j y_j} \cdot Q_\Lambda(N+1+j-y_1 - \cdots - y_{N+1}) \, dy_1 \cdots dy_{N+1}.$$

Apply now inductively formula (72) to the last expression. $\square$

Acknowledgments. The first author acknowledges the support of the Fulbright commission while the second author thanks the Alexander von Humboldt-Stiftung for supporting him in the framework of the Feodor-Lynen-program.

References

1. de Boor, C., R. DeVore and A. Ron, On the construction of multi-variate (pre)wavelets, Constr. Approx. **9** (1993), 123–166.

2. de Boor, C., R. DeVore and A. Ron, Approximation from shift-invariant subspaces of $L_2(\mathbb{R}^d)$. Trans. Amer. Math. Soc. **341** (1994), 787–806.

3. Chui, C. K., *An Introduction to Wavelets*, Academic Press, Boston 1992.

4. Dyn, N. and A. Ron, Recurrence relations for Tchebycheffian B-splines, Journal d'analyse Mathem. **51** (1988), 118–138.

5. Dyn, N. and A. Ron, Cardinal translation invariant Tchebycheffian Bsplines, Approximation Theory and its Applications **6** (1990), 1–12.

6. Kounchev, O. I., Definition and basic properties of polysplines, I and II, C.R. Acad. Bulg. Sci. **44** (1991), No. 7 and 8, 9–11, 13–16.

7. Kounchev, O. I., *Multivariate Polysplines. Applications to Numerical and Wavelet Analysis*, Academic Press, Boston 2001.

8. Kounchev, O. I. and H. Render, Multivariate cardinal splines via spherical harmonics, submitted.

9. Kounchev, O. I. and H. Render, The interpolation problem for cardinal splines, submitted.

10. Kounchev, O. I. and H. Render, Symmetry of interpolation polysplines and L−splines, in *Trends in Approximation Theory*, K. Kopotun, T. Lyche, and M. Neamtu (eds.), Vanderbilt University Press, Nashville, 2001, 191–202.

11. Lyche, T. and L. L. Schumaker, L−spline wavelets, in *Wavelets: Theory, Algorithms, and Applications*, C. Chui, L. Montefusco, and L. Puccio (eds.), Academic Press, New York, 1994, 197–212.

12. Micchelli, C. A., Oscillation matrices and cardinal spline interpolation, in *Studies in Spline Functions and Approximation Theory*, S. Karlin, C. Micchelli, A. Pinkus, and I. Schoenberg (eds.), Academic Press, New York, 1976, 163–202.

13. Micchelli, C. A., Cardinal L−spline, in *Studies in Spline Functions and Approximation Theory*, S. Karlin, C. Micchelli, A. Pinkus, and I. Schoenberg (eds.), Academic Press, New York, 1976, 203–250.

14. Schumaker, L. L., *Spline Functions: Basic Theory*, J. Wiley, New York, 1981.

15. Stein, E. M. and G. Weiss, *Introduction to Fourier Analysis on Euclidean Spaces*, Princeton University Press, Princeton, 1971.

16. Vekua, I. N., *New methods for Solving Elliptic Equations*, NH Publ. Co., John Wiley & Sons, Inc., New York, 1967.

Ognyan Kounchev
Institute of Mathematics
Bulgarian Academy of Sciences
Bonchev Str. 9
1143 Sofia, Bulgaria
kounchev@bas.bg

Hermann Render
Fachbereich Mathematik
Gerhard-Mercator Universität Duisburg
D-47048 Duisburg, Germany
render@math.uni-duisburg.de

Discrete GB-Splines and Their Properties

Boris I. Kvasov

Abstract. Discrete generalized splines are continuous piecewise defined functions which meet some smoothness conditions for the first and second divided differences at the knots. Direct algorithms and recurrence relations are proposed for constructing discrete generalized B-splines (discrete GB-splines for short). Properties of discrete GB-splines and their series are studied. It is shown that discrete GB-splines form weak Chebyshev systems and that series of discrete GB-splines have a variation diminishing property.

§1. Introduction

The tools of generalized splines and GB-splines are widely used in solving problems of shape preserving approximation (e.g., see [7]). By introducing various parameters into the spline structure, one can preserve characteristics of the initial data such as positivity, monotonicity, convexity, presence of linear sections, etc. Here, the main challenge is to develop algorithms that choose parameters automatically. Recently, in [2] a difference method for constructing shape preserving hyperbolic splines as solutions of multipoint boundary value problems was developed. Such an approach avoids the computation of hyperbolic functions and has substantial other advantages. However, the extension of a mesh solution will be a discrete hyperbolic spline. In this paper we consider more general constructions of discrete generalized splines and discrete GB-splines, and investigate their main properties.

§2. Discrete Generalized Splines

Let a partition $\Delta : a = x_0 < x_1 < \cdots < x_N = b$ of the interval $[a, b]$ be given. For fixed $\tau_j^{L_i} > 0$ and $\tau_j^{R_i} > 0$, $j = i, i+1$, and a function S

Approximation Theory X: Wavelets, Splines, and Applications   355
Charles K. Chui, Larry L. Schumaker, and Joachim Stöckler (eds.), pp. 355–368.
Copyright Ⓞ 2002 by Vanderbilt University Press, Nashville, TN.
ISBN 0-8265-1416-2.

which is defined and continuous on the real line $\mathbb{R}$ we introduce the linear difference operators

$$D_{i,1}S(x) = (\lambda_i^{R_i} S[x - \tau_i^{L_i}, x] + \lambda_i^{L_i} S[x, x + \tau_i^{R_i}])(1 - t)$$
$$+ (\lambda_{i+1}^{R_i} S[x - \tau_{i+1}^{L_i}, x] + \lambda_{i+1}^{L_i} S[x, x + \tau_{i+1}^{R_i}])t,$$

$$D_{i,2}S(x) = 2S[x - \tau_i^{L_i}, x, x + \tau_i^{R_i}](1 - t) + 2S[x - \tau_{i+1}^{L_i}, x, x + \tau_{i+1}^{R_i}]t,$$

$$x \in [x_i, x_{i+1}], \quad i = 0, \ldots, N - 1,$$

where $\lambda_j^{R_i} = 1 - \lambda_j^{L_i} = \tau_j^{R_i}/(\tau_j^{L_i} + \tau_j^{R_i})$, $j = i, i + 1$ and $t = (x - x_i)/h_i$, $h_i = x_{i+1} - x_i$. The square parentheses denote the usual first and second divided differences of the function S.

We associate to Δ a system of functions $\{1, x, \Phi_i, \Psi_i\}$, $i = 0, \ldots, N - 1$, which are defined and continuous on $\mathbb{R}$ and for given i are linearly independent on the interval $[x_i, x_{i+1}]$. The functions Φ_i and Ψ_i are subject to the constraints

$$\Phi_i(x_{i+1} - \tau_{i+1}^{L_i}) = \Phi_i(x_{i+1}) = \Phi_i(x_{i+1} + \tau_{i+1}^{R_i}) = 0, \quad D_{i,2}\Phi_i(x_i) = 1,$$
$$\Psi_i(x_i - \tau_i^{L_i}) = \Psi_i(x_i) = \Psi_i(x_i + \tau_i^{R_i}) = 0, \qquad D_{i,2}\Psi_i(x_{i+1}) = 1. \tag{1}$$

Any element S_i of the linear space Υ_i spanned by the four functions 1, x, Φ_i, Ψ_i can be uniquely written as

$$S_i(x) = S_i(x_i)(1 - t) + S_i(x_{i+1})t + D_{i,2}S_i(x_i)[\Phi_i(x) - \Phi_i(x_i)(1 - t)]$$
$$+ D_{i,2}S_i(x_{i+1})[\Psi_i(x) - \Psi_i(x_{i+1})t]. \tag{2}$$

Definition 1. *A function $S : [a, b] \to \mathbb{R}$ is called a* discrete generalized spline *if:*

(i) *for any integer i, $0 \le i \le N - 1$, there exists a unique function $S_i \in \Upsilon_i$ such that*

$$S(x) \equiv S_i(x), \quad x \in [x_i, x_{i+1}]; \tag{3}$$

(ii) *for all integers $i = 1, \ldots, N - 1$, S satisfies the following smoothness conditions*

$$S_{i-1}(x_i) = S_i(x_i),$$
$$D_{i-1,1}S_{i-1}(x_i) = D_{i,1}S_i(x_i), \tag{4}$$
$$D_{i-1,2}S_{i-1}(x_i) = D_{i,2}S_i(x_i).$$

The set of discrete generalized splines satisfying Definition 1 will be denoted by S_4^{DG}. The usual operations of addition of elements from S_4^{DG} and their multiplication by real numbers give again elements in the set S_4^{DG} which hence is a linear space.

Definition 1 generalizes the notion of discrete cubic splines in [8]. If $\tau_j^{L_i} \to 0$, $\tau_j^{R_i} \to 0$, $j = i, i + 1$ for all i, then as the limiting case we obtain generalized splines in [6]. If $\tau_i^{L_j} = \tau_i^L$ and $\tau_i^{R_j} = \tau_i^R$, $j = i - 1, i$, then according to smoothness conditions (4), the values of the functions S_{i-1} and S_i at the three consecutive points $x_i - \tau_i^L$, x_i, $x_i + \tau_i^R$ coincide. Setting $\tau_j^{L_i} = \tau_j^{R_i} = \tau_i$, $j = i, i+1$, we obtain $D_{i,1}S(x) = S[x - \tau_i, x + \tau_i]$ and $D_{i,2}S(x) = 2S[x - \tau_i, x, x + \tau_i]$, which is the case discussed in [2].

According to the conditions (4), the discrete generalized spline S defined by (2) and (3) can be written as

$$
\begin{aligned}
S(x) =& S(x_i)(1 - t) + S(x_{i+1})t + M_i[\Phi_i(x) - \Phi_i(x_i)(1 - t)] \\
& + M_{i+1}[\Psi_i(x) - \Psi_i(x_{i+1})t],
\end{aligned}
\tag{5}
$$

for $x \in [x_i, x_{i+1}]$ and $i = 0, \ldots, N - 1$, where $M_j = D_{i,2}S_i(x_j)$, $j = i, i+1$.

The functions Φ_i and Ψ_i depend on tension parameters which influence the behaviour of S fundamentally. We call them the **defining functions**. In practice one takes Φ_i to depend on a parameter p_i, and Ψ_i to depend on a parameter q_i, $0 \leq p_i, q_i < \infty$. In the limiting case when $p_i, q_i \to \infty$ we require that $\lim_{p_i \to \infty} \Phi_i(x) = 0$, $x \in (x_i, x_{i+1}]$ and $\lim_{q_i \to \infty} \Psi_i(x) = 0$, $x \in [x_i, x_{i+1})$ so that the function S in formula (5) turns into a linear function. Additionally, we require that if $p_i = q_i = 0$ for all i, then we get a discrete cubic spline.

§3. Construction of Discrete GB-Splines

Let us construct a basis for the space of discrete generalized splines S_4^{DG} by using functions which have local supports of minimum length. Since $\dim(S_4^{DG}) = N + 3$ we extend the grid Δ by adding the points x_j, $j = -3, -2, -1, N+1, N+2, N+3$, such that $x_{-3} < x_{-2} < x_{-1} < a$, $b < x_{N+1} < x_{N+2} < x_{N+3}$. As in Section 2, for each interval $[x_i, x_{i+1}]$, $i = -3, -2, -1, N, N+1, N+2$, we introduce the linear space Υ_i. This permits us to define the discrete generalized spline S on the extended interval $[x_{-3}, x_{N+3}]$.

We demand that the discrete GB-splines $B_{-3}, \ldots, B_{N-1}$ have the properties

$$
\begin{aligned}
& B_i(x) > 0, \quad x \in (x_i + \tau_i^{R_i}, x_{i+4} - \tau_{i+4}^{L_{i+3}}), \tag{6} \\
& B_i(x) \equiv 0, \quad x \notin (x_i, x_{i+4}),
\end{aligned}
$$

$$
\sum_{j=-3}^{N-1} B_j(x) \equiv 1, \quad x \in [a, b]. \tag{7}
$$

According to (5), on the interval $[x_j, x_{j+1}]$, $j = i, \ldots, i + 3$, for each $i = -3, \ldots, N - 1$ the discrete GB-spline B_i has the form

$$
B_i(x) \equiv \bar{B}_{i,j}(x) = P_{i,j}(x) + \Phi_j(x)M_{j,B_i} + \Psi_j(x)M_{j+1,B_i}, \tag{8}
$$

where $P_{i,j}$ is a linear polynomial and $M_{l,\mathrm{B}_i} = D_{j,2}\bar{\mathrm{B}}_{i,j}(x_l)$, $l = j, j+1$ are constants to be determined.

The smoothness conditions (4) together with the constraints (1) give the relations

$$P_{i,j}(x_j) = P_{i,j-1}(x_j) + z_j M_{j,\mathrm{B}_i},$$
$$D_{j,1}P_{i,j}(x_j) = D_{j-1,1}P_{i,j-1}(x_j) + c_{j-1,2}M_{j,\mathrm{B}_i},$$

where

$$z_j \equiv z_j(x_j) = \Psi_{j-1}(x_j) - \Phi_j(x_j),$$
$$c_{j-1,2} = D_{j-1,1}\Psi_{j-1}(x_j) - D_{j,1}\Phi_j(x_j).$$

Thus

$$P_{i,j}(x) = P_{i,j-1}(x) + [z_j + c_{j-1,2}(x - x_j)]M_{j,\mathrm{B}_i}. \qquad (9)$$

According to (4), the condition $\mathrm{B}_i(x) \equiv 0$ for $x \notin (x_i, x_{i+4})$ is satisfied if and only if

$$\bar{\mathrm{B}}_{i,i}(x_i) = D_{i,1}\bar{\mathrm{B}}_{i,i}(x_i) = D_{i,2}\bar{\mathrm{B}}_{i,i}(x_i) = 0,$$
$$\bar{\mathrm{B}}_{i,i+3}(x_{i+4}) = D_{i+3,1}\bar{\mathrm{B}}_{i,i+3}(x_{i+4}) = D_{i+3,2}\bar{\mathrm{B}}_{i,i+3}(x_{i+4}) = 0.$$

Due to (8) and (1), the latter relations are equivalent to

$$P_{i,i} \equiv 0, \quad M_{i,\mathrm{B}_i} = 0 \quad \text{and} \quad P_{i,i+3} \equiv 0, \quad M_{i+4,\mathrm{B}_i} = 0.$$

Therefore, by repeated use of (9) we obtain

$$P_{i,j}(x) = \sum_{l=i+1}^{j} [z_l + c_{l-1,2}(x - x_l)]M_{l,\mathrm{B}_i} = - \sum_{l=j+1}^{i+3} [z_l + c_{l-1,2}(x - x_l)]M_{l,\mathrm{B}_i}.$$

In particular, the following identity is valid,

$$\sum_{j=i+1}^{i+3} [z_j + c_{j-1,2}(x - x_j)]M_{j,\mathrm{B}_i} \equiv 0,$$

from which one obtains the equalities

$$\sum_{j=i+1}^{i+3} c_{j-1,2}y_j^r M_{j,\mathrm{B}_i} = 0, \quad r = 0, 1, \quad y_j = x_j - \frac{z_j}{c_{j-1,2}}. \qquad (10)$$

Thus the formula for the discrete GB-spline B_i takes the form

$$\mathrm{B}_i(x) = \begin{cases} \Psi_i(x)M_{i+1,\mathrm{B}_i}, & x \in [x_i, x_{i+1}), \\ (x - y_{i+1})c_{i,2}M_{i+1,\mathrm{B}_i} + \Phi_{i+1}(x)M_{i+1,\mathrm{B}_i} + \Psi_{i+1}(x)M_{i+2,\mathrm{B}_i}, \\ \qquad x \in [x_{i+1}, x_{i+2}), \\ (y_{i+3} - x)c_{i+2,2}M_{i+3,\mathrm{B}_i} + \Phi_{i+2}(x)M_{i+2,\mathrm{B}_i} + \Psi_{i+2}(x)M_{i+3,\mathrm{B}_i}, \\ \qquad x \in [x_{i+2}, x_{i+3}), \\ \Phi_{i+3}(x)M_{i+3,\mathrm{B}_i}, & x \in [x_{i+3}, x_{i+4}), \\ 0, & \text{otherwise}. \end{cases}$$

$$(11)$$

After substituting formula (11) into the normalization condition (7) written for $x \in [x_i, x_{i+1}]$, $i = 0, \ldots, N-1$, we obtain

$$\sum_{j=i-3}^{i} B_j(x) = \Phi_i(x) \sum_{j=i-3}^{i-1} M_{i,B_j} + \Psi_i(x) \sum_{j=i-2}^{i} M_{i+1,B_j}$$
$$+ (y_{i+1} - x)c_{i,2} M_{i+1,B_{i-2}} + (x - y_i)c_{i-1,2} M_{i,B_{i-1}} \equiv 1.$$

Due to the linear independence of functions 1, x, Φ_i, and Ψ_i on $[x_i, x_{i+1}]$, the latter relation is satisfied if and only if

$$\sum_{j=i-3}^{i-1} M_{i,B_j} = \sum_{j=i-2}^{i} M_{i+1,B_j} = 0, \tag{12}$$

$$y_{i+1}c_{i,2} M_{i+1,B_{i-2}} - y_i c_{i-1,2} M_{i,B_{i-1}} = 1,$$
$$c_{i,2} M_{i+1,B_{i-2}} - c_{i-1,2} M_{i,B_{i-1}} = 0. \tag{13}$$

In particular, from (13) we derive the identity

$$(y_{i+1} - x)c_{i,2} M_{i+1,B_{i-2}} + (x - y_i)c_{i-1,2} M_{i,B_{i-1}} \equiv 1.$$

Solving system (13) and using (10) or (12), we obtain

$$M_{j,B_i} = \frac{y_{i+3} - y_{i+1}}{c_{j-1,2}\omega'_{i+1}(y_j)}, \qquad j = i+1, i+2, i+3,$$
$$\omega_{i+1}(x) = (x - y_{i+1})(x - y_{i+2})(x - y_{i+3})$$

or with the notation $c_{j,3} = y_{j+2} - y_{j+1}$, $j = i, i+1$,

$$M_{i+1,B_i} = \frac{1}{c_{i,2}c_{i,3}},$$
$$M_{i+2,B_i} = -\frac{1}{c_{i+1,2}}\left(\frac{1}{c_{i,3}} + \frac{1}{c_{i+1,3}}\right), \tag{14}$$
$$M_{i+3,B_i} = \frac{1}{c_{i+2,2}c_{i+1,3}}.$$

§4. Properties of Discrete GB-Splines

The functions $B_{-3}, \ldots, B_{N-1}$ possess many of the properties inherent in the usual discrete polynomial B-splines. To provide inequality (6), in what follows we need to impose additional conditions on the functions Φ_j and Ψ_j which, as the reader may readily check, are satisfied by all the defining functions given in Section 8. The proofs of the following four assertions repeat those given in [5].

Lemma 1. *Let the conditions*

$$0 < 2\Phi_j(x_j) < -h_i D_{j,1}\Phi_j(x_j), \quad 0 < 2\Psi_j(x_{j+1}) < h_j D_{j,1}\Psi_j(x_{j+1}),$$

$j = i+1, i+2, i+3$, *be satisfied. Then in (14)*

$$c_{j,k} > 0, \quad j = i, \dots, i+4-k; \quad k = 2, 3,$$

and, therefore,

$$(-1)^{j-i-1} M_{j,B_i} > 0, \quad j = i+1, i+2, i+3.$$

Theorem 1. *Let the conditions of Lemma 1 be satisfied, the functions* Φ_j *and* Ψ_j *be convex and* $D_{j,2}\Phi_j$ *and* $D_{j,2}\Psi_j$ *be strictly monotone on the interval* $[x_j, x_{j+1}]$ *for all* j. *Then the functions* $B_{-3}, \dots, B_{N-1}$ *have the following properties:*

(a) $B_j(x) > 0$ *for* $x \in (x_j + \tau_j^{R_j}, x_{j+4} - \tau_{j+4}^{L_{j+3}})$, *and* $B_j(x) \equiv 0$ *if* $x \notin (x_j, x_{j+4})$;

(b) B_j *satisfies the smoothness conditions (4)*;

(c) $\Phi_j(x) = c_{j-1,2} c_{j-2,3} B_{j-3}(x)$, $\Psi_j(x) = c_{j,2} c_{j,3} B_j(x)$ *for* $x \in [x_j, x_{j+1}]$, $j = 0, \dots, N-1$, *and*

$$\sum_{j=-3}^{N-1} y_{j+2}^r B_j(x) \equiv x^r, \quad r = 0, 1 \quad \text{for} \quad x \in [a, b]. \tag{15}$$

Lemma 2. *The functions* $B_{-3}, \dots, B_{N-1}$ *are splines from* S_4^{DG} *with finite supports of minimal length.*

Theorem 2. *The functions* $B_{-3}, \dots, B_{N-1}$ *are linearly independent on* $[a, b]$ *and form a basis of the space of discrete generalized splines* S_4^{DG}.

§5. Local Approximation by Discrete GB-Splines

According to Theorem 2, any discrete generalized spline $S \in S_4^{DG}$ can be uniquely written in the form

$$S(x) = \sum_{j=-3}^{N-1} b_j B_j(x), \quad x \in [a, b] \tag{16}$$

for some coefficients b_j.

If the coefficients b_j in (16) are known, then by virtue of formula (11) we can write out an expression for the discrete generalized spline S on the interval $[x_i, x_{i+1}]$, which is convenient for calculations,

$$S(x) = b_{i-2} + b_{i-1}^{(1)}(x - y_i) + b_{i-1}^{(2)}\Phi_i(x) + b_i^{(2)}\Psi_i(x), \tag{17}$$

where

$$b_j^{(k)} = \frac{b_j^{(k-1)} - b_{j-1}^{(k-1)}}{c_{j,4-k}}, \quad k = 1, 2; \quad b_j^{(0)} = b_j. \tag{18}$$

The representations (16) and (17) allow us to find a simple and effective way to approximate a given continuous function f from its samples.

Theorem 3. *Let a continuous function* f *be given by its samples* $f(y_j)$, $j = -1, \ldots, N + 1$, *where* y_j *is defined in* (10). *Then for* $b_j = f(y_{j+2})$, $j = -3, \ldots, N-1$, *formula* (16) *is exact for linear polynomials and provides a formula for local approximation.*

Proof: It suffices to employ the identities (15). Inserting the coefficients $b_{j-2} = 1$ and $b_{j-2} = y_j$ in formula (16), and using the identities (15), we prove the first assertion of the theorem.

For $b_{j-2} = f(y_j)$, formula (17) can be rewritten as

$$S(x) = f(y_i) + f[y_i, y_{i+1}](x - y_i) + (y_{i+1} - y_{i-1})f[y_{i-1}, y_i, y_{i+1}]c_{i-1,2}^{-1}\Phi_i(x)$$
$$+ (y_{i+2} - y_i)f[y_i, y_{i+1}, y_{i+2}]c_{i,2}^{-1}\Psi_i(x), \qquad x \in [x_i, x_{i+1}].$$

This is the formula of local approximation. The theorem is thus proved. □

Corollary 1. *Let a continuous function* f *be given by its samples* $f_j = f(x_j)$, $j = -2, \ldots, N + 2$. *Then by setting*

$$b_{j-2} = f_j - \left(\Psi_{j-1}(x_j)f[x_j, x_{j+1}] - \Phi_j(x_j)f[x_{j-1}, x_j]\right)c_{j-1,2}^{-1} \qquad (19)$$

in (16), *we obtain a formula of three-point local approximation, which is exact for linear polynomials.*

Proof: It suffices to check the result for the monomials 1 and x. Then according to (19), we obtain $b_{j-2} = 1$ and $b_{j-2} = y_j$, and it only remains to make use of the identities (15). This proves the corollary. □

Equation (17) permits us to write the coefficients of the spline S in its representation (16) in the form

$$b_{j-2} = \begin{cases} S(y_j) - D_{j-1,2}S(x_{j-1})\Phi_{j-1}(y_j) - D_{j,2}S(x_j)\Psi_{j-1}(y_j), & y_j < x_j, \\ S(y_j) - D_{j,2}S(x_j)\Phi_j(y_j) - D_{j+1,2}S(x_{j+1})\Psi_j(y_j), & y_j \geq x_j. \end{cases}$$

According to this formula we have $b_{j-2} = S(y_j) + O(\overline{h}_j^2)$, $\overline{h}_j = \max(h_{j-1}, h_j)$. Hence it follows that the control polygon (e.g., see [3]) converges quadratically to the function f when $b_{j-2} = f(y_j)$, or if the formula (19) is used.

§6. Recurrence Formulae for Discrete GB-Splines

Let us define functions

$$B_{j,2}(x) = \begin{cases} D_{j,2}\Psi_j(x), & x \in [x_j, x_{j+1}), \\ D_{j+1,2}\Phi_{j+1}(x), & x \in [x_{j+1}, x_{j+2}], \quad j = i, i+1, i+2. \quad (20) \\ 0, & \text{otherwise}, \end{cases}$$

We assume that the functions $D_{j,2}\Psi_j$ and $D_{j+1,2}\Phi_{j+1}$ are strictly monotone on $[x_j, x_{j+1})$ and $[x_{j+1}, x_{j+2}]$, respectively. The splines $\mathrm{B}_{j,2}$ are a generalization of the "hat-functions" for polynomial B-splines. They are nonnegative, and furthermore, $\mathrm{B}_{j,2}(x_j) = \mathrm{B}_{j,2}(x_{j+2}) = 0$, $\mathrm{B}_{j,2}(x_{j+1}) = 1$. Let us denote

$$D_1 S(x) \equiv D_{i,1} S_i(x), \\ D_2 S(x) \equiv D_{i,2} S_i(x), \quad x \in [x_i, x_{i+1}], \quad i = 0, \ldots, N-1;$$

then from (4) $D_1 S$ and $D_2 S$ are well defined if $S \in S_4^{DG}$. With the previous notation, according to (11), (14) and (20) we obtain

$$D_2 \mathrm{B}_i(x) = \sum_{j=i+1}^{i+3} M_{j,\mathrm{B}_i} \mathrm{B}_{j-1,2}(x)$$

$$= \frac{1}{c_{i,3}} \left(\frac{\mathrm{B}_{i,2}(x)}{c_{i,2}} - \frac{\mathrm{B}_{i+1,2}(x)}{c_{i+1,2}} \right) - \frac{1}{c_{i+1,3}} \left(\frac{\mathrm{B}_{i+1,2}(x)}{c_{i+1,2}} - \frac{\mathrm{B}_{i+2,2}(x)}{c_{i+2,2}} \right). (21)$$

In addition, the function $D_1 \mathrm{B}_i$ satisfies the relation

$$D_1 \mathrm{B}_i(x) = \frac{\mathrm{B}_{i,3}(x)}{c_{i,3}} - \frac{\mathrm{B}_{i+1,3}(x)}{c_{i+1,3}}, \tag{22}$$

where

$$\mathrm{B}_{j,3}(x) = \begin{cases} \dfrac{D_{j,1}\Psi_j(x)}{c_{j,2}}, & x \in [x_j, x_{j+1}), \\[2mm] 1 + \dfrac{D_{j+1,1}\Phi_{j+1}(x)}{c_{j,2}} - \dfrac{D_{j+1,1}\Psi_{j+1}(x)}{c_{j+1,2}}, & x \in [x_{j+1}, x_{j+2}), \\[2mm] -\dfrac{D_{j+2,1}\Phi_{j+2}(x)}{c_{j+1,2}}, & x \in [x_{j+2}, x_{j+3}), \\[2mm] 0, & \text{otherwise.} \end{cases}$$

$$\tag{23}$$

Using formula (23), it is easy to show that the functions $\mathrm{B}_{-2,3}, \ldots,$ $\mathrm{B}_{N-1,3}$ satisfy the first and second smoothness conditions in (4), have supports of minimum length, are linearly independent and form a partition of unity:

$$\sum_{j=1}^{N-1} \mathrm{B}_{j,3}(x) \equiv 1, \quad x \in [a, b].$$

Figures 1 and 2 show the graphs of the discrete GB-splines $\mathrm{B}_{j,2}$, $\mathrm{B}_{j,3}$, and B_j (from left to right) on a uniform mesh with step size $h_i = 1$ and with $\tau_j^{L_i} = \tau_j^{R_i} = \tau$, $j = i, i+1$, for all i. We have chosen discretization

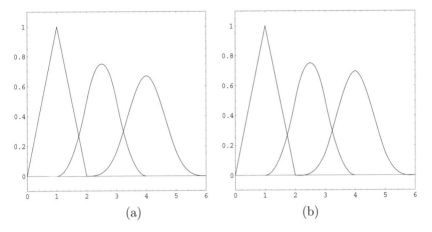

Fig. 1. The discrete GB-splines $B_{j,k}$, $k = 2, 3, 4$ (from left to right) on a uniform mesh with step size $h_i = 1$, no tension and discretization parameter $\tau = 0.1$ (a) and $\tau = 0.33$ (b).

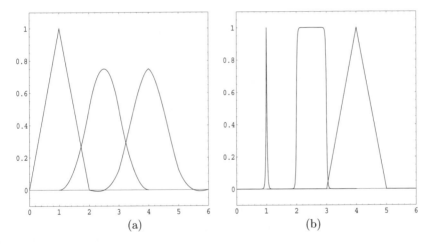

Fig. 2. Same as Figure 1, but with discretization parameter $\tau = 0.5$ (a) and with tension parameters $q_i = 50$, for all i (b).

parameter $\tau = 0.1$ (Fig. 1(a) and Fig. 2(b)), $\tau = 0.33$ (Fig. 1(b)) and $\tau = 0.5$ (Fig. 2(a)) for

$$\Psi_i(x) = \psi_i(q_i, t)h_i^2 = \frac{\hat{\tau}_i \sinh q_i t - t \sinh(q_i \hat{\tau}_i)}{\frac{4}{\hat{\tau}_i} \sinh^2 \frac{q_i \hat{\tau}_i}{2} \sinh q_i} h_i^2, \quad \hat{\tau}_i = \frac{\tau}{h_i},$$

$$\Phi_i(x) = \psi_i(q_i, 1 - t)h_i^2.$$

This is a special case of Example 4 in Section 8. In Figures 1 and 2(a) we

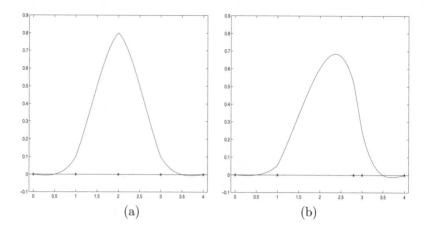

Fig. 3. The discrete GB-splines $B_{j,4}$ on a uniform mesh (a) and on a non-uniform mesh (b). The asterisk $*$ denotes the x_i. For both plots $q_i = 2$ and $\tau = 0.5$.

have parameters $q_i = 0$, i.e. we have conventional discrete cubic B-splines (e.g., see [8]). Visually, the presence of intervals where the B-spline B_j is negative is more visible with growing discretization parameter τ. In Figure 2(b) the tension parameters are $q_i = 50$ for all i, whence the shape of the graphs is practically unchanged when τ increases from 0.1 to 0.5. As the limit for $q_i \to \infty$ we obtain the pulse function for $B_{j,2}$, the "step-function" for $B_{j,3}$ and the "hat-function" for B_j (all of height 1).

Figure 3 shows the graphs of discrete GB-splines $B_{j,4}$ on a uniform mesh (left) and on a nonuniform mesh (right), where the asterisk $*$ denotes the x_i. For both plots $q_i = 2$ and $\tau = 0.5$.

Applying formulae (21) and (22) to the representation (16), we also obtain

$$D_1 S(x) = \sum_{j=-2}^{N-1} b_j^{(1)} B_{j,3}(x), \quad D_2 S(x) = \sum_{j=-1}^{N-1} b_j^{(2)} B_{j,2}(x), \quad (24)$$

where $b_j^{(k)}$, $k = 1, 2$ are defined in (18).

§7. Series of Discrete GB-Splines (Uniform Case)

Let us suppose that each step size $h_i = x_{i+1} - x_i$ of the mesh $\Delta : a = x_0 < x_1 < \cdots < x_N = b$ is an integer multiple of the same tabulation step, τ, of some uniform mesh refinement on $[a, b]$. For $\theta \in \mathbb{R}$, $\tau > 0$ define $\mathbb{R}_{\theta\tau} = \{\theta + i\tau : i \text{ is an integer}\}$ and let $\mathbb{R}_{\theta 0} = \mathbb{R}$. For any $a, b \in \mathbb{R}$ and $\tau > 0$ let $[a, b]_\tau = [a, b] \cap \mathbb{R}_{a\tau}$.

The functions $B_{j,2}$, $B_{j,3}$, and B_j with $\tau_j^{L_i} = \tau_j^{R_i} = \tau$, $j = i, i+1$ for all i are nonnegative on the discrete interval $[a, b]_\tau$. This permits us to reprove the main results for discrete polynomial splines in [9] for series of discrete generalized splines. In particular, if in (16) and (24) we have coefficients $b_j^{(k)} > 0$, $k = 0, 1, 2$, $j = -3 + k, \ldots, N - 1$, then the spline S will be a positive, monotonically increasing and convex function on $[a, b]_\tau$.

Denote by $\operatorname{supp}_\tau B_i = \{x \in \mathbb{R}_{a,\tau} | B_i(x) > 0\}$ the discrete support of the spline B_i, i.e. the discrete set $(x_i + \tau, x_{i+4} - \tau)_\tau$.

Theorem 4. *Assume that $\zeta_{-3} < \zeta_{-2} < \cdots < \zeta_{N-1}$ are prescribed points on the discrete line $\mathbb{R}_{a,\tau}$. Then*

$$D = \det\left(B_i(\zeta_j)\right) \geq 0, \quad i, j = -3, \ldots, N - 1$$

and strict positivity holds if and only if

$$\zeta_i \in \operatorname{supp}_\tau B_i, \quad i = -3, \ldots, N - 1. \tag{25}$$

The proof of this theorem repeats that of Theorem 8.66 in [9, p. 355]. The following three statements follow immediately from Theorem 4.

Corollary 2. *The system of discrete GB-splines $B_{-3}, \ldots, B_{N-1}$ associated with knots on $\mathbb{R}_{a,\tau}$ is a weak Chebyshev system according to the definition given in [9, p. 36], i.e. for any $\zeta_{-3} < \zeta_{-2} < \cdots < \zeta_{N-1}$ in $\mathbb{R}_{a,\tau}$ we have $D \geq 0$ and $D > 0$ if and only if condition (25) is satisfied. In the latter case the discrete generalized spline $S(x) = \sum_{j=-3}^{N-1} b_j B_j(x)$ has no more than $N + 2$ zeros.*

Corollary 3. *If the conditions of Theorem 4 are satisfied, then the solution of the interpolation problem*

$$S(\zeta_i) = f_i, \quad i = -3, \ldots, N - 1, \quad f_i \in \mathbb{R} \tag{26}$$

exists and is unique.

Let $A = \{a_{ij}\}$, $i = 1, \ldots, m$, $j = 1, \ldots, n$, be a rectangular $m \times n$ matrix with $m \leq n$. The matrix A is said to be totally nonnegative (totally positive) (e.g., see [4]) if the minors of all order of the matrix are nonnegative (positive), i.e. for all $1 \leq p \leq m$, we have

$$\det\left(a_{i_k j_l}\right) \geq 0 \ (> 0) \quad \text{for all} \quad \begin{array}{l} 1 \leq i_1 < \cdots < i_p \leq m, \\ 1 \leq j_1 < \cdots < j_p \leq n. \end{array}$$

Corollary 4. *For arbitrary integers $-3 \leq \nu_{-3} < \cdots < \nu_{p-4} \leq N - 1$ and $\zeta_{-3} < \zeta_{-2} < \cdots < \zeta_{p-4}$ in $\mathbb{R}_{a,\tau}$ we have*

$$\bar{D}_p = \det\{B_{\nu_i}(\zeta_j)\} \geq 0, \quad i, j = -3, \ldots, p - 4$$

and strict positivity holds if and only if

$$\zeta_i \in \mathrm{supp}_\tau B_{\nu_i}, \quad i = -3, \dots, p-4$$

i.e. the matrix $\{B_j(\zeta_i)\}_{i,j=-3,\dots,N-1}$ is totally nonnegative.

The last statement is proved by induction based on Theorem 4 and the recurrence relations for the minors of the matrix $\{B_j(\zeta_i)\}$. The proof does not differ from that of Theorem 8.67 described in [9, p. 356].

Since the supports of discrete GB-splines are finite, the matrix of system (26) is banded and has seven nonzero diagonals in general. The matrix is tridiagonal if $\zeta_i = x_{i+2}$, $i = -3, \dots, N-1$.

An important particular case of the problem in which $S'(x_i) = f'_i$, $i = 0, N$, can be obtained by passing to the limit as $\zeta_{-3} \to \zeta_{-2}$, $\zeta_{N-1} \to \zeta_{N-2}$.

de Boor and Pinkus [1] proved that linear systems with totally nonnegative matrices can be solved by Gaussian elimination without pivoting. Thus, the system (26) can be solved effectively by the conventional Gauss method.

Denote by $S^-(\mathbf{v})$ the number of sign changes (variations) in the sequence of components of the vector $\mathbf{v} = (v_1, \cdots, v_n)$, with zeros being neglected. Karlin [4] showed that if a matrix A is totally nonnegative, then it decreases the variation, i.e.

$$S^-(A\mathbf{v}) \le S^-(\mathbf{v}).$$

By Corollary 4, the totally nonnegative matrix $\{B_j(\zeta_i)\}_{i,j=-3,\dots,N-1}$ formed by discrete GB-splines decreases the variation.

For a bounded real function f, let $S^-(f)$ be the number of sign changes of the function f on the real axis $\mathbb{R}$, without taking into account the zeros

$$S^-(f) = \sup_n S^-[f(\zeta_1), \dots, f(\zeta_n)], \quad \zeta_1 < \zeta_2 < \cdots < \zeta_n.$$

Theorem 5. *The discrete generalized spline $S(x) = \sum_{j=-3}^{N-1} b_j B_j(x)$ is a variation diminishing function, i.e. the number of sign changes of S does not exceed that in the sequence of its coefficients:*

$$S^- \left(\sum_{j=-3}^{N-1} b_j B_j \right) \le S^-(\mathbf{b}), \quad \mathbf{b} = (b_{-3}, \dots, b_{N-1}).$$

The proof of this statement is the same as that of Theorem 8.68 for discrete polynomial B-splines in [9, p. 356].

§8. Examples of Defining Functions

Let us give some choices of the defining functions Φ_i and Ψ_i for discrete generalized splines that conform to the sufficiency conditions derived earlier in the paper. Putting

$$\Psi_i(x) = \psi_i(t)h_i^2 = \psi(q_i, \hat{\tau}_i^{L_i}, \hat{\tau}_i^{R_i}, t)h_i^2, \quad \Phi_i(x) = \psi(p_i, \hat{\tau}_{i+1}^{R_i}, \hat{\tau}_{i+1}^{L_i}, 1-t)h_i^2,$$

$$\hat{\tau}_j^{L_i} = \tau_j^{L_i}/h_i, \quad \hat{\tau}_j^{R_i} = \tau_j^{R_i}/h_i; \quad j = i, i+1; \quad 0 \le p_i, q_i < \infty,$$

we consider some possibilities for choosing the functions ψ_i which, due to the constraints (1), satisfy the conditions

$$\psi_i(-\hat{\tau}_i^{L_i}) = \psi_i(0) = \psi_i(\hat{\tau}_i^{R_i}) = 0, \quad D_{i+1,2}\psi_i(1) = h_i^{-2}. \tag{27}$$

1) Discrete rational spline with linear denominator:

$$\psi_i(t) = C_i \frac{(t + \hat{\tau}_i^{L_i})t(t - \hat{\tau}_i^{R_i})}{1 + q_i(1-t)}.$$

2) Discrete rational spline with quadratic denominator:

$$\psi_i(t) = C_i \frac{(t + \hat{\tau}_i^{L_i})t(t - \hat{\tau}_i^{R_i})}{1 + q_i t(1-t)}.$$

3) Discrete exponential spline:

$$\psi_i(t) = C_i(t + \hat{\tau}_i^{L_i})t(t - \hat{\tau}_i^{R_i})exp\big(-q_i(1-t)\big).$$

4) Discrete hyperbolic spline:

$$\psi_i(t) = C_{i,1}\Big[\sinh q_i t - t\frac{\sinh q_i \hat{\tau}_i^{R_i}}{\hat{\tau}_i^{R_i}}\Big] + C_{i,2}\Big[\cosh q_i t - 1 - t\frac{\cosh q_i \hat{\tau}_i^{R_i} - 1}{\hat{\tau}_i^{R_i}}\Big].$$

5) Discrete cubic spline with additional knots:

$$\psi_i(t) = \frac{1}{2}\frac{(t - \beta_i + \hat{\tau}_i^{L_i})(t - \beta_i)_+(t - \beta_i - \hat{\tau}_i^{R_i})}{3(1 - \beta_i) + \hat{\varepsilon}_{i+1} - \hat{\varepsilon}_i},$$

$$\hat{\varepsilon}_j = \hat{\tau}_j^{R_i} - \hat{\tau}_j^{L_i}, \quad j = i, i+1; \quad \beta_i = 1 - (1 + q_i)^{-1}, \quad E_+ = \max(0, E).$$

The points $x_i + \alpha_i h_i$ ($\alpha_i = (1 + p_i)^{-1}$) and $x_i + \beta_i h_i$ fix the position of two additional knots of the spline on the interval $[x_i, x_{i+1}]$. By moving these knots one can perform a transfer from a discrete cubic spline to piecewise linear interpolation.

6) Discrete spline of variable order:

$$\psi_i(t) = C_i(t + \hat{\tau}_i^{L_i})t^{k_i}(t - \hat{\tau}^{R_i}), \quad k_i = 1 + q_i.$$

The constants C_i in the expressions for the function ψ_i above are calculated from the condition (27) for the second divided difference of ψ_i. To find $C_{i,k}$, $k = 1, 2$, one needs additionally to use the condition $\psi_i(-\hat{\tau}_i^{L_i}) = 0$. It is easy to check that in all cases we get the corresponding defining functions in [5] by setting $\hat{\tau}_j^{L_i} = \hat{\tau}_j^{R_i} = 0$, $j = i, i+1$.

Acknowledgments. The author was supported by the Thailand Research Fund under grant number BRG/08/2543. He also wishes to express his gratitude to both the referee and the editor for their invaluable remarks and suggestions for improvements.

References

1. de Boor, C., and A. Pinkus, Backward error analysis for totally positive linear systems, Numer. Math. **27** (1977), 485–490.

2. Costantini, P., B. I. Kvasov, and C. Manni, On discrete hyperbolic tension splines, Advances in Computational Mathematics **11** (1999), 331–354.

3. Hoschek, J., and D. Lasser, *Fundamentals of Computer Aided Geometric Design*, A K Peters, Welesley, Massachusetts, 1993.

4. Karlin, S., *Total Positivity, Vol. 1*, Stanford University Press, Stanford, CA, 1968.

5. Kvasov, B. I., Local bases for generalized cubic splines, Russ. J. Numer. Anal. Math. Modelling **10** (1995), 49–80.

6. Kvasov, B. I., GB-splines and their properties, Annals of Numerical Mathematics **3** (1996), 139–149.

7. Kvasov, B. I., *Methods of Shape-Preserving Spline Approximation*, World Scientific Publ. Co. Pte. Ltd., Singapore, 2000.

8. Lyche, T., Discrete cubic spline interpolation, BIT **16** (1976), 281–290.

9. Schumaker, L. L., *Spline Functions: Basic Theory*, Wiley, New York, 1981.

Boris I. Kvasov
School of Mathematics
Suranaree University of Technology
111 University Avenue
Nakhon Ratchasima 30000, Thailand
boris@math.sut.ac.th

Parameterizations of Univariate Orthogonal Wavelets with Short Support

Ming-Jun Lai and David W. Roach

Abstract. In this note, we give a complete and simple parameterization for the length-six one-dimensional filters which satisfy the necessary conditions for orthogonality. This parameterization includes all compactly supported univariate scaling functions contained within the interval $[0,5]$ using dilation factor 2. These formulae are a convenient way to generate a continuum of wavelets with varying approximation properties ranging from scaling functions which only reproduce constants to ones which reproduce linears and quadratics. In addition, solutions for the length-eight and length-ten filters are given where the parameters have transcendental constraints. We conclude this note with some interesting numerical experiments comparing the parameterized wavelets of length six with the standard Haar, D4, and D6 wavelets in an image compression scheme.

§1. Introduction

The use of compactly supported orthonormal wavelets is common in various applications. The Daubechies orthonormal wavelets are an important class which have been used extensively for applications such as image compression. The construction of wavelets usually begins with a scaling function ϕ which satisfies various properties such as refinability and orthogonality. Numerous necessary and sufficient conditions have been developed to aid in the construction of wavelets (see [1,6], and many others). One method for constructing compactly supported scaling functions is to begin with the dilation equation

$$\phi(x) = \sum_k h_k \phi(2x - k)$$

Approximation Theory X: Wavelets, Splines, and Applications   369
Charles K. Chui, Larry L. Schumaker, and Joachim Stöckler (eds.), pp. 369–384.
Copyright © 2002 by Vanderbilt University Press, Nashville, TN.
ISBN 0-8265-1416-2.

with a finite number of coefficients. In order for a function of this type to
be orthonormal to its integer shifts, the coefficients in the dilation equation
must satisfy

$$m(1) = 1$$
$$|m(z)|^2 + |m(-z)|^2 = 1, \quad z = e^{i\omega},$$

where $m(z) = \frac{1}{2}\sum_k h_k z^k$ is the associated trigonometric polynomial. For
scaling functions with support contained in $[0,5]$ these necessary conditions suggest a simple technique for parameterizing their solution. A different approach in [8] gives the explicit parameterization of the length-four
filters and a constructive technique for longer filters. The method shown
here is the same technique we used to parameterize compactly supported
bivariate scaling functions(see [2,3,4,5]). Another approach for dealing
with the bivariate case can be seen in [7].

   As will be shown, the necessary conditions for orthogonality yield a
set of linear and nonlinear equations which necessarily imply that various sums of dilation coefficients form perfect squares. This leads to the
introduction of the free parameters. In Section 2, we give the parameterization of the length-four filters for completeness. In Section 3, we derive
the explicit parameterization for the length-six filters. Section 4 gives the
constrained solutions for the length-eight and length-ten filters. Finally,
we conclude in Section 5 with an interesting numerical experiment comparing the standard Haar, D4, and D6 wavelets with the parameterized
wavelets of length-six in an image compression scheme.

§2. Length-Four Solution

The formulas for the parameterization of the length-four filters were given
in [8], but we include them here for completeness. Let $H_4(z) = a_0 + b_0 z +
a_1 z^2 + b_1 z^3$.

Lemma 1. $H_4(z)$ satisfies $H_4(1) = 1$ and

$$|H_4(z)|^2 + |H_4(-z)|^2 = 1, \qquad \forall z = e^{i\omega} \text{ and } \omega \in \mathbb{R}$$

if and only if

$$a_0 = \frac{1}{4} + \frac{1}{2\sqrt{2}}\cos\alpha, \quad b_0 = \frac{1}{4} + \frac{1}{2\sqrt{2}}\sin\alpha,$$

$$a_1 = \frac{1}{4} - \frac{1}{2\sqrt{2}}\cos\alpha, \quad b_1 = \frac{1}{4} - \frac{1}{2\sqrt{2}}\sin\alpha,$$

for any $\alpha \in \mathbb{R}$.

Proof: First, we have

$$|H_4(z)|^2 = a_0^2 + b_0^2 + a_1^2 + b_1^2$$
$$+ (a_0 b_0 + b_0 a_1 + a_1 b_1)(z + z^{-1})$$
$$+ (a_0 a_1 + b_0 b_1)(z^2 + z^{-2}) + a_0 b_1(z^3 + z^{-3}).$$

Then $|H_4(z)|^2 + |H_4(-z)|^2 = 1$ implies that

$$a_0^2 + a_1^2 + b_0^2 + b_1^2 = \tfrac{1}{2}$$
$$a_0 a_1 + b_0 b_1 = 0. \tag{1}$$

It follows from $H_4(1) = 1$ that $a_0 + a_1 + b_0 + b_1 = 1$. Also, $H(-1) = 0$ implies that $a_0 + a_1 - b_0 - b_1 = 0$. Thus, we have

$$a_0 + a_1 = b_0 + b_1 = \frac{1}{2}. \tag{2}$$

Moreover, the two nonlinear equations from (1) imply

$$(a_0 - a_1)^2 + (b_0 - b_1)^2 = \frac{1}{2}$$

or

$$a_0 - a_1 = \frac{1}{\sqrt{2}} \cos \alpha, \quad b_0 - b_1 = \frac{1}{\sqrt{2}} \sin \alpha.$$

Thus, after combining these equations with equation (2), we have the formulas as stated in the lemma.

On the other hand, using the expressions for the a_i's and b_i's with a straightforward calculation shows that in fact

$$|H_4(z)|^2 + |H_4(-z)|^2 = 1.$$

This completes the proof. □

Example 1. When $\alpha = \frac{\pi}{4}$, we get $H_4(z) = \frac{1+z}{2}$, which is associated with the Haar wavelet.

Example 2. When $\alpha = \frac{5\pi}{12}$, we get $a_0 = \frac{1+\sqrt{3}}{8}$ and $a_1 = \frac{3-\sqrt{3}}{8}$ as well as $b_0 = \frac{3+\sqrt{3}}{8}$, $b_1 = \frac{1-\sqrt{3}}{8}$. Then $H_4(z)$ is associated with the Daubechies D4 wavelet.

Next, we look for choices of α where $H_4(z)$ has a second-order vanishing moment. That is, $H_4(z) = \left(\frac{1+z}{2}\right)^2 p(z)$ where $p(z)$ is some trigonometric polynomial.

In this case $\frac{d}{dz}H(z)|_{z=-1} = 0$ which is equivalent to $b_0 - 2a_1 + 3b_1 = 0$. That is,

$$\begin{aligned}
0 &= b_0 - 2a_1 + 3b_1 \\
&= \frac{1}{4} + \frac{1}{2\sqrt{2}} \sin \alpha - \frac{1}{2} + \frac{1}{\sqrt{2}} \cos \alpha + \frac{3}{4} - \frac{3}{2\sqrt{2}} \sin \alpha \\
&= \frac{1}{2} + \frac{1}{\sqrt{2}} (\cos \alpha - \sin \alpha) \\
&= \frac{1}{2} + \cos \left(\alpha + \frac{\pi}{4} \right),
\end{aligned}$$

which is only satisfied by $\alpha = \frac{5\pi}{12}$ or $\alpha = \frac{13\pi}{12}$. Both are associated with Daubechies' D4 wavelet. Thus, D4 is the only member of the family with two vanishing moments, which is a well-known fact.

§3. Length-Six Solution

Let $H_6(z) = a_0 + b_0 z + a_1 z^2 + b_1 z^3 + a_2 z^4 + b_2 z^5$ be a trigonometric polynomial of $z = e^{i\omega}$.

Lemma 2. $H_6(z)$ satisfies $H_6(1) = 1$ and

$$|H_6(z)|^2 + |H_6(-z)|^2 = 1, \forall z = e^{i\omega}, \qquad \omega \in \mathbb{R}$$

if and only if

$$a_0 = \frac{1}{8} + \frac{1}{4\sqrt{2}} \cos\alpha + \frac{p}{2} \cos\beta$$

$$a_1 = \frac{1}{4} - \frac{1}{2\sqrt{2}} \cos\alpha$$

$$a_2 = \frac{1}{8} + \frac{1}{4\sqrt{2}} \cos\alpha - \frac{p}{2} \cos\beta$$

$$b_0 = \frac{1}{8} + \frac{1}{4\sqrt{2}} \sin\alpha + \frac{p}{2} \sin\beta$$

$$b_1 = \frac{1}{4} - \frac{1}{2\sqrt{2}} \sin\alpha$$

$$b_2 = \frac{1}{8} + \frac{1}{4\sqrt{2}} \sin\alpha - \frac{p}{2} \sin\beta,$$

where

$$p = \frac{1}{2}\sqrt{1 + \sin(\alpha + \frac{\pi}{4})}$$

for any $\alpha, \beta \in \mathbb{R}$.

Proof: First of all, $H_6(1) = 1$ implies that $\sum_{i=0}^{2} a_i + b_i = 1$. Also, $H_6(-1) = 0$ implies that

$$\sum_{i=0}^{2} a_i - \sum_{i=0}^{2} b_i = 0.$$

It follows that

$$a_0 + a_1 + a_2 = b_0 + b_1 + b_2 = \frac{1}{2}. \tag{3}$$

Next, $|H_6(z)|^2 + |H_6(-z)|^2 = 1$ implies that

$$a_0^2 + a_1^2 + a_2^2 + b_0^2 + b_1^2 + b_2^2 = \frac{1}{2}$$

$$a_0 a_1 + a_1 a_2 + b_0 b_1 + b_1 b_2 = 0 \tag{4}$$

$$a_0 a_2 + b_0 b_2 = 0.$$

From these nonlinear equations, it follows that

$$(a_0 - a_1 + a_2)^2 + (b_0 - b_1 + b_2)^2 = \frac{1}{2},$$

and

$$a_0 - a_1 + a_2 = \frac{1}{\sqrt{2}} \cos \alpha, \quad b_0 - b_1 + b_2 = \frac{1}{\sqrt{2}} \sin \alpha. \qquad (5)$$

Combining (5) with (3), we have

$$a_1 = \frac{1}{4} - \frac{1}{2\sqrt{2}} \cos \alpha, \qquad b_1 = \frac{1}{4} - \frac{1}{2\sqrt{2}} \sin \alpha$$

$$\qquad\qquad\qquad\qquad\qquad\qquad\qquad\qquad (6)$$

$$a_0 + a_2 = \frac{1}{4} + \frac{1}{2\sqrt{2}} \cos \alpha, \quad b_0 + b_2 = \frac{1}{4} + \frac{1}{2\sqrt{2}} \cos \alpha.$$

Moreover,

$$
\begin{aligned}
(a_0 - a_2)^2 + (b_0 - b_2)^2 &= \frac{1}{2} - a_1^2 - b_1^2 \\
&= \frac{1}{2} - (\frac{1}{4} - \frac{1}{4\sqrt{2}}(\cos \alpha + \sin \alpha)) \\
&= \frac{1}{4} + \frac{1}{4\sqrt{2}}(\cos \alpha + \sin \alpha) \\
&= \frac{1}{4} + \frac{1}{4} \sin(\alpha + \frac{\pi}{4}),
\end{aligned}
$$

which implies that

$$a_0 - a_2 = \frac{1}{2}\sqrt{1 + \sin(\alpha + \frac{\pi}{4})} \cos \beta$$

$$b_0 - b_2 = \frac{1}{2}\sqrt{1 + \sin(\alpha + \frac{\pi}{4})} \sin \beta.$$

The result follows by adding these equations to those from (6).

On the other hand, a routine calculation establishes that $H_6(z)$ with these coefficients satisfies $H(1) = 1$ and $|H_6(z)|^2 + |H_6(-z)|^2 = 1$. This completes the proof. $\square$

Example 3. If $\alpha = \frac{\pi}{4}$ and $\beta = \frac{\pi}{4}$, then $H_6(z)$ is associated with the Haar wavelet.

Example 4. If $\alpha = \frac{5\pi}{12}$ and $\beta = \frac{\pi}{3}$, then $H_6(z)$ is associated with the Daubechies D4 wavelet.

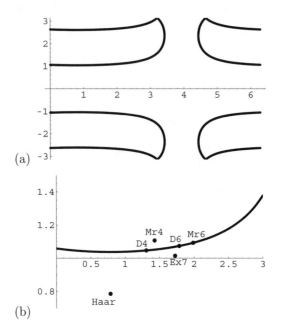

Fig. 1. (a) A plot of the ordered pairs (α, β) for which $H_6(z)$ has two vanishing moments, (b) a zoomed in plot of the above graph showing the positions of Haar, D4, D6, Example 7 (Ex7), and the most-regular length-four (Mr4) and length-six (Mr6) scaling functions.

Example 5. *If*

$$\cos\alpha = -\frac{1}{4}\sqrt{8 + \sqrt{15 + 12\sqrt{10}}}, \quad \sin\alpha = \frac{1}{4}\sqrt{8 - \sqrt{15 + 12\sqrt{10}}},$$

$$\cos\beta = \frac{1}{4}\sqrt{8 - 8\sqrt{-25 + 8\sqrt{10}}}, \quad \sin\beta = \frac{1}{4}\sqrt{8 + 8\sqrt{-25 + 8\sqrt{10}}},$$

then $H_6(z)$ is the filter associated with Daubechies D6 wavelet.

Example 6. *To see when $H_6(z)$ has two vanishing moments, we require* $H'(-1) = 0$ *which implies*

$$\begin{aligned}
0 =& H'(-1) = b_0 - 2a_1 + 3b_1 - 4a_2 + 5b_2 \\
=& 1/2 + 2b_1 + 4b_2 - 2a_1 - 4a_2 \\
=& 1/2 - \sqrt{2(1 + \sin(\alpha + \pi/4))}\sin(\beta - \pi/4)
\end{aligned}$$

or

$$\sin(\alpha + \pi) = \frac{1}{8\sin^2(\beta - \pi/4)} - 1.$$

When the ordered pair (α, β) lies along the curves given in Figure 1(a), $H_6(z)$ *has two vanishing moments.*

Example 7. When $\alpha = 1.725080699801023$ and $\beta = 1.01424683724616$, the scaling function is as shown in Figure 2(b). This scaling function performed better than Haar, D4, and D6 in our numerical experiments as shown in the last Section. It should be noted that this scaling function does not have a second-order vanishing moment as can be seen in Figure 1(b).

Example 8. With $\alpha = 1.4288992721907328$, $\beta = 1.1071487177940904$, $H_6(z)$ is associated with the most-regular length-four filter as given in [1]. When $\alpha = 1.9886461158096038$ and $\beta = 1.0934936891036087$, $H_6(z)$ is associated with the most-regular length-six filter as given in [1]. Each of these scaling functions are shown in Figure 2(c) and 2(d).

§4. Constrained Solutions of Lengths Eight and Ten

The length-eight solution has 4 parameters and a transcendental constraint, and the length-ten solution has 5 parameters and a transcendental constraint as well. We will need the following lemma in the proofs of Lemmas 4 and 5.

Lemma 3. *Suppose* $a, b, c,$ *and* $d \in \mathbb{R}$. *Then* $a^2 + b^2 + c^2 + d^2 = 1$ *if and only if*

$$a = \cos\beta\cos\gamma, \quad b = \cos\beta\sin\gamma, \quad c = \sin\beta\cos\theta, \quad d = \sin\beta\sin\theta$$

for $\beta, \gamma,$ *and* $\theta \in \mathbb{R}$.

Let $H_8(z) = a_0 + b_0 z + a_1 z^2 + b_1 z^3 + a_2 z^4 + b_2 z^5 + a_3 z^6 + b_3 z^7$ be a trigonometric polynomial of $z = e^{i\omega}$.

Lemma 4. $H_8(z)$ *satisfies* $H_8(1) = 1$ *and*

$$|H_8(z)|^2 + |H_8(-z)|^2 = 1, \forall z = e^{i\omega}, \qquad \omega \in \mathbb{R}$$

if and only if

$$a_0 = \frac{1}{8} + \frac{1}{4\sqrt{2}}\cos\alpha + \frac{1}{2\sqrt{2}}\cos\beta\cos\gamma$$

$$a_1 = \frac{1}{8} - \frac{1}{4\sqrt{2}}\cos\alpha + \frac{1}{2\sqrt{2}}\cos\beta\sin\gamma$$

$$a_2 = \frac{1}{8} + \frac{1}{4\sqrt{2}}\cos\alpha - \frac{1}{2\sqrt{2}}\cos\beta\cos\gamma$$

$$a_3 = \frac{1}{8} - \frac{1}{4\sqrt{2}}\cos\alpha - \frac{1}{2\sqrt{2}}\cos\beta\sin\gamma$$

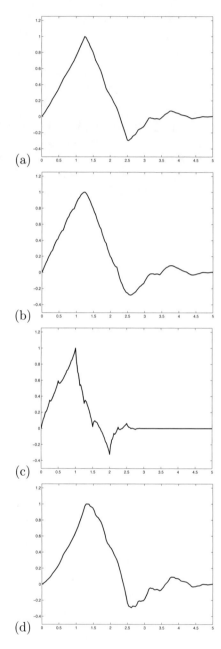

Fig. 2. Graphs of four different scaling functions: (a) D6 scaling function with three vanishing moments from Example 5, (b) H6 scaling function from Example 7, (c) most-regular length-four scaling function from Example 8, (d) most-regular length-six scaling function from Example 8.

$$b_0 = \frac{1}{8} + \frac{1}{4\sqrt{2}} \sin\alpha + \frac{1}{2\sqrt{2}} \sin\beta\cos\theta$$

$$b_1 = \frac{1}{8} - \frac{1}{4\sqrt{2}} \sin\alpha + \frac{1}{2\sqrt{2}} \sin\beta\sin\theta$$

$$b_2 = \frac{1}{8} + \frac{1}{4\sqrt{2}} \sin\alpha - \frac{1}{2\sqrt{2}} \sin\beta\cos\theta$$

$$b_3 = \frac{1}{8} - \frac{1}{4\sqrt{2}} \sin\alpha - \frac{1}{2\sqrt{2}} \sin\beta\sin\theta,$$

where $\alpha, \beta, \theta,$ and $\gamma \in \mathbb{R}$ satisfy

$$\sqrt{2}\cos\theta\sin\beta - 2\cos\theta\sin\alpha\sin\beta + \sqrt{2}\cos\beta(\cos\gamma - \sin\gamma)$$
$$- 4\cos^2\beta\cos\gamma\sin\gamma - 2\cos\alpha\cos\beta(\cos\gamma + \sin\gamma)$$
$$- \sqrt{2}\sin\beta\sin\theta - 2\sin\alpha\sin\beta\sin\theta - 4\cos\theta\sin^2\beta\sin\theta = 0.$$

Proof: As before, $H_8(1) = 1$ implies that $\displaystyle\sum_{i=0}^{3} a_i + b_i = 1$ and $H_8(-1) = 0$ implies that

$$\sum_{i=0}^{3} a_i - \sum_{i=0}^{3} b_i = 0.$$

It follows that

$$a_0 + a_1 + a_2 + a_3 = b_0 + b_1 + b_2 + b_3 = \frac{1}{2}. \tag{7}$$

Again, $|H_8(z)|^2 + |H_8(-z)|^2 = 1$ implies the following set of nonlinear equations:

$$a_0^2 + a_1^2 + a_2^2 + a_3^2 + b_0^2 + b_1^2 + b_2^2 + b_3^2 = \frac{1}{2}$$
$$a_0 a_1 + a_1 a_2 + a_2 a_3 + b_0 b_1 + b_1 b_2 + b_2 b_3 = 0 \tag{8}$$
$$a_0 a_2 + a_1 a_3 + b_0 b_2 + b_1 b_3 = 0$$
$$a_0 a_3 + b_0 b_3 = 0.$$

By adding all the equations together from (8), we see that

$$(a_0 - a_1 + a_2 - a_3)^2 + (b_0 - b_1 + b_2 - b_3)^2 = \frac{1}{2}.$$

So,

$$a_0 - a_1 + a_2 - a_3 = \frac{1}{\sqrt{2}}\cos\alpha, \quad b_0 - b_1 + b_2 - b_3 = \frac{1}{\sqrt{2}}\sin\alpha,$$

which implies that

$$a_0 + a_2 = \frac{1}{4} + \frac{1}{2\sqrt{2}}\cos\alpha, \quad a_1 + a_3 = \frac{1}{4} - \frac{1}{2\sqrt{2}}\cos\alpha$$

$$b_0 + b_2 = \frac{1}{4} + \frac{1}{2\sqrt{2}}\sin\alpha, \quad b_1 + b_3 = \frac{1}{4} - \frac{1}{2\sqrt{2}}\sin\alpha.$$

Continuing along this line, by subtracting the third equation from the first equation of (8), we have

$$(a_0 - a_2)^2 + (a_1 - a_3)^2 + (b_0 - b_2)^2 + (b_1 - b_3)^2 = \frac{1}{2}.$$

Using Lemma 3,

$$a_0 - a_2 = \frac{1}{\sqrt{2}} \cos \beta \cos \gamma, \quad a_1 - a_3 = \frac{1}{\sqrt{2}} \cos \beta \sin \gamma$$

$$b_0 - b_3 = \frac{1}{\sqrt{2}} \sin \beta \cos \theta, \quad b_1 - b_3 = \frac{1}{\sqrt{2}} \sin \beta \sin \theta.$$

Combining these four equations with the previous group of four, we arrive at the necessary formulae as listed in the lemma.

Plugging these formulae into the nonlinear system of equations reveals that the first and third equations from (8) are satisfied, but the second and fourth equations produce a single transcendental constraint as stated in the lemma. □

Example 9. *Let* $\alpha = 2.2400754386946162, \beta = 0.7535419996522459,$ $\gamma = 0.9614024467911164,$ *and* $\theta = -0.02541300114737489.$ *Then* $H_8(z)$ *is associated with Daubechies scaling function D8.*

Example 10. *In order for* $H_8(z)$ *to have a second order vanishing moment, we require* $H'(-1) = 0.$ *It follows that*

$$a_1 + 2a_2 + 3a_3 = b_1 + 2b_2 + 3b_3 + \frac{1}{4}.$$

Using the solution formulae in Lemma 4, we have

$$\frac{1}{2} - \sin(\alpha - \frac{\pi}{4}) + 2 \cos \beta \sin(\gamma + \frac{\pi}{4}) - 2 \sin \beta \sin(\theta + \frac{\pi}{4}) = 0.$$

Combining this with the transcendental constraint in Lemma 4, we can find many solutions for $H_8(z)$ *with the second order vanishing moment. For example,*

$$\alpha = \frac{5\pi}{12}, \beta = -\frac{\pi}{2}, \gamma = -\frac{\pi}{4}, \text{ and } \theta = -\frac{\pi}{4},$$

and

$$\alpha = \frac{5\pi}{12}, \beta = \pi, \gamma = -\frac{\pi}{4}, \text{ and } \theta = -\frac{\pi}{4}.$$

Let $H_{10}(z) = a_0 + b_0 z + a_1 z^2 + b_1 z^3 + a_2 z^4 + b_2 z^5 + a_3 z^6 + b_3 z^7 + a_4 z^8 + b_4 z^9$ be a trigonometric polynomial of $z = e^{i\omega}$.

Lemma 5. $H_{10}(z)$ satisfies $H_{10}(1) = 1$ and

$$|H_{10}(z)|^2 + |H_{10}(-z)|^2 = 1, \forall z = e^{i\omega}, \qquad \omega \in \mathbb{R}$$

if and only if

$$a_0 = \frac{1}{16} + \frac{1}{8\sqrt{2}}\cos\alpha + \frac{1}{4\sqrt{2}}\cos\beta\cos\gamma + \frac{r}{2}\cos\delta$$

$$a_1 = \frac{1}{8} - \frac{1}{4\sqrt{2}}\cos\alpha + \frac{1}{2\sqrt{2}}\cos\beta\sin\gamma$$

$$a_2 = \frac{1}{8} + \frac{1}{4\sqrt{2}}\cos\alpha - \frac{1}{2\sqrt{2}}\cos\beta\cos\gamma$$

$$a_3 = \frac{1}{8} - \frac{1}{4\sqrt{2}}\cos\alpha - \frac{1}{2\sqrt{2}}\cos\beta\sin\gamma$$

$$a_4 = \frac{1}{16} + \frac{1}{8\sqrt{2}}\cos\alpha + \frac{1}{4\sqrt{2}}\cos\beta\cos\gamma - \frac{r}{2}\cos\delta$$

$$b_0 = \frac{1}{16} + \frac{1}{8\sqrt{2}}\sin\alpha + \frac{1}{4\sqrt{2}}\sin\beta\cos\theta + \frac{r}{2}\sin\delta$$

$$b_1 = \frac{1}{8} - \frac{1}{4\sqrt{2}}\sin\alpha + \frac{1}{2\sqrt{2}}\sin\beta\sin\theta$$

$$b_2 = \frac{1}{8} + \frac{1}{4\sqrt{2}}\sin\alpha - \frac{1}{2\sqrt{2}}\sin\beta\cos\theta$$

$$b_3 = \frac{1}{8} - \frac{1}{4\sqrt{2}}\sin\alpha - \frac{1}{2\sqrt{2}}\sin\beta\sin\theta$$

$$b_4 = \frac{1}{16} + \frac{1}{8\sqrt{2}}\sin\alpha + \frac{1}{4\sqrt{2}}\sin\beta\cos\theta - \frac{r}{2}\sin\delta,$$

where

$$r = \sqrt{\frac{1}{2} - a_1^2 - a_2^2 - a_3^2 - b_1^2 - b_2^2 - b_3^2}$$

and $\alpha, \beta, \theta, \gamma$, *and* $\delta \in \mathbb{R}$ *satisfy*

$$\cos\beta\left(\cos\gamma(\sqrt{2} - 2\cos\alpha) - 8\sqrt{2}r\cos\delta\sin\gamma\right) +$$

$$\sin\beta\left(\cos\theta(\sqrt{2} - 2\sin\alpha) - 8\sqrt{2}r\sin\delta\sin\theta\right) = 0.$$

Proof: Along the same lines, we have the following nonlinear equations to solve:

$$a_0^2 + a_1^2 + a_2^2 + a_3^2 + a_4^2 + b_0^2 + b_1^2 + b_2^2 + b_3^2 + b_4^2 = \frac{1}{2}$$

$$a_0a_1 + a_1a_2 + a_2a_3 + a_3a_4 + b_0b_1 + b_1b_2 + b_2b_3 + b_3b_4 = 0$$

$$a_0a_2 + a_1a_3 + a_2a_4b_0b_2 + b_1b_3 + b_2b_4 = 0 \qquad (9)$$

$$a_0a_3 + a_1a_4 + b_0b_3 + b_1b_4 = 0$$

$$a_0a_4 + b_0b_4 = 0,$$

as well as the linear equation $\sum_{i=0}^{4} a_i = \sum_{i=0}^{4} b_i$. Using various combinations of these equations as we have before, it can be seen that

$$a_0 + a_1 + a_2 + a_3 + a_4 = \frac{1}{2}, \qquad b_0 + b_1 + b_2 + b_3 + b_4 = \frac{1}{2}$$

$$a_0 - a_1 + a_2 - a_3 + a_4 = \frac{1}{\sqrt{2}} \cos \alpha, \qquad b_0 - b_1 + b_2 - b_3 + b_4 = \frac{1}{\sqrt{2}} \sin \alpha$$

$$a_0 - a_2 + a_4 = \frac{1}{\sqrt{2}} \cos \beta \cos \gamma, \qquad b_0 - b_2 + b_4 = \frac{1}{\sqrt{2}} \sin \beta \cos \theta$$

$$a_1 - a_3 = \frac{1}{\sqrt{2}} \cos \beta \sin \gamma, \qquad b_1 - b_3 = \frac{1}{\sqrt{2}} \sin \beta \sin \theta.$$

So, we can solve for a_1, a_2, a_3, b_1, b_2, and b_3 as

$$a_1 = \frac{1}{8} - \frac{1}{4\sqrt{2}} \cos \alpha + \frac{1}{2\sqrt{2}} \cos \beta \sin \gamma$$

$$a_2 = \frac{1}{8} + \frac{1}{4\sqrt{2}} \cos \alpha - \frac{1}{2\sqrt{2}} \cos \beta \cos \gamma$$

$$a_3 = \frac{1}{8} - \frac{1}{4\sqrt{2}} \cos \alpha - \frac{1}{2\sqrt{2}} \cos \beta \sin \gamma$$

$$b_1 = \frac{1}{8} - \frac{1}{4\sqrt{2}} \sin \alpha + \frac{1}{2\sqrt{2}} \sin \beta \sin \theta$$

$$b_2 = \frac{1}{8} + \frac{1}{4\sqrt{2}} \sin \alpha - \frac{1}{2\sqrt{2}} \sin \beta \cos \theta$$

$$b_3 = \frac{1}{8} - \frac{1}{4\sqrt{2}} \sin \alpha - \frac{1}{2\sqrt{2}} \sin \beta \sin \theta.$$

Using the first and last equation from (9), we have

$$(a_0 - a_4)^2 + a_1^2 + a_2^2 + a_3^2 + (b_0 - b_4)^2 + b_1^2 + b_2^2 + b_3^2 = \frac{1}{2}.$$

Hence, it follows that

$$a_0 - a_4 = r \cos \delta$$
$$b_0 - b_4 = r \sin \delta$$
$$r = \sqrt{\frac{1}{2} - a_1^2 - a_2^2 - a_3^2 - b_1^2 - b_2^2 - b_3^2}.$$

Thus, the necessary formulae from the lemma follow. Upon substituting these formulae back into the nonlinear equations, we find that the first, third, and last equations are satisfied whereas the second and fourth produce a single transcendental equation. $\square$

Example 11. Let $\alpha = 2.6829477415207257$, $\beta = 0.714939482206344$, $\gamma = 1.028362000753886$, $\theta = 0.2225085841811395$, and suppose $\delta = 1.2907830472783552$. Then $H_{10}(z)$ is associated with the Daubechies D10 wavelet.

Example 12. *In order for $H_{10}(z)$ to satisfy the second order vanishing moment condition, we need $H'_{10}(-1) = 0$, i.e.*

$$H'_{10}(-1) = b_0 + 3b_1 + 5b_2 + 7b_3 + 9b_4 - 2a_1 - 4a_2 - 6a_3 - 8a_4$$
$$= \frac{1}{2} + 4r\cos\delta - 4r\sin\delta + \sqrt{2}\cos\beta\sin\gamma - \sqrt{2}\sin\beta\sin\theta$$
$$= 0.$$

There are numerous possibilities. For example, when

$$\alpha = 2.8975168508124955, \beta = \frac{3\pi}{4}, \gamma = \frac{\pi}{6}, \theta = 0, \text{ and } \delta = \frac{\pi}{4},$$

then both the transcendental condition and the second moment condition are satisfied.

§5. Numerical Experiment

We have implemented an image compression scheme as a means of comparing the parameterized solutions of length six with the standard wavelets of Haar, D4, and D6. (We are still working on the comparison using H_8 and H_{10}. The results will be reported elsewhere.) The scheme consists of the following:

- Decomposing the gray-scale values of various images with size 512×512 to a maximum number of levels.

- Encoding the decomposed image using an embedded zero-tree encoder to a specified file size. All of the images in this experiment have a file size of 262,159 bytes which is approximately one byte per pixel. The actual compressed file sizes are 32,793 bytes (8:1), 16,409 bytes (16:1), 8,217 bytes (32:1), and 4,121 bytes (64:1).

- Decoding the compressed file.

- Reconstructing the image using the wavelet transform and rounding the values to the nearest integer.

- Calculating the peak signal to noise ratio (PSNR) which is a measure of the root mean squared error (RMSE) in the sense that

$$RMSE = \frac{1}{512^2} \sum_{i,j=1}^{512} (p_{i,j} - \hat{p}_{i,j})^2$$
$$PSNR = 20\log_{10}(\frac{255}{RMSE})$$

where $p_{i,j}$ is the original gray-scale value and $\hat{p}_{i,j}$ is the value after reconstruction.

Image: Lena 512×512				
Wavelet	8:1	16:1	32:1	64:1
Haar	36.2258	32.6462	29.5685	27.5420
D4	38.4440	34.9209	31.6733	28.8185
D6	38.7819	35.3234	32.0479	29.0727
H6(Ex7)	38.8167	35.4208	32.1585	29.1588

Image: Barbara 512×512				
Wavelet	8:1	16:1	32:1	64:1
Haar	30.4954	26.8119	24.6329	22.7409
D4	32.8675	28.6364	25.6853	23.3821
D6	33.4311	29.0735	25.9431	23.4715
H6(Ex7)	33.6441	29.2804	26.1074	23.5789

Image: Boat 512×512				
Wavelet	8:1	16:1	32:1	64:1
Haar	34.7720	30.7103	27.5762	25.4130
D4	35.6517	31.5910	28.5165	26.0746
D6	35.8593	31.8088	28.6402	26.1880
H6(Ex7)	35.9301	31.9080	28.7187	26.2733

Image: Finger-print 512×512				
Wavelet	8:1	16:1	32:1	64:1
Haar	32.4415	29.9961	28.5828	27.3396
D4	33.3224	30.8632	29.4448	27.9762
D6	33.8077	31.2097	29.8049	28.2562
H6(Ex7)	33.9236	31.3387	29.9687	28.3928

Image: Marmousi 512×512				
Wavelet	8:1	16:1	32:1	64:1
Haar	35.7244	31.1309	27.7014	25.0277
D4	41.3471	36.2831	31.8262	27.5415
D6	43.6310	37.7401	33.0496	28.9160
H6(Ex7)	45.5190	38.6440	33.9109	29.5047

Image: Crowd 512×512				
Wavelet	8:1	16:1	32:1	64:1
Haar	33.3109	29.2032	26.1346	23.7222
D4	34.7214	30.6307	27.4101	24.8170
D6	35.1740	31.0444	27.7601	25.0348
H6(Ex7)	35.2893	31.1590	27.8690	25.1465

Tab. 1. PSNR comparison of the standard Daubechies wavelets and the parameterized wavelet of length six from Example 7.

Figure 3 shows the images used in the compression scheme, and Tab. 1 gives the PSNR values for the standard wavelets versus a parameterized wavelet for each image at various compression ratios. The parameterized wavelet was selected based upon having the largest PSNR values for the parameter values near D6. The search for the parameterized wavelet was limited, and a more exhaustive search may yield better results. Note that a higher PSNR value is better. In this case, we were able to find a single

Fig. 3. Test images: Lena, Finger-print, Barbara, Marmousi, Boat, and Crowd.

length six parameterized filter given in Example 7 which out-performed all the others tested on all six test images and at each compression ratio.

Acknowledgments. This project was partially funded by a research enhancement grant from the Kentucky NSF EPSCoR Foundation.

References

1. Daubechies, I., *Ten Lectures on Wavelets*, SIAM, Philadelphia, 1992.

2. He, W. and M. J. Lai, Examples of bivariate nonseparable compactly supported orthonormal continuous wavelets, in *Wavelet Applications in Signal and Image Processing IV*, Proceedings of SPIE, 3169, 303–314, 1997, also appears in IEEE Transactions on Image Processing **9** (2000), 949–953.

3. Lai, M. J. and D. W. Roach, Construction of bivariate symmetric orthonormal wavelets with short support, Univ. of Georgia Math. Preprint Series No. 22 (1999) and No. 20, (2000).

4. Lai, M. J. and D. W. Roach, The nonexistence of bivariate symmetric wavelets with short support and two vanishing moments, in *Trends in Approximation Theory*, K. Kopotun, T. Lyche, and M. Neamtu (eds.), Vanderbilt University Press, Nashville, 2001, 213–223.

5. Lai, M. J. and D. W. Roach, Nonseparable symmetric wavelets with short support, Proceedings of SPIE Conference on Wavelet Applications in Signal and Image Processing VII, Vol. 3813, July 1999, 132–146.

6. Lawton, W.,   Necessary and sufficient conditions for constructing orthonormal wavelet bases, J. Math. Phys. **32** (1991), 57–61.

7. Tian, Jun and R. O. Wells, Algebraic structures of orthogonal wavelet spaces, J. Applied and Computational Harmonic Analysis **8** (2000), 223–248.

8. Wells, R. O., Jr., Parameterizing smooth compactly supported wavelets, Trans. Amer. Math. Soc. **338** (1993), 919–931.

Ming-Jun Lai
Mathematics Department
University of Georgia
Athens, GA 30602
mjlai@math.uga.edu

David W. Roach
Mathematics and Statistics Department
Murray State University
Murray, KY 42071
david.roach@murraystate.edu

Bivariate C^1 Cubic Splines
for Exterior Biharmonic Equations

Ming-Jun Lai, Paul Wenston, and Lung-an Ying

Abstract. In this paper we deal with numerical solutions of exterior biharmonic equations. We apply bivariate C^1 cubic splines to solve exterior biharmonic equations using the infinite element method which was pioneered in Ying [7]. A numerical algorithm is derived and implemented in MATLAB. Some results from numerical experiments are included to demonstrate the convergence of the method.

§1. Introduction

In this paper we are interested in numerically solving the following exterior biharmonic equation using bivariate spline functions:

$$\begin{cases} \triangle^2 u = f & \text{outside of } \Omega \\ \frac{\partial}{\partial n} u = g & \text{on } \partial\Omega \\ u = h & \text{on } \partial\Omega, \end{cases} \tag{1.1}$$

where $\triangle$ is the usual Laplace operator, $\Omega \subset \mathbf{R}^2$ is a polygonal domain, $f \in L^2(\mathbf{R}^2\backslash\Omega) \cap C_0(\mathbf{R}^2)$, and g and h are two functions defined on the boundary of Ω which are compatible in the following sense: there exists a $u_0 \in C_0(\mathbf{R}^2) \cap H^2(\mathbf{R}^2\backslash\Omega)$ satisfying

$$\begin{cases} \frac{\partial}{\partial n} u_0 = g & \text{on } \partial\Omega \\ u_0 = h & \text{on } \partial\Omega. \end{cases} \tag{1.2}$$

The biharmonic equation in (1.1) is associated with the 2D Stokes equations for exterior flow when

$$\int_{\partial\Omega} g\, ds = 0, \tag{1.3}$$

Approximation Theory X: Wavelets, Splines, and Applications

Charles K. Chui, Larry L. Schumaker, and Joachim Stöckler (eds.), pp. 385–404.

Copyright ℗ 2002 by Vanderbilt University Press, Nashville, TN.

ISBN 0-8265-1416-2.

and in fact, is the stream function formulation of the 2D Stokes equations:

$$
\begin{cases}
-\nu\Delta\mathbf{u} + \nabla p = \mathbf{f}, & (x,y) \in \Omega \\
\mathrm{div}\,\mathbf{u} = 0, & (x,y) \in \Omega \\
\mathbf{u} = \mathbf{g}, & (x,y) \in \partial\Omega,
\end{cases}
\tag{1.4}
$$

where $\mathbf{u} = (u_1, u_2)$ is the velocity vector, p the pressure function and $\mathbf{f} = (f_1, f_2)$ is an external force. Here we have assumed that

$$
\int_{\partial\Omega} \mathbf{n}\cdot\mathbf{g}\,ds = 0.
$$

Since $\mathrm{div}\,\mathbf{u} = 0$, there exists a stream function u such that

$$
u_1 = \frac{\partial}{\partial y}u, \quad u_2 = -\frac{\partial}{\partial x}u
$$

(cf. Girault and Raviart [3]). By taking derivatives of the first two equations in (1.4) and subtracting the resulting second equation from the first equation, we obtain a biharmonic equation with $f = \dfrac{-1}{\nu}\left(\dfrac{\partial}{\partial y}f_1 - \dfrac{\partial}{\partial x}f_2\right)$, $g = \mathbf{n}\cdot\mathbf{g}$ with $\mathbf{n}$ being the normal direction of $\partial\Omega$, and h an anti-derivative of $\tau\cdot\mathbf{g}$ with τ being the tangential direction of $\partial\Omega$.

In the rest of the paper we concentrate on the numerical solution of the biharmonic equation (1.1) over the exterior domain of a given polygon Ω. Let $A(u,v) = \int_{\mathbf{R}^2\backslash\Omega} \Delta u \Delta v\,dxdy$ be a bilinear form, and let $|u|_2$ be the seminorm on the Sobolev space $H^2(\mathbf{R}^2\backslash\Omega)$. Set

$$
H_0 := \overline{C_0^\infty(\mathbf{R}^2\backslash\overline{\Omega})}^{|\cdot|_2}
$$

and

$$
H := \overline{C_0^\infty(\mathbf{R}^2\backslash\Omega)}^{|\cdot|_2}.
$$

It is easy to see that both H_0 and H are Hilbert spaces, and that the inner product in H_0 is equivalent to $A(\cdot,\cdot)$.

We now introduce the **weak solution** of exterior biharmonic equations. We say that $u = w + u_0$ is a weak solution of (1.1) if $w \in H_0$ and is such that

$$
A(w,v) = \int_{\mathbf{R}^2\backslash\Omega} fv\,dxdy - \int_{\mathbf{R}^2\backslash\Omega} \Delta u_0 \Delta v\,dxdy, \forall v \in H_0.
\tag{1.5}
$$

By Lax-Milgram's theorem, there exists a unique $w \in H_0$ satisfying (1.5). Thus, the weak solution u of (1.1) exists.

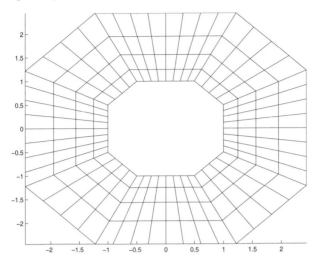

Fig. 1. First few rings of a quadrangulation of the exterior of an octagonal domain.

To find the weak solution numerically, we shall use the so-called *infinite element method* (cf.[Ying [7,8]]) which is a special method for dealing with exterior domain problems.

We will use the bivariate C^1 cubic spline space which consists of the well-known Fraeijs de Veubeke and Sander elements (cf. [2,6]) or [4]) to construct infinite elements. For simplicity, we first consider the case where Ω is a star-shaped polygon in $\mathbb{R}^2$. We will discuss the case where Ω is not such a polygon in §7.

Suppose that Ω is a star-shaped polygon in $\mathbb{R}^2$ and that the origin $(0,0) \in \Omega$. Partition the boundary $\partial\Omega$ into pieces by choosing (x_i, y_i), $i = 1, \ldots, N$ on $\partial\Omega$ which include all the corner points of $\partial\Omega$. Draw rays from the origin $(0,0)$ passing through (x_i, y_i) which divide $\mathbb{R}^2\backslash\Omega$ into N sections. Let $\xi > 1$ be a real number. Let $(x_{i,k}, y_{i,k})$ be the point on the ray passing through (x_i, y_i) defined by the equation $r(x_{i,k}, y_{i,k}) = \xi^k r(x_i, y_i)$, with $r(x, y) =$ the distance of (x, y) from the origin. Connect all $(x_{i,k} y_{i,k})$, $i = 1, \ldots, N$, by piecewise linear segments to get a ring Γ_k. Let Ω_k be the domain between Γ_k and Γ_{k-1} for $k = 1, \cdots$, with $\Gamma_0 = \partial\Omega$.

As illustrated in Fig. 1 the rays through (x_i, y_i) and the rings Γ_k quadrangulate the exterior of Ω.

We let $\Diamond$ be the special triangulation obtained from this quadrangulation by adding both diagonals of each quadrilateral. Let

$$S_3^1(\Diamond) = \{S \in C^1(\mathbb{R}^2\backslash\Omega), S \in P_3, \forall t \in \Diamond\}$$

be the space of C^1 piecewise cubic splines with respect to this triangulation.

Using the Lax-Milgram theorem easily yields

Theorem 1. *There exists a unique approximate weak solution* $u_s \in S_3^1(\diamondsuit) \cap H$ *such that*

$$A(u_s, v) = \int_{\mathbf{R}^2 \setminus \Omega} fv dx dy, \qquad \forall v \in S_3^1(\diamondsuit) \cap H_0 \qquad (1.6)$$

and $\frac{\partial}{\partial n} u_s = g_I$, $u = h_I$, *on* $\partial\Omega$. *Here,* g_I *and* h_I *are approximations of* g *and* h *on* $\partial\Omega$ *respectively.*

The infinite element method has been applied to elliptic and parabolic equations in exterior domains and problems with singular solutions (cf. Ying [8]). Using the spline functions in $S_3^1(\diamondsuit)$ greatly reduces the complexity of the analysis of nonconforming finite elements for biharmonic equations and speeds up the convergence of the numerical solution of the exterior biharmonic equation.

The paper is organized as follows: We first give some preliminary results explaining the construction of infinite elements using C^1 cubic splines in §2. Then we derive an algorithm for computing the numerical solution of homogeneous exterior biharmonic equations in §3. In §4 we present some examples illustrating the numerical performance of the algorithm derived in the previous section. In §5 we derive a numerical algorithm for nonhomogeneous exterior biharmonic equations. Convergence of these algorithms is studied in §6. Finally, we describe how to treat non-star-shaped domains and how to deal with singularities of the solutions around the corners of polygonal domains.

§2. Preliminaries

We first construct locally supported basis functions (infinite elements) for $S_3^1(\diamondsuit)$. Write $\overline{x}_{i,k} = \frac{1}{2}(x_{i,k} + x_{i+1,k})$, $\overline{y}_{i,k} = \frac{1}{2}(y_{i,k} + y_{i+1,k})$ for $i = 1, \ldots, N$, with $x_{N+1,k} = x_{1,k}$ and $y_{N+1,k} = y_{1,k}$ and $\hat{x}_{i,k} = \frac{1}{2}(x_{i,k}+x_{i,k+1})$, $\hat{y}_{i,k} = \frac{1}{2}(y_{i,k} + y_{i,k+1})$, for $i = 1, \cdots, N$. Suppose $\varphi_{i,k}$, $\psi_{x,i,k}$, $\psi_{y,i,k}$, $\eta_{h,i,k}$, $\eta_{v,i,k}$ in $S_3^1(\diamondsuit)$ satisfy the following conditions. Let $\varphi_{i,k}$ satisfy

$$\begin{cases} \varphi_{i,k}(x_{j,\ell}, y_{j,\ell}) = \begin{cases} \xi^k r(x_i, y_i), & \text{if } j = i, \ell = k \\ 0 & \text{otherwise} \end{cases} \\ \frac{\partial}{\partial x}\varphi_{i,k}(x_{j,\ell}, y_{j,\ell}) = \frac{\partial}{\partial y}\varphi_{i,k}(x_{j,\ell}, y_{j,\ell}) = 0 \\ \frac{\partial}{\partial n}\varphi_{i,k}(\overline{x}_{j,\ell}, \overline{y}_{j,\ell}) = \frac{\partial}{\partial n}\varphi_{i,k}(\hat{x}_{j,\ell}, \hat{y}_{j,\ell}) = 0 \end{cases}$$

for all $j = 1, \ldots, N$ and $\ell \geq 0$, where $\dfrac{\partial}{\partial n}$ denotes the normal derivative.

Let $\psi_{x,i,k}$ satisfy

$$\begin{cases} \psi_{x,i,k}(x_{j,\ell}, y_{j,\ell}) = \frac{\partial}{\partial y}\psi_{x,j,k}(x_{j,\ell}, y_{j,\ell}) = 0 \\ \frac{\partial}{\partial x}\psi_{x,i,k}(x_{j,\ell}, y_{j,\ell}) = \begin{cases} 1 & \text{if } j = i, \ell = k \\ 0 & \text{otherwise} \end{cases} \\ \frac{\partial}{\partial n}\psi_{x,i,k}(\overline{x}_{j,\ell}, \overline{y}_{j,\ell}) = \frac{\partial}{\partial n}\psi_{x,j,n}(\hat{x}_{j,\ell}, \hat{y}_{j,\ell}) = 0 \end{cases}$$

for all $j = 1, \ldots n$ and $\ell \geq 0$. Similarly we require that $\psi_{y,i,k}$ satisfy

$$\begin{cases} \psi_{y,i,k}(x_{j,\ell}, y_{j,\ell}) = \frac{\partial}{\partial x}\psi_{y,j,k}(x_{j,\ell}, y_{j,\ell}) = 0 \\ \frac{\partial}{\partial y}\psi_{y,i,k}(x_{j,\ell}, y_{j,\ell}) = \begin{cases} 1 & \text{if } j = i, \ell = k \\ 0 & \text{otherwise} \end{cases} \\ \frac{\partial}{\partial n}\psi_{y,i,k}(\overline{x}_{j,\ell}, \overline{y}_{j,\ell}) = \frac{\partial}{\partial n}\psi_{y,j,n}(\hat{x}_{j,\ell}, \hat{y}_{j,\ell}) = 0 \end{cases}$$

for all $j = 1, \ldots n$ and $\ell \geq 0$.

Finally, let $\eta_{v,i,k}$ satisfy

$$\begin{cases} \eta_{v,i,k}(x_{j,\ell}, y_{j,\ell}) = \frac{\partial}{\partial x}\eta_{v,j,k}(x_{j,\ell}, y_{j,\ell}) = 0 \\ \frac{\partial}{\partial y}\eta_{v,i,k}(x_{j,\ell}, y_{j,\ell}) = \frac{\partial}{\partial n}\eta_{v,i,k}(\overline{x}_{j,\ell}, \overline{y}_{j,\ell}) = 0, \\ \frac{\partial}{\partial n}\eta_{v,i,k}(\hat{x}_{j,\ell}, \hat{y}_{j,\ell}) = \begin{cases} 1 & \text{if } j = i, \ell = k \\ 0 & \text{otherwise} \end{cases} \end{cases}$$

and $\eta_{h,i,k}$ satisfy

$$\begin{cases} \eta_{h,i,k}(x_{j,\ell}, y_{j,\ell}) = \frac{\partial}{\partial x}\eta_{h,j,k}(x_{j,\ell}, y_{j,\ell}) = 0 \\ \frac{\partial}{\partial y}\eta_{h,i,k}(x_{j,\ell}, y_{j,\ell}) = \frac{\partial}{\partial n}\eta_{h,i,k}(\hat{x}_{j,\ell}, \hat{y}_{j,\ell}) = 0, \\ \frac{\partial}{\partial n}\eta_{h,i,k}(\overline{x}_{j,\ell}, \overline{y}_{j,\ell}) = \begin{cases} 1 & \text{if } j = i, \ell = k \\ 0 & \text{otherwise} \end{cases} \end{cases}$$

for all $j = 1, \ldots n$ and $\ell \geq 0$.

By the construction theory of FVS elements or C^1 cubic splines over $\Diamond$, it is known that such spline functions exist. We note the following important property:

Lemma 1. *For all $k \geq 0$,*

$$\varphi_{i,k+1}(x, y) = \xi\varphi_{i,k}\left(\frac{x}{\xi}, \frac{y}{\xi}\right),$$

$$\psi_{x,i,k+1}(x, y) = \xi\psi_{x,i,k}\left(\frac{x}{\xi}, \frac{y}{\xi}\right), \qquad \psi_{y,i,k+1}(x, y) = \xi\psi_{y,i,k}\left(\frac{x}{\xi}, \frac{y}{\xi}\right),$$

$$\eta_{v,i,k+1}(x, y) = \xi\eta_{v,i,k}\left(\frac{x}{\xi}, \frac{y}{\xi}\right), \qquad \eta_{h,i,k+1}(x, y) = \xi\eta_{h,i,k}\left(\frac{x}{\xi}, \frac{y}{\xi}\right).$$

Proof: These relations may be proved by using the B-form representation for these functions. $\square$

Thus, any spline function $s \in S_3^1(\diamondsuit)$ can be written as

$$s = \sum_{k=0}^{\infty} \sum_{i=1}^{N} \left(a_{i,k}\varphi_{i,k} + b_{i,k}\psi_{x,i,k} + c_{i,k}\psi_{y,i,k} + d_{i,k}\eta_{h,i,k} + e_{i,k}\eta_{v,i,k} \right).$$

For simplicity we group $a_{i,k}, b_{i,k}, c_{i,k}, d_{i,k}$ together into one 4N–vector a_k. Also we denote by b_k the vector of $e_{i,k}$, $i = 1, \cdots, N$. Thus, for $u, v \in S_3^1(\diamondsuit)$

$$
\begin{aligned}
A(u, v) &= \int_{\mathbf{R}^2 \backslash \Omega} \triangle u \triangle v \, dx dy = \sum_{k=1}^{\infty} \int_{\Omega_k} \triangle u \triangle v \, dx dy \\
&= \sum_{k=1}^{\infty} (\tilde{a}_{k-1}^T, \tilde{b}_{k-1}^T, \tilde{a}_k^T) B_k \begin{pmatrix} a_{k-1} \\ b_{k-1} \\ a_k \end{pmatrix},
\end{aligned}
\tag{2.1}
$$

where u is associated with vectors $a_0, b_0, a_1, b_1, \cdots$ while v is associated with vectors $\tilde{a}_0, \tilde{b}_0, \tilde{a}_1, \tilde{b}_1, \cdots$. Here B_k is the bending matrix whose entries consist of

$$\int_{\Omega_k} \triangle\varphi_{i,k} \triangle\varphi_{j,k} \, dx dy, \qquad \int_{\Omega_k} \triangle\varphi_{i,k} \triangle\psi_{x,j,k} \, dx dy,$$

etc.. With Lemma 1 we are able to prove

Lemma 2. $B_{k+1} = B_k$ for all $k \geq 1$.

Proof: Consider a typical entry of B_{k+1}:

$$
\begin{aligned}
\int_{\Omega_{k+1}} \triangle\varphi_{i,k+1} \triangle\varphi_{j,k+1} \, dx dy &= \int_{\Omega_k} \frac{1}{\xi^2} \triangle\xi\varphi_{i,k} \frac{1}{\xi^2} \triangle\xi\varphi_{j,k} \xi dx' \xi dy' \\
&= \int_{\Omega_k} \triangle\varphi_{i,k} \triangle\varphi_{j,k} \, dx dy,
\end{aligned}
$$

where we have used Lemma 1 and the substitution $x = \xi x'$ and $y = \xi y'$. Clearly, that is an entry of B_k. A similar calculation applies to the other entries. $\square$

From now on we shall write B for all B_k's.

§3. Numerical Solutions of Homogeneous Exterior Biharmonic Equations

Recall that $u \in S_3^1(\diamondsuit) \cap H$ is a weak solution of the homogeneous exterior biharmonic if for all $v \in S_3^1(\diamondsuit) \cap H_0$

$$A(u, v) = \sum_{k=1}^{\infty} (\tilde{a}_{k-1}^T, \tilde{b}_{k-1}^T, \tilde{a}_k^T) B \begin{pmatrix} a_{k-1} \\ b_{k-1} \\ a_k \end{pmatrix} = 0. \tag{3.1}$$

Note that $v \in S_3^1(\diamondsuit) \cap H_0$ implies that $\tilde{a}_0 = 0$ while the remaining coefficients of v can individually take on arbitrary values. Writing $B = \begin{pmatrix} B_{11} & B_{12} & B_{13} \\ B_{21} & B_{22} & B_{23} \\ B_{31} & B_{32} & B_{33} \end{pmatrix}$ in the block structure according to the partition of the vectors $(\tilde{a}_{k-1}^T, \tilde{b}_{k-1}^T, \tilde{a}_k^T)$ and $\begin{pmatrix} a_{k-1} \\ b_{k-1} \\ a_k \end{pmatrix}$, by (3.1) we have the following infinite system of linear equations:

$$B_{21} a_0 + B_{22} b_0 + B_{23} a_1 = 0,$$

$$B_{31} a_0 + B_{32} b_0 + (B_{33} + B_{11}) a_1 + B_{12} b_1 + B_{13} a_2 = 0,$$

and for $k = 1, 2, \cdots$,

$$B_{21} a_{k-1} + B_{22} b_{k-1} + B_{23} a_k = 0,$$

$$B_{31} a_{k-1} + B_{32} b_{k-1} + (B_{33} + B_{11}) a_k + B_{12} b_k + B_{13} a_{k+1} = 0.$$

It follows that

$$b_{k-1} = -B_{22}^{-1} B_{21} a_{k-1} - B_{22}^{-1} B_{23} a_k$$

and

$$C a_{k-1} + D a_k + C^T a_{k+1} = 0$$

for $k = 1, 2, \ldots$, where

$$\begin{cases} C = B_{31} - B_{32} B_{22}^{-1} B_{21} \\ D = B_{33} + B_{11} - B_{32} B_{22}^{-1} B_{23} - B_{12} B_{22}^{-1} B_{21}. \end{cases} \tag{3.2}$$

By Theorem 1 the vector a_1 is uniquely determined by the boundary conditions, i.e., a_0. That is, there exists a matrix X such that $a_1 = X a_0$. It is easy to see that $a_2 = X a_1$ for the same matrix X. In general we have

$$a_k = X a_{k-1} = \cdots = X^k a_0.$$

It follows that
$$(C + DX + C^T X^2)a_0 = 0$$
for any given boundary conditions a_0. Thus we obtain the matrix equation
$$C + DX + C^T X^2 = 0.$$

In order to construct such a matrix X, we consider
$$R_1 = \begin{bmatrix} D & C \\ I & 0 \end{bmatrix}, \qquad R_2 = \begin{bmatrix} -C^T & 0 \\ 0 & I \end{bmatrix}, \tag{3.3}$$
where I denotes the identity matrix of the same size as D. Note that
$$R_1 \begin{bmatrix} \lambda_i g_i \\ g_i \end{bmatrix} = \lambda_i R_2 \begin{bmatrix} \lambda_i g_i \\ g_i \end{bmatrix}$$
for any eigenpair (λ_i, g_i) of X. That is, $[\lambda_i g_i^T, g_i^T]$ is an eigenvector of the generalized eigenvalue problem $R_1 v = \lambda R_2 v$.

Since all eigenvalues of X are eigenvalues of $R_1 v = \lambda R_2 v$, we shall identify which eigenvalues of $R_1 v = \lambda R_2 v$ are also an eigenvalue of X. We have

Lemma 3. *All eigenvalues λ of X which are not equal to 1 have a magnitude less than 1. When $\lambda = 1$, λ is a double eigenvalue and its eigenvectors correspond to $u = x$ and $u = y$.*

Proof: Let λ be an eigenvalue of X with an associated eigenvector g_λ. Suppose that $|\lambda| \geq 1$. By Theorem 1, for g_λ, there is a unique spline $u_\lambda \in S_3^1(\diamondsuit) \cap H$ which is the approximate solution of the homogeneous biharmonic equation and satisfies the boundary conditions corresponding to the vector g_λ. We note that the semi-norm
$$|u_\lambda|_2^2 = \int_{\mathbf{R}^2 \setminus \Omega} \left[\left(\frac{\partial^2}{\partial x^2} u_\lambda \right)^2 + 2 \left(\frac{\partial^2}{\partial xy} u_\lambda \right)^2 + \left(\frac{\partial^2}{\partial y^2} u_\lambda \right)^2 \right] dxdy$$
is bounded. Similar to (2.1) and using Lemma 2, we have
$$|u_\lambda|_2^2 = \sum_{k=1}^{\infty} (a_{k-1}^T, b_{k-1}^T, a_{k-1}^T) \tilde{B} \begin{pmatrix} a_{k-1} \\ b_{k-1} \\ a_k \end{pmatrix}$$
for some matrix $\tilde{B}$. However, using the expression for b_{k-1} in terms of a_{k-1} and a_k and noting that $a_k = \lambda^k a_0$ with $|\lambda| \geq 1$, we have, for another matrix $\hat{B}$,
$$|u_\lambda|_2^2 = \sum_{k=1}^{\infty} (a_{k-1}^T a_k^T) \hat{B} \begin{pmatrix} a_{k-1} \\ a_k \end{pmatrix}$$
$$= \sum_{k=1}^{\infty} (\lambda^{k-1})^2 (g_\lambda^T, \lambda g_\lambda^T) \hat{B} \begin{pmatrix} g_\lambda \\ \lambda g_k \end{pmatrix} = \infty$$

if $(g_\lambda^T, \lambda g_\lambda^T) = \hat{B} \begin{pmatrix} g_\lambda \\ \lambda g_\lambda \end{pmatrix} \neq 0.$

Otherwise, $|u_\lambda|_2 = 0$ which implies that $u_\lambda = x$ or $u_\lambda = y$ or $u_\lambda =$ constant. It can be seen that $u_\lambda = 1$ corresponds to $\lambda = \frac{1}{\xi} < 1$ by checking the coefficients corresponding to $\phi_{i,k}$. If $u_\lambda \neq x$ and $u_\lambda \neq y$, then the magnitude of λ has to be less than 1. This completes the proof. $\square$

If (λ, v) is an eigenpair of $R_1 v = \lambda R_2 v$ with $|\lambda| < 1$, then writing $v = [v_1, v_2]^T$ with $v_1 = \lambda v_2$, (λ, v_2) is an eigenpair of X. With these eigenvalues and eigenvectors, in addition to $\lambda = 1$, $u_\lambda = x$ and $u_\lambda = y$, we have $X = E\Lambda E^T$ with Λ being a diagonal matrix with eigenvalues on its diagonal and E containing eigenvectors. The above discussion leads to the following

Algorithm 1. (Numerical Solution of Exterior Homogeneous Biharmonic Equations)

Step 1: Construct a bending matrix $B = \begin{pmatrix} B_{11} & B_{12} & B_{13} \\ B_{21} & B_{22} & B_{23} \\ B_{31} & B_{32} & B_{33} \end{pmatrix}$ over Ω_1.

Step 2: Compute all eigenvalues and associated eigenvectors for the generalized eigenvalue problem

$$R_1 v = \lambda R_2 v$$

with R_1 and R_2 defined in (3.2) and (3.3).

Step 3: Find all eigenvalues whose magnitude is less than 1 and whose eigenvectors v can be decomposed to $v = \begin{bmatrix} \lambda v_1 \\ v_1 \end{bmatrix}$. In addition, find the two eigenvectors v corresponding to the eigenvalue 1. They can be constructed using the interpolation properties of the basis functions $\varphi_{i,k}, \psi_{x,i,k}, \psi_{y,i,k},$ $\eta_{v,i,k}, \eta_{h,i,k}$. Assemble $X = E\Lambda E^{-1}$.

Step 4: Compute $a_k = E\Lambda^k E^{-1} a_0$ and

$$b_k = -B_{22}^{-1} B_{21} a_{k-1} - B_{22}^{-1} B_{23} a_k$$

for $k = 1, 2, \cdots$.

§4. Numerical Experiments

We report on some numerical experiments on the convergence of the approximate weak solution in $S_3^1(\diamondsuit) \cap H$ to the weak solution in H. We have implemented bivariate C^1 cubic splines in MATLAB to numerically solve homogeneous biharmonic equations over exterior domains. Our programs are able to solve exterior homogeneous biharmonic equation over any star-shaped polygonal domains.

Consider the solution of the following exterior problem:

$$\begin{cases} \Delta^2 u = 0, & \text{exterior of } \Omega \\ u = g, & \text{on } \partial\Omega \\ \frac{\partial}{\partial n} u = h, & \text{on } \partial\Omega, \end{cases}$$

where Ω is the octagon as shown in Figure 1. Recall that the general solution of the homogeneous biharmonic equation is given by

$$u(x,y) = ar^{-n} \sin(n\theta) + br^{2-m} \cos(m\theta) \tag{4.1}$$

for any real numbers a and b and any nonnegative integer m and n. Here, (r,θ) is the polar coordinate of (x,y).

We use the function (4.1) with $m = 3$, $n = 2$, $a = b = 1$ as the exact solution of an exterior homogeneous biharmonic equation. We then input the boundary of a star-shaped polygonal domain, $\xi > 1$ and the boundary values of $u(x,y)$ into our MATLAB programs and compare the numerical solution against the exact solution. In the following table we list the maximum relative error of the first order derivatives of the numerical solution as compared with the exact solution. We list the relative errors over first few rings. Starting with 24 points on the boundary $\partial\Omega$ and letting $\xi = 1.25$, we refine the triangulations by inserting midpoints between any two consecutive points on $\partial\Omega$ and choose $\xi = 1.125$. Then we refine the triangulations once more by inserting midpoints and choose $\xi = 1.0625$. Thus, we list the errors based on the three different triangulations. From Tables 1–3, we can see that the relative errors are reduced when the triangulation is refined and ξ decreases.

§5. Numerical Solutions of Nonhomogeneous Exterior Biharmonic Equations

We now discuss the solution of the nonhomogeneous biharmonic equation. We begin with several lemmas.

Lemma 4. If $u \in S_3^1(\diamondsuit) \cap H$ is an approximate weak solution of the homogeneous biharmonic equation, then

$$A(u,v) = \tilde{a}_0 K_z a_0, \qquad \forall v \in S_3^1(\diamondsuit) \cap H$$

where $K_z = B_{11} - B_{12}B_{22}^{-1}B_{21} + (B_{13} - B_{12}B_{22}^{-1}B_{23})X$. Here, a_0 and $\tilde{a}_0$ are vectors consisting of the boundary function and partial derivatives values of u and v, respectively.

Tab. 1. The maximum relative error of the first order derivatives D_x and D_y over rings R_k with $\xi = 1.25$.

Rings	D_x	D_y
k=1	0.1608	0.1869
k=2	0.1323	0.2207
k=3	0.0656	0.1312
k=4	0.0664	0.1382
k=5	0.0648	0.1295

Tab. 2. The maximum relative error of the first order derivatives D_x and D_y over rings R_k with $\xi = 1.125$.

Rings	D_x	D_y
k=1	0.1074	0.0892
k=2	0.0531	0.1389
k=3	0.0676	0.1161
k=4	0.0505	0.1156
k=5	0.0526	0.1096

Tab. 3. The maximum relative error of the first order derivatives D_x and D_y over rings R_k with $\xi = 1.0625$.

Rings	D_x	D_y
k=1	0.0541	0.0567
k=2	0.0333	0.0747
k=3	0.0363	0.0747
k=4	0.0371	0.0734
k=5	0.0411	0.0783
k=6	0.0376	0.0757
k=7	0.0399	0.0732
k=8	0.0376	0.0739

Proof: Since $A(u,v) = \displaystyle\sum_{k=1}^{\infty}(\tilde{a}_{k-1}^T, \tilde{b}_{k-1}^T, \tilde{a}_k^T)B\begin{pmatrix} a_{k-1} \\ b_{k-1} \\ a_k \end{pmatrix}$ for any $v \in S_3^1(\diamondsuit) \cap H$, and since $A(u,v) = 0$ for all $v \in S_3^1(\diamondsuit) \cap H_0$, we may assume that v is such that $\tilde{b}_0 = \tilde{a}_1 = \tilde{b}_1 = \cdots = 0$. We thus have

$$A(u,v) = \tilde{a}_0^T(B_{11}a_0 + B_{12}b_0 + B_{13}a_1)$$
$$= \tilde{a}_0^T(B_{11} - B_{12}B_{22}^{-1}(B_{21} + B_{23}X) + B_{13}X)a_0 = \tilde{a}_0^T K_z a_0. \quad \square$$

We may write

$$A(u,v) = (\tilde{a}_0^T, \tilde{b}_0^T, \tilde{a}_1^T) B \begin{pmatrix} a_0 \\ b_0 \\ a_1 \end{pmatrix} + \tilde{a}_1 K_z a_1.$$

For $\tilde{a}_0 = 0 = \tilde{b}_0$ and $\tilde{a}_1 \neq 0$, we have

$$\tilde{a}_1^T (B_{31} a_0 + B_{32} b_0 + B_{33} a_1) + \tilde{a}_1 K_z a_1 = 0.$$

It follows that

$$(B_{33} + K_z - B_{32} B_{22}^{-1} B_{23}) a_1 + (B_{31} - B_{32} B_{22}^{-1} B_{21}) a_0 = 0.$$

Thus, we have

Lemma 5. $X = -(B_{33} + K_z - B_{32} B_{22}^{-1} B_{23})^{-1} (B_{31} - B_{32} B_{22}^T B_{21}).$

For the nonhomogeneous biharmonic equation the approximate weak solution $u \in S_3^1(\diamondsuit) \cap H$ satisfies the boundary conditions

$$\frac{\partial u_s}{\partial n}\Big|_{\partial \Omega} = g_I \text{ and } u_s|_{\partial \Omega} = h_I$$

and the variational equation

$$A(u,v) = \langle f, v \rangle, \qquad \forall v \in S_3^1(\diamondsuit) \cap H_0$$

where $\langle f, v \rangle = \int_{\mathbf{R}^2 \setminus \Omega} f v \, dx dy$. Let $A_k(u,v)$ be defined by the equation

$$A_k(u,v) = \sum_{\ell=k+1}^{\infty} \int_{\Omega_\ell} \triangle u \triangle v \, dx dy$$

and $\langle f, v \rangle_k$ by the equation

$$\langle f, v \rangle_k = \sum_{\ell=k+1}^{\infty} \int_{\Omega_\ell} f v \, dx dy.$$

As before, we use $(a_0, b_0, a_1, b_1, \dots,)^T$ to denote the vector associated with the approximate weak solution u and $(\tilde{a}_0, \tilde{b}_0, \tilde{a}_1, \tilde{b}_1, \dots)^T$, with $\tilde{a}_0 = 0$ to denote the vector associated with a test function $v \in S_3^1(\diamondsuit) \cap H_0$. The following result is easily seen:

Lemma 6. *For all $k \geq 0$,*

$$
A_k(u, v) - \langle f, v \rangle_k = (\tilde{a}_k^T, \tilde{b}_k^T, \tilde{a}_{k+1}^T) B \begin{pmatrix} a_k \\ b_k \\ a_{k+1} \end{pmatrix}
$$

$$
- (\tilde{a}_k^T, \tilde{b}_k^T, \tilde{a}_{k+1}^T) \begin{pmatrix} f_k \\ g_k \\ f_{k+1} \end{pmatrix} + A_{k+1}(u, v) - \langle f, v \rangle_{k+1}
$$

where f_k, g_k, f_{k+1} is a vector whose entries consist of $\int_{\Omega_k} f \varphi_{i,k} dx dy$, etc.

Let u^* be a particular solution satisfying $A(u^*, v) = \langle f, v \rangle$, where u^* may not satisfy the given boundary conditions in (1.1). Then $w = u - u^*$ is a solution of the homogeneous biharmonic equation. If we use $(c_0, d_0, c_1, d_1, \cdots)$ to denote the vector associated with w, then the above implies

$$
c_k = X^k c_0, d_{k-1} = -B_{22}^{-1}(B_{21}c_{k-1} + B_{23}c_k).
$$

Thus, $a_0 = c_0 + a_0^*$ or $c_0 = a_0 - a_0^*$, and

$$
a_1 = c_1 + a_1^* = X c_0 + a_1^* = X a_0 + q_1.
$$

In general, we have $a_k = X a_{k-1} + q_k$. Also,

$$
b_0 = d_0 + b_0^* = -B_{22}^{-1} B_{21} c_0 - B_{22}^{-1} B_{23} c_1 + b_0^*
$$

$$
= -B_{22}^{-1} B_{21} a_0 - B_{22}^{-1} B_{23} a_1 + \tilde{b}_0^*
$$

for some $\tilde{b}_0^*$. Moreover for $v \in S_3^1(\diamondsuit)$ with $\tilde{b}_0 = \tilde{a}_1 = \tilde{b}_1 = \ldots = 0$, we have

$$
\begin{aligned}
A(u,v) &- \langle f, v \rangle \\
&= \tilde{a}_0^T (B_{11} a_0 + B_{12} b_0 + B_{13} a_1) - \tilde{a}_0^T f_0 \\
&= \tilde{a}_0^T ((B_{11} - B_{12} B_{22}^{-1} B_{21}) a_0 + (B_{13} - B_{12} B_{22}^{-1} B_{23}) a_1 + B_{12} \tilde{b}_0^*) - \tilde{a}_0^T f_0 \\
&= \tilde{a}_0^T K_z a_0 - \tilde{a}_0^T h_0
\end{aligned}
$$

for some vector h_0. In general, we have

$$
A_k(u, v) - \langle f, v \rangle_k = \tilde{a}_k^T K_z a_k - \tilde{a}_k^T h_k
$$

for $k = 0, 1, 2, \ldots$. We are now able to derive recurrence formulas for a_k and b_k which are also dependent on h_k. For this purpose we need to derive a recurrence formula for h_k. These derivations will be based upon the following restatement of Lemma 6:

$$A_k(u, v) - \langle f, v \rangle_k =$$

$$(\tilde{a}_k^T, \tilde{b}_k^T, \tilde{a}_{k+1}^T) \begin{pmatrix} B_{11} & B_{12} & B_{13} \\ B_{21} & B_{22} & B_{23} \\ B_{31} & B_{32} & B_{33} \end{pmatrix} \begin{pmatrix} a_k \\ b_k \\ a_{k+1} \end{pmatrix} - (\tilde{a}_k^T, \tilde{b}_k^T, \tilde{a}_{k+1}^T) \begin{pmatrix} f_k \\ g_k \\ f_{k+1} \end{pmatrix}$$

$$+ A_{k+1}(u, v) - \langle f, v \rangle_{k+1}.$$

If we choose v such that all coefficients $\tilde{a}_k$'s and $\tilde{b}_k$'s of v are zero except for one vector $\tilde{b}_k \neq 0$, then $A(u, v) - \langle f, v \rangle = 0$ implies that $A_k(u, v) - \langle f, v \rangle_k = 0$. By using the above formula, we have

$$B_{21} a_k + B_{22} b_k + B_{23} a_{k+1} - g_k = 0,$$

or $b_k = B_{22}^{-1}(g_k - B_{21} a_k - B_{23} a_{k+1})$, for all $k \geq 0$. If we choose $v \in S_3^1(\diamondsuit) \cap H_0$ such that all coefficients of v are zero except for $\tilde{a}_{k+1} \neq 0$, then $A_k(u, v) - \langle f, v \rangle_k = 0$ which implies

$$B_{31} a_k + B_{32} b_k + B_{33} a_{k+1} - f_{k+1} + K_z a_{k+1} - h_{k+1} = 0.$$

That is,

$$(B_{33} + K_z - B_{32} B_{22}^{-1} B_{23}) a_{k+1}$$
$$= h_{k+1} + f_{k+1} - B_{31} a_k - B_{32} B_{22}^{-1} g_k + B_{32} B_{22}^{-1} B_{21} a_k.$$

By Lemma 5, we have

$$a_{k+1} = X a_k + (B_{33} + K_z - B_{32} B_{22}^{-1} B_{23})^{-1}(h_{k+1} + f_{k+1} - B_{32} B_{22}^{-1} g_k).$$

Finally, if we choose $v \in S_3^1(\diamondsuit) \cap H_0$ with all coefficients zero except for $\tilde{a}_k \neq 0$ for some $k \geq 1$, then

$$K_z a_k - h_k = (B_{11}, B_{12}, B_{13}) \begin{pmatrix} a_k \\ B_{22}^{-1}(g_k - B_{21} a_k - B_{23} a_{k+1}) \\ a_{k+1} \end{pmatrix} - f_k.$$

After simplification, we have

$$(K_z - B_{11} + B_{12} B_{22}^{-1} B_{21}) a_k$$
$$= (B_{13} - B_{12} B_{22}^{-1} B_{23}) a_{k+1} + B_{12} B_{22}^{-1} g_k + h_k - f_k.$$

Using Lemma 5, the expression for K_z in Lemma 4 and the expression for a_{k+1}, we have

$$h_k - f_k = - B_{12} B_{22}^{-1} g_k - (B_{13} - B_{12} B_{22}^{-1} B_{23}) \times$$
$$(B_{33} + K_z - B_{32} B_{22}^{-1} B_{23})^{-1}(h_{k+1} + f_{k+1} - B_{32} B_{22}^{-1} g_k).$$

Therefore, we have

Algorithm 2. (Numerical Solution of Exterior Nonhomogeneous Biharmonic Equations)

Step 1: Start with $h_\infty = 0$ or $h_\ell = 0$ for sufficiently large integer ℓ. Compute

$$h_k = f_k - B_{12}B_{22}^{-1}g_k - (B_{13} - B_{12}B_{22}^{-1}B_{23}) \times$$
$$(B_{33} + K_z - B_{32}B_{22}^{-1}B_{23})^{-1}(h_{k+1} + f_{k+1} - B_{32}B_{22}^{-1}g_k) \tag{5.1}$$

for $k = \ell - 1, \ell - 2, \cdots, 1$.

Step 2: Compute

$$a_{k+1} = Xa_k + (B_{33} + K_z - B_{32}B_{22}^{-1}B_{23})^{-1}(h_{k+1} + f_{k+1} - B_{32}B_{22}^{-1}g_k) \tag{5.2}$$

for $k = 0, 1, 2, \cdots, \ell$.

Step 3: Compute

$$b_k = B_{22}^{-1}(g_k - B_{21}a_k - B_{23}a_{k+1}), \tag{5.3}$$

for $k = 0, 1, 2, \cdots, \ell$.

We have recently implemented the above algorithm in MATLAB and the numerical experiments will be reported elsewhere.

§6. Convergence Analysis

In this section we discuss the convergence of the approximate weak solution in $S_3^1(\diamondsuit) \cap H$ to the weak solution in H. Let u be the weak solution in H satisfying

$$A(u, v) = \langle f, v \rangle, \qquad \forall v \in H_0$$

and $\frac{\partial}{\partial n}u|_{\partial\Omega} = g$, $u|_{\partial\Omega} = h$. Let u_s be the approximate weak solution in $S_3^1(\diamondsuit) \cap H$ satisfying

$$A(u_s, v) = \langle f, v \rangle, \qquad \forall v \in H_0 \cap S_3^1(\diamondsuit)$$

and $\frac{\partial}{\partial n}u_s|_{\partial\Omega} = g$, $u_s|_{\partial\Omega} = h$. It is standard that

$$|u - u_s|_{2,\mathbf{R}^2\setminus\Omega}^2 = \int_{\mathbf{R}^2\setminus\Omega} |\triangle(u - u_s)|^2 dxdy$$

$$= \int_{\mathbf{R}^2\setminus\Omega} \triangle(u - u_s)\triangle(u - u_s - s)dxdy$$

$$\leq |u - u_s|_{2,\mathbf{R}^2\setminus\Omega}|u - u_s - s|_{2,\mathbf{R}^2\setminus\Omega}$$

for any $s \in S_3^1(\diamondsuit) \cap H_0$. That is;

$$|u - u_s|_{2,\mathbf{R}^2 \setminus \Omega} \leq \inf_{s \in H_0 \cap S_3^1(\diamondsuit)} |u - u_s - s|_{2,\mathbf{R}^2 \setminus \Omega}.$$

In particular we have

$$|u - u_s|_{2,\mathbf{R}^2 \setminus \Omega} \leq |u - u_I|_{2,\mathbf{R}^2 \setminus \Omega}$$

where u_I is a quasi-interpolant of u in $S_3^1(\diamondsuit)$ satisfying the boundary conditions. By [1] or [5],

$$|u - u_I|_{2,\Omega_k} \leq C|\diamondsuit|_{\Omega_k}^2 |u|_{4,\Omega_k}.$$

Note that the mesh size of $\diamondsuit \cap \Omega_1$ is

$$\overline{h} = \max\{(\xi - 1)r_0, \ \mathrm{dist}((x_i, y_i), (x_{i+1}, y_{i+1}))\}$$

where $r_0 = \max\{r(x, y), (x, y) \in \partial\Omega\}$. Hence

$$|u - u_I|_{2,\mathbf{R}^2 \setminus \Omega}^2 = \sum_{k=1}^{\infty} |u - u_I|_{2,\Omega_k}^2$$

$$\leq C \sum_{k=1}^{\infty} (\xi^k \overline{h})^4 |u|_{4,\Omega_k}^2 \leq C\overline{h}^4 \sum_{k=1}^{\infty} (\xi^{2k} |u|_{4,\Omega_k})^2.$$

Therefore, we have obtained the following

Theorem 2. *Suppose that $u \in H$ and that*

$$r^2 \left(\frac{\partial}{\partial x}\right)^{\alpha} \left(\frac{\partial}{\partial y}\right)^{\beta} u \in L_2(\mathbb{R}^2 \setminus \Omega)$$

for all (α, β) satisfying $\alpha + \beta = 4$. Then

$$|u - u_s|_{2,\mathbf{R}^2 \setminus \Omega} \leq C\overline{h}^2 |r^2 D^4 u|_{\mathbf{R}^2 \setminus \Omega},$$

where

$$|r^2 D^4 u|_{\mathbf{R}^2 \setminus \Omega} = \max_{\alpha + \beta = 4} \left(\int_{\mathbf{R}^2 \setminus \Omega} \left|r(x,y)^2 \left(\frac{\partial}{\partial x}\right)^{\alpha} \left(\frac{\partial}{\partial y}\right)^{\beta} u(x,y)\right|^2 dx dy\right)^{\frac{1}{2}}.$$

§7. Remarks

We now discuss how to treat exterior biharmonic equations outside a polygon $\Omega \subset \mathbf{R}^2$ which is not star-shaped. First of all, let Ω_0 be the convex hull of Ω. As in the previous section, we have

$$\tilde{A}(u,v) - \langle f, v \rangle_{\mathbf{R}^2 \setminus \Omega_0} = \int_{\mathbf{R}^2 \setminus \Omega_0} \triangle u \triangle v \, dx dy - \int_{\mathbf{R}^2 \setminus \Omega_0} f v \, dx dy$$

$$= \tilde{a}_0^T K_z a_0 + \tilde{a}_0^T h_0,$$

where $a_0 = (a_{01}, a_{02})^T$ with $a_{0,1}$ being the spline coefficients associated with $\partial \Omega_0 \cap \partial \Omega$ and a_{02} associated with $\partial \Omega_0 \setminus \partial \Omega$. Here, u is the approximate weak solution in $S_3^1(\diamondsuit) \cap H$ satisfying the boundary conditions on $\partial \Omega_0 \cap \partial \Omega$.

According to the split of $a_0 = (a_{0,1}, a_{0,2})^T$, K_z can be written as

$$K_z = \begin{pmatrix} K_z^{11} & K_z^{12} \\ K_z^{21} & K_z^{22} \end{pmatrix}.$$

Next we subdivide $\Omega_0 \setminus \Omega$ into triangulated quadrilaterals $\hat{\diamondsuit}$ and let $S_3^1(\hat{\diamondsuit})$ be the C^1 cubic spline space over $\hat{\diamondsuit}$. We can write

$$\hat{A}(u,v) - \langle f, v \rangle_{\Omega_0 \setminus \Omega} = \int_{\Omega_0 \setminus \Omega} \triangle u \triangle v \, dx dy - \int_{\Omega_0 \setminus \Omega} f v \, dx dy dy$$

$$= \tilde{a}_0^T K_y c_0 + \tilde{c}_0 \tilde{h}_0$$

for $u, v \in S_3^1(\hat{\diamondsuit})$, where c_0 is a vector consisting of (a_{02}, a_{03}, a_{04}) with a_{03} being the coefficients corresponding to the given boundary conditions on $\partial \Omega \setminus \partial \Omega_0$, and a_{04} the coefficients of the spline functions supported inside $\Omega_0 \setminus \Omega$. Accordingly, K_y may be written as

$$K_y = \begin{pmatrix} K_y^{11} & K_y^{12} & K_y^{13} \\ K_y^{21} & K_y^{22} & K_y^{23} \\ K_y^{31} & K_y^{32} & K_y^{33} \end{pmatrix}.$$

It follows from $A(u,v) - \langle f, v \rangle = \tilde{A}(u,v) - \langle f, v \rangle_{\mathbf{R}^2 \setminus \Omega_0} + \hat{A}(u,v) - \langle f, v \rangle_{\Omega_0 \setminus \Omega}$ that

$$A(u,v) - \langle f, v \rangle$$

$$= (\tilde{a}_{01}^T, \tilde{a}_{02}^T, \tilde{a}_{03}^T, \tilde{a}_{04}^T) \begin{pmatrix} K_z^{11} & K_z^{12} & 0 & 0 \\ K_z^{21} & K_z^{22} - K_y^{11} & -K_y^{12} & -K_y^{13} \\ 0 & -K_y^{21} & -K_y^{22} & -K_y^{23} \\ 0 & -K_y^{31} & -K_y^{32} & -K_y^{33} \end{pmatrix} \begin{pmatrix} a_{01} \\ a_{02} \\ a_{03} \\ a_{04} \end{pmatrix}$$

$$- (\tilde{a}_{01}^T, \tilde{a}_{02}^T, \tilde{a}_{03}^T, \tilde{a}_{04}^T) \begin{pmatrix} h_{01} \\ h_{02} + \tilde{h}_{02} \\ \tilde{h}_{03} \\ \tilde{h}_{04} \end{pmatrix},$$

where we have written $h_0 = (h_{01}^T, h_{02}^T)$ and $\tilde{h}_0 = (\tilde{h}_{02}^T, \tilde{h}_{03}^T, \tilde{h}_{04}^T)$.

For all test functions $v \in H_0 \cap (S_3^1(\diamondsuit) \oplus S_3^1(\diamondsuit))$, we have $\tilde{a}_{01} = 0$ and $\tilde{a}_{03} = 0$, where both $\tilde{a}_{02}$ and $\tilde{a}_{04}$ are arbitrary. It thus follows that

$$\begin{pmatrix} K_z^{22} - K_y^{11} & K_y^{13} \\ -K_y^{31} & -K_y^{33} \end{pmatrix} \begin{pmatrix} a_{02} \\ a_{04} \end{pmatrix}$$
$$= -\begin{pmatrix} K_z^{21} & K_y^{1-2} \\ 0 & -K_y^{32} \end{pmatrix} \begin{pmatrix} a_{01} \\ a_{03} \end{pmatrix} + \begin{pmatrix} h_{02} + \tilde{h}_{02} \\ \tilde{h}_{04} \end{pmatrix}. \tag{7.1}$$

Hence we have the following algorithm:

Algorithm 3. (Numerical Solution of Exterior Biharmonic Equations for a Non-star Shaped Domain)

Step 1: Start with $h_\ell = 0$ for sufficiently large ℓ. Compute h_k using the formula in (5.1) for $k = \ell - 1, \ell - 2, \ldots, 1, 0$.

Step 2: Solve the linear system (7.1).

Step 3: Compute a_{k+1} recursively using (5.2) for $k = 0, 1, \ldots$.

Step 4: Compute b_k using formula (5.3).

Finally, we remark on how to treat the singularities of the weak solutions near the corners of a polygonal domain Ω. We first subdivide $\mathbb{R}^2 \backslash \Omega$ as before. For each corner point (x_i, y_i) of Ω, we further subdivide the two quadrilaterals Q_i, Q_i' which share the edge $\langle (x_i, y_i), (x_{i,1}, y_{i,1}) \rangle$ using a parameter $\xi_{in} < 1$ in the same way as we subdivide $\mathbb{R}^2 \backslash \Omega$. More precisely we choose $(\tilde{x}_{j,k}, \tilde{y}_{j,k}) = (x_i, y_i) + \xi_{in}^k r(x_j - x_i, y_j - y_i)$, for each $(x_j, y_j) \in \{(x_{i-1}, y_{i-1}), (x_{i-1,1}, y_{i-1,1}), (x_{i,1}, y_{i,1}), (x_{i+1,1}, y_{i+1,1}), (x_{i+1}, y_{i+1})\}$. We connect these $(\tilde{x}_{j,1}, \tilde{y}_{j,1})$ together to form two sub-quadrilaterals Q_1 and Q_1' which are similar to Q and Q'. We repeat this construction to get Q_2 and Q_2' inside Q_1 and Q_1', and so on. By adding the two diagonals in each quadrilateral, we obtain a triangulated quadrangulation $\diamondsuit_i$ associated with (x_i, y_i). Let $S_3^1(\diamondsuit_i)$ be the C^1 cubic spline space over $\diamondsuit_i$. As before, we construct basic functions $\varphi_{j,k}$, $\psi_{x,j,k}$ $\psi_{y,j,k}$, $\eta_{h,j,k}$, $\eta_{v,j,k}$ using ξ_{in} instead of ξ. Let

$$A^{(i)}(u, v) - \langle f, v \rangle^{(i)} = \int_{\Omega_1 \cap \diamondsuit_1} (\triangle u \triangle v - fv) dx dy$$
$$= (\tilde{a}_0^{(i)})^T K_z^{(i)} a_0^{(i)} + (\tilde{a}_0^{(i)})^T h^{(i)}$$

for appropriate vectors $\tilde{a}_0^{(i)}$, $a_0^{(i)}$ and $h^{(i)}$. Then the values of the approximate weak solution u at $(\tilde{x}_{j,k}, \tilde{y}_{j,k})$ is $\dfrac{a_k^{(i)}}{\xi_{in}^k r(x_j - x_i, y_j - y_i)}$.

Since $a_k^{(i)} = (X^{(i)})^k a_0^{(i)} = c^{(i)}(\Gamma^{(i)})^k(c^{(i)})^{-1} a_0^{(i)}$ for some matrix $X^{(i)}$, $c^{(i)}$ and the diagonal matrix $\Gamma^{(i)}$ which contains the eigenvalues $\lambda^{(i)}$'s of $X^{(i)}$, we know that

$$u_s(\tilde{x}_{j,k}, \tilde{y}_{j,k}) \approx c_j(\lambda_{largest}^{(i)})^k(\xi_{in}^k r(x_j - x_i, y_j - y_i)$$

$$\approx c_j r(\tilde{x}_{j,k} - x_i, \tilde{y}_{j,k} - y_i)^{\frac{\log \lambda_{largest}^{(i)}}{\log \xi_{in}} + 1},$$

where $\lambda_{largest}^{(i)}$ denotes the largest eigenvalue of $X^{(i)}$ in absolute value, and c_j is a constant. As $k \to \infty$, $r(\tilde{x}_{jk} - x_i, \tilde{y}_{jk} - y_i) \to 0$ and

$$\frac{\partial^2}{\partial x^2} u_s(x_i, y_i) \cong \frac{\partial^2}{\partial x^2} u_s(\tilde{x}_{j,k}, \tilde{y}_{j,k})$$

$$\cong c_j r(\tilde{x}_{jk} - x_i, \tilde{y}_{jk} - y_i)^{\frac{\log \lambda_{largest}^{(i)}}{\log \xi_{in}} - 1}.$$

If $c_j \neq 0$ and $\frac{\log|\lambda_{largest}^{(i)}|}{\log \xi_{in}} < 1$ then u has a singularity at (x_i, y_i).

The above argument provides a numerical method to determine the singularity at all corner points of $\partial\Omega$.

Acknowledgments. The first author of this paper would like to thank the National Science Foundation for its support under grant #DMS-9870187.

References

1. Ciavaldini, J. F. and J. C. Nedelec, Sur l'élément de Fraeijs de Veubeke et Sander, Rev. Francaise Automat. Informat. Rech. Opér., Anal. Numer., R2 (1974), 29–45.

2. Fraeijs de Veubeke, B, A conforming finite element for plate bending, J. Solids Structures **4** (1968), 95–108.

3. Girault, V. and P.A. Raviart, *Finite Element Methods for Navier-Stokes Equations*, Springer-Verlag, 1986.

4. Lai, M. J., Scattered data interpolation and approximation by using bivariate C^1 piecewise cubic polynomials, Comput. Aided Geom. Design **13** (1996), 81–88.

5. Lai, M. J. and L. L. Schumaker, On the approximation power of splines on triangulated quadrangulations, SIAM Numer. Anal. **36** (1999), 143–159.

6. Sander, G., Bornes supérieures et inférieures dans l'analyse matricielle des plaques en flexion-torsion, Bull. Soc. Royale Sciences Liège **33** (1964), 456–494.

7. Ying, L. A., The infinite similar element method for calculating stress intensity factors, Scientia Sinica **21** (1978), 19–43.

8. Ying, L. A., *Infinite Element Method*, Peking Univ. Press and Vieweg Publishing, 1995.

Ming-Jun Lai
Department of Mathematics
The University of Georgia
Athens, GA 30602
mjlai@math.uga.edu

Paul Wenston
Department of Mathematics
The University of Georgia
Athens, GA 30602
paul@math.uga.edu

Lung-an Ying
School of Mathematical Sciences
Peking University
Beijing, China
yingla@pku.edu.cn

Lagrange Interpolation by C^1 Cubic Splines on Triangulations of Separable Quadrangulations

Günther Nürnberger, Larry L. Schumaker, and Frank Zeilfelder

Abstract. We construct a local Lagrange interpolation method based on C^1 cubic splines on certain triangulations obtained from a separable quadrangulation by adding one or two diagonals. The method provides optimal order approximation of smooth functions. We illustrate the efficiency of the method by applying it to fit a large scattered data set corresponding to real world terrain.

§1. Introduction

Recently there has been considerable interest in spaces of polynomial splines defined on triangulations obtained from a quadrangulation by inserting one or both diagonals in each quadrangle, see [11,12,14,17,21,29]. At the same time, there has also been considerable interest in finding C^1 spline spaces which are suitable for Lagrange interpolation, see [7,18–27]. In this paper we consider the following related problem:

Problem 1. Let $\mathcal{V} := \{\xi_i\}_{i=1}^n$ be a set of points in the plane, and let $\Diamond$ be a quadrangulation with vertices at the points of $\mathcal{V}$. Find a triangulation $\triangle$ of $\Diamond$ and a set of additional points $\{\xi_i\}_{i=n+1}^N$ such that for every choice of the data $\{z_i\}_{i=1}^N$, there is a unique C^1 cubic spline $s \in \mathcal{S}_3^1(\triangle)$ satisfying

$$s(\xi_i) = z_i, \qquad i = 1, \ldots, N. \tag{1}$$

We call $P := \{\xi_i\}_{i=1}^N$ and $\mathcal{S}_3^1(\triangle)$ a Lagrange interpolation pair.

Approximation Theory X: Wavelets, Splines, and Applications   405
Charles K. Chui, Larry L. Schumaker, and Joachim Stöckler (eds.), pp. 405–424.
Copyright ⊕ 2002 by Vanderbilt University Press, Nashville, TN.
ISBN 0-8265-1416-2.

In particular, we are interested in constructing Lagrange interpolation pairs in such a way that the interpolating spline depends only locally on the data. One of the main difficulties of constructing such pairs is that the known local Hermite interpolation methods [1,2,6,9–12,15–17,28,29] cannot be directly transformed to local Lagrange interpolation schemes. Recently, a local Lagrange interpolation method was developed in [21] for a special class of checkerboard quadrangulations. In this paper we extend those results to a much larger class of so-called *separable quadrangulations* defined in Section 2 below. In particular, we give explicit algorithms for creating the Lagrange interpolation pair $\mathcal{S}_3^1(\triangle)$ and P in such a way that for any given data, the interpolating spline depends locally on the data and can be constructed in $\mathcal{O}(N)$ operations. We also show that the method produces optimal order approximation of smooth functions, even though several quadrilaterals only contain one diagonal.

While not directly applicable to interpolation of scattered data, Lagrange interpolating pairs nevertheless provide a useful tool for *fitting* scattered data. For example, they can be integrated into a two-stage method (cf. [30] and Sect. 7 below) where the first stage involves interpolating the data with a linear spline surface s_1 over a triangulation of the data points, and where the second stage involves interpolation of s_1 by a C^1 cubic spline s_2 on a much coarser triangulation $\triangle$, and based on samples of s_1 at the points of P.

The paper is organized as follows. In Section 2 we define separable quadrangulations, and describe an algorithm for classifying quadrilaterals which will be used in Section 3 to define an associated triangulation and corresponding Lagrange interpolation set of points P. In Section 4 we compute the dimension of the spline space $\mathcal{S}_3^1(\triangle)$, and illustrate it with two examples in Section 5. In Section 6 we introduce the cardinal basis splines associated with P, and show that they form a stable local basis for $\mathcal{S}_3^1(\triangle)$. A corresponding interpolation operator is defined in Section 6, and is shown to provide optimal order approximation of smooth functions. In Section 7 we illustrate the performance of the two-stage method mentioned above to the modelling of some very complicated terrain based on a very large set of scattered data. We conclude with several remarks.

§2. Decomposing Separable Quadrangulations

Throughout this paper we assume that $\Diamond$ is a quadrangulation of a simply connected set Ω, and that the largest angle in any quadrilateral $Q \in \Diamond$ is less than π.

Definition 2. *A quadrangulation $\Diamond$ is said to be* **separable** *provided there exists a set $\Diamond_0$ of quadrilaterals in $\Diamond$ such that for every interior vertex v of $\Diamond$, there is a unique quadrilateral $Q \in \Diamond_0$ with vertex at v.*

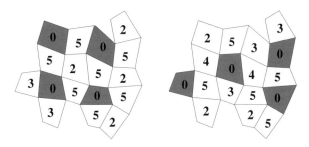

Fig. 1. Separable quadrangulations where every interior vertex is of degree four.

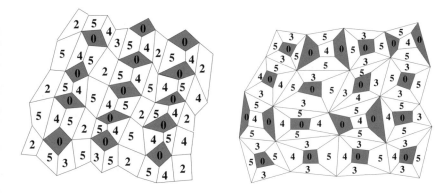

Fig. 2. Separable quadrangulations which contain interior vertices of odd degree.

Fig. 3. A quadrangulation which is not separable.

Figures 1 and 2 show several examples of separable quadrangulations, where the quadrilaterals in $\Diamond_0$ are shown in black. As shown in Figure 1, for a given quadrangulation $\Diamond$, there may be more than one way to choose $\Diamond_0$. The class of separable quadrangulations is strictly larger than the class of checkerboard quadrangulations treated in [21], as shown by the quadrangulations in Figure 2. On the other hand, it is easy to see that not all quadrangulations are separable; an example is given in Figure 3. In the following, we say that two quadrilaterals in $\Diamond$ are **neighbors** provided they have a common edge. They are said to **touch** if they have a common vertex but no common edge.

As a first step towards constructing a triangulation $\triangle$ of $\diamond$ and a corresponding set $P := \{\xi_i\}_{i=1}^N$ of Lagrange interpolation points, we now describe an algorithm for decomposing $\diamond$ into six classes of quadrilaterals:

$$\diamond := \diamond_0 \cup \diamond_1 \cup \diamond_2 \cup \diamond_3 \cup \diamond_4 \cup \diamond_5. \tag{2}$$

For convenience of notation, we write the partial unions as

$$\diamond_{0i} := \diamond_0 \cup \cdots \cup \diamond_i, \qquad i = 1, \ldots, 5.$$

Algorithm 3. *(Construction of $\diamond_1$, $\diamond_2$ and $\diamond_3$). Suppose $\diamond$ is a separable quadrangulation, and that $\diamond_0$ is a subset of quadrilaterals as in Definition 2.*

1) *Start with $\diamond_1 = \emptyset$, and repeat as often as possible: if there exists Q in $\diamond$ which does not touch any quadrilateral in $\diamond_{01}$, add Q to $\diamond_1$.*

2) *Start with $\diamond_2 = \emptyset$, and repeat as often as possible: if there exists Q which is not in $\diamond_{02}$ and is not a neighbor of any quadrilateral in $\diamond_{02}$, add Q to $\diamond_2$.*

3) *Start with $\diamond_3 = \emptyset$, and repeat as often as possible: if there exists Q which is not in $\diamond_{03}$ but which has exactly one neighbor in $\diamond_{02}$ and no neighbor in $\diamond_3$, add Q to $\diamond_3$.*

We note that after carrying out Algorithm 3, some of the sets $\diamond_i$ with $0 \leq i \leq 3$ can be empty. We say that a collection $\mathcal{C} := \{Q_1, \ldots, Q_m\}$ of quadrilaterals in $\diamond$ is a **connected component** provided that Q_i and Q_{i+1} are neighbors for each $i = 1, \ldots, m-1$. We call a connected component $\mathcal{C}$ a **chain** if $|j - i| > 1$ implies Q_i and Q_j are not neighbors (except that we allow Q_m and Q_1 to be neighbors, in which case we have a **closed** chain).

Lemma 4. *The set $\diamond^* = \diamond \setminus \diamond_{03}$ consists of a finite number of chains.*

Proof: First we observe that any $Q \in \diamond^*$ has at most two neighbors in $\diamond^*$. Indeed, if it had four neighbors in $\diamond^*$, it would have been selected for $\diamond_1$ or $\diamond_2$. If it had three neighbors in $\diamond^*$, it would have been selected for $\diamond_3$ if it also had one neighbor in $\diamond_1 \cup \diamond_2$, and would have been selected for $\diamond_2$, otherwise. It is now clear that any connected component of $\diamond^*$ must be a chain. $\square$

Figure 4 shows the four different types of chains in $\diamond^*$. Since $\diamond$ is a separable quadrangulation, it follows that no chain in $\diamond^*$ is a cell of quadrilaterals, i.e., a closed chain where all its quadrilaterals have a common vertex. We have marked the quadrilaterals in Figure 4 to illustrate the construction of the sets $\diamond_4$ and $\diamond_5$ as described in the following algorithm.

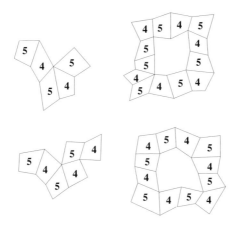

Fig. 4. Four types of chains can appear in $\diamondsuit^*$.

Algorithm 5. *(Construction of $\diamondsuit_4$ and $\diamondsuit_5$).*

1) *For every chain $\mathcal{C} = \{Q_1, \ldots, Q_m\}$, put the even numbered quadrilaterals in $\diamondsuit_4$.*

2) *Put all remaining quadrilaterals in $\diamondsuit_5$.*

We illustrate the decomposition (2) in Figures 1 and 2, where the quadrilaterals in $\diamondsuit_i$ are marked with the corresponding integer, and where the quadrilaterals in $\diamondsuit_0$ are colored black to clearly show how they cover the interior vertices of $\diamondsuit$. For later use, we note that if $Q \in \diamondsuit_4$, then m of its edges with $2 \leq m \leq 3$ lie in $\diamondsuit_{03}$, while if $Q \in \diamondsuit_5$, then m of its edges with $2 \leq m \leq 4$ lie in $\diamondsuit_{04}$.

§3. Construction of a Lagrange Interpolation Pair

Suppose $\diamondsuit$ is a separable quadrangulation which has been decomposed as described in the previous section. In this section we present an algorithm which will triangulate $\diamondsuit$ and also produce a set of points P such that P and $\mathcal{S}_3^1(\triangle)$ form a Lagrange interpolation pair. In creating the algorithm, we have taken special care to insure that the resulting interpolation method is both stable and local, see below.

The points of P will be chosen as a subset of the set $\mathcal{D}$ of **domain points** associated with the Bernstein-Bézier representation of $\mathcal{S}_3^1(\triangle)$, cf. [3,12–14]. As is well known, $\mathcal{D}$ contains one point at each vertex v of $\triangle$, two points on each edge $e := \langle u, v \rangle$ located at $(2u + v)/3$ and $(u + 2v)/3$, and one point at the center $\xi_T := (u + v + w)/3$ of each triangle $T := \langle u, v, w \rangle$. For each $v \in \triangle$, we define the **disk of radius 1 around** v to be the set $D(v)$ containing v and the closer of the two domain points in the interior of each edge attached to v.

Algorithm 6. *(Constructing a Lagrange interpolation pair).* Let $\diamondsuit$ be a separable quadrangulation. Initialize $P = \emptyset$.

1) For each $Q := \langle v_1, v_2, v_3, v_4 \rangle$ in $\diamondsuit_0 \cup \diamondsuit_1$: Divide Q into two triangles $T_Q := \langle v_1, v_2, v_3 \rangle$ and $\tilde{T}_Q := \langle v_1, v_3, v_4 \rangle$. For each $i = 1, \ldots, 4$, add the domain points in $D(v_i) \cap Q$ to P and mark v_i. Add the point $(v_1 + v_2 + v_3)/3$.

2) For $m = 3, \ldots, 0$, repeat as often as possible: If there exists $Q := \langle v_1, v_2, v_3, v_4 \rangle$ in $\diamondsuit_2 \cup \diamondsuit_3$ with m unmarked vertices $v_1, \ldots, v_m$, add the domain points $D(v_i) \cap Q$ and mark v_i for $i = 1, \ldots, m$. Divide Q into two triangles by inserting the diagonal $\langle v_1, v_3 \rangle$. If $Q \in \diamondsuit_2$, add the point $(v_1 + v_2 + v_3)/3$.

3) For each $Q \in \diamondsuit_4$: Divide Q into four triangles by inserting both diagonals of Q. Add the points in $D(v) \cap Q$ for all unmarked vertices v of Q. Add the point w_Q where the diagonals cross. If Q shares only two edges with quadrilaterals in $\diamondsuit_{03}$, add the point $(2w_Q + v)/3$, where v is a vertex of one of the shared edges but not both. Mark the vertices of Q.

4) For each $Q \in \diamondsuit_5$: Divide Q into four triangles by inserting both diagonals of Q. Add the points in $D(v) \cap Q$ for all unmarked vertices v of Q. Let $\mathcal{M}_5$ be the collection of quadrilaterals in $\diamondsuit_5$ which have already been dealt with. If Q shares three or fewer edges with quadrilaterals in $\diamondsuit_{04} \cup \mathcal{M}_5$, add the point w_Q where the diagonals cross. If Q shares only two edges, add the point $(2w_Q + v)/3$, where v is a vertex of one of the shared edges but not both. Mark the vertices of Q and put Q in $\mathcal{M}_5$.

The choices of interpolation points in steps 1) – 3) of Algorithm 6 are illustrated in Figures 5 – 7, where the Lagrange interpolation points are marked with black and grey dots. Figure 6 shows the various cases of step 2), depending on the number of unmarked vertices, where the marked vertices are indicated with an open disk. Here, the point marked with a grey dot is only chosen as an interpolation point if the quadrilateral is in $\diamondsuit_2$. Figure 7 shows the various choices of interpolation points in step 3) of the algorithm. In this step the choice of points depends on the number of edges (drawn with thick lines) which are shared with quadrilaterals in $\diamondsuit_{03}$.

Figures 8 and 9 show the results of applying Algorithm 6 to the quadrangulations on the left in Figures 1 and 2. The added diagonals are drawn with thick lines, and the points in the local Lagrange interpolation set P are marked with black dots. The construction shows that the quadrilaterals in $\diamondsuit_{03}$ contain exactly one diagonal, while the remaining quadrilaterals in $\diamondsuit$ contain both diagonals.

Fig. 5. Choice of interpolation points in 1) of Algorithm 6.

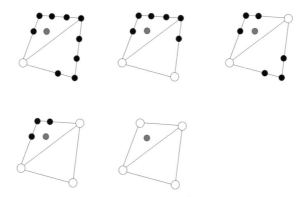

Fig. 6. Choices of interpolation points in 2) of Algorithm 6.

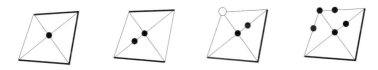

Fig. 7. Choices of interpolation points in 3) of Algorithm 6.

Theorem 7. *Let $\triangle$ and $P := \{\xi_i\}_{i=1}^{N}$ be the triangulation and point set constructed by Algorithm 6. Then P and $\mathcal{S}_3^1(\triangle)$ form a Lagrange interpolation pair.*

Proof: Given $z := \{z_i\}_{i=1}^{N}$, we need to prove that there exists a unique $s \in \mathcal{S}_3^1(\triangle)$ satisfying the interpolation conditions (1). Assuming s is represented in Bernstein-Bézier form (cf. [3,12–14]), we now show how to compute the corresponding B-coefficients step by step, working through the quadrilaterals in groups. We treat all of the quadrilaterals in one group before moving on to the next group. We note that for each vertex v, once we have determined the coefficients corresponding to v and two additional (noncollinear) points in $D(v)$, then using the C^1 smoothness conditions we can compute all other coefficients corresponding to domain points in $D(v)$. These in turn determine its gradient at v. We maintain a list $\mathcal{V}_G$ of such vertices, starting with $\mathcal{V}_G = \emptyset$.

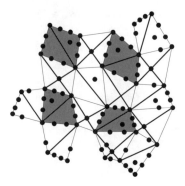

Fig. 8. The Lagrange interpolation pair for Example 9.

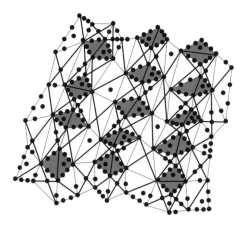

Fig. 9. The Lagrange interpolation pair for Example 10.

1) $Q \in \Diamond_0 \cup \Diamond_1$. Since the set P contains all of the domain points on the boundary of Q, we can solve the Lagrange interpolation problem at the four points on each edge to uniquely determine the associated coefficients. Thus we can put all four vertices of Q into $\mathcal{V}_G$. Now there remains just one undetermined coefficient of $s|_{T_Q}$, namely the one associated with the point at the center of T_Q. But this is uniquely determined by the interpolation condition at that point. The remaining coefficient of $s|_{\tilde{T}_Q}$ can then be uniquely determined from one C^1 smoothness condition across the edge between T_Q and $\tilde{T}_Q$.

2) $Q \in \Diamond_2 \cup \Diamond_3$. We first deal with all such Q which correspond to $m = 3$ in step 2 of Algorithm 6. This means that (numbering the vertices of Q appropriately), we know that the set P contains the points $D(v_i) \cap Q$ for $i = 1, 2, 3$, and that the fourth vertex v_4 belongs to $\mathcal{V}_G$ – i.e., the coefficients in $D(v_4) \cap Q$ are already known or can be computed from

smoothness conditions from previously computed coefficients in $D(v_4) \cap T$ for some triangle T attached to v_4. Then using Hermite interpolation on each of the edges of Q, we can find all of the coefficients of $g := s|_Q$ corresponding to domain points on the edges of Q. If $Q \in \diamondsuit_2$, we then compute the coefficient of g corresponding to the center of one of the triangles in Q by interpolation at that point, while if $Q \in \diamondsuit_3$, we can compute that coefficient by a smoothness condition across the edge which joins Q to a quadrilateral in $\diamondsuit_{02}$. The coefficient at the center of the other triangle can now be computed by a C^1 smoothness condition across the edge between the triangles. We now repeat this process for quadrilaterals Q corresponding to $m = 2$, then $m = 1$, and finally $m = 0$.

3) $Q \in \diamondsuit_4$. In this case Q is split into four subtriangles by the two diagonals of Q which we assume cross at w_Q. Let $T_i := \langle w_Q, v_i, v_{i+1} \rangle$, $i = 1, \ldots, 4$, be the four subtriangles of Q ($v_5 := v_1$). At least three vertices of Q lie in $\mathcal{V}_G$. First we treat all quadrilaterals with three such vertices, then all quadrilaterals with four such vertices. In both cases we can use univariate Hermite interpolation to compute the coefficients of $g := s|_Q$ corresponding to all domain points on the boundary of Q. The computation of the remaining coefficients depends on how many edges of Q lie in $\diamondsuit_{03}$. If there are three such edges, then using the C^1 smoothness conditions determines the B-coefficients of the polynomial pieces associated with the domain points at the centers of the corresponding three triangles which we may assume without loss of generality are T_1, T_2, T_3. Then a C^1 smoothness condition across the edge $\langle w_Q, v_2 \rangle$ determines the B-coefficient corresponding to the domain point $(2w_Q + v_2)$. The coefficient corresponding to the domain point $(2w_Q + v_3)$ is determined in a similar way. Since interpolation at w_Q determines the associated coefficient, we now know the gradient of s at w_Q, and all remaining coefficients of $s|_Q$ are uniquely determined.

The situation is slightly different if only two edges of Q lie in $\diamondsuit_{03}$. First suppose these two edges are opposite. Without loss of generality we may assume they are the edges $\langle v_1, v_2 \rangle$ and $\langle v_3, v_4 \rangle$. Then C^1 smoothness across these edges determines the coefficients of s associated with the centers of the triangles T_1 and T_3. Now P also contains w_Q as well as the point $(2w_Q + v_j)/3$ for some vertex v_j of Q, and univariate Hermite interpolation uniquely determines s on the edge $e := \langle w_Q, v_j \rangle$ and thus the coefficient corresponding to $(2w_Q + v_j)/3$. At this point we can use C^1 smoothness across e to determine the coefficient corresponding to the center point of the neighboring triangle T, then C^1 smoothness across the edge $\tilde{e}$ leading to the next triangle to get another coefficient corresponding to a domain point in T of the form $(2w_Q + v_k)/3$. This determines the gradient of s at w_Q and subsequently all remaining coefficients of $s|_Q$. Finally, suppose the two edges lying in $\diamondsuit_{03}$ are adjoining. Without loss

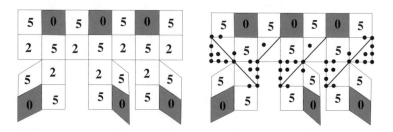

Fig. 10. Choice of points for quadrilaterals in $\Diamond_2$ near the boundary.

of generality, we may assume they are the edges $\langle v_1, v_2 \rangle$ and $\langle v_2, v_3 \rangle$. In this case C^1 smoothness across the edge $\langle w_Q, v_2 \rangle$ determines the coefficient corresponding to $(2w_Q + v_2)/3$. The coefficient at w_Q and one other coefficient in $D(w_Q)$ are determined by interpolation as before.

4) $Q \in \Diamond_5$. This case is almost the same as case 3, except that now we distinguish various subcases according to how many edges Q shares with $\Diamond_{04}$. The other difference is that now it is possible that all four edges are shared. In this case we use C^1 smoothness across all four of the edges to determine the coefficients at the centers of each of the triangles $T_1, \ldots, T_4$. Then using C^1 smoothness across each of the edges $\langle w_Q, v_i \rangle$, we can determine all of the coefficients associated with the domain points $(2w_Q + v_i)/3$ for $i = 1, \ldots, 3$, thus giving the gradient of s at w_Q. All other coefficients follow as before. $\square$

The process of computing B-coefficients described in Theorem 7 is local in the sense that at each step we compute coefficients associated with just one quadrilateral. It follows that the complexity of the overall computation is $\mathcal{O}(N)$. It is also local in the sense that the values of the computed coefficients in a given quadrilateral depend only on data in nearby quadrilaterals (how far away is made more precise in Section 6 below). The desire to make the method as local as possible is the reason for the relatively complex structure of Algorithm 6, in particular as regards the choice of points in quadrilaterals in $\Diamond_2$ which lie near the boundary of $\Diamond$. In particular, it is important to deal with all quadrilaterals in $\Diamond_2$ with 3 unmarked vertices before doing those with 2 unmarked vertices, and so on. Figure 10 shows how this works in a simple case.

We also claim that the process of computing the B-coefficients of s in Theorem 7 is stable in the sense that the maximum absolute value of the computed coefficients is bounded by $K\|z\|$, where $\|z\| := \max_{1 \le i \le N} |z_i|$ and K is a constant which depends only on the smallest angle $\theta_\triangle$ in the triangulation $\triangle$, see Remark 6. Indeed, it is well-known that computing coefficients from smoothness conditions is stable in this sense, see [13]. On the other hand, computing the coefficient c_{111} of a cubic polynomial in Bernstein-Bézier form from an interpolation condition at the point ξ_{111} at the center of the triangle is also stable because the corresponding 1×1 matrix corresponds to $B_{111}^3(\xi_{111}) = 2/9$, which is independent of the geometry.

§4. Dimension of $\mathcal{S}_3^1(\triangle)$

If $\mathcal{S}_3^1(\triangle)$ and P form a Lagrange interpolation pair, then the fact that interpolation at the points of P uniquely determines an $s \in \mathcal{S}_3^1(\triangle)$ immediately implies that $\dim \mathcal{S}_3^1(\triangle) = \#P = N$. In this section we give an explicit formula for N which is useful for comparing sizes of spline spaces.

Let n_V and n_E denote the number of vertices and edges of $\Diamond$, respectively, and let

$$n_1 := \# \Diamond_{03}.$$

This is the number of quadrilaterals for which Algorithm 6 inserts just one diagonal in constructing $\triangle$.

Theorem 8. *Suppose $\triangle$ and P are a Lagrange interpolation pair produced by Algorithm 6 applied to a separable quadrangulation Q. Then*

$$\dim \mathcal{S}_3^1(\triangle) = 3n_V + n_E - 3n_1. \tag{3}$$

Proof: Suppose ϕ is the triangulation obtained from Q by inserting both diagonals in each quadrilateral. Note that by definition, it is impossible for two quadrilaterals in $\Diamond_3$ to be neighbors. Moreover, the set $\Diamond_{03}$ does not contain a closed chain of quadrilaterals. It is shown in [12] that $\dim \mathcal{S}_3^1(\phi) = 3n_V + n_E$. But it is easy to see that if we remove one diagonal from a quadrilateral, then the dimension goes down by 3. If this process is repeated n_1 times, the dimension is reduced by $3n_1$, since the quadrilaterals of $\Diamond_{03}$ are isolated in the above sense. $\square$

§5. Examples

To illustrate the above results, we give two examples.

Example 9. *Let $\Diamond$ be the quadrangulation shown in Figure 1.*

Discussion: The triangulation and point set produced by Algorithm 6 are shown in Figure 8, where the points of P are marked with black dots. In this case $V_B = 22$, $V_I = 7$ and (cf. [14]),

$$n_Q = \frac{V_B + 2V_I - 2}{2} = 17, \qquad n_E = \frac{3V_B + 4V_I - 4}{2} = 45.$$

Since $n_1 = 10$, Theorem 8 asserts that $N = \dim \mathcal{S}_3^1(\triangle) = 3 \cdot 29 + 45 - 3 \cdot 10 = 102$. $\square$

Example 10. Let $\Diamond$ be the quadrangulation shown in Figure 2.

Discussion: The triangulation and point set produced by Algorithm 6 are shown in Figure 9, where the points of P are marked with black dots. In this case $V_B = 30$, $V_I = 50$, $n_Q = 64$, $n_E = 143$, and $n_1 = 28$. Theorem 8 gives $N = \dim \mathcal{S}_3^1(\triangle) = 299$. $\square$

§6. Bounding the Error of Interpolation

As a tool for obtaining an error bound for our interpolation method, we now introduce the **cardinal basis splines** associated with the Lagrange interpolation pair $P := \{\xi_i\}_{i=1}^N$ and $\mathcal{S}_3^1(\triangle)$. For each $1 \leq i \leq N$, let B_i be the unique spline in $\mathcal{S}_3^1(\triangle)$ such that

$$B_i(\xi_j) = \delta_{ij}, \qquad i, j = 1, \ldots, N. \tag{4}$$

Theorem 11. *The set of splines* $\{B_i\}_{i=1}^N$ *forms a stable local basis for* $\mathcal{S}_3^1(\triangle)$.

Proof: The proof of Theorem 7 shows that all B-coefficients of B_i are uniquely and stably determined for $i = 1, \ldots, N$. Moreover, we claim

$$\operatorname{supp}(B_i) \subseteq \operatorname{star}^8(Q), \qquad i = 1, \ldots, N, \tag{5}$$

where $\operatorname{star}(Q)$ is the union of all quadrilaterals which intersect with Q in at least one point, and $\operatorname{star}^m(Q) := \operatorname{star}^{m-1}(\operatorname{star}(Q))$. To verify (5), first consider the case where B_i corresponds to a point $\xi_i \in P$ in the center of a subtriangle of a quadrilateral $Q \in \Diamond_2 \cup \Diamond_3$. In this case P contains enough points to force B_i and its gradient to be zero at all vertices (and thus along all edges) of $\Diamond$. But using C^1 smoothness conditions across edges to compute other coefficients inside of triangles, the nonzero value of the coefficient corresponding to ξ_i can **propagate** to the centers of neighboring triangles. Such propagation can occur only within a chain of quadrilaterals. Moreover, by the choice of points associated with quadrilaterals in $\Diamond_4$, it is easy to see that the maximal chain of propagation is of the form $Q_1, \ldots, Q_5$, where $Q_i \in \Diamond_{i+1}$ for $i = 1, \ldots, 4$ and $Q_5 \in \Diamond_5$. (Two quadrilaterals in $\Diamond_5$ can be neighbors in an odd circular chain). This case is illustrated in Figure 11, where ξ_i is symbolized by a black dot and the dotted lines indicate odd circular chains. It follows that the support of B_i is at most $\operatorname{star}^4(Q)$ in this case. The case where $\xi_i \in P$ is the center of a subtriangle of a quadrilateral in $\Diamond_{01}$ is analogous.

We now consider the case where B_i corresponds to a point $\xi_i \in P$ in a disk $D(v) \cap Q$ for some vertex v of a quadrilateral Q. In this case, propagation can take place along an edge $e := \langle v, w \rangle$ so that a nonzero value or gradient at v can lead to a nonzero value or gradient at w. This is of concern when v and w are not in the same quadrilateral. We claim, however, that such propagation is quite limited. Clearly, propagation to w is not possible if w is a vertex of a quadrilateral in $\Diamond_{01}$. Thus, any extended propagation would have to occur along a path of edges passing through boundary vertices. Such paths can be quite complex (see Figures 10 and 12). However, for propagation to occur, e must contain the point $(2w + v)/3$ but not the point $(v + 2w)/3$. Making use of the order in which the quadrilaterals are treated in step 2 of Algorithm 6, it can be verified that propagation along edges is limited to going from one quadrilateral Q to another $\tilde{Q}$ in star$^4(Q)$. But a nonzero gradient at w can lead to a nonzero center coefficient in a quadrilateral in $\Diamond_2 \cup \Diamond_3$ with vertex at w, where it can then propagate along a chain as discussed in the previous paragraph. Combining these observations, we conclude that the support of B_i is at most star$^8(Q)$. $\square$

Figure 12 shows an example of a maximal chain of propagation for a basis spline B_i which corresponds to a point $\xi_i \in P$ in a disk $D(v) \cap Q$ for some vertex v of a quadrilateral Q. In this figure the open disks indicate the vertices that are marked after step 1) of Algorithm 6. Here, we assume that in step 2) of the algorithm, the quadrilaterals in $\Diamond_2$ are considered from the left to the right. The small numbers inside these quadrilaterals indicate the number of unmarked vertices, and the black dots are the chosen interpolation points. The dotted line indicates an odd circular chain.

Although by Theorem 11 in some cases there may be basis functions in $\mathcal{S}_3^1(\triangle)$ with rather large supports, such cases are actually quite rare, and most of the time the basis splines have supports which are much smaller than star$^8(Q)$ for some Q. Indeed, the supports frequently consist only of a small chain of quadrilaterals.

In view of (4), it is clear that for every $f \in C(\Omega)$, the spline $s \in \mathcal{S}_3^1(\triangle)$ which interpolates f at the points of P is given by the formula

$$\mathcal{I}f := \sum_{i=1}^{N} f(\xi_i)\, B_i. \tag{6}$$

We can regard $\mathcal{I}$ as an operator mapping functions in $C(\Omega)$ into splines in $\mathcal{S}_3^1(\triangle)$. It should be emphasized that we have introduced the cardinal basis splines and this formula purely as a theoretical tool. We do not make use of this representation for computing or storing an interpolating spline – for that purpose, the Bernstein-Bézier representation is much better suited. But it is useful for proving the following theorem concerning the approximation of functions in the Sobolev space $W_\infty^{m+1}(\Omega)$. Let $|\cdot|_{m+1,\infty}$ be the standard Sobolev seminorm, and let $|\triangle|$ be the maximum of the diameters of the triangles in $\triangle$.

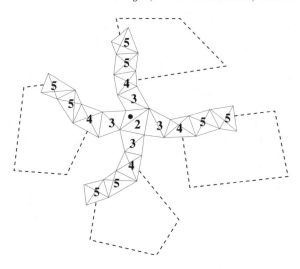

Fig. 11. Maximal chains of propagation in the interior.

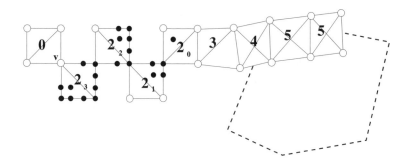

Fig. 12. Maximal chains of propagation along boundary edges.

Theorem 12. *There exists a constant C depending only on the smallest angle in $\triangle$ such that if f is in the Sobolev space $W_\infty^{m+1}(\Omega)$ with $0 \leq m \leq 3$,*

$$\|D_x^\alpha D_y^\beta (f - \mathcal{I}f)\|_{\infty,\Omega} \leq C \, |\triangle|^{m+1-\alpha-\beta} \, |f|_{m+1,\infty,\Omega}, \qquad (7)$$

for all $0 \leq \alpha + \beta \leq m$.

Proof: We apply Theorem 5.1 of [13]. Clearly, $\mathcal{I}p = p$ for all cubic polynomials. The hypothesis (5.3) of that theorem is trivial since $|f(\xi)| \leq \|f\|_{T_\xi}$, where T_ξ is the triangle which contains ξ. □

The analog of this error bound also holds for the p-norm, $1 \leq p < \infty$. For $p = \infty$, this result can also be established using the Bramble-Hilbert lemma, or by using the weak-interpolation methods described in [7,9,18,22,26,27].

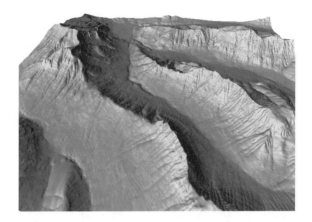

Fig. 13. A piecewise linear interpolating spline based on 1,142,239 points.

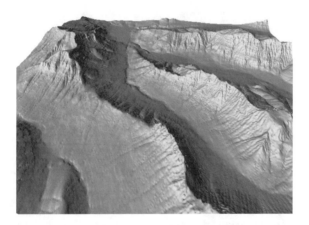

Fig. 14. A cubic spline interpolating the surface of Fig. 13 at 199,692 points.

§7. Numerical Example

In this section we illustrate the application of our method to the problem of fitting a smooth surface to a very large set of real world scattered data. We begin with elevation values of a terrain surface at 1,142,239 points in a rectangular region H. We apply the two-stage method mentioned in the introduction; *i.e.*, we first interpolate the data with a C^0 linear spline based on a triangulation of the data points, and then fit the resulting surface with a C^1 cubic spline based on the checkerboard quadrangulation obtained by imposing a $n \times n$ grid on H, where $n = 258$. In this case we have an interpolation pair involving 199,692 interpolation points.

Fig. 15. The interpolating spline of Figure 14 with texture.

Figure 13 shows the C^0 piecewise linear spline which interpolates the original set of data points, while Figure 14 shows the corresponding C^1 cubic spline surface. In terms of number of data needed to represent the surfaces, this corresponds to a compression rate of about 5.7. Despite this relatively high compression rate, the figures show that the cubic spline provides a realistic reconstruction of the original surface. The maximal error is about 10 meters.

The implementation of the interpolation method is straightforward, and as discussed above has $\mathcal{O}(N)$ complexity. In particular, for this example the construction of the cubic interpolating spline took about 3 minutes on a standard PC. Figure 15 shows the results of applying some texture to the spline of Figure 14.

Our numerical tests with complicated real world surfaces indicate that the best compression rates can be obtained by replacing the interpolation conditions at interior points of quadrilaterals containing only one diagonal by a condition forcing the surface to reproduce quadratic polynomials. These methods are under further study.

Numerical tests for synthetic data show the optimal approximation order of the methods. The interested reader may consult [21,25–27] for experiments with related local Lagrange interpolation methods.

§8. Remarks

Remark 1. We do not know of any way to characterize the set of separable quadrangulations in terms of elementary properties of quadrangulations such as numbers of edges attached to each vertex, etc. Given a quadrangulation $\Diamond$, we can always start with an arbitrary quadrilateral, mark its vertices, then recursively look for additional quadrilaterals with four unmarked vertices. The process always stops after at most $O(n/4)$ steps, where n is the number of vertices of $\Diamond$. If this process does not reveal whether $\Diamond$ is separable, we can start with a different quadrilateral and try again, but there is no guarantee of getting a definitive answer without looking at all possible starting quadrilaterals, and searching through all possible choices along the way (which would be a very expensive process). We are currently developing an alternative approach to solving Problem 1 which will work for arbitrary quadrangulations.

Remark 2. If we are given Hermite data at a set of points $\mathcal{V}$, then for any triangulation $\triangle$ with vertices at the points of $\mathcal{V}$, there are several well-known local interpolation methods based on C^1 splines. In particular, we can use the classical quintic macro-element, or if we are willing to split the triangles of $\triangle$, either of the Clough-Tocher [6] or Powell-Sabin [28] macro-elements, which make use of cubic and quadratic splines, respectively. A generalization of these methods to C^r splines was given in [1,2,15,16]. There are also methods based on quartic splines, see [4,5,10].

Remark 3. There is also a simple local interpolation method for Hermite data given at the vertices of an arbitrary quadrangulation $\Diamond$. It is based on C^1 cubic splines defined on the triangulation $\oplus$ obtained by adding both diagonals to each quadrilateral, see [11,12,29].

Remark 4. The situation is much more complicated for *Lagrange* interpolation since Hermite interpolation methods cannot be directly converted to Lagrange interpolation schemes. In particular, the interpolation points near the boundary of $\Diamond$ have to be chosen carefully in order to guarantee that the method is local. Local Lagrange interpolation methods based on C^1 splines on triangulations have been obtained only recently, see [21,25–27]. The constructions involve coloring the triangles. Results on non-local Lagrange interpolation and Hermite interpolation by bivariate splines on classes of triangulations were given in [7–9,18–20,22–24].

Remark 5. The interpolation method described here is based on C^1 cubic splines on triangulations which are obtained from a given quadrangulation by inserting one diagonal in some quadrilaterals and two diagonals in others. On these triangulations, the spline space $\mathcal{S}_3^1(\triangle)$ has optimal order approximation power. This should be contrasted with the case of C^1 cubic splines defined on triangulations obtained from a given quadrangulation by inserting just one diagonal in each quadrilateral. It is well known [3] that these spaces do not possess optimal approximation power.

Remark 6. It is easy to see that the angles of a quadrangulation can be well-behaved in the sense that the largest angle is bounded by a constant $\kappa < \pi$, but the smallest angle $\theta_\triangle$ in the associated triangulation can still be arbitrarily small. However, conversely, clearly $2\theta_\triangle \leq \kappa \leq \pi - 2\theta_\triangle$.

Acknowledgments. The second author was supported by the Army Research Office under grant DAAD-19-99-1-0160.

References

1. Alfeld, P. and L. L. Schumaker, Smooth macro-elements based on Clough-Tocher triangle splits, Numer. Math., to appear.

2. Alfeld, P. and L. L. Schumaker, Smooth macro-elements based on Powell-Sabin triangle splits, Advances in Comp. Math., to appear.

3. Boor, C. de and K. Höllig, Approximation order from bivariate C^1-cubics: a counterexample, Proc. Amer. Math. Soc. **87** (1983), 649–655.

4. Chui, C. K. and D. Hong, Construction of local C^1 quartic spline elements for optimal-order approximation, Math. Comp. **65** (1996), 85–98.

5. Chui, C. K. and D. Hong, Swapping edges of arbitrary triangulations to achieve the optimal order of approximation, SIAM J. Numer. Anal. **34** (1997), 1472–1482.

6. Clough, R. W. and J. L. Tocher, Finite element stiffness matrices for analysis of plates in bending, Proc. Conf. on Matrix Methods in Structural Mechanics (1965), Wright Patterson A.F.B., Ohio.

7. Davydov, O., G. Nürnberger, and F. Zeilfelder, Approximation order of bivariate spline interpolation for arbitrary smoothness, J. Comp. Appl. Math. **90** (1998), 117–134.

8. Davydov, O., G. Nürnberger, and F. Zeilfelder, Cubic spline interpolation on nested polygon triangulations, in *Curve and Surface Fitting: Saint Malo 1999*, A. Cohen et. al. (eds.), Vanderbilt University Press, Nashville, 1999, 161–170.

9. Davydov, O., G. Nürnberger, and F. Zeilfelder, Bivariate spline interpolation with optimal approximation order, Constr. Approx. **17** (2001), 181–208.

10. Davydov, O. and L. L. Schumaker, Approximation and interpolation with C^1 quartic bivariate splines, SIAM J. Numer. Anal., to appear.

11. Fraeijs de Veubeke, B., A conforming finite element for plate bending, J. Solids Structures **4** (1968), 95–108.

12. Lai, M.-J., Scattered data interpolation and approximation using bivariate C^1 piecewise cubic polynomials, Comput. Aided Geom. Design **13** (1996), 81–88.

13. Lai, M.-J. and L. L. Schumaker, On the approximation power of bivariate splines, Advances in Comp. Math. **9** (1998), 251–279.

14. Lai, M.-J. and L. L. Schumaker, On the approximation power of splines on triangulated quadrangulations, SIAM J. Numer. Anal. **36** (1999), 143–159.

15. Lai, M.-J. and L. L. Schumaker, Macro-elements and stable bases for splines on Clough-Tocher triangulations, Numer. Math. **88** (2001), 105–119.

16. Lai, M.-J. and L. L. Schumaker, Macro-elements and stable bases for splines on Powell-Sabin triangulations, Math. Comp., to appear.

17. Lai, M.-J. and L. L. Schumaker, Quadrilateral macro-elements, SIAM J. Math. Anal., to appear.

18. Nürnberger, G., Approximation order of bivariate spline interpolation, J. Approx. Theory **87** (1996), 117–136.

19. Nürnberger, G. and T. Rießinger, Lagrange and Hermite interpolation by bivariate splines, Numer. Funct. Anal. Optim. **13** (1992), 75-96.

20. Nürnberger, G. and T. Rießinger, Bivariate spline interpolation at grid points, Numer. Math. **71** (1995), 91-119.

21. Nürnberger, G., L. L. Schumaker, and F. Zeilfelder, Local Lagrange interpolation by bivariate C^1 cubic splines, in *Mathematical Methods in CAGD: Oslo 2000*, T. Lyche and L. L. Schumaker (eds.), Vanderbilt University Press, Nashville, 2001, 393–404.

22. Nürnberger, G. and G. Walz, Error analysis in interpolation by bivariate C^1-splines, IMA J. Numer. Anal. **18** (1998), 485–508.

23. Nürnberger, G. and F. Zeilfelder, Interpolation by spline spaces on classes of triangulations, J. Comput. Appl. Math. **119** (2000), 347–376.

24. Nürnberger, G. and F. Zeilfelder, Developments in bivariate spline interpolation, J. Comput. Appl. Math. **121** (2000), 125–152.

25. Nürnberger, G. and F. Zeilfelder, Local Lagrange interpolation by cubic splines on a class of triangulations, in *Trends in Approximation Theory*, K. Kopotun, T. Lyche, and M. Neamtu, (eds.), Vanderbilt University Press (Nashville), 2001, 341–350.

26. Nürnberger, G. and F. Zeilfelder, Local Lagrange interpolation on Powell-Sabin triangulations and terrain modelling, in *Recent Progress in Multivariate Approximation*, W. Haussmann, K. Jetter, and M. Reimer, (eds.), Birkhäuser Verlag (Basel), 2001, 227–244.

27. Nürnberger, G. and F. Zeilfelder, Lagrange interpolation by bivariate C^1 splines with optimal approximation order, manuscript, 2001.

28. Powell, M. J. D. and M. A. Sabin, Piecewise quadratic approximations on triangles, ACM Trans. Math. Software **3** (1977), 316–325.

29. Sander, G., Bornes supérieures et inférieures dans l'analyse matricielle des plaques en flexion-torsion, Bull. Soc. Royale Sciences Liège **33** (1964), 456–494.

30. Schumaker, L.L., Fitting Surfaces to Scattered Data, in *Approximation Theory II*, G.G. Lorentz, et. al., (eds.), 1976, 203–268.

Günther Nürnberger
Institut für Mathematik
Universität Mannheim
D-68131 Mannheim
GERMANY
nuernberger@euklid.math.uni-mannheim.de

Larry L. Schumaker
Center for Constructive Approximation
Department of Mathematics
Vanderbilt University
Nashville, TN 37205
USA
s@mars.cas.vanderbilt.edu

Frank Zeilfelder
Institut für Mathematik
Universität Mannheim
D-68131 Mannheim
GERMANY
zeilfeld@euklid.math.uni-mannheim.de

Framelets with Many Vanishing Moments

Alexander Petukhov

Abstract. We study *tight wavelet frames* associated with given refinable functions which are obtained with *the unitary extension principles*. All possible polynomial symbols generating refinable functions and admitting framelets with a given number of vanishing moments are found. We also consider constructions of (anti)symmetric framelets generated by a symmetric refinable function. We give a few examples of symmetric framelet systems with two and three generators.

§1. Introduction

We consider compactly supported *wavelet frames* (or *framelets*) generated by a refinable function. The main goal of our paper is to present a general construction of such frames with an arbitrary number of vanishing moments.

In this paper, we shall consider only functions of one variable in the space $L^2(\mathbb{R})$ with the inner product

$$\langle f, g \rangle = \int_{-\infty}^{\infty} f(x)\overline{g(x)}\, dx.$$

As usual, $\hat{f}(\omega)$ denotes the Fourier transform of $f(x) \in L^2(\mathbb{R})$, defined by

$$\hat{f}(\omega) = \int_{-\infty}^{\infty} f(x)e^{-ix\omega}\, dx.$$

A tight frame in $L^2(\mathbb{R})$ is defined to be a collection of functions $\{f_j\}$ such that the Parseval equality

$$\|f\|^2 = \sum_j |\langle f, f_j \rangle|^2$$

Approximation Theory X: Wavelets, Splines, and Applications   425
Charles K. Chui, Larry L. Schumaker, and Joachim Stöckler (eds.), pp. 425–432.
Copyright Ⓒ 2002 by Vanderbilt University Press, Nashville, TN.
ISBN 0-8265-1416-2.

holds for any $f \in L^2(\mathbb{R})$. A tight frame

$$\psi^j_{k,m}(x) := 2^{k/2} \psi^j(2^k x - m), \ k, m \in \mathbb{Z}, \ j = 1, \ldots, n$$

is called a **wavelet (or affine) tight frame**. The functions ψ^j are known as **framelets**. Orthonormal wavelets [3] give a special case of a framelet system.

Suppose a real-valued function $\varphi \in L^2(\mathbb{R})$ satisfies the following conditions:

(a) $\hat{\varphi}(2\omega) = m_0(\omega)\hat{\varphi}(\omega)$, where m_0 is an essentially bounded 2π-periodic function;

(b) $\lim_{\omega \to 0} \hat{\varphi}(\omega) = (2\pi)^{-1/2}$,

then the function φ is called **refinable** or **scaling**, m_0 is called a **symbol** of φ, and the relation in item (a) is called a **refinement equation**.

In spite of the fact that in most practically important cases the refinable function can be easily reconstructed from its symbol, the problem of existence of a scaling function satisfying a refinement equation with the given symbol is not completely solved. Here we will not discuss the problem of recovering the function φ from its symbol. So in what follows the notion of a refinable function is basic for us, and a symbol is only an attribute of a refinable function.

Every refinable function generates a **multiresolution analysis (MRA)** of the space $L^2(\mathbb{R})$, i.e., a nested sequence

$$\ldots \subset V^{-1} \subset V^0 \subset V^1 \subset \ldots \subset V^j \subset \ldots$$

of closed linear subspaces of $L^2(\mathbb{R})$ such that

- $\cap_{j \in \mathbb{Z}} V^j = \{0\}$;
- $\overline{\cup_{j \in \mathbb{Z}} V^j} = L^2(\mathbb{R})$;
- $f(x) \in V^j \Leftrightarrow f(2x) \in V^{j+1}$.

The MRA is defined by letting V^j be the closure of the linear span of the functions $\{\varphi(2^j x - n)\}_{n \in \mathbb{Z}}$.

§2. Constructing Tight Frames

The most popular approach to the design of orthonormal wavelets is based on the construction of an MRA for the space $L^2(\mathbb{R})$ generated with a given refinable function $\phi \in V^0$ whose shifts $\{\phi(x-n)\}_{n \in \mathbb{Z}}$ form an orthonormal basis of V^0. It is well known that the problem of finding orthonormal wavelet bases generated by a scaling function can be reduced to solving the matrix equation

$$M(\omega)M^*(\omega) = I, \tag{1}$$

where

$$M(\omega) = \begin{pmatrix} m_0(\omega) & m_1(\omega) \\ m_0(\omega + \pi) & m_1(\omega + \pi) \end{pmatrix},$$

and $m_i(\omega)$ are essentially bounded functions with $m_i(-\omega) = \overline{m_i(\omega)}$, i.e., the Fourier series of these functions have real coefficients. It is known (see [3]), that (1) is necessary for any scaling function $\varphi(x)$ and the associated wavelet $\psi(x)$, $\hat{\psi}(2\omega) = m_1(\omega)\hat{\varphi}(\omega)$ to generate an orthogonal wavelet basis. Any refinable function φ whose symbol m_0 is a solution to (1) generates a tight frame (see [6] for the case when m_0 is a polynomial, the general case was proved in [1]).

To find all solutions to (1) we need to find a solution of the equation

$$|m_0(\omega)|^2 + |m_0(\omega + \pi)|^2 = 1, \tag{2}$$

and then all possible functions m_1 can be represented in the form

$$m_1(\omega) = \alpha(\omega)\overline{e^{i\omega}m_0(\omega + \pi)}, \tag{3}$$

where $\alpha(\omega)$ is an arbitrary π-periodic function, satisfying $|\alpha(\omega)| = 1$, $\alpha(-\omega) = \overline{\alpha(\omega)}$.

Now suppose we have an arbitrary refinable function $\varphi(\omega)$ with the symbol m_0 which does not satisfy (2), $\varphi_m(x) := \varphi(x - m)$, $m \in \mathbb{Z}$. Our goal is to construct a collection of functions $\{\psi^j\}_{j=1}^n \subset V^1$ with masks $\{m_j\}_{j=1}^n$ such that the sets

$$\varphi_m, \psi_{k,m}^j, \quad m \in \mathbb{Z}, k \geq 0, j = 1, \ldots, n, \tag{4}$$

and

$$\psi_{k,m}^j, \quad m, k \in \mathbb{Z}, j = 1, \ldots, n, \tag{5}$$

constitute tight frames of $L^2(\mathbb{R})$.

Let φ be a refinable function with a symbol m_0,

$$\hat{\psi}^k(\omega) = m_k(\omega/2)\hat{\varphi}(\omega/2) \in V^1,$$

where each symbol m_k is a 2π-periodic and essentially bounded function for $k = 1, 2, \ldots, n$. It is well known that for constructing practically important tight frames, the matrix

$$\boldsymbol{M}(\omega) = \begin{pmatrix} m_0(\omega) & m_1(\omega) & \ldots & m_n(\omega) \\ m_0(\omega + \pi) & m_1(\omega + \pi) & \ldots & m_n(\omega + \pi) \end{pmatrix}$$

plays an important role. In [7], we proved the following statement:

Theorem 1. *Systems of functions (4) and (5) form tight frames of $L^2(\mathbb{R})$ if and only if their symbols satisfy the equation*

$$\boldsymbol{M}(\omega)\boldsymbol{M}^*(\omega) = I. \tag{6}$$

A solution to (6) exists if and only if

$$|m_0(\omega)|^2 + |m_0(\omega + \pi)|^2 \leq 1. \tag{7}$$

All possible solutions to (6) can be represented in the form

$$\begin{pmatrix} m_1(\omega) \\ \dots \\ m_n(\omega) \end{pmatrix} = P(\omega) \begin{pmatrix} \overline{\left(\dfrac{e^{i\omega} m_0(\omega + \pi)}{B(\omega)} \right)} \\ A(\omega)\dfrac{m_0(\omega)}{B(\omega)} \end{pmatrix}, \tag{8}$$

where $A(\omega)$, $B(\omega)$ satisfy the relations

$$|A(\omega)|^2 = 1 - |m_0(\omega)|^2 - |m_0(\omega + \pi)|^2,$$
$$|B(\omega)|^2 = |m_0(\omega)|^2 + |m_0(\omega + \pi)|^2,$$

and $P(\omega)$ is an arbitrary $n \times 2$ matrix with π-periodic elements whose columns are orthonormal. For those ω for which $B(\omega) = 0$, in the right-hand side of (8), we take an arbitrary vector with unit norm.

In particular, Theorem 1 claims the existence of a solution with 2 generators ψ^1, ψ^2 for any refinable function φ satisfying (7).

Remark. Equation (6) (the *unitary extension principle*) appeared in [10]. In that paper under some additional decay assumption for $\hat{\varphi}$, A. Ron and Z. Shen proved that (6) implies that the corresponding system forms a tight frame.

By the Riesz lemma on factorization of positive trigonometric polynomials, for the rational function m_0 we can choose framelets with rational symbols. The problem of finding polynomial solutions for the polynomial m_0 can be reduced to an appropriate choice of the matrix $P(\omega)$. This problem was completely solved in [7]. A different proof of the existence of polynomial solutions can be found in [2].

§3. Framelets with Many Vanishing Moments

One of our goals is to describe all possible symbols m_0 admitting framelets with a given number of vanishing moments. In this case, the corresponding framelets can be found, for example, by the methods from [7]. A slightly different approach to constructing compactly supported framelets with many vanishing moments can be found in [11]. In what follows we will use the substitution $x = \sin^2 \frac{\omega}{2}$.

Theorem 2. *The refinable function φ admits framelets with N vanishing moments if and only if*

$$|m_0(\omega)|^2 = (1 - x)^N p(x), \tag{9}$$

where

$$p(x) = \sum_{j=0}^{N-1} \binom{N+j-1}{j} x^j - x^N q(x) \geq 0, \quad 0 \leq x \leq 1,$$

q is an algebraic polynomial, $q(x) + q(1 - x) \geq 0$ for $0 \leq x \leq 1$.

Substituting a sufficiently small constant for q in the theorem, we obtain the following result.

Corollary. *For any $N > 0$, there exists a polynomial symbol m_0 of degree $2N$ admitting framelets with N vanishing moments.*

Proof: We note that the choice $q \equiv 0$ gives Daubechies' compactly supported orthogonal wavelet basis. Since columns of the matrix $P(\omega)$ from Theorem 1 are orthonormal, by (8), elements of any of its columns cannot vanish simultaneously. So the minimal number of vanishing moments for the framelets ψ^j is equal to the minimum order of $\omega = 0$ as a zero of the polynomials $m_0(\omega + \pi)$ and $A(\omega)$.

Thus, if a refinable function $m_0(\omega + \pi)$ has a root of order N at the origin, we necessarily have (9). Let us introduce the polynomial

$$r(x) = -p(x) + \sum_{j=0}^{N-1} \binom{N+j-1}{j} x^j.$$

Then

$$|A(\omega)|^2 = 1 - (1 - x)^N \left(\sum_{j=0}^{N-1} \binom{N+j-1}{j} x^j - r(x) \right)$$

$$- x^N \left(\sum_{j=0}^{N-1} \binom{N+j-1}{j} (1 - x)^j - r(1 - x) \right)$$

$$= (1 - x)^N r(x) + x^N r(1 - x).$$

Therefore, $\omega = 0$ is a root of the polynomial $|A(\omega)|^2$ with order $2N$ if and only if $r(x) = x^N q(x)$. The necessity of the condition $q(x) + q(1 - x) \geq 0$ follows from the positivity of $|A(\omega)|$. The necessity is proved. The sufficiency can be checked by direct substitution. $\square$

§4. Symmetry

In most applied problems (anti)symmetry of the basic functions is very desirable. C. Chui and W. He [2] proved that for a symmetric refinable function φ a framelet system generated with at most three (anti)symmetric framelets exists. The problem of the existence of two (anti)symmetric framelets was solved in our paper [8].

Theorem 3. *[8] If a symmetric refinable function φ satisfies inequality (7), then an (anti)symmetric framelet system generated by 2 framelets exists if and only if all roots of the polynomial $\mathcal{A}(z) := A(\omega)$, where $z = e^{-i\omega}$, have even multiplicity.*

By Theorems 2 and 3, a refinable function φ satisfying (7) admits two framelets with N vanishing moments if and only if the following conditions hold:

- $q(x) + q(1 - x) \geq 0$;
- $p(x) = \sum_{j=0}^{N-1} \binom{N+j-1}{j} x^j - x^N q(x) \geq 0$;
- all roots $(x_0 \neq 0, x_0 \neq 1)$ of the polynomial $p(x)$ have even multiplicity;
- all roots of the polynomial $q(x) + q(1 - x)$ $(x_0 \neq 0, x_0 \neq 1)$ have even multiplicity.

We don't know how to find the general solution satisfying the above properties for an arbitrary N. So we just give a few simple examples. We denote by $h_k(z) := m_k(\omega)$ the z-transforms of the functions m_k, $z = e^{-i\omega}$.

The symbols of the (anti)symmetric framelet system with 2 vanishing moments generated by two framelets with the shortest support are

$$h_0(z) = \frac{(-3z^{-1} + 14 - 3z^1)(1 + z)^3 z^{-1}}{64},$$

$$h_1(z) = -0.18154(z^{-2} + z^3) + 0.30258(z^{-1} + z^2) - 0.12103(1 + z),$$

$$h_2(z) = \frac{3(z^{-2} - z^3) - 5(z^{-1} - z^2)}{16}.$$

The shortest symbol of the refinable function admitting three (anti) symmetric framelets with three vanishing moments is

$$h_0(z) = \frac{z^{-2}(1 + z)^4(3z^{-2} - 18z^{-1} + 38 - 18z^1 + 3z^2)}{128}.$$

Despite the fact that compactly supported systems are the most popular in applications, wavelets and wavelet frames with rational symbols (see, for example, [4,5,9]) have the same computational efficiency and

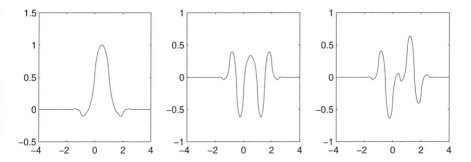

Fig. 1. Framelets with two vanishing moments. $\deg m_0 = 5$.

deserve consideration due to their flexibility and variety of choice. A one-parameter family of refinable functions with rational symbols of least degree admitting two (anti)symmetric framelets with two vanishing moments is

$$h_0(z) = \frac{P(z)}{Q(z^2)},$$

where

$$P(z) = -az^{-2} + 0.25z^{-1} + (0.5 + 2a) + 0.25z - az^2,$$

$$Q(z) = bz^{-1} + (1 - 2b) + bz,$$

with

$$b = \frac{1}{16} - a, \quad \frac{1}{16(1 - \sqrt{2})} \le a \le \frac{1}{16(1 + \sqrt{2})}.$$

The end points $a = (16(1 \pm \sqrt{2}))^{-1}$ give orthonormal wavelet bases. The case $a = (16(1 - \sqrt{2}))^{-1}$ was considered in [5], whereas the case $a = (16(1 + \sqrt{2}))^{-1}$ was unknown. For $a = -1/48$ the corresponding tight frame consists of piecewise *linear splines*. Since the parameter a is very small, the refinable function "almost coincides" with the linear B-spline.

References

1. Bownik, M., Tight frames of multidimensional wavelets, J. Fourier Anal. Appl. **3** (1997), 525–542.

2. Chui, C. K. and W. He, Compactly supported tight frames associated with refinable functions, Appl. Comput. Harm. Anal. **8** (2000), 293–319.

3. Daubechies, I., *Ten Lectures on Wavelets*, CBMS conference series in Applied Mathematics, Vol **61**, SIAM, Philadelphia, 1992.

4. Heising, G., D. Marpe, H. L. Cycon, and A. P. Petukhov, Wavelet-based very low bit-rate video coding using image warping and overlapped block motion compensation, IEE Proc. Vision, Image and Signal Processing **148** (2001), #2, 93–102.

5. Herley, C. and M. Vetterli, Wavelets and recursive filter banks, IEEE Trans. on Signal Processing **41** (1993), 2536–2556.

6. Lawton, W., Tight frames of compactly supported affine wavelets, J. Math. Phys. **31** (1990), 1898–1901.

7. Petukhov, A., Explicit construction of framelets, Appl. Comput. Harm. Anal. **11** (2001), 313–327.

8. Petukhov, A., Symmetric framelets, Constr. Approx. (to appear). See also the preprint of IMI of the University of South Carolina (2000), # 15. http://www.math.sc.edu/~imip/00papers/0015.ps.

9. Petukhov, A., Biorthogonal wavelet bases with rational masks and their applications, Trudy St. Petersburg Mat. Ob. **7** (1999), 168–193.

10. Ron, A. and Z. Shen, Affine systems in $L_2(\mathbb{R}^d)$: the analysis of the analysis operator, J. Func. Anal. **148** (1997), 408–447.

11. Selesnick, I., Smooth wavelet tight frames with zero moments, Appl. Comput. Harm. Anal. **10** (2001), 163–181.

Alexander Petukhov
Department of Mathematics
University of South Carolina
Columbia, SC 29208
petukhov@math.sc.edu

Stability of Radial Basis Function Interpolants

Robert Schaback

Abstract. The stability of the linear systems arising from scattered data interpolation problems with radial basis functions is analysed in full generality. Since lower bounds for the smallest eigenvalue of the coefficient matrix yield upper bounds for the absolute error of the RBF coefficients in terms of the absolute errors in the data, we then focus on a new and short proof of such bounds.

§1. Introduction

We shall study the stability of multivariate interpolation by conditionally positive definite radial functions of order $m \geq 0$.

Definition 1. *A univariate function*

$$\phi \; : \; \mathbb{R}_{\geq 0} \to \mathbb{R}$$

is called conditionally positive definite of order *m on $\mathbb{R}^d$, if for all possible choices of sets*

$$X = \{x_1, \ldots, x_N\} \subset \mathbb{R}^d$$

of N distinct points, the quadratic form induced by the $N \times N$ matrix

$$A = (\phi(\|x_j - x_k\|_2))_{1 \leq j, k \leq N} \tag{1}$$

is positive definite on the subspace

$$V := \left\{ \alpha \in \mathbb{R}^N \; : \; \sum_{j=1}^{N} \alpha_j p(x_j) = 0 \text{ for all } p \in \mathbb{P}_m^d \right\},$$

Approximation Theory X: Wavelets, Splines, and Applications
Charles K. Chui, Larry L. Schumaker, and Joachim Stöckler (eds.), pp. 433–440.
Copyright Ⓒ 2002 by Vanderbilt University Press, Nashville, TN.
ISBN 0-8265-1416-2.

where $\mathbb{P}_m^d$ stands for the space of d-variate polynomials of order not exceeding m.

Note that $m = 0$ implies $V = \mathbb{R}^N$ because of $\mathbb{P}_m^d = \{0\}$, and then the matrix A in (1) is positive definite. The most prominent examples of conditionally positive definite radial basis functions of order m on $\mathbb{R}^d$ are

$$
\begin{array}{llll}
\phi(r) & = & (-1)^{\lceil \beta/2 \rceil} r^\beta & \beta > 0, \beta \notin 2\mathbb{N}_0 & m \geq \lceil \beta/2 \rceil \\
\phi(r) & = & (-1)^{k+1} r^{2k} \log(r) & k \in \mathbb{N} & m \geq k+1 \\
\phi(r) & = & (c^2 + r^2)^{\beta/2} & \beta < 0 & m \geq 0 \\
\phi(r) & = & (-1)^{\lceil \beta/2 \rceil}(c^2 + r^2)^{\beta/2} & \beta > 0, \beta \notin 2\mathbb{N}_0 & m \geq \lceil \beta/2 \rceil \\
\phi(r) & = & e^{-\alpha r^2} & \alpha > 0 & m \geq 0 \\
\phi(r) & = & (1 - r)_+^4 (1 + 4r) & d \leq 3 & m \geq 0
\end{array}
$$

See e.g. [10] for a comprehensive derivation of the properties of these functions.

Interpolation of real values $f_1, \ldots, f_N$ on a set $X = \{x_1, \ldots, x_N\}$ of N distinct scattered points of $\mathbb{R}^d$ by a radial basis function ϕ is done by solving the $(N + Q) \times (N + Q)$ system

$$
\begin{array}{ccccc}
A\alpha & + & P\beta & = & f \\
P^T \alpha & + & 0 & = & 0,
\end{array}
\tag{2}
$$

where $Q = \dim \mathbb{P}_m^d$ and

$$
P = (p_i(x_j))_{1 \leq j \leq N, 1 \leq i \leq Q}
$$

for a basis $p_1, \ldots, p_Q$ of $\mathbb{P}_m^d$. In fact, if the additional assumption

$$
\mathrm{rank}\,(P) = Q \leq N
\tag{3}
$$

holds, then the system (2) is uniquely solvable. The resulting interpolant has the form

$$
s(x) = \sum_{j=1}^{N} \alpha_j \phi(\|x_j - x\|_2) + \sum_{i=1}^{Q} \beta_i p_i(x)
\tag{4}
$$

with the additional condition $\alpha \in V$.

§2. Stability

To investigate the numerical stability of the system (2), we replace α, β, f by perturbations of the original quantities and get

$$
\begin{array}{ccccc}
(\Delta\alpha)^T A(\Delta\alpha) & + & 0 & = & (\Delta\alpha)^T \Delta f \\
P^T (\Delta\alpha) & + & 0 & = & 0.
\end{array}
\tag{5}
$$

Since A is positive definite on the subspace $V = \ker P^T$, there are positive eigenvalues $\Lambda \geq \lambda$ such that

$$\Lambda \|\alpha\|_2^2 \geq \alpha^T A \alpha \geq \lambda \|\alpha\|_2^2 \text{ for all } \alpha \in V = \ker P^T. \tag{6}$$

We can insert this into (5) to get

$$\|\Delta \alpha\|_2 \leq \frac{1}{\lambda} \|\Delta f\|_2$$

to bound the absolute error $\Delta \alpha$ of α by the absolute error Δf in the data vector f. If the solution α is nonzero, we have that $f - P\beta$ is nonzero, and a similar argument combining (2) and (6) yields

$$\frac{\|\Delta \alpha\|_2}{\|\alpha\|_2} \leq \frac{\Lambda}{\lambda} \frac{\|\Delta f\|_2}{\|f - P\beta\|_2}$$

for the relative error. Thus, the standard L_2 theory of numerical stability applies to the RBF part of (2). The condition is given by the ratio Λ/λ, while stability of the absolute error is dominated by λ.

The stability of the calculation of β follows the lines of the stability theory for discrete L_2 polynomial approximation, because we have

$$\beta = (P^T P)^{-1} P^T (f - A\alpha). \tag{7}$$

This means that the polynomial part can be calculated from the residuals of the RBF data via the standard operator $(P^T P)^{-1} P^T$ of discrete L_2 polynomial approximation. The absolute error of the residuals can be bounded by

$$\|\Delta f - A\Delta \alpha\|_2 \leq \|\Delta f\|_2 (1 + \frac{\Lambda}{\lambda}),$$

and we can see how the condition of the RBF part enters into the stability theory for the polynomial part: the upper bounds have to be multiplied by $1 + \frac{\Lambda}{\lambda}$.

To derive stability bounds for practical use, one needs upper bounds for Λ and lower bounds for λ.

§3. Upper Bounds for Eigenvalues

For bounded radial basis functions one can get crude upper bounds for Λ via Gerschgorin's theorem. In fact, if we normalize ϕ to satisfy

$$1 = \phi(0) \geq \phi(r) \text{ for all } r \in [0, \infty),$$

then

$$|1 - \Lambda| \leq N - 1,$$

which is not too bad for standard applications and compared to the bad behavior of λ to become apparent later. In particular, this bound is independent of the data locations and the smoothness of ϕ, which have a strong influence on lower bounds for λ.

A somewhat more general argument works for cases in which we have a convolution representation

$$\phi(\|x - y\|_2) = \int_{\mathbf{R}^d} \Psi(x - t)\Psi(y - t)dt. \tag{8}$$

This is actually true for all smooth unconditionally positive definite and Fourier–transformable functions, because the Fourier transform of $\Psi(x)$ can be obtained via the square root of the (nonnegative) Fourier transform of $\phi(\| \cdot \|_2)$. If we assume (8), then

$$\alpha^T A\alpha = \int_{\mathbf{R}^d} \left(\sum_{j=1}^{N} \alpha_j \Psi(x_j - t) \right)^2 dt$$

$$\leq \left(\sum_{j=1}^{N} \alpha_j^2 \right) \int_{\mathbf{R}^d} \sum_{j=1}^{N} \Psi(x_j - t)^2 dt$$

$$= N\phi(0) \sum_{j=1}^{N} \alpha_j^2.$$

For general positive definite functions $\Phi(x, y)$, we know that

$$\Phi(x, y) = (\Phi(x, \cdot), \Phi(y, \cdot))_{\mathcal{H}}$$

holds, where $\mathcal{H}$ is the native Hilbert space for Φ (see e.g. [9] for details). Then the above argument takes the form

$$\alpha^T A\alpha = \left\| \sum_{j=1}^{N} \alpha_j \Phi(x_j, \cdot) \right\|_{\mathcal{H}}^2 \leq \left(\sum_{j=1}^{N} \alpha_j^2 \right) \left(\sum_{j=1}^{N} \Phi(x_j, x_j) \right). \tag{9}$$

The case of unbounded radial basis functions is somewhat more complicated. It comprises the radial basis functions with positive minimal order of conditional positive definiteness, for instance the multiquadrics $\phi(r) = \sqrt{r^2 + c^2}$ or thin–plate splines $\phi(r) = r^2 \log r$. In contrast to the above upper bound, which did not use the additional condition $\alpha \in V = \ker P^T$, we now have to rely on the latter. If we insert (7) into the first equation in (2), we get

$$(A - P(P^T P)^{-1} P^T A)\alpha = f - P(P^T P)^{-1} P^T f.$$

Due to $P(P^T P)^{-1} P^T \alpha = 0$, this system can be written as

$$RAR^T \alpha = Rf,$$

where $R := I - P(P^T P)^{-1} P^T = R^T$ is the operator that maps discrete data into their residuals after least–squares approximation by polynomials from $\mathbb{P}_m^d$. The matrix $B := RAR^T$ will now have a significantly better behavior than A, as far as upper bounds are concerned, because it can be interpreted as an interpolation matrix of an unconditionally positive definite non–radial function (see §6 of [9]). In particular, we can apply (9) to this new function, and we get a numerically computable upper bound on the largest eigenvalue. Further details are suppressed here. We summarize:

Theorem 2. *For any positive definite radial basis function ϕ, the largest eigenvalue Λ of the matrix A as defined in (1) is bounded above by $N\phi(0)$. For non–radial positive definite basis functions the bound takes the form*

$$\sum_{j=1}^{N} \Phi(x_j, x_j),$$

and the case of positive order of conditional positive definiteness can be reduced to the positive definite non–radial case by matrix transformations.

§4. Lower Bounds for Eigenvalues

The work of Ball [1,2], Narcowich, Sivakumar, and Ward [5,6] already contains lower bounds for the smallest eigenvalue λ of A, and these bounds are near–optimal due to [7]. However, the proofs are complicated, and we want to provide a much shorter though less general argument, which can be transferred to expansion kernels [4]. It relies on the existence of positive definite functions with compact support, which were not available before 1995 due to Wu [13] and Wendland [12].

The idea is to perturb the matrix A on the diagonal by subtracting from a conditionally positive definite radial function ϕ of order m some positive definite radial function ψ with small support, such that $\phi - \psi$ still is conditionally positive (semi)definite of order m. If we write A_ϕ when A in (1) is based on ϕ, we then get

$$\alpha^T A_\phi \alpha = \alpha^T A_{\phi-\psi} \alpha + \alpha^T A_\psi \alpha \geq \alpha^T A_\psi \alpha = \psi(0)\|\alpha\|_2^2 \text{ for all } \alpha \in V$$

if the support of ϕ is smaller than the minimal distance

$$q := \min_{1 \leq i < j \leq N} \|x_i - x_j\|_2 \tag{10}$$

between two different data points, and then $A_\psi = \psi(0)I$.

Theorem 3. Let ϕ be a conditionally positive definite function of order m. If ψ is a positive definite radial basis function with support in $[0, q]$ with q from (10) such that $\phi - \psi$ is conditionally positive definite of order at least m, then $\psi(0)$ is a lower bound for the smallest eigenvalue of A_ϕ as defined in (1).

Thus we get $\psi(0)$ as a lower bound for λ, whenever we can find a ψ with support in $[0, q]$ such that ψ and $\phi - \psi$ are positive definite. Of course, we would like to take the maximal $\psi(0)$ under these conditions, but the corresponding optimization problem still is an open challenge.

To be more specific, we confine ourselves to conditionally positive definite radial basis functions with a radial generalized Fourier transform $\widehat{\phi}$ satisfying

$$\widehat{\phi}(r) \geq c_\infty r^{-d-\beta} \text{ for all } r > 1. \tag{11}$$

Note that this places an upper bound on the smoothness of ϕ, and thus it rules out infinitely differentiable cases like the Gaussian and the multiquadrics. Furthermore, it implies by arguments from [11] that the standard $L_\infty(\Omega)$ error bounds for interpolation of functions in the native space $\mathcal{H}$ cannot be better than of order $\mathcal{O}(h^{\beta/2})$, where h is the data density

$$h := \sup_{y \in \Omega} \min_{1 \leq j \leq N} \|x_j - y\|_2.$$

By the Uncertainty Principle in [8], the optimal lower bounds of eigenvalues λ have the form $\mathcal{O}(q^\beta)$ for small q, and this is what we want to recover by our new technique.

Theorem 4. Let ϕ be a conditionally positive definite radial basis function whose Fourier transform has at most the decay (11). Then the smallest eigenvalues of the matrices A in (1) have a lower bound of the form $\lambda \geq cq^\beta$ for all data sets with $q \leq 1$.

Proof: For convenience of notation, we add

$$0 < c_0 \leq \widehat{\phi}(r) \text{ for all } r \leq 1$$

to (11). From Wendland's supply of arbitrarily smooth compactly supported positive definite radial basis functions, we can find some σ with support on $[0, 1]$ satisfying $\sigma(0) = 1$ and having a positive radial Fourier transform $\widehat{\sigma}$ with

$$\begin{aligned} \widehat{\sigma}(r) &\leq C_\infty r^{-d-\beta} & r > 0 \\ \widehat{\sigma}(r) &\leq C_0 & r \leq 1. \end{aligned}$$

We now take $\psi(\cdot) = \epsilon\sigma(\cdot/q)$ to squeeze the support of ψ into $[0, q]$, and we maximize ϵ under the constraint

$$\widehat{\phi}(r) \geq \widehat{\psi}(r) = \epsilon q^d \widehat{\sigma}(rq) \text{ for all } r \geq 0 \tag{12}$$

which still makes $\phi - \psi$ conditionally positive semidefinite of at least the same order as ϕ, because this order is related to the order of the singularity of $\widehat{\phi}$ at zero.

We first treat the case $r > 1$, in which it suffices to guarantee (12) by

$$\widehat{\psi}(r) = \epsilon q^d \widehat{\sigma}(rq) \le \epsilon q^d C_\infty(rq)^{-d-\beta} = \epsilon q^{-\beta} C_\infty r^{-d-\beta} \le c_\infty r^{-d-\beta} \le \widehat{\phi}(r)$$

by picking

$$\epsilon \le \frac{c_\infty}{C_\infty} q^\beta.$$

The case $r \le 1$ has $rq \le 1$ and we can satisfy

$$\widehat{\psi}(r) = \epsilon q^d \widehat{\sigma}(rq) \le \epsilon q^d C_0 \le \epsilon C_0 \le c_0 \le \widehat{\phi}(r)$$

by taking

$$\epsilon \le \frac{c_0}{C_0}. \quad \square$$

Note that we could incorporate infinitely differentiable cases like the Gaussian and the multiquadrics, if we had a sufficient supply of infinitely differentiable radial basis functions with small compact supports. The case of expansion kernels suffered also from lack of positive definite functions with arbitrarily small support, but this was overcome in [4]. An application to non–radial basis functions with varying scales and shapes is in [3].

Acknowledgments. Special thanks go to Holger Wendland for proof-reading.

References

1. Ball, K., Eigenvalues of Euclidean distance matrices, J. Approx. Theory **68** (1992), 74–82.

2. Ball, K., N. Sivakumar, and J.D. Ward, On the sensitivity of radial basis interpolation to minimal data separation distance. Constr. Approx. **8** (1992), 401–426.

3. Bozzini, M., L. Lenarduzzi, M. Rossini, and R. Schaback, Interpolation by basis functions of different scales and shapes, preprint 2001.

4. Narcowich, F. J., R. Schaback, and J. D. Ward, Stability estimates for interpolation by positive definite kernels, preprint, 2001.

5. Narcowich, F. J., N. Sivakumar, and J. D. Ward, On condition numbers associated with radial–function interpolation, J. Math. Anal. Appl. **186** (1994), 457–485.

6. Narcowich, F. J., and J. D. Ward, Norm estimates for the inverses of a general class of scattered–data radial–function interpolation matrices, J. Approx. Theory **69** (1992), 84–109.

7. Schaback, R., Lower bounds for norms of inverses of interpolation matrices for radial basis functions, J. Approx. Theory **79** (1994), 287–306.

8. Schaback, R., Error estimates and condition numbers for radial basis function interpolation, Advances in Comp. Math. **3** (1995), 251–264.

9. Schaback, R., Native Hilbert spaces for radial basis functions, in *New Developments in Approximation Theory*, M. D. Buhmann, D. H. Mache, M. Felten, and M. W. Müller (eds.), Birkhäuser Verlag, International Series of Numerical Mathematics **132** (1999), 255–282.

10. Schaback, R. and H. Wendland, Characterization and construction of radial basis functions, in *Multivariate Approximation and Applications*, Dyn, N., D. Leviatan, D. Levin, and A. Pinkus (eds.), Cambridge University Press 2000.

11. Schaback, R. and H. Wendland, Inverse and saturation theorems for radial basis function interpolation, Math. Comp., to appear.

12. Wendland, H., Piecewise polynomial, positive definite and compactly supported radial functions of minimal degree, Advances in Comp. Math. **4** (1995), 389–396.

13. Wu, Z., Multivariate compactly supported positive definite radial functions, Advances in Comp. Math. **4** (1995), 283–292.

Prof. Dr. R. Schaback
Institut für Numerische und Angewandte Mathematik
Zentrum für Informatik
Lotzestraße 16–18
D-37083 Göttingen
Germany
schaback@math.uni-goettingen.de

Extremal Properties of Conditionally Positive Definite Functions

Hans Strauss

Abstract. In this paper we consider extremal properties of conditionally positive definite functions. We study extremal problems for interpolation and approximation problems which are closely related. General theorems of extremal properties are obtained which contain results of the literature. Moreover, some new classes of positive definite functions are given.

§1. Introduction

It is well known that spline functions and related functions are solutions of extremal problems. In [1] several results of this type are already given. These investigations have been extended to more general problems, e.g., to smoothing splines which are an important tool in data fitting. In multivariate approximation, thin plate splines satisfy certain extremal problems. These problems can be considered as special problems for interpolation with conditionally positive definite functions.

Here, we want to give general results concerning extremal properties of conditionally positive definite functions. We consider interpolation and approximation problems, and also show that there is a relationship between these problems. Moreover, we also consider some new classes of positive definite functions.

In Section 2 we give some basic definitions and results. In Section 3 we first define some interpolation and approximation problems, and show that these problems have solutions which can be obtained by solving similar linear systems. Then we consider extremal properties, and obtain general theorems which contain several results from the literature. Moreover, the close relationship of the extremal properties of certain interpolation and

Approximation Theory X: Wavelets, Splines, and Applications   441
Charles K. Chui, Larry L. Schumaker, and Joachim Stöckler (eds.), pp. 441–452.
Copyright ℗ 2002 by Vanderbilt University Press, Nashville, TN.
ISBN 0-8265-1416-2.

approximation problems follows from our results. In Section 4 we give some new classes of positive definite functions. In the univariate case we get the functions via an integration of positive definite functions. These functions then have a higher order of differentiability. In order to get multivariate positive definite functions, we consider products of positive definite functions.

§2. Definitions

Let S be a subset of $\mathbb{R}^d$ with $d \geq 1$, and let $F(S)$ be the class of real–valued functions $f : S \to \mathbb{R}$. Assume that $U = \text{span}\,\{u_1, \ldots, u_m\}$ is an m–dimensional subspace of $F(S)$. For a given finite set of distinct points $X = \{x_1, \ldots, x_n\}$, $m \leq n$, in S we define

$$U_X^\perp = \left\{ (a_1, \ldots, a_n)^T \in \mathbb{R}^n : \sum_{i=1}^n a_i u_j(x_i) = 0, \ j = 1, \ldots, m \right\}.$$

Definition 2.1. Let S and U be defined as above and $P : S \times S \to \mathbb{R}$.

(a) The function P is called conditionally positive definite (CPD) w.r.t. U if the matrix $M = (P(x_i, x_j))_{i,j=1,\ldots,n}$ is conditionally nonnegative definite w.r.t. $U_X^\perp$, i.e. $\sum_{i,j=1}^n c_i c_j P(x_i, x_j) \geq 0$ for all $\mathbf{c} = (c_1, \ldots, c_n)$ in $U_X^\perp$ and for all choices of finite subsets of distinct points $X = \{x_1, \ldots, x_n\} \subseteq S$.

(b) The function P is called conditionally strictly positive definite (CSPD) w.r.t. U if strict inequality holds for all $\mathbf{c} \in U_X^\perp \setminus \{\mathbf{0}\}$ and for all choices of finite subsets of distinct points $X = \{x_1, \ldots, x_n\} \subseteq S$. Then M is called a conditionally positive definite matrix w.r.t. $U_X^\perp$.

(c) The function P is called (strictly) positive definite (PD (SPD)) if $U = \{\mathbf{0}\}$.

Let S and U be given as above. Moreover, let $H \subseteq F(S)$ be a linear subspace endowed with a semi–inner product $(\ ,\ )$ such that $U \subseteq H$ and $(h, h) = 0$ if and only if $h \in U$, i.e. U is the null space of $(\ ,\ )$.

Remark. It can easily be shown that $(f, u) = 0$ for all $f \in H$ and $u \in U$.

Let $K : S \times S \to \mathbb{R}$ be a reproducing kernel of H w.r.t. U. For a definition see [4,13]. The reproducing kernel K w.r.t. U satisfies the following properties for all choices of finite subsets of distinct points $X = \{x_1, \ldots, x_n\} \subseteq S$ and all $(a_1, \ldots, a_n)^T \in U_X^\perp$:

(a) $\sum_{i=1}^n a_i K(., x_i) \in H$;

(b) $(f, \sum_{i=1}^n a_i K(., x_i)) = \sum_{i=1}^n a_i f(x_i)$.

Using the aforementioned properties of reproducing kernels, the following result can be shown (see [12]).

Proposition 2.2. *Let $K : S \times S \to \mathbb{R}$ be a reproducing kernel of the space H w.r.t. U and let $X = \{x_1, \ldots, x_n\}$ be a set of distinct points in S.*

(a) *Then the matrix $(K(x_i, x_j))_{i,j=1}^n$ is a conditionally nonnegative matrix w.r.t. $U_X^\perp$.*

(b) *The matrix $(K(x_i, x_j))_{i,j=1}^n$ is a conditionally positive definite matrix w.r.t. $U_X^\perp$ if and only if $\sum_{i=1}^n a_i K(., x_i) \notin U$ for all $\mathbf{a} = (a_1, \ldots, a_n)^T \in U_X^\perp \setminus \{\mathbf{0}\}$.*

Remark. In order to get CPD functions, we start here with a reproducing kernel space. It is also possible to start with a CPD function and then define a reproducing kernel Hilbert space which is the native space w.r.t. its reproducing kernel (see [7]).

§3. Extremal Problems

Let $H \subset F(S)$ be a semi–inner product space which has a reproducing kernel K w.r.t. U, where U is the null space of the semi–inner product. We first consider an approximation problem. Let distinct points $X = \{x_1, \ldots, x_n\}$ in S be given, where $n \geq m$. Moreover, let W_n be an $n \times n$–matrix and assume that W_n is a nonnegative definite matrix. We now define the functional

$$d_X(f, \mathbf{a}, \mathbf{b}) = \frac{1}{2}\left(\left\| f - \sum_{i=1}^n a_i K(\cdot, x_i) - \sum_{i=1}^m b_i u_i \right\|^2 + \lambda \mathbf{a}^T W_n \mathbf{a}\right) \quad (1)$$

for all $\mathbf{a} = (a_1, \ldots, a_n)^T \in \mathbb{R}^n$, $\mathbf{b} = (b_1, \ldots, b_m)^T \in \mathbb{R}^m$ and some given $\lambda > 0$. Using the notation

$$C_X = (u_i(x_j))_{i=1, j=1}^{m,n},$$

where C_X is an $m \times n$– matrix, we obtain

$$U_X^\perp = \{\mathbf{a} \in \mathbb{R}^n : C_X \mathbf{a} = \mathbf{0}\}.$$

We now consider the following problem.

Problem 3.1. *Given $f \in H$, minimize $d_X(f, \mathbf{a}, \mathbf{b})$ over all $\mathbf{a} \in \mathbb{R}^n, \mathbf{b} \in \mathbb{R}^m$ subject to*

$$C_X \mathbf{a} = \mathbf{0}.$$

It has been shown in [12] that this approximation problem can be transformed into a minimization problem of a quadratic functional with side conditions. If we use the notation

$$\mathbf{f}_X := (f(x_1) \ldots f(x_n))^T$$
$$K_X = (K(x_i, x_j))_{i,j=1}^n,$$

then the following is true:

Proposition 3.2. *Let $f \in H$ be given. Then for all $\mathbf{a} \in U_X^{\perp}$ and $\mathbf{b} \in \mathbb{R}^m$ we have*

$$d_X(f, \mathbf{a}) := d_X(f, \mathbf{a}, \mathbf{b}) = \frac{1}{2} \mathbf{a}^T (K_X + \lambda W_n) \mathbf{a} - \mathbf{f}_X^T \mathbf{a} + \frac{1}{2}(f, f).$$

In order to solve Problem 3.1 we shall need solutions of the following problem.

Problem 3.3. *Determine vectors $\mathbf{a}^* \in \mathbb{R}^n$ and $\mathbf{d}^* \in \mathbb{R}^m$ which solve the linear system*

$$(K_X + \lambda W_n)\mathbf{a} + C_X^T \mathbf{d} = \mathbf{y},$$
$$C_X \mathbf{a} = \mathbf{0}$$

for a given $\mathbf{y} \in \mathbb{R}^n$.

It is well known that there is a unique solution of Problem 3.3 if $K_X + \lambda W_n$ is a conditionally positive definite matrix w.r.t. $U_X^{\perp}$ and C_X has rank m.

In the following, we always denote by s^* the function

$$s^* = \sum_{i=1}^{n} a_i^* K(\cdot, x_i) + \sum_{i=1}^{m} d_i^* u_i, \qquad (2)$$

where $\mathbf{a}^* = (a_1^*, \ldots, a_n^*)^T$ and $\mathbf{d}^* = (d_1^*, \ldots, d_m^*)^T$ are a solution of Problem 3.3.

Proposition 3.4. *Let $f \in H$ and $\mathbf{y} = \mathbf{f}_X$. Assume that the vectors $\mathbf{a}^* \in \mathbb{R}^n$ and $\mathbf{d}^* \in \mathbb{R}^m$ are a solution of Problem 3.3, where $K_X + \lambda W_n$ is a conditionally positive definite matrix. Then*

$$d_X(f, \mathbf{a}^*) \le d_X(f, \mathbf{a})$$

for all $\mathbf{a} \in U_X^{\perp}$.

Remark. (a) The matrix W_n defines a regularisation term of the approximation problem. Let K_X and λW_n be positive definite matrices. Then the linear system

$$(K_X + \lambda W_n)\mathbf{a} = \mathbf{f}_X$$

determines a solution of a problem which can be considered as a regularisation of the least squares problem of minimizing

$$\left\| f - \sum_{i=1}^{n} a_i K(\cdot, x_i) \right\|$$

for $\mathbf{a} = (a_1, ..., a_n)^T \in \mathbb{R}^n$. This is called a Tichonov regularisation. The matrix λW_n can be used to improve the condition of the linear system.

(b) It is well known that solutions (2) of Problem 3.3 also appear in interpolation with smoothing splines (see [8,14]). In this paper the relationship between Problem 3.1 and interpolation with smoothing splines is also studied.

Extremal properties.
In order to consider extremal properties, we now use a matrix W_n of the form

$$W_n = \begin{pmatrix} P & 0 \\ 0 & 0 \end{pmatrix} \tag{3}$$

where P is a regular $(r \times r)$–matrix, $0 \le r \le n$. Since W_n is a nonnegative definite matrix, it follows that P is a positive definite matrix. The pseudoinverse of W_n is

$$W_n^{\perp} = \begin{pmatrix} P^{-1} & 0 \\ 0 & 0 \end{pmatrix}$$

(see [5]). The matrix $W_n^{\perp}$ is also a nonnegative definite matrix.

Remark 3.5. Let the matrix W_n be of the form (3), let $P = ((p_{ij})_{i,j=1}^{r})$ and the function s^* be defined as in (2). Then

$$s^*(x_i) + \lambda \sum_{j=1}^{r} p_{ij} a_j^* = y_i, \ i = 1, ..., r$$

and

$$s^*(x_i) = y_i, \quad i = r + 1, \ldots, n.$$

This means that we solve an interpolation problem for the points x_i, $i = r + 1, ..., n$ only. In the points x_i, $i = 1, ..., n$, we do not interpolate, since we may have noisy data.

In the next theorem we obtain a general result comparing the seminorms of special interpolating functions in H.

Theorem 3.6. Let W_n be defined as in (3), and let $f \in H, \mathbf{y} \in \mathbb{R}^n$ be given satisfying $f(x_i) = y_i$, $i = r + 1, \ldots, n$. Assume that s^* is defined as in (2). Then

$$\lambda(s^*, s^*) + 2(\mathbf{f}_X - \mathbf{s}_X^*)W_n^{\perp}(\mathbf{y} - \mathbf{s}_X^*) \le \lambda(f, f).$$

Proof: It follows from Proposition 3.2 that

$$d_X(s^*, \mathbf{a}^*) = \frac{1}{2}(\mathbf{a}^*)^T(K_X + \lambda W_n)\mathbf{a}^* - (\mathbf{s}_X^*)^T\mathbf{a}^* + \frac{1}{2}(s^*, s^*)$$

and for all $f \in H$

$$d_X(f, \mathbf{a}^*) = \frac{1}{2}(\mathbf{a}^*)^T(K_X + \lambda W_n)\mathbf{a}^* - \mathbf{f}_X^T\mathbf{a}^* + \frac{1}{2}(f, f).$$

We conclude from the definition of d_X that

$$d_X(s^*, \mathbf{a}^*) = \frac{1}{2}\lambda(\mathbf{a}^*)^T W_n\mathbf{a}^*$$

$$\leq \frac{1}{2}\|f - s^*\|^2 + \frac{1}{2}\lambda(\mathbf{a}^*)^T W_n\mathbf{a}^* = d_X(f, \mathbf{a}^*).$$

Hence, we obtain

$$(s^*, s^*) + 2(\mathbf{f}_X - \mathbf{s}_X^*)^T\mathbf{a}^* \leq (f, f). \tag{4}$$

It follows from Remark 3.5 that

$$\lambda W_n\mathbf{a}^* = \mathbf{y} - \mathbf{s}_X^*. \tag{5}$$

Let

$$\tilde{I} := \begin{pmatrix} I_r & 0 \\ 0 & 0 \end{pmatrix} = W_n^{\perp}W_n,$$

where I_r is an $(r \times r)$ – identity matrix. Since $f(x_i) = s^*(x_i) = y_i$, $i = r+1, \ldots, n$, we have together with (5) that

$$(\mathbf{f}_X - \mathbf{s}_X^*)^T\mathbf{a}^* = (\mathbf{f}_X - \mathbf{s}_X^*)^T\tilde{I}\mathbf{a}^* =$$

$$(\mathbf{f}_X - \mathbf{s}_X^*)^T W_n^{\perp}W_n\mathbf{a}^* = \frac{1}{\lambda}(\mathbf{f}_X - \mathbf{s}_X^*)^T W_n^{\perp}(\mathbf{y} - \mathbf{s}_X^*).$$

If we insert this expression in (4) the theorem is proved. $\square$

Remark 3.7. If we consider the proof of Theorem 3.6 for a special case, we can easily show the relationship between two famous extremal properties which have already been considered in [1] for the case of polynomial splines. Using the theory of CPD functions, we can consider more general situations and also have very simple proofs, as shown next.

We let $\mathbf{y} \in \mathbb{R}^n$ and define the set

$$H_{X,\mathbf{y}} = \{f \in H : \mathbf{f}_X = \mathbf{y}\}.$$

Let W_n be a null matrix and let s^* be the function in (2). Then it follows from Remark 3.5 that $s^* \in H_{X,\mathbf{y}}$. It also follows from Proposition 3.2 that

$$d_X(f, \mathbf{a}^*) = \frac{1}{2}(\mathbf{a}^*)^T K_X\mathbf{a}^* - \mathbf{y}^T\mathbf{a}^* + \frac{1}{2}(f, f).$$

Since

$$0 = d_X(s^*, \mathbf{a}^*) \leq d_X(f, \mathbf{a}^*),$$

we get

$$(s^*, s^*) \leq (f, f).$$

We have two famous properties: Given $f \in H_{X,\mathbf{y}}$ and s^* as in (2),

(a) $\|f - s^*\| \leq \|f - s\|$ for all $s \in G$, where $G = \operatorname{span}\{u_1, \ldots, u_m, K(\cdot - x_1), \ldots, K(\cdot - x_n)\}$.

(b) $\|s^*\| \leq \|f\|$.

Remark. In the literature much more attention has been paid to the last inequality. But we see from Remark 3.7 that these inequalites are more or less the same thing.

We now define a functional and derive extremal properties from Theorem 3.6. Let $f \in H$ and $\mathbf{y} \in \mathbb{R}^n$. Moreover, let W_n be defined as in (2). We define the functional

$$L(f) = (\mathbf{f}_X - \mathbf{y})^T W_n^\perp (\mathbf{f}_X - \mathbf{y}) + \lambda(f, f).$$

Theorem 3.8. *Let $f \in H$ and $\mathbf{y} \in \mathbb{R}^n$ satisfy $f(x_i) = y_i$, $i = r+1, ..., n$. Assume that s^* is defined as in (2). Then*

$$L(s^*) \leq L(s^*) + (\mathbf{f}_X - \mathbf{s}_X^*)^T W_n^\perp (\mathbf{f}_X - \mathbf{s}_X^*) \leq L(f).$$

Proof: It follows from Theorem 3.6 that

$$\lambda(s^*, s^*) + 2(\mathbf{f}_X - \mathbf{s}_X^*)^T W_n^\perp (\mathbf{y} - \mathbf{s}_X^*) \leq \lambda(f, f)$$

for $f \in H$ satisfying $f(x_i) = y_i$, $i = r+1 \ldots n$. Then

$$\begin{aligned}
L(f) &= \lambda(f, f) + (\mathbf{f}_X - \mathbf{y})^T W_n^\perp (\mathbf{f}_X - \mathbf{y}) \\
&\geq \lambda(s^*, s^*) + 2(\mathbf{f}_X - \mathbf{s}_X^*)^T W_n^\perp (\mathbf{y} - \mathbf{s}_X^*) \\
&\quad + (\mathbf{f}_X - \mathbf{s}_X^*)^T W_n^\perp (\mathbf{f}_X - \mathbf{s}_X^*) + 2(\mathbf{f}_X - \mathbf{s}_X^*)^T W_n^\perp (\mathbf{s}_X^* - \mathbf{y}) \\
&\quad + (\mathbf{s}_X^* - \mathbf{y})^T W_n^\perp (\mathbf{s}_X^* - \mathbf{y}) \\
&= L(s^*) + (\mathbf{f}_X - \mathbf{s}_X^*)^T W_n^\perp (\mathbf{f}_X - \mathbf{s}_X^*) \geq L(s^*).
\end{aligned}$$

The last inequality follows from the fact that $W_n^\perp$ is also a nonnegative definite matrix. $\square$

Special cases.

Let $W_n = \text{diag}(w_1, \ldots, w_r, 0, \ldots, 0)$, $0 \le r \le n$, $w_1 > 0, \ldots, w_r > 0$ and let $\mathbf{y} \in \mathbb{R}^n$. We define the functional

$$L(f) = \sum_{i=1}^{r} \frac{1}{w_i} (f(x_i) - y_i)^2 + \lambda(f, f).$$

Assume that s^* is a solution of (2). Then we obtain from Theorem 3.8 that

$$L(s^*) + \sum_{i=1}^{r} \frac{1}{w_i} (f(x_i) - s^*(x_i))^2 \le L(f)$$

for all $f \in H$ satisfying $f(x_i) = y_i$, $i = r+1, \ldots, n$.

If we set $W_n = \text{diag}(n, \ldots, n)$, then we obtain a well known property of smoothing splines, see also [8,14],

$$L(s^*) + \frac{1}{n} \sum_{i=1}^{n} (f(x_i) - s^*(x_i))^2 \le L(f).$$

§4. Positive Definite Functions

In a famous theorem of Bochner the following is shown: Let g be a complex function on $\mathbb{R}^d$. Then the function $P(x, y)$ where $P(x, y) = g(x - y)$ is a continuous PD function if and only if g is the Fourier transform of a finite nonnegative Borel measure μ. Moreover, if the carrier of μ is not a set of Lebesgue measure zero, then P is an SPD function (see [3]).

Many classes of PD and SPD functions are well known (see [2,3,6,7,9]). In this section we want to construct some new classes of PD and SPD functions.

Univariate functions.

In the case $d = 1$ there are many classes of such functions. In particular we obtain PD functions using the Polya criterium: P is a PD function if g is a real, even continuous function which is convex on $(0, \infty)$ and $\lim_{x \to \infty} g(x) = 0$ (see [9]). The drawback of such functions is that they are not differentiable at 0. In the following we want to construct PD functions via integration of PD functions.

Let $M : \mathbb{R} \to \mathbb{R}$ be given, such that $M(x) = M(-x)$ for all $x \ge 0$ and $M(x) = 0$ for $x \ge a$ for some given positive real number a. Moreover, assume that M satisfies the condition

$$\int_{-a}^{a} M(t) \cos(xt) \, dt \ge 0,$$

for all $x \in \mathbb{R}$. Then it follows that M is a PD function.

By employing the function M, we want to construct new functions which are also PD functions and have a higher order of differentiability. Let a, b be real numbers that satisfy $0 < b < a$, and let $c = a + b$. Set $g(u) = M(u + b) - M(u)$ and $G(t) = \int_{-c}^{t} g(u)du$. Then we define

$$B(x) = \int_{-c}^{x} (G(t) - G(t - b))dt.$$

Theorem 4.1. *Let $M \in C^m(\mathbb{R})$ be defined as above, where m is a nonnegative integer. Then the following is true:*

(a) *The function B is an even, nonnegative function in $C^{m+2}(\mathbb{R})$ with support $[-c, c]$.*

(b) *B is a PD function on $\mathbb{R}$.*

Proof: (a) It follows from the definition that G has support $[-c, a]$, and hence $G(\cdot - b)$ has support $[-a, c]$. Then it follows that B has support $[-c, -c]$. Since $M \in C^m(\mathbb{R})$, we obtain that $B \in C^{m+2}(\mathbb{R})$. It also follows from the construction that B is a nonnegative function.

(b) We shall show that

$$F(x) = \int_{\mathbb{R}} B(t)e^{ixt}dt = \int_{-c}^{c} B(t)\cos(xt)dt \geq 0$$

for all $x \in \mathbb{R}$. Then B is a PD function. We distinguish two cases:

(i) Let $x \neq 0$. Since $B \in C^{m+2}(\mathbb{R})$ and B has support $[-c, c]$, we have

$$B(-c) = B(c) = B'(-c) = B'(c) = 0.$$

Then integration by parts yields

$$\begin{aligned}
F(x) &= \int_{-c}^{c} B(t)\cos(xt)dt \\
&= \frac{1}{x}B(t)\sin(xt) \mid_{-c}^{c} - \frac{1}{x} \int_{-c}^{c} B'(t)\sin(xt)dt \\
&= \frac{1}{x^2}B'(t)\cos(xt) \mid_{-c}^{c} - \frac{1}{x^2} \int_{-c}^{c} B''(t)\cos(xt)dt \\
&= -\frac{1}{x^2} \int_{-c}^{c} (g(t) - g(t - b))\cos(xt)dt \\
&= -\frac{1}{x^2} \int_{-c}^{c} (M(t + b) - 2M(t) + M(t - b))\cos(xt)dt.
\end{aligned}$$

Moreover, we have

$$-x^2 F(x) = \int_{-c}^{a-b} M(t+b)\cos(xt)dt - 2\int_{-a}^{a} M(t)\cos(xt)dt$$

$$+ \int_{-a+b}^{c} M(t-b)\cos(xt)dt$$

$$= \int_{-a}^{a} M(t)\big(\cos(x(t-b)) - 2\cos(xt) + \cos(x(t+b))\big)\,dt$$

$$= \int_{-a}^{a} M(t)(2\cos(xt)\cos(xb) - 2\cos(xt))dt$$

$$= 2(\cos(xb) - 1)\int_{-a}^{a} M(t)\cos(xt)dt.$$

It follows from the assumptions that $\int_{-a}^{a} M(t)\cos(xt)dt \geq 0$ for $x \in \mathbb{R}$ and therefore

$$F(x) = \frac{2}{x^2}(1 - \cos(xb))\int_{-a}^{a} M(t)\cos(xt)dt \geq 0$$

for all $x \in \mathbb{R}$.

(ii) Let $x = 0$. Then

$$F(0) = \int_{-c}^{c} B(t)dt > 0$$

since $B(t) > 0, t \in (-c, c)$.

This proves the theorem. $\square$

We have obtained a recursion formula which defines PD and SPD functions with increasing order of differentiability. In the following we want to define multivariate SPD functions as products of univariate functions.

Products of functions.

We now consider functions of the form

$$f(x) = \int_{-\infty}^{\infty} e^{itx} d\mu(t),$$

where μ is a finite nonnegative Borel measure. As above it follows that $P(x, y) = f(x - y)$ is a PD function. We consider tensor–products of such functions.

If we consider Bochner's result and the fact that the multivariate Fourier transform of a tensor product function is the tensor product of the univariate Fourier transforms, we obtain the following result.

Theorem 4.2. Let $\mu_k, k = 1, 2$, be finite nonnegative Borel measures on $\mathbb{R}$, and let

$$f_k(x) = \int_{-\infty}^{\infty} e^{itx} d\mu_k(t), \quad k = 1, 2.$$

Assume that the carrier of μ_1 and μ_2 has positive Lebesgue measure. Let

$$f(x, y) = f_1(x)f_2(y).$$

Then, for every set of distinct points $X = \{((x_i, y_i))_{i=1}^n\}$ in $\mathbb{R}^2$, the matrix

$$A = (f(x_j - x_k, y_j - y_k))_{j,k=1}^n$$

is a positive definite matrix, i.e. the function

$$f_1(x - u)f_2(y - v)$$

where $(x, y), (u, v) \subset \mathbb{R}^2$ is an SPD function on $\mathbb{R}^2$.

Remark. This theorem shows that we can construct multivariate SPD functions using products of univariate SPD functions. For products of reproducing kernel functions see also [10,11].

References

1. Ahlberg, J. H., E. N. Nilson and J. L. Walsh, *The Theory of Splines and Their Applications*, Academic Press, New York, 1967.

2. Cheney, E. W., Approximation using positive definite functions, in *Approximation Theory VIII, Vol. 1: Approximation and Interpolation*, Charles K. Chui and Larry L. Schumaker (eds.), World Scientific Publishing Co., Inc., Singapore, 1995, 145–168.

3. Cheney, E. W. and W. Light, *A Course in Approximation Theory*, Brooks/Cole Publishing Company, Pacific Grove/U.S.A., 1999.

4. Duchon, J., Splines minimizing rotation-invariant semi-norms in Sobolev spaces, in *Constructive Theory of Functions of Several Variables*, W. Schempp and K. Zeller (eds.), Springer–Verlag, Berlin, 1977, 85–100.

5. Lawson, C. L. and R. J. Hanson, *Solving Least Squares Problems*, SIAM, Philadelphia, 1995.

6. Saitoh, S. *Integral Transforms, Reproducing Kernels and Their Applications*, Longman, Essex, 1997.

7. Schaback, R., Native Hilbert spaces for radial basis functions I, in *New Developments in Approximation Theory*, M. W. Müller, M. D. Buhmann, D. H. Mache and M. Felten (eds.), Birkhäuser Verlag, Basel, 1999, 255–282.

8. Schumaker, L. L. and F. Utreras, Asymptotic properties of complete smoothing splines and applications, SIAM J. Sci. Statist. Comput. **9** (1988), 24–38.

9. Stewart, J., Positive definite functions and generalizations, an historical survey, Rocky Mountain J. Math. **6** (1976), 409–434.

10. Strauss, H., On interpolation with products of positive definite functions, Numerical Algorithms **15** (1997), 153–165.

11. Strauss, H., On products of positive definite functions, in *Approximation Theory IX, Vol. 1: Theoretical Aspects*, Charles K. Chui and Larry L. Schumaker (eds.), Vanderbilt University Press, Nashville, 1998, 317–324.

12. Strauss, H., Approximation with constraints by conditionally positive definite functions, preprint.

13. Utreras, F. I., Cross validation techniques for smoothing spline functions in one or two dimensions, in *Smoothing Techniques for Curve Estimation*, M. Rosenblatt and Th. Gasser (eds.), Springer Verlag, Berlin, 1979, 196–232.

14. Wahba, G., *Spline Models for Observational Data*, CBMS-NSF Regional Conference Series in Applied Mathematics 59, SIAM, Philadelphia, PA, 1990.

Hans Strauss
Institut für Angewandte Mathematik
Universität Erlangen–Nürnberg
Martensstr. 3
D–91058 Erlangen
Germany
strauss@am.uni-erlangen.de

On Linear Independence of Generators of FSI Distribution Spaces on ℝ

Jianzhong Wang

Abstract. A distribution space is called finitely shift invariant (FSI) if it is generated by a vector-valued distribution (generator) and its shifts. The linear independence of shifts of a generator is important in many applications. This paper provides a characterization for linear independence of the shifts of a generator of an FSI distribution space and presents a way to find the generators with linear independent shifts. The generators of some special FSI distribution spaces are discussed in detail.

§1. Introduction

Let $\phi = (\phi_1, \phi_2, \cdots, \phi_r)^T$ be a compactly supported vector-valued distribution on ℝ. We call r the length of ϕ and denote it by $\text{Length}(\phi)$. When $\text{Length}(\phi) = 1, \phi$ is also called a scalar distribution. Let $(l)^r$ denote the space containing all r-vector sequences of complex numbers, and let $(l_0)^r$ denote the subspace of $(l)^r$, which contains all compactly supported sequences in $(l)^r$. For $\boldsymbol{u} \in (l)^r$, its kth term is the row vector $\boldsymbol{u}(k) = (u_1(k), \cdots, u_r(k)) \in \mathbb{C}^r$ and its jth component is the scalar sequence $\boldsymbol{u}_j = (u_j(k))_{k \in \mathbb{Z}}$. The semi-convolution of ϕ and $\boldsymbol{u} \in (l)^r$ is defined by

$$\phi *' \boldsymbol{u} := \sum_{k \in \mathbb{Z}} \boldsymbol{u}(k)\phi(\cdot - k), \quad \boldsymbol{u} \in (l)^r, \tag{1}$$

where $\boldsymbol{u}(k)\phi(\cdot - k) = \sum_{j=1}^r u_j(k)\phi_j(\cdot - k)$. Let

$$S := S(\phi) := \{\phi *' \boldsymbol{u}; \quad \boldsymbol{u} \in (l)^r\}$$

and

$$S_0 := S_0(\phi) := \{\phi *' \boldsymbol{u}; \quad \boldsymbol{u} \in (l_0)^r\}.$$

Approximation Theory X: Wavelets, Splines, and Applications
Charles K. Chui, Larry L. Schumaker, and Joachim Stöckler (eds.), pp. 453–472.
Copyright Ⓞ 2002 by Vanderbilt University Press, Nashville, TN.
ISBN 0-8265-1416-2.

Then S and S_0 both are finitely shift invariant (FSI) spaces generated by ϕ. We call ϕ their generator. Set

$$K_0(\phi) = \{\boldsymbol{u} \in (l_0)^r; \quad \phi*'\boldsymbol{u} = 0\}$$

and

$$K(\phi) = \{\boldsymbol{u} \in (l)^r; \quad \phi*'\boldsymbol{u} = 0\}.$$

If $K_0(\phi) = \{0\}$, we say that ϕ has finitely linearly independent shifts. In the case that $K_0(\phi) \neq \{0\}$, the FSI space $S(\phi)$ may be generated by a generator with a smaller length. Let

$$\Psi = \{\psi; \quad \text{supp}\,\psi < \infty \text{ and } S(\psi) = S(\phi)\}.$$

Then the number $\eta = \min(\text{Length}(\psi); \psi \in \Psi)$ is called the cardinal number of ϕ and denoted by $\#(\phi)$. If $K(\phi) = \{0\}$, we say that ϕ has (global) linearly independent shifts and call ϕ an LI generator of $S(\phi)$. It is obvious that if ψ is an LI generator of $S(\phi)$, then $\text{Length}(\psi) = \#(\phi)$.

Many papers in the literature discussed linear independence of shifts of functions (or distributions). Dahmen and Micchelli [5] and Ron [23] discussed shifts of scalar distributions. Jia and Micchelli [17] generalized the results of [5] and [23] to vector-valued distributions. Linear independence of shifts of refinable distributions is also well studied. Jia and Wang [19] showed a characterization of linear independence of shifts of a refinable scalar distribution. (Also see [24] and [27]). The generalization of their results to refinable vector-valued distributions (of single variable) was made by Wang [26], Hogan [11], and Plonka [21]. Jia and Hogan [12] gave a characterization for multivariate refinable vector-valued distributions. Other papers on this aspect can be found in the references of the papers mentioned above.

The problem of cardinality of FSI spaces is first raised by de Boor, DeVore, and Ron in their paper [4]. Compared to the discussion of linear independence, fewer papers are concerned with this problem. In this paper, we give a uniform treatment for both cardinality and linear independence. In Section 2, we discuss the representation of compactly supported distributions and their properties. In Section 3, we derive a characterization of cardinality and linear independence for vector-valued distributions, and show how to obtain an LI generator for an FSI space. Using the results in Section 3, we study the spline-type generators in Section 4, where several examples are also given to illustrate the usage of the results in the section. The discussion of the refinable generators is included in Section 5.

§2. Compactly Supported Distributions

Before we study linear independence of shifts of compactly supported vector-valued distributions, we introduce some notions of compactly supported distributions.

Let $\Omega \subset \mathbb{R}$ be an open set. Let $C^k(\Omega)$ be the linear space of all functions which have continuous derivatives up to (and including) $k(\geq 0)$, and $C^\infty(\Omega)$ be the set of all infinitely differentiable functions. Let $C_0^k(\Omega) \subset C^k(\Omega)$ be the space that contains all compactly supported functions in $C^k(\Omega)$, and $C_0^\infty(\Omega) \subset C^\infty(\Omega)$ be the space that contains all compactly supported functions in $C^\infty(\Omega)$. We write $C(\Omega) = C^0(\Omega)$ and $C_0(\Omega) = C_0^0(\Omega)$. For any compact support $K \subset \Omega$, let $\mathcal{D}_K(\Omega)$ be the set of all functions $f \in C_0^\infty(\Omega)$ such that $\mathrm{supp}\,(f) \subset K$. Define a family of semi-norms on $\mathcal{D}_K(\Omega)$ by

$$p_{K,m}(f) = \sup_{s \leq m, x \in K} \|D^s f\|_\infty, \quad 0 \leq m < \infty, \quad K \subset \Omega. \tag{2}$$

$\mathcal{D}_K(\Omega)$ is a locally convex (linear topological) space and $C_0^\infty(\Omega)$ equipped with the strict inductive limit of $\mathcal{D}_K(\Omega)$'s, where K ranges over all compact subsets of Ω, is the locally convex linear topological space $\mathcal{D}(\Omega)$. A continuous linear functional defined on $\mathcal{D}(\Omega)$ is called a (Schwartz) **distribution**. The space of all distributions is denoted by $\mathcal{D}(\Omega)'$. Corresponding to (Schwartz) distributions, $\mathcal{D}(\Omega)$ is called their **base space**, and functions in $\mathcal{D}(\Omega)$ are called **test functions**. The value of a distribution φ at a test function f is denoted by (φ, f). Let $\mathcal{S}$ be a locally convex linear topological space such that $\mathcal{D}(\Omega) \subset \mathcal{S}$. If a distribution φ on $\mathcal{D}(\Omega)$ can be extended to a continuous linear functional on $\mathcal{S}$, then we call φ a distribution on (the base space) $\mathcal{S}$, and all distributions on $\mathcal{S}$ form the **dual space** of $\mathcal{S}$, which is denoted by $\mathcal{S}'$. In this paper, we mainly consider compactly supported distributions. A point x_0 is called a **support point** of a distribution φ if, for any $\epsilon > 0$, there is a test function f with $\mathrm{supp}\,f \subset (x_0 - \epsilon, x_0 + \epsilon)$ such that $(\varphi, f) \neq 0$. The set of all support points of φ is called the **support** of φ and denoted by $\mathrm{supp}\,\varphi$. It is obvious that $\mathrm{supp}\,\varphi$ is a closed set. The locally convex space $\mathcal{K}(\Omega)$ is defined as the space $C^\infty(\Omega)$ equipped with the semi-norms (2). Similarly, the space $\mathcal{K}^k(\Omega)$ is defined as the space $C^k(\Omega)$ equipped with the semi-norms

$$p_{K,m}(f) = \sup_{s \leq m, x \in K} \|D^s f\|_\infty, \quad 0 \leq m \leq k, \quad K \subset \Omega.$$

It is known that a compactly supported distribution φ on $\mathcal{D}(\Omega)$ can be extended in one and only one way to a distribution $\tilde{\varphi}$ on $\mathcal{K}(\Omega)$ such that $(\tilde{\varphi}, f) = 0$ if f vanishes in a neighborhood of $\mathrm{supp}\,\varphi$. For convenience, we shall continue to denote this unique extension (of φ) by φ.

The following result gives a representation of a compactly supported distribution.

Theorem 1. ([7], Vol. 2., Chapter 2, Section 4.4). *Let φ be a compactly supported distribution on $\mathcal{D}(\mathbb{R})$ with $\operatorname{supp}\varphi \subset [a,b]$. Then there is a positive integer s such that, for any $\epsilon > 0$, there exist continuous functions $\varphi_{q\epsilon} \in C_0((a-\epsilon, b+\epsilon)), 0 \le q \le s$, such that*

$$\varphi = \sum_{q=0}^{s} D^q \varphi_{q\epsilon},$$

where D is the generalized derivative operator. That is, for any $f \in \mathcal{K}(\mathbb{R})$,

$$(\varphi, f) = \sum_{q=0}^{s}(-1)^q \int_{a-\epsilon}^{b+\epsilon} \varphi_{q\epsilon}(x)D^q f(x)dx. \tag{3}$$

The positive integer s in Theorem 1 is called the **order** of the distribution φ and denoted by $\operatorname{ord}(\varphi)$. By (3), any compactly supported distribution has a finite order. In numerical analysis and in wavelet theory, we often need to deal with the spaces of finitely differentiable functions. Hence, we often have to (continuously) extend distributions on these spaces, such as $\mathcal{K}^s(\mathbb{R})$. By (3), we can see that any s-order distribution with $\operatorname{supp}\varphi \subset [a,b]$ can be uniquely extended on $\mathcal{K}^s(\mathbb{R})$, (*i.e.*, $\varphi \in \mathcal{K}^s(\mathbb{R})'$). For an integer $s > 0$, we define the distribution space

$$\mathcal{F}_s([a,b]) = \{\varphi \in \mathcal{K}^s(\mathbb{R})'; \quad \operatorname{supp}\varphi \subset [a,b] \ \& \operatorname{ord}(\varphi) \le s\}.$$

To identify functions in $\mathcal{F}_s([a,b])$, we introduce the following notion. A set of functions $\{f_\lambda\}_{\lambda \in \Lambda} \subset \mathcal{K}^s(\mathbb{R})$ is called a **determinative set** of $\mathcal{F}_s([a,b])$ if, for $\varphi \in \mathcal{F}_s([a,b])$,

$$(\varphi, f_\lambda) = 0, \quad \forall \lambda \in \Lambda, \quad \Longrightarrow \varphi = 0.$$

Let

$$C_0^s([\alpha, \beta]) = \{f \in C_0^s(\mathbb{R}); \quad \operatorname{supp} f \subset [\alpha, \beta]\}.$$

The Banach space $\mathcal{K}_0^s([\alpha, \beta])$ is the space $C_0^s([\alpha, \beta])$ equipped with the norm $||f|| = \max_{0 \le i \le s} ||D^i f||_\infty$. It is clear that $\mathcal{K}_0^s([\alpha, \beta]) \subset \mathcal{K}^s(\mathbb{R})$. By Theorem 1, we have the following.

Corollary 2. *For a $\delta > 0$, let $\{f_\lambda\}_{\lambda \in \Lambda}$ span a dense set of $\mathcal{K}_0^s([a-\delta, b+\delta])$. Then $\{f_\lambda\}_{\lambda \in \Lambda}$ is a determinative set of $\mathcal{F}_s([a,b])$.*

Proof: Assume that $\varphi \in \mathcal{F}_s([a,b])$ satisfies $(\varphi, f_\lambda) = 0, \forall \lambda \in \Lambda$. We shall prove $\varphi = 0$. Since $\{f_\lambda\}_{\lambda \in \Lambda}$ spans a dense set in $\mathcal{K}_0^s([a-\delta, b+\delta])$, we have $(\varphi, f) = 0 \quad \forall f \in \mathcal{K}_0^s([a-\delta, b+\delta])$. This implies $(\varphi, f) = 0$ for all test functions with supports in $[a-\delta, b+\delta]$. Since $\operatorname{supp}\varphi \subset [a,b]$, we have $(\varphi, f) = 0$ for all test function with supports in $\mathbb{R} \setminus [a,b]$. Thus $(\varphi, f) = 0$.

Now assume f is an arbitrary test function. We select a positive $\epsilon < \delta$, and let $h_\epsilon \in C_0^\infty(\mathbb{R})$ be a function that is 1 on $(a - \epsilon, b + \epsilon)$ and vanishes on $\mathbb{R} \setminus [a - \delta, b + \delta]$. We now have

$$f = h_\epsilon f + (1 - h_\epsilon)f,$$

where supp $((1 - h_\epsilon)f) \subset \mathbb{R} \setminus (a - \epsilon, b + \epsilon)$ and supp $(h_\epsilon f) \subset [a - \delta, b + \delta]$. Thus $(\varphi, f) = (\varphi, h_\epsilon f) + (\varphi, (1 - h_\epsilon)f) = 0$. Hence $\varphi = 0$. $\square$

In many applications, Sobolev spaces on an open set $\Omega \subset \mathbb{R}$ are often needed. Let n be a non-negative integer. For $1 \leq p \leq \infty$, we define

$$W^{n,p}(\Omega) = \{f; \quad D^j f \in L_p(\Omega), \quad 0 \leq j \leq n\},$$

which is equipped with the norm $||f||_{n,p} = \max_{0 \leq i \leq n} ||D^i f||_p$. When $p = 2$, it becomes a Hilbert space, which is often denoted by $H^n(\Omega)$. It is clear that an s-order compactly supported distribution is a distribution on $W^{s,p}(\mathbb{R})$ for any p, $1 \leq p \leq \infty$.

The homogeneous Sobolev space $W_0^{n,p}(\Omega)$ is a subspace of $W^{n,p}(\Omega)$ defined by $W_0^{n,p}(\Omega) = W^{n,p}(\Omega) \cap C_0^{n-1}(\Omega)$. We also define $W_0^{n,p}([a,b]) = W^{n,p}((a,b)) \cap C_0^{n-1}([a,b])$. Hence, $\mathcal{K}_0^n(\Omega)$ can be embedded into $W_0^{n,p}(\Omega)$, and $\mathcal{K}_0^n([a,b])$ can be embedded into $W_0^{n,p}([a,b])$. By Corollary 2, if $\{f_\lambda\}_{\lambda \in \Lambda}$ spans a dense set of $W_0^{s,p}([a - \delta, b + \delta])$ for some $p, 1 \leq p \leq \infty$, then $\{f_\lambda\}_{\lambda \in \Lambda}$ is a determinative set of $\mathcal{F}_s([a,b])$.

The restriction of a distribution $\varphi \in \mathcal{D}(\mathbb{R})'$ on an open interval (α, β) is well-defined. It can be described as follows. Let $\tilde{\varphi}$ be a distribution on $\mathcal{D}((\alpha, \beta))$ defined by

$$(\tilde{\varphi}, f) = (\varphi, f), \quad \forall f \in \mathcal{D}((\alpha, \beta)).$$

Since $\mathcal{D}((\alpha, \beta))$ can be embedded into $\mathcal{D}(\mathbb{R})$, we can extend $\tilde{\varphi}$ to a distribution on $\mathcal{D}(\mathbb{R})$ in such a way that $(\tilde{\varphi}, f) = 0$ for any f vanishing on $[\alpha, \beta]$. This extension is the restriction of φ on (α, β) and denoted by $\varphi|_{(\alpha,\beta)}$. Without loss of generality, for any interval $[\alpha, \beta]$, we can assume that a compactly supported distribution φ has supp $\varphi \subset [a, b]$, where a and b are selected such that $(\alpha, \beta) \subset [a, b]$. By Theorem 1, for a fixed $\epsilon > 0$, there is an $s > 0$ and continuous functions $\varphi_{q\epsilon}$ with supp $\varphi_{q\epsilon} \subset [a - \epsilon, b + \epsilon]$ such that $\varphi = \sum_{q=0}^s D^q \varphi_{q\epsilon}$. Then $\varphi|_{(\alpha,\beta)}$ can be represented as

$$(\varphi|_{(\alpha,\beta)}, f) = \sum_{q=0}^s (-1)^q \int_\alpha^\beta \varphi_{q\epsilon}(x) D^q f(x) dx, \quad \forall f \in \mathcal{K}^s(\mathbb{R}). \quad (4)$$

Let

$$\mathcal{F}_s((\alpha, \beta)) = \{\varphi|_{(\alpha,\beta)}; \quad \varphi \in \mathcal{K}(\mathbb{R})', \quad \mathrm{ord}\,(\varphi) \leq s\}.$$

By (4), we can prove the following.

Lemma 3. *If* $\{f_\lambda\}_{\lambda \in \Lambda}$ *spans a dense set of* $\mathcal{K}_0^s([\alpha, \beta])$, *(or of* $W_0^{s,p}([\alpha, \beta])$, $1 \leq p \leq \infty$,*) then* $\{f_\lambda\}_{\lambda \in \Lambda}$ *is a determinative set of* $\mathcal{F}_s((\alpha, \beta))$.

The proof of the lemma is similar to Corollary 2. We skip it. The restriction of φ on a half-open interval $[\alpha, \beta)$ can be defined by

$$\varphi|_{[\alpha,\beta)} = \varphi|_{(\gamma,\beta)} - \varphi|_{(\gamma,\alpha)}, \tag{5}$$

where γ can be any number satisfying $\gamma < \alpha$. Note that the definition of $\varphi|_{[\alpha,\beta)}$ is independent of γ. The distribution $\varphi|_{[\alpha,\beta)} - \varphi|_{(\alpha,\beta)}$ is supported at the point α, which is a finite combination of Dirac's distribution and its derivatives at α. Let $\mathcal{F}_s([\alpha, \beta)) = \{\varphi|_{[\alpha,\beta)}; \ \varphi \in \mathcal{K}(\mathbb{R})', \ \mathrm{ord}\,(\varphi) \leq s\}$. It is clear that we have the following.

Lemma 4. *Let* $\gamma < \alpha$. *If* $\{f_\lambda\}_{\lambda \in \Lambda}$ *spans a dense set of* $\mathcal{K}_0^s([\gamma, \beta])$, *(or of* $W_0^{s,p}([\gamma, \beta])$, $1 \leq p \leq \infty$,*) then* $\{f_\lambda\}_{\lambda \in \Lambda}$ *is a determinative set of* $\mathcal{F}_s([\alpha, \beta))$.

A distribution φ is called **regular** if, for any $f \in \mathcal{D}(\mathbb{R})$,

$$(\varphi, f) = \int_{\mathbb{R}} \varphi(x) f(x) dx.$$

A compactly supported regular distribution φ with $\mathrm{supp}\,\varphi \subset [a, b]$ is a function in $L_1([a, b])$. Then $\varphi|_{(\alpha,\beta)} = \varphi \chi_{(\alpha,\beta)}$, where $\chi_{(\alpha,\beta)}$ is the characteristic function of the interval (α, β). When φ is regular, $\varphi|_{(\alpha,\beta)} = \varphi|_{[\alpha,\beta)}$.

§3. LI Generators of FSI Spaces

In this section, we study linear independence of shifts of compactly supported vector-valued distributions. We also discuss how to obtain LI generators of an FSI space from its non-LI generators.

For a compactly supported distribution φ, its **integer support** is defined as an integer interval $[M, N)$ such that $\mathrm{supp}\,\varphi \subset [M, N)$, but $\varphi|_{[M,M+1)} \neq 0$ and $\varphi|_{[N-1,N)} \neq 0$. If φ is regular, then the condition $\mathrm{supp}\,\varphi \subset [M, N)$ can be relaxed to $\mathrm{supp}\,\varphi \subset [M, N]$. We denote the integer support of φ by $\mathrm{isupp}\,\varphi$. For a compactly supported vector-valued distribution ϕ, we define $\mathrm{isupp}\,\phi = \{\,\mathrm{isupp}\,\phi_1, \cdots, \mathrm{isupp}\,\phi_r\}$. Let $\mathrm{isupp}\,\phi_i = [M_i, N_i), 1 \leq i \leq r$. Since we consider linear independence of shifts of φ, without loss of generality, we can assume $M_i = 0, 1 \leq i \leq r$. We now define

$$\Lambda = \cup_{j=1}^r \Lambda_j, \tag{6}$$

where $\Lambda_j = \{(j, k); \ 0 \leq k \leq N_j - 1\}$. Write $\lambda = |\Lambda|$. For any $(j, k) \in \Lambda$, define $\phi_j^k = \phi_j(\cdot + k)|_I$, where $I = [0, 1)$. Thus, the space $S(\phi)|_I$ is a finite dimensional space spanned by

$$\Phi := (\phi_j^k)_{(j,k) \in \Lambda}. \tag{7}$$

Let $b = (b_1, \cdots, b_m)^T$ be a basis of $S(\phi)|_I$. Then there exists a $\lambda \times m$ matrix M such that

$$\Phi = Mb,$$

where bi-indices (j, k) are used for the rows of M. We call M a representing matrix of ϕ (with respect to b). Let e^i be the ith column of M. The $(j, k)^{th}$ component of e^i is denoted by $e_j^i(k)$. Let $N = \max_{1 \leq i \leq r} N_i$ and

$$\hat{e}_j^i(k) = \begin{cases} e_j^i(k), & (j, k) \in \Lambda \\ 0, & (j, k) \notin \Lambda. \end{cases}$$

Defining the $r \times m$ matrix

$$\boldsymbol{M}(k) = (\hat{e}_j^i(k))_{i,j=1}^{m,r}, \quad 0 \leq k \leq N - 1,$$

we have

$$\phi = \sum_{k=0}^{N-1} \boldsymbol{M}(k) b(\cdot - k). \tag{8}$$

Write $\phi^k = (\phi_1^k, \cdots, \phi_r^k)^T$. It is obvious that $\boldsymbol{M}(k) b = \phi^k$. The Laurent polynomial matrix $M(z) := \sum \boldsymbol{M}(k) z^k$ is called a derived matrix of M. We say that a Laurent polynomial matrix is fundamental if it has full rank for all $z \in \mathbb{C} \setminus \{0\}$. Although a representing matrix M of ϕ is dependent of the selection of the basis of $S(\phi)|_I$, all representing matrices of ϕ are equivalent under elementary column transforms. In fact, if b_1 and b_2 are two bases of $S(\phi)|_I$, then there is an invertible matrix T such that $b_1 = Tb_2$. Let M_1 and M_2 be the representing matrices of ϕ with respect to b_1 and b_2 respectively. Then we have $M_2 = M_1 T$, *i.e.*, M_1 and M_2 are equivalent. Representing matrices of a vector-valued distribution have the following property.

Lemma 5. *The rank of representing matrices of a vector-valued distribution φ is equal to the dimension of the space $S(\phi)|_I$. Besides, $\phi_j^k = 0$ if and only if the $(j, k)^{th}$ row of a representing matrix is 0.*

Proof: Let $b = (b_1, \cdots, b_m)^T$ be a basis of $S(\phi)|_I$ and M be the corresponding representing matrix of φ. If the rank of $M = [e^1, e^2, \cdots, e^m]$ is less than m, then the matrix M can be decomposed into

$$M = [\hat{e}^1, \cdots, \hat{e}^{m-1}, 0] C,$$

where C is an $m \times m$ invertible matrix. Let $\tilde{b} = Cb$. Then $\tilde{b} \subset S(\phi)|_I$ and $\phi = [\hat{e}^1, \cdots, \hat{e}^{m-1}, 0] \tilde{b}$. Thus, span $\{\tilde{b}_1, \cdots, \tilde{b}_{m-1}\} = S(\phi)|_I$. This is a contradiction of b being a basis of $S(\phi)|_I$. Hence rank $M = m$. We have $\phi_j^k = \sum_{i=1}^m e_j^i(k) b_i$. Since b is a basis, $\phi_j^k = 0$ if and only if $(e_j^1(k), \cdots, e_j^m(k)) = 0$, *i.e.*, the $(j, k)^{th}$ row of M is 0. $\square$

We say the shifts of ϕ are linearly independent on I if all non-vanishing components of Φ form a basis of $S(\phi)|_I$. By Lemma 5, we have the following.

Corollary 6. *The shifts of ϕ are linearly independent on I if and only if there are m nonzero rows in a representing matrix of ϕ.*

Proof: Let M be a representing matrix of ϕ. If M has m non-zero rows, the non-vanishing components in Φ have the same indices as these non-zero rows. Since all nonzero rows of M form an invertible matrix, these components form a basis of $S(\phi)|_I$. The proof of the reverse is trivial. □

By (8), we have, for any $\boldsymbol{a} \in (l)^r$,

$$\phi*'\boldsymbol{a} = b *' (\mathcal{C}_M\boldsymbol{a}), \tag{9}$$

where $\mathcal{C}_M$ is the convolution operator : $(l)^r \to (l)^m$ defined by

$$\boldsymbol{v}(l) := (\mathcal{C}_M\boldsymbol{a})(l) = \sum_{k \in \mathbb{Z}} \boldsymbol{a}(k)\boldsymbol{M}(l - k). \tag{10}$$

Taking the z-transform of (10), we have $v(z) = a(z)M(z)$, where $a(z) = \sum \boldsymbol{a}(k)z^k$, $v(z) = \sum \boldsymbol{v}(k)z^k$. It is obvious that $\mathcal{C}_M$ in also a linear operator $(l_\infty)^r \to (l_\infty)^m$ and $(l_0)^r \to (l_0)^m$. Denote its kernel on $(l)^r$, on $(l_\infty)^r$, and on $(l_0)^r$ by $\ker\mathcal{C}_M, \ker_b\mathcal{C}_M$, and $\ker_0\mathcal{C}_M$ respectively. Let $\Gamma = \{z \in \mathbb{C}, \ |z| = 1\}$.

We now characterize linear independence of shifts of ϕ in terms of its representing matrix M.

Theorem 7. *Let ϕ be a vector-valued distribution and b be an arbitrary basis of the space $S(\phi)|_I$. Let M be the representing matrix of ϕ with respect to b, and let $M(z)$ be the derived matrix of M. Then the following statements hold:*

(1) $K(\phi) = \{0\}$ *if and only if $M(z)$ is fundamental. Moreover, $c \in K(\phi)$ if and only if $\mathcal{C}_M\boldsymbol{c} = 0$.*

(2) $K_0(\phi) = \{0\}$ *if and only if $\operatorname{rank} M(z) = r$. Moreover, $\#(\phi)$ is equal to $\operatorname{rank} M(z)$.*

(3) *Let $M(z) = L(z)B_c(z)R(z)$, where $B_c(z)$ is Smith's canonical form of $M(z)$, and matrices $L(z)$ and $R(z)$ both are fundamental. Then*

$$\psi(\cdot) = \sum_{k=0}^{N-1} \boldsymbol{R}(k)b(\cdot - k) \tag{11}$$

is an LI generator of $S(\phi)$.

Proof: It is obvious that $K(b) = \{0\}$. By (9), we have $K(\phi) = \ker(\mathcal{C}_M)$ and $K_0(\phi) = \ker_0(\mathcal{C}_M)$. According to [17], $K(\phi) = \{0\}$ if and only if there exists a row vector $c \in \mathbb{C}^r\backslash\{0\}$ and a complex $w \in (\mathbb{C}\backslash\{0\})^r$ such that

$$\sum_{k \in \mathbb{Z}^s} w^{-k}c\phi(x - k) = 0,$$

which is equivalent to

$$\sum_{k\in\mathbb{Z}}\sum_{l\in\mathbb{Z}} w^{-k}c\boldsymbol{M}(l-k)b(x-l) = 0. \tag{12}$$

Since $\{b(x-l)\}_{l\in\mathbb{Z}}$ is linearly independent, (12) holds if and only if

$$\sum_{k\in\mathbb{Z}} w^{-k}c\boldsymbol{M}(l-k) = 0 \quad \forall l \in \mathbb{Z},$$

which is equivalent to $c M(w) = 0$, *i.e.*, rank $M(w) < r$. Hence, $K(\phi) = \{0\}$ if and only if rank $M(w) = r$ for all $w \in (\mathbb{C}\backslash\{0\})^s$, *i.e.*, $M(z)$ is fundamental. Similarly, $K_0(\phi) \neq \{0\}$ if and only if there is a non-zero $\boldsymbol{a} \in (l_0)^r$ such that $\phi *' \boldsymbol{a} = 0$, which is equivalent to $\mathcal{C}_M \boldsymbol{a} = 0$, *i.e.*, rank $M(z) < r$. We now assume that rank $M(z) = \gamma(< r)$. Then $M(z)$ has Smith's canonical decomposition

$$M(z) = L(z)B_c(z)R(z),$$

where $B_c(z) = \begin{pmatrix} B_\gamma & 0 \\ 0 & 0 \end{pmatrix}$ and $B_\gamma = \mathrm{diag}\,(d_1(z),\cdots,d_\gamma(z))$, $d_j(z) \neq 0$, $j = 1,\cdots,\gamma$. Let $G_\gamma(z)$ be the $r \times \gamma$ matrix containing the first γ columns of $L(z)B_c(z)$ and $R_\gamma(z)$ be the $\gamma \times m$ matrix containing the first γ rows of the matrix $R(z)$. We now define the vector-valued distribution ψ by $\psi(x) = \sum R_\gamma(k)b(x-k)$. It is obvious that shifts of ψ are linearly independent and Length$(\psi) = \gamma$. From $\phi(x) = \sum G_\gamma(k)\psi(x-k)$, we have $\phi \subset S(\psi)$. On the other hand, since $L(z)$ is fundamental, so is its inverse $L^{-1}(z)$. Let $H_\gamma(z)$ be the $\gamma \times r$ matrix containing the first γ rows of $L^{-1}(z)$. Since $H_\gamma(z)$ is a polynomial matrix, $M_\gamma(z) = B_\gamma^{-1}(z)H_\gamma(z)$ is a rational matrix. We have $\psi(x) = \sum M_\gamma(k)\phi(x-k)$, *i.e.*, $\psi \in S(\phi)$. Hence, ψ is an LI generator of $S(\phi)$ and $\#(\phi) = \gamma$. $\square$

Corollary 8. *If ϕ is a vector-valued distribution in $W^{n,p}(\mathbb{R})'$ and ψ is an LI generator of $S(\phi)$, then $\psi \subset W^{n,p}(\mathbb{R})'$.*

Proof: If $\phi \subset W^{n,p}(\mathbb{R})'$, then the basis b (of $S(\phi)|_I$) $\subset W^{n,p}(\mathbb{R})'$. By (11), $\psi \subset W^{n,p}(\mathbb{R})'$. $\square$

From the last part of the proof of Theorem 7, we have the following result.

Corollary 9. *Let ϕ be a vector-valued distribution, M be a representing matrix of ϕ, and $M(z)$ be the derived matrix of M. Assume that $M(z) = L(z)B_c(z)R(z)$ is Smith's canonical decomposition of $M(z)$, where $B_c(z) = \begin{pmatrix} B_\gamma & 0 \\ 0 & 0 \end{pmatrix}$ and $B_\gamma = \mathrm{diag}\,(d_1(z),\cdots,d_\gamma(z))$, $d_j(z) \neq 0$ for*

$j = 1, \cdots, \gamma$. Let $H_\gamma(z)$ be the $\gamma \times r$ matrix containing the first γ rows of $L^{-1}(z)$. Write $M_\gamma(z) = B_\gamma^{-1}(z)H_\gamma(z)$. Then $\psi(x) = \sum M_\gamma(k)\phi(x - k)$ is an LI generator of the space $S(\phi)$.

From Theorem 7, we know that if we want to find an LI generator of $S(\phi)$, we only need to find a representing matrix of ϕ. We now discuss how to find it. Let $F = (f_1, \cdots, f_m)^T$ be a vector-valued distribution and $G = (g_1, \cdots, g_n)$ be a vector-valued function in the base space. Then we define $(F, G) = ((f_i, g_j))_{i,j=1}^{m,n}$, which is an $m \times n$ matrix. If F is regular, we sometimes use the notation $\int_{\mathbb{R}} F(x)G(x)dx$ instead of (F, G).

Theorem 10. *Assume the compactly supported vector-valued distribution ϕ is regular and $\phi \subset L_p(\mathbb{R})$. Define the index set Λ by (6) and Φ by (7). Assume $\{g_j\}_{j \in J}$ spans a dense set in $L_q(I), \frac{1}{p} + \frac{1}{q} = 1$. Denote $\boldsymbol{v}_j = \int_I \Phi(x)g_j(x)dx$, which is a vector in $\mathbb{R}^\lambda$. Then any linearly independent system in $\{\boldsymbol{v}_j\}_{j \in J}$ forms a representing matrix of ϕ. Similarly, assume ϕ is a vector-valued distribution of order s and $\{g_j\}_{j \in J} \subset W_0^{s,p}([0 - \epsilon, 1])$ spans a dense set in $W_0^{s,p}([0 - \epsilon, 1])$ for an $\epsilon > 0$. Let $\boldsymbol{v}_j = (\Phi, g_j) \in \mathbb{R}^\lambda$. Then any linearly independent system in $\{\boldsymbol{v}_j\}_{j \in J}$ forms a representing matrix of ϕ.*

Proof: We first assume that $\phi \subset L_p(\mathbb{R})$ and $\{g_j\}_{j \in J}$ spans a dense set in $L_q(I)$. Let all columns of $b = [b_1, \cdots, b_m]$ form the basis of $S(\phi)|_I$ and M be the representing matrix of ϕ with respect to b. Write $M = [e^1, \cdots, e^m]$. From $\Phi = Mb$, we have

$$\boldsymbol{v}_j = \int_I \Phi(x)g_j(x)dx = M \int_I b(x)g_j(x)dx := M\boldsymbol{d}_j. \qquad (13)$$

Then $\text{span}\{\boldsymbol{v}_j\}_{j \in J} \subset \text{span}\{e^1, \cdots, e^m\}$. Let $d_j(k) = \int_I b_k(x)g_j(x)dx$ and $\boldsymbol{d}(k) = (d_j(k))_{j \in J}$, which is an infinite row vector. We now claim that the m sequences $\boldsymbol{d}(1), \cdots, \boldsymbol{d}(m)$ are linearly independent. To verify this, we assume the vector $c = (c_1 \cdots, c_m)$ satisfies $\sum_{i=1}^m c_i\boldsymbol{d}(i) = 0$. It follows that

$$\int_I \sum_{i=1}^m c_ib_i(x)g_j(x)dx = 0, \quad \forall j \in J.$$

Note that the distribution $\sum_{i=1}^m c_ib_i \in L_p(I)$. Since $\{g_j\}_{j \in J}$ spans a dense set in $L_q(I)$, we have $\sum_{i=1}^m c_ib_i(x) = 0$. From the fact that b is a linearly independent system of $L_p(I)$, we have $c = 0$, which implies that the m sequences $\boldsymbol{d}(1), \cdots, \boldsymbol{d}(m)$ are linearly independent. We now have, by (13),

$$\text{span}\{\boldsymbol{v}_j\}_{j \in J} = \text{span}\{e^1, \cdots, e^m\},$$

i.e., a linearly independent system in $\{\boldsymbol{v}_j\}_{j \in J}$ forms a representing matrix of ϕ. When ϕ is a vector-valued distribution of order s and $\{g_j\}_{j \in J} \subset$

$W_0^{s,p}([0-\epsilon,1])$, it follows that any set that spans a dense set of $W_0^{s,p}([0-\epsilon,1])$ is a determinative set of $\mathcal{K}_0^s(I)$, and we can get the proof in a similar way. $\square$

§4. On Spline-type Distributions

Let

$$\pi_I := \{x_0, \cdots, x_N\}, \quad 0 = x_0 < \cdots < x_N = 1,$$

be a finite partition of I. We denote $\Delta_i = [x_{i-1}, x_i)$.Then $I = \cup_{i=1}^N \Delta_i$. Shifting π_I by an integer k, we have a partition on $I_k = [k-1,k)$, which is denoted by $\pi_{I_k} (= \pi_I + k)$. Thus, $I_k = \cup_{i=1}^N \Delta_i^k$, where $\Delta_i^k = [x_{i-1} + k - 1, x_i + k - 1)$. Then $\pi = \cup_{k\in\mathbb{Z}} \pi_{I_k}$ defines a partition of $\mathbb{R}$. A vector-valued distribution ϕ is called a **spline-type distribution** on π, if there exists a finite linearly independent distribution system $P = \{p_1, p_2, \cdots, p_n\}$ such that each $\phi|_{\Delta_i^k}, 1 \leq i \leq N, k \in \mathbb{Z}$, is a linear combination of P. We say $S(\phi)$ is a **spline subspace** on π if any distribution $f \in S(\phi)$ is a spline-type distribution on π (with respect to P). Let ϕ be a compactly supported distribution. If each $\phi|_j^k$ is a spline-type distribution on I (with respect to P), then $S(\phi)$ is a spline subspace. The set P is called a **foundation** of ϕ. Most useful generators of FSI spaces are spline-type distributions. We now discuss linear independence of their shifts. In this section, we keep the assumption that the distributions we considered are compactly supported.

Let $Q = \{q_1, \cdots, q_l\}$ be a set of algebraic linear functionals on P. As we defined in the previous section, Q is called a determinative set of P, if $(Q, f) = 0$ for all $f \in \mathrm{span}\, P$ implies $f = 0$. It is obvious that Q is a determinative set of P if and only if the matrix $(Q, P) = ((q_i, p_j))_{i,j=1}^{l,n}$ has full column rank. A determinative set Q is called a **dual set** of P if $(Q, P) = I_n$. The dual set of P is denoted by $P^* = \{p_1^*, \cdots, p_n^*\}$. For example, a polynomial spline of order n is a linear combination of $\{1, x, \cdots, x^{n-1}\}$ on each subinterval Δ_i^k. One of the dual sets of $\{1, x, \cdots, x^{n-1}\}$ is $\{\delta, D\delta, \cdots, \frac{D^{n-1}\delta}{(n-1)!}\}$, where δ is Dirac's distribution. For spline-type distributions, we have the following.

Theorem 11. *Assume ϕ is a spline-type vector-valued distribution and its foundation is $P = \{p_1, \cdots, p_n\}$. Assume $\phi_j^k = \sum_{i=1}^n c_{i,l}^{j,k} p_i$ on Δ_l. Let $c_{i,l} = \left(c_{i,l}^{j,k}\right)_{(j,k)\in\Lambda}$. Then any basis of $V_p = \mathrm{span}\,\{c_{i,l} : i = 1, \cdots, n, l = 1, \cdots, N\}$ forms a representing matrix of ϕ.*

Proof: Let $b = (b_1, \cdots, b_m)^T$ be a basis of $S(\phi)|_I$. Then there is a representing matrix M of ϕ such that $\Phi = Mb$. Let P be the foundation of ϕ.

Note that, on each Δ_l, b_t is a linear combination of P :

$$b_t = \sum_{s=1}^{n} d_{s,l}^{t} p_s, \quad \text{on } \Delta_l, \ 1 \leq t \leq m.$$

Let $\{p_1^*, \cdots, p_n^*\}$ be a dual set of P. It follows that

$$(p_w^*, b_t) = \sum_{s=1}^{n} d_{w,l}^{t} (p_w^*, p_s) = d_{w,l}^{t} \quad \text{on } \Delta_l.$$

Then we have

$$(p_w^*, \Phi_j^k) = (p_w^*, \sum_{i=1}^{n} c_{i,l}^{j,k} p_i) = c_{w,l}^{j,k} \quad \text{on } \Delta_l.$$

It follows that

$$(p_w^*, \phi|_{\Delta_l}) = \boldsymbol{c}_{w,l}.$$

On the other hand, we have

$$(p_w^*, \phi|_{\Delta_l}) = (p_w^*, Mb|_{\Delta_l}) = M\boldsymbol{d}_{w,l},$$

where $\boldsymbol{d}_{w,l} = (d_{w,t}^{1}, \cdots, d_{w,l}^{m})^T$. Hence,

$$\boldsymbol{c}_{w,l} = M\boldsymbol{d}_{w,l}.$$

Let C be the matrix whose columns are $\{\boldsymbol{c}_{w,l}\}_{w,l=1}^{n,N}$, and D be the matrix whose columns are $\{\boldsymbol{d}_{w,l}\}_{w,l=1}^{n,N}$. We claim that D has full row rank. Note that the tth row of D is $\boldsymbol{d}^t = (d_{w,l}^{t})_{w,l=1}^{n,N}$. Assume that there is a vector $\boldsymbol{a} = (a_1 \cdots, a_m)$ such that $\sum_{t=1}^{m} a_t \boldsymbol{d}^t = 0$. It follows that

$$\sum_{t=1}^{m} a_t d_{w,l}^{t} = 0, \quad \forall w = 1, \cdots, n, \quad l = 1, \cdots, N.$$

This implies $\sum_{t=1}^{m} a_t b^t = 0$ on I, which is a contradiction of $\{b_i\}_{i=1}^{m}$ being a basis of $S(\phi)|_I$. Since now D has full row rank, the column space of C is equal to the column space of M. However, M has full column rank. Hence the basis of the column space of C (which is V_p) forms a representing matrix of ϕ. $\square$

In some applications, the locally linear independence of shifts of a distribution is required. It is defined as follows. Let $O \subset \mathbb{R}$ be an open set. Then the set $O \cap I$ is called a **relative open set** of I and denoted by O_I. If for any relative open set O_I, the non-vanishing distributions in $\{\phi_j^k|_{O_I}\}_{(j,k)\in\Lambda}$ are linearly independent, then we say the shifts of ϕ are **locally linearly independent**. We have the following result.

Theorem 12. *Let ϕ be a spline-type vector-valued distribution such that its foundation contains only entire functions. Assume that the partition corresponding to ϕ is created by $\pi_I = \{0, 1\}$. If the number of all nonzero rows of its representing matrix M is equal to the number of columns of M, then the shifts of ϕ are locally linearly independent.*

Proof: We know that M has full column rank. If the number of all nonzero rows of its representing matrix M is equal to the number of columns of M, then non-vanishing components in the set $\{\phi_j^k\}_{(j,k) \in \Lambda}$ form a basis of $S(\phi)|_I$. Thus, they are linearly independent on any relative open set in I since all $p_i, 1 \le i \le n$, are entire functions. Hence, the shifts of ϕ are locally linearly independent. $\square$

We now give some examples to illustrate Theorem 11 and Theorem 12.

Example 1. The n-th order cardinal B-spline can be represented by

$$B_n(x) = \sum_{i=0}^{n} (-1)^i \binom{n}{i} (x - i)_+^{n-1}.$$

We now apply Theorem 12 to prove that the shifts of B_n are locally linearly independent. The partition corresponding to B_n is $\pi_I : \{0, 1\}$. We have

$$\phi^k := B_n(\cdot + k)|_I = \sum_{i=0}^{k} (-1)^i \binom{n}{i} (x - i)^{n-1}, \quad 0 \le k \le n - 1.$$

Its foundation is $p = (1, x, \cdots, x^{n-1})^T$. Write

$$\boldsymbol{d}_l = (-1)^l \binom{n}{l} \left[\binom{n-1}{0}, \binom{n-1}{1}(-l)^1, \cdots, \binom{n-1}{n-1}(-l)^{n-1} \right]^T$$

and $\boldsymbol{c}_i = \boldsymbol{d}_1 + \cdots + \boldsymbol{d}_i$. Then the vector $\Phi = (\phi^0, \cdots, \phi^{n-1})^T$ and the matrix $C = [\boldsymbol{c}_1, \cdots, \boldsymbol{c}_n]$ have the relation $\Phi = Cp$. Let $\boldsymbol{a} = (0, \cdots, -(n - 1))$. Denote the Vandermonde matrix of $\boldsymbol{a}$ by $V_{\boldsymbol{a}}$. We have

$$|\det C| = \prod_{l=0}^{n} \binom{n}{l} \prod_{l=0}^{n-1} \binom{n-1}{l} \det V_{\boldsymbol{a}} \ne 0.$$

Hence, C is non-singular. By Theorem 12, the shifts of B_n are locally linearly independent.

Example 2. Let $\phi = (\phi_1, \phi_2)^T$, where $\phi_1 = B_1$ and $\phi_2 = B_2$. The equation $\Phi = Cp$ can be represented as

$$\begin{pmatrix} \phi_1^0 \\ \phi_1^1 \\ \phi_2^0 \\ \phi_2^1 \end{pmatrix} = \begin{pmatrix} 1 & 0 \\ 0 & 0 \\ 0 & 1 \\ 2 & -1 \end{pmatrix} \begin{pmatrix} 1 \\ x \end{pmatrix}.$$

The matrix C has rank 2. Hence, it is a representing matrix of ϕ. We have

$$C(z) = \begin{pmatrix} 1 & 0 \\ 0 & 1 \end{pmatrix} + \begin{pmatrix} 0 & 0 \\ 2 & -1 \end{pmatrix} z = \begin{pmatrix} 1 & 0 \\ 2z & 1-z \end{pmatrix},$$

which has rank 1 at $z = 1$. Hence the shifts of ϕ are linearly dependent. In fact, we have $\sum_{k \in \mathbb{Z}} (\phi_1(x-k) - \phi_2(x-k)) = 0$. If we set $\phi_1 = B_1 + \delta$, where δ is Dirac's distribution at 0. Then we have

$$\begin{pmatrix} \phi_1^0 \\ \phi_1^1 \\ \phi_2^0 \\ \phi_2^1 \end{pmatrix} = \begin{pmatrix} 1 & 1 & 0 \\ 0 & 0 & 0 \\ 0 & 0 & 1 \\ 0 & 2 & -1 \end{pmatrix} \begin{pmatrix} \delta \\ 1 \\ x \end{pmatrix}.$$

The matrix $C(z)$ becomes $\begin{pmatrix} 1 & 1 & 0 \\ 0 & 2z & 1-z \end{pmatrix}$. Then $C(z)$ has full rank for all $z \in \mathbb{C} \setminus \{0\}$. Hence, the shifts of ϕ become linearly independent.

Example 3. Let $\phi = (\phi_1, \phi_2)^T$, where $\phi_1 = \frac{1}{2}B_2(2x)$ and $\phi_2 = B_2(x)$. The partition corresponding to ϕ is $\pi_I = \{0, 1/2, 1\}$. On $[0, 1/2)$, we have

$$\begin{pmatrix} \phi_1^0 \\ \phi_1^1 \\ \phi_2^0 \\ \phi_2^1 \end{pmatrix} = \begin{pmatrix} 0 & 1 \\ 0 & 0 \\ 0 & 1 \\ 2 & -1 \end{pmatrix} \begin{pmatrix} 1 \\ x \end{pmatrix}.$$

On $[1/2, 1)$, we have

$$\begin{pmatrix} \phi_1^0 \\ \phi_1^1 \\ \phi_2^0 \\ \phi_2^1 \end{pmatrix} = \begin{pmatrix} 1 & -1 \\ 0 & 0 \\ 0 & 1 \\ 2 & -1 \end{pmatrix} \begin{pmatrix} 1 \\ x \end{pmatrix}.$$

Then

$$C = \begin{pmatrix} 0 & 1 & 1 & -1 \\ 0 & 0 & 0 & 0 \\ 0 & 1 & 0 & 1 \\ 2 & -1 & 2 & -1 \end{pmatrix}.$$

We can select the first three columns as a basis of the column space of C. These columns form a representing matrix of ϕ. Since the number of its non-zero rows is equal to the number of its columns, the shifts of ϕ are linearly independent on I. Hence they are also linearly independent. However, we cannot get the local linear independence of them by Theorem 12, for the condition of $\pi_I = \{0, 1\}$ is not valid. In fact, the shifts of ϕ are locally linearly dependent, which can be verified directly on the interval $[0, 1/2)$.

Example 4. This example is borrowed from Example 3 of [12]. Let

$$\phi_1(x) = \begin{cases} 24x^2 + 9x^2, & x \in [0,1) \\ 69 - 42x + 6x^2, & x \in [1,2) \\ 81 - 54x + 9x^2, & x \in [2,3) \\ 0, & \text{otherwise} \end{cases}$$

and

$$\phi_2(x) = \begin{cases} 27x^2 + \frac{27}{4}x^2, & x \in [0,1) \\ 72 - 45x + \frac{27}{4}x^2, & x \in [1,2) \\ 81 - 54x + 9x^2, & x \in [2,3) \\ 0, & \text{otherwise} \end{cases}.$$

Then $\phi = (\phi_1, \phi_2)^T$ satisfies the refinement equation

$$\phi = \sum_{k=0}^{4} M(k)\phi(2\cdot -k),$$

where

$$M(z) = \begin{pmatrix} \frac{-1}{4} & \frac{2}{3} \\ \frac{-9}{16} & 1 \end{pmatrix} + \begin{pmatrix} \frac{-11}{4} & \frac{32}{9} \\ \frac{-51}{16} & \frac{9}{4} \end{pmatrix} z + \begin{pmatrix} \frac{-61}{12} & \frac{16}{3} \\ \frac{-21}{4} & \frac{11}{2} \end{pmatrix} z^2$$
$$+ \begin{pmatrix} \frac{-21}{4} & \frac{16}{3} \\ \frac{-21}{4} & \frac{16}{3} \end{pmatrix} z^3 + \begin{pmatrix} -2 & 2 \\ -2 & 2 \end{pmatrix} z^4.$$

The authors of [12] proved the (global) linear independence of the shifts of ϕ using the refinement property of ϕ. We now show their linear independence by Theorem 11. We have

$$\begin{pmatrix} \phi_1^0 \\ \phi_1^1 \\ \phi_1^2 \\ \phi_2^0 \\ \phi_2^1 \\ \phi_2^2 \end{pmatrix} = \begin{pmatrix} 0 & 24 & 9 \\ 69 & -42 & 6 \\ 81 & -54 & 9 \\ 0 & 27 & \frac{27}{4} \\ 72 & -45 & \frac{27}{4} \\ 81 & -54 & 9 \end{pmatrix} \begin{pmatrix} 1 \\ x \\ x^2 \end{pmatrix}.$$

The matrix has full rank. Hence it is a representing matrix of ϕ. We have

$$C(z) = \begin{pmatrix} 69z + 81z^2 & 24 - 42z - 54z^2 & 9 + 6z^2 + 9z^2 \\ 72z + 81z^2 & 27 - 45z - 54z^2 & \frac{27}{4} + \frac{27}{4}z + 9z^2 \end{pmatrix}.$$

The submatrix $\begin{pmatrix} 69z + 81z^2 & 24 - 42z - 54z^2 \\ 72z + 81z^2 & 27 - 45z - 54z^2 \end{pmatrix}$ is singular at $z = 0, -1\pm$

$2\sqrt{6}/3$, and the submatrix $\begin{pmatrix} 69z + 81z^2 & 9 + 6z^2 + 9z^2 \\ 72z + 81z^2 & \frac{27}{4} + \frac{27}{4}z + 9z^2 \end{pmatrix}$ is singular at

-1 and $27/5$. Hence $C(z)$ has hull rank on $\mathbb{C} \setminus \{0\}$. Therefore, the shifts of ϕ are linearly independent.

§5. On Refinable Distributions

In this section we assume $\phi = (\phi_1, \cdots, \phi_r)^T$ is refinable, *i.e.*, ϕ satisfies the refinement equation

$$\phi(\cdot) = \sum_{k=0}^{M} \boldsymbol{P}(k)\phi(2 \cdot -k), \tag{14}$$

where $\boldsymbol{P} = (\boldsymbol{P}(k))_{k=0}^{M}$ (called the mask of ϕ) is a sequence of $r \times r$ matrices. The vector-valued distribution ϕ in (14) is often called a scaling vector, and the corresponding space $S(\phi)$ $(S_0(\phi))$ is called a refinable finitely shift invariant (RFSI) space. Let

$$P = 2^{-1} \sum_{k=0}^{M} \boldsymbol{P}(k).$$

We assume the matrix P has eigenvalue 1, but has no eigenvalues 2^k, $k = 1, 2, \cdots$. Under these conditions, for any right 1-eigenvector of P, say u, there exists a unique solution of (14) such that $\hat{\phi}(0) = u$. The smoothness of ϕ can be determined via its mask. (See [1,10,9,15,13,25], and references therein.)

If ϕ satisfies the refinement equation (14), then isupp $\phi_i \subset [0, M + 1), 1 \leq i \leq r$. We define two matrices $\boldsymbol{A}_0$ and $\boldsymbol{A}_1$ by

$$\boldsymbol{A}_0 = \begin{pmatrix} \boldsymbol{P}(0) & 0 & 0 & \cdots & 0 \\ \boldsymbol{P}(2) & \boldsymbol{P}(1) & \boldsymbol{P}(0) & \cdots & 0 \\ \boldsymbol{P}(4) & \boldsymbol{P}(3) & \cdots & \cdots & \cdots \\ \vdots & \vdots & \vdots & \vdots & \vdots \\ \boldsymbol{P}(2M) & \boldsymbol{P}(2M-1) & \cdots & \cdots & \boldsymbol{P}(M) \end{pmatrix} \tag{15}$$

and

$$\boldsymbol{A}_1 = \begin{pmatrix} \boldsymbol{P}(1) & \boldsymbol{P}(0) & 0 & \cdots & 0 \\ \boldsymbol{P}(3) & \boldsymbol{P}(2) & \boldsymbol{P}(1) & \cdots & 0 \\ \boldsymbol{P}(5) & \boldsymbol{P}(4) & \cdots & \cdots & \cdots \\ \vdots & \vdots & \vdots & \vdots & \vdots \\ \boldsymbol{P}(2M+1) & \boldsymbol{P}(2M) & \cdots & \cdots & \boldsymbol{P}(M+1) \end{pmatrix}. \tag{16}$$

Let Φ be defined by (7). From (14), we derive

$$\Phi|_{[0,1/2)} = \boldsymbol{A}_0 \Phi(2\cdot)|_{[0,1/2)},$$
$$\Phi|_{[1/2,1)} = \boldsymbol{A}_1 \Phi(2 \cdot -1)|_{[1/2,1)}.$$

To obtain a representing matrix of ϕ, we introduce r-regular compactly supported generators of a multiresolution analysis.

Definition 13. *A nested sequence of subspaces of $L_2(\mathbb{R})$*

$$\cdots \subset V_{-1} \subset V_0 \subset V_1 \subset \cdots$$

is called a multiresolution analysis (MRA) *of $L_2(\mathbb{R})$, if it has the following properties:*

(1) $\cap_{j \in \mathbb{Z}} V_j = \{0\}$, $\overline{\cup_{j \in \mathbb{Z}} V_j} = L_2(\mathbb{R})$.

(2) $\forall f \in L_2(\mathbb{R})$, $\forall j \in \mathbb{Z}$, $\quad f(x) \in V_j \iff f(2x) \in V_{j+1}$.

(3) $\forall f \in L_2(\mathbb{R})$, $\forall k \in \mathbb{Z}$, $\quad f(x) \in V_j \iff f(x-k) \in V_j$.

(4) *There exists a function $g \in V_0$ such that $\{g(x-k)\}_{k \in \mathbb{Z}}$ is a basis of V_0. The function g is called a* generator *of the MRA.*

Note that in this paper the generator is only required to generate a basis of V_0, which may not be a Riesz basis.

Definition 14. *A function $g \in L_2(\mathbb{R})$ is said to be an r-regular generator of an MRA if $g \in H^r(\mathbb{R})$ and*

$$|D^j g(x)| \le C_m (1 + |x|)^{-m} \ a.e., \quad 0 \le j \le r, \quad \forall m \in \mathbb{Z}.$$

Hence, if a generator $g \in H^r(\mathbb{R})$ is compactly supported, then it is r-regular.

Theorem 15. *Assume ϕ is an s-order refinable vector-valued distribution and g is an s-regular compactly supported generator of an MRA. Let $\boldsymbol{u}_k = (\Phi, g(\cdot - k))$, and let τ be the index-shift operator defined by $\tau(\boldsymbol{u}_k) = \boldsymbol{u}_{k-1}$. Let $\boldsymbol{A}_0$ and $\boldsymbol{A}_1$ be the matrices defined by (15) and (16). Let*

$$V = \operatorname{span} \{(\boldsymbol{A}_0 + \boldsymbol{A}_1 \tau)^m \tau^n \boldsymbol{u}_0, \quad m = 0, 1, \cdots, \quad n \in \mathbb{Z}\}.$$

Then any basis of the space V forms a representing matrix of ϕ.

Proof: Write $g_{jk} = g(2^j \cdot -k)$. Then the set $\{g_{jk}\}_{j \ge 0, k \in \mathbb{Z}}$ spans a dense set of $H^s(\mathbb{R})$. Without loss of generality, we assume $\operatorname{supp} g \subset [0, L]$, where L is a natural number. Let $\Gamma = \{(j, k), \ \operatorname{supp} g_{jk} \subset [-L, L]\}$. Then $\{g_{jk}\}_{(j,k) \in \Gamma}$ spans a dense set in $H_0^s([-L, L])(\supset H_0^s([-1, 1]))$. It follows that $\{g_{jk}\}_{(j,k) \in \Gamma}$ is a determinative set of $\mathcal{F}_0^s(I)$. Let

$$\hat{V} = \operatorname{span} \{(\Phi, g_{jk})\}_{(j,k) \in \Gamma}.$$

By Theorem 10, the basis of $\hat{V}$ is a representing matrix of ϕ. Note that the support of ϕ is I. Hence,

$$\hat{V} = \operatorname{span} \{(\Phi, g_{mn}); \quad m = 0, 1, \cdots, \quad n \in \mathbb{Z}\}.$$

We now prove $V = \hat{V}$. In fact,

$$
\begin{aligned}
(\Phi, g_{1n}) &= A_0(\Phi(2\cdot), g_{0n}(2\cdot)) + A_1(\Phi(2\cdot - 1), g_{0n}(2\cdot)) \\
&= \frac{1}{2}\left(A_0(\Phi(\cdot), g_{0n}(\cdot)) + A_1(\Phi(\cdot), g_{0n}(\cdot + 1))\right) \\
&= \frac{1}{2}(A_0 u_n + A_1 u_{n-1}) \\
&= \frac{1}{2}(A_0 + A_1 \tau)\tau^{-n} u_0.
\end{aligned}
$$

By mathematical induction, we have $(\Phi, g_{mn}) = \frac{1}{2^m}(A_0 + A_1\tau)^m \tau^{-n} u_0$. The theorem is proved. $\square$

If ϕ is a regular vector-valued distribution, then we have the following.

Corollary 16. *Assume that ϕ is a regular refinable vector-valued distribution. Let $u = \int_0^1 \Phi(x)dx$ and*

$$
V = \mathrm{span}\,\{\prod_{j=0}^{m} A_{\epsilon_j} u, \quad m = 0, 1, 2, \cdots\},
$$

where $\epsilon_j \in \{0, 1\}$. Then any basis of the space V forms a representing matrix of ϕ.

Proof: Since ϕ is regular, any set which spans a dense set of $L_\infty(I)$ is a determinative set of ϕ. We now let $g = \chi_I$. Then the set $\{g_{mn}\}$, where $m = 0, 1, \cdots$ and $0 \le n \le 2^m - 1$, is a dense set in $L_\infty(I)$. Note that $\int_0^1 \Phi(x)g(x)dx = u$ and $\int_0^1 \Phi(x)g(x - k)dx = 0, k \ne 0$. The corollary follows from Theorem 15. $\square$

Remark. If ϕ is a compactly supported continuous vector-valued function, then we can evaluate ϕ at any real number. In this case, the vector u in Corollary 16 can be replaced by $u = \Phi(0)$. [8] uses it to discuss local linear independence of refinable continuous vector-valued functions.

Finally, we briefly discuss local linear independence of shifts of refinable distributions. It is clear that if the shifts of ϕ are locally linearly independent, then they are linearly independent on I. We know from Corollary 6 that ϕ has linearly independent shifts on I if and only if the number of nonzero rows in its representing matrix M is equal to the rank of M. Hence, we only need to study the conditions which ensure linear independence on I implying local linear independence.

Theorem 17. *Let ϕ be a refinable vector-valued distribution. Assume the shifts of ϕ are linearly independent on I. Then they are locally linearly independent if and only if the non-zero rows of any matrix $\prod_{j=0}^{m} A_{\epsilon_j}, m = 0, 1, \cdots$, are linearly independent. In particular, if both A_0 and A_1 have no zero rows, the conditions become that A_0 and A_1 both are invertible.*

Proof: The proof is similar to the proof of Theorem 3.1 in [8]. $\square$

References

1. Cohen, A., I. Daubechies, and G. Plonka, Regularity of refinable function vectors, J. Fourier Anal. Appl. **3** (1997), 295–324.

2. Chui, C. K. and J. Z. Wang, On compactly supported spline wavelets and a duality principle, Trans. Amer. Math. Soc. **330** (1992), 903–916.

3. de Boor, C., R. DeVore, and A. Ron, Approximation from shift-invariant subspaces of $L^2(\mathbb{R}^d)$, J. Funct. Anal. **119** (1994), 37–78.

4. de Boor, C., R. DeVore, and A. Ron, Approximation orders of FSI spaces in $L_2(\mathbb{R}^d)$, Constr. Approx. **14** (1998), 631–652.

5. Dahmen, W. and C. A. Micchelli, Translates of multivariate splines, Linear Algebra Appl. **52** (1983), 217–234.

6. Dahmen, W. and C. A. Micchelli, Biorthogonal wavelet expansions, Constr. Approx. **13** (1997), 293–328.

7. Gelfand, I. M. and L. E. Shilov, *Generalized Functions*, Vol 2, 1958.

8. Goodman, T. N. T., R. Q. Jia, and D. X. Zhou, Local linear independence of refinable vectors of functions, Proc. of the Royal Society of Edinburgh **130**(2000), 813–826.

9. Han, B., and R. Q. Jia, Multivariate refinement equations and subdivision schemes, SIAM J. Math. Anal. **29** (1998), 1177–1199.

10. Heil, C., and D. Collela, Matrix refinement equations: Existence and uniqueness, J. Fourier Anal. Appl. **2** (1996), 363–377.

11. Hogan, T. A., Stability and independence of the shifts of finitely many refinable functions, J. Fourier Anal. Appl. **3**(1997), 757–774.

12. Hogan, T. A. and R. Q. Jia, Dependency relations among the shifts of a multivariate refinable distribution, Constr. Approx. **17** (2001), 19–37.

13. Jiang, Q. T., and Z. W. Shen, On existence and weak stability of matrix refinable functions, Constr. Approx. **15** (1999), 337–353.

14. Jia, R. Q., Shift-invariant spaces on the real line, Proc. Amer. Math. Soc. **125** (1997), 785–793.

15. Jia, R. Q., Subdivision schemes in L_p spaces, Advances in Computational Mathematics **3** (1995), 390–341.

16. Jia, R. Q., The subdivision and transition operators associated with a refinement equation, in *Advanced Topics in Multivariate Approximation*, F. Fontanella, K. Jetter and P. J. Laurent (eds.), 1996, 1–16.

17. Jia, R. Q., and C. A. Micchelli, On linear independence for integer translates of a finite number of functions, Proc. Edinburgh Math. Soc. **36** (1992), 69–85.

18. Jia, R. Q., S. D. Riemenschneider, and D. X. Zhou, Approximation by multiple refinable functions, Canad. J. Math. **49** (1997), 944–962.

19. Jia, R. Q., and J. Z. Wang, Stability and linear independence associated with wavelet decompositions, Proc. Amer. Math. Soc. **117** (1993), 1115–1124.

20. Micchelli, C. A., Regularity of multiwavelets, Adv. Comput. Math. **7** (1997), 455–545.

21. Plonka, G., Necessary and sufficient conditions for orthonormality of scaling vectors, in *Multivariate Approximation and Splines*, G. Nürnberger, J. W. Schmidt, and G. Walz (eds.), Birkhäuser Verlag, Basel (1997), 205–218.

22. Plonka, G., and V. Strela, Construction of multi-scaling functions with approximation and symmetry, SIAM J. Math. Anal. **29** (1998), 481–510.

23. Ron, A., A necessary and sufficient condition for the linear independence of the integer translates of a compactly supported distribution, Constr. Approx. **5** (1989), 297–308.

24. Ron, A., Characterizations of linear independence and stability of the shifts of a univariate refinable function in terms of its refinement mask, CMS Technical Report # 93-3, Madison WI, September 1992.

25. Shen, Z. W., Refinable function vectors, SIAM J. Math. Anal. **29** (1998), 235–250.

26. Wang, J. Z., Stability and linear independence associated with scaling vectors, SIAM J. Math. Anal. **29** (1998), 1140–1156.

27. Zhou, D. X., Stability of refinable functions, multiresolution analysis and Haar bases, SIAM J. Math. Anal. **27** (1996), 891–904.

Jianzhong Wang
Sam Houston State University
Huntsville, TX 77341-2206
USA
mth_jxw@shsu.edu

Fast Evaluation of Radial Basis Functions: Methods Based on Partition of Unity

Holger Wendland

Abstract. We combine the theory of radial basis function interpolation with a partition of unity method to solve large-scale, scattered data problems. We analyze the computational complexity and pay special attention to the underlying data structure. Finally, we give a numerical example.

§1. Radial Basis Functions

Radial basis functions are nowadays a popular choice for solving interpolation problems, where the data sites $X = \{x_1, \ldots, x_N\}$ are scattered points in $\mathbb{R}^d$ with $d \geq 2$. The idea is to choose a fixed conditionally positive definite function Φ of a certain order m and to form the interpolant to the data $f_1, \ldots, f_N$ by

$$s_{f,X} = \sum_{j=1}^{N} \alpha_j \Phi(\cdot - x_j) + p, \tag{1}$$

where p is a polynomial of degree $m - 1$. Additionally to the interpolation conditions $s_{f,X}(x_j) = f_j$, $1 \leq j \leq N$, the coefficients have to satisfy the relation $\sum \alpha_j q(x_j) = 0$ for all polynomials q of degree less than m. It is well known that under certain mild conditions on the location of the data sites, this interpolation problem is uniquely solvable for a wide range of basis functions, including thin plate splines, multiquadrics and Gaussians (see [7] for a recent overview). These globally supported basis functions have the drawback of being unfavourably expensive concerning computational complexity. To be more precise, direct methods for computing the interpolant $s_{f,X}$ need $\mathcal{O}(N^3)$ operations and each evaluation of (1) takes

Approximation Theory X: Wavelets, Splines, and Applications   473
Charles K. Chui, Larry L. Schumaker, and Joachim Stöckler (eds.), pp. 473–483.
Copyright Ⓒ 2002 by Vanderbilt University Press, Nashville, TN.
ISBN 0-8265-1416-2.

another $\mathcal{O}(N)$ operations. This is obviously too much if problems with more than $N = 100,000$ points have to be dealt with and many people consider radial basis functions only to be useful for small problems. But this is definitely wrong. Much work in the direction of fast solvers for the interpolation problem and in the direction of fast evaluation of the inter- polant (1) has been done in recent years (cf. [2,3,4,5,8]). All these methods retain the initial approximation space and allow to decrease the computa- tional complexity dramatically. But since everything has its price, these methods need some insight into multipole expansions, which are rather delicate to implement.

Hence, we want to propose another method, which is actually not a new one. The idea is simply to solve a large number of small, local problems instead of one large-scale problem and to put the local solutions together by a partition of unity method. As a matter of fact, this is a very natural way to deal with radial basis functions, which are actually local methods. They are local in the following sense. If we investigate the approximation property of our interpolant (1) at a point x, we only need those data points x_j that are close to x. We will come back to this in a later section.

This paper is organized as follows. The next section is devoted to a general description of partition of unity approximation methods. In the third section we will use radial basis functions as the local spaces for the partition of unity method. This allows us to show that this method inherits the approximation properties of the global radial basis function interpolant. In the fourth section we discuss data structures and compu- tational complexity. In the fifth section we shortly comment on stability of the procedure. In the final section we provide a numerical example.

§2. Partition of Unity

The idea of a partion of unity method is the following. We start with a mildly overlapping covering $\{\Omega_j\}_{j=1}^M$ of the region Ω. We will make the term "mildly overlapping" more precise in just a moment. Associated with this covering we choose a partition of unity, i.e. a family of compactly supported, continuous functions $\{w_j\}$ with $\sum_{j=1}^M w_j = 1$ and $\mathrm{supp}(w_j) \subseteq \overline{\Omega_j}$. Moreover, for every cell Ω_j we choose an approximation space V_j. Then, a function f is approximated on each cell by a local approximant $s_j \in V_j$ and the local approximants are put together by forming

$$s_f(x) = \sum_{j=1}^M s_j(x) w_j(x). \tag{2}$$

To be more precise, we introduce

Definition 1. *Let $\Omega \subseteq \mathbb{R}^d$ be a bounded set. Let $\{\Omega_j\}_{j=1}^M$ be an open and bounded covering of Ω. This means all Ω_j are open and bounded and Ω is contained in their union. Set $\delta_j = \mathrm{diam}(\Omega_j) = \sup_{x,y\in\Omega_j} \|x-y\|_2$. We call a family of functions $\{w_j\}_{j=1}^M$ with $w_j \in C^k(\mathbb{R}^d)$* a k-stable partition of unity with respect to the covering $\{\Omega_j\}$, *if*

1) $\mathrm{supp}(w_j) \subseteq \overline{\Omega_j}$,

2) $\displaystyle\sum_{j=1}^M w_j \equiv 1$ *on* Ω,

3) *for every $\alpha \in \mathbb{N}_0^d$ with $|\alpha| \leq k$ there exists a constant $C_\alpha > 0$ such that*

$$\|D^\alpha w_j\|_{L_\infty(\Omega_j)} \leq C_\alpha/\delta_j^{|\alpha|}$$

for all $1 \leq j \leq M$.

So far, we did not make any further assumptions on the covering $\{\Omega_j\}$, but for efficiency it is necessary that the cardinality of

$$I(x) := \{j : x \in \Omega_j\}$$

is uniformly bounded on Ω. Nonetheless, even without this assumption we can give a first convergence result. But without this assumption we will lose convergence orders.

Theorem 2. *Let $\Omega \subseteq \mathbb{R}^d$ be bounded. Suppose $\{\Omega_j\}_{j=1}^M$ is an open and bounded covering of Ω, and $\{w_j\}_{j=1}^M$ is a k-stable partition of unity. Let $f \in C^k(\Omega)$ be the function to be approximated. Let $V_j \subseteq C^k(\Omega_j)$ be given. Assume that the local approximation spaces V_j have the following approximation property: On each patch $\Omega_j \cap \Omega$, f can be approximated by a function $s_j \in V_j$ such that*

$$\|D^\alpha f - D^\alpha s_j\|_{L_\infty(\Omega\cap\Omega_j)} \leq \varepsilon_j(\alpha).$$

Then the function (2) satisfies

$$|(D^\alpha f - D^\alpha s_f)(x)| \leq \sum_{j\in I(x)} \sum_{\beta\leq\alpha} \binom{\alpha}{\beta} C_{\alpha-\beta} \delta_j^{|\beta|-|\alpha|} \varepsilon_j(\beta) \qquad (3)$$

for all $|\alpha| \leq k$.

Proof: The proof is straightforward. We simply use Leibniz' rule and the fact that the $\{w_j\}$ form a partition of unity to derive

$$(D^\alpha f - D^\alpha s_f)(x) = D^\alpha \sum_{j=1}^M w_j(x)(f(x) - s_j(x))$$

$$= \sum_{j\in I(x)} \sum_{\beta\leq\alpha} \binom{\alpha}{\beta} D^{\alpha-\beta} w_j(x) D^\beta (f - s_j)(x).$$

The assumed bounds on the derivatives of w_j and on $D^\beta(f - s_j)$ now yield the stated result. $\square$

§3. Radial Basis Functions as the Local Method

It is now our goal to use radial basis functions to define the local approximation spaces. To this end and for the reader's convenience, we review some details on local error estimates for radial basis function interpolation, which can be found in [11].

To every conditionally positive definite function Φ and every region $\Omega \subseteq \mathbb{R}^d$, there exists a natural function space $\mathcal{N}_\Phi(\Omega)$, the native Hilbert space. This space can be defined in various ways, and we assume that the reader is familiar with the concept (see [10]). In many cases the native space turns out to be a classical smoothness space like Sobolev or Beppo Levi space. Here, we need only few properties. The first one is that if $\Phi \in C^k(\mathbb{R}^d)$, its smoothness is inherited by the native space via $\mathcal{N}_\Phi(\Omega) \subseteq C^{\lfloor k/2 \rfloor}(\Omega)$. Hence, we might naturally ask for bounds on $D^\alpha f - D^\alpha s_{f,X}$ for $f \in \mathcal{N}_\Phi(\Omega)$ and $\alpha \in \mathbb{N}_0^d$ with $|\alpha| \le k/2$. The second property we need is the extension property. Every function $f \in \mathcal{N}_\Phi(\Omega)$ has a natural, norm preserving extension to a function $Ef \in \mathcal{N}_\Phi(\mathbb{R}^d)$.

As usual we will measure the approximation error in terms of the fill distance

$$h_{X,\Omega} = \sup_{x \in \Omega} \min_{x_j \in X} \|x - x_j\|_2.$$

We will state our convergence results only in case of (conditionally) positive definite functions with a finite number of continuous derivatives. But it should be clear from the proofs that if Gaussians or multiquadrics are used, convergence orders are spectral again. To get the full approximation order, we use the idea of Hölder continuity, but we need only a weak form. Hence we define the space $C_\nu^k(\mathbb{R}^d)$ to be the space of all functions $f \in C^k(\mathbb{R}^d)$ such that their derivatives of order k satisfy $D^\alpha f(x) = \mathcal{O}(\|x\|_2^\nu)$ for $\|x\|_2 \to 0$.

The local version for error estimates on radial basis functions is taken from [11]. It can be formulated as

Theorem 3. *Suppose* $\Phi \in C_\nu^k(\mathbb{R}^d)$ *is conditionally positive definite of order* m. *Let* $\tilde\theta \in (0, \pi/2)$ *and* $h > 0$ *be given. Let* $\tilde\Omega \subseteq \mathbb{R}^d$ *satisfy an interior cone condition with angle* $\tilde\theta$ *and radius* $\tilde r = C_s h$. *Suppose further that* $\tilde X = \{\tilde x_1, \ldots, \tilde x_M\} \subseteq \tilde\Omega$ *satisfies* $h_{\tilde X, \tilde\Omega} \le h$. *Then there exist constants* $C, C_0 > 0$ *depending only on* d, k, m, $\tilde\theta$, *and* Φ, *but not on* $h_{X,\Omega}$, *such that*

$$\left\| D^\alpha f - D^\alpha s_{f,X} \right\|_{L_\infty(\tilde\Omega)} \le C h_{\tilde X, \tilde\Omega}^{\frac{k+\nu}{2} - |\alpha|} |f|_{\mathcal{N}_\Phi(\Omega)}$$

for all $f \in \mathcal{N}_\Phi(\Omega)$ *and all* $|\alpha| \le k/2$, *provided that* $C_s \ge C_0$ *and* $h \le h_0$.

Now, let us come back to the partition of unity method. As mentioned before, we want to use the radial basis function interpolant as the local approximant. Hence, we set $X_j = X \cap \Omega_j$ and $s_j = s_{f,X_j}$, so that the global approximant becomes

$$s_f(x) = \sum_{j=1}^{M} s_{f,X_j}(x) w_j(x).$$

Since the partition of unity is given by compactly supported functions, and since we interpolate on each cell, it is clear that the resulting function s_f interpolates the data as well.

To use Theorem 3 in this context, we have to make some additional assumptions on the covering $\{\Omega_j\}$.

Definition 4. *Suppose $\Omega \subseteq \mathbb{R}^d$ is bounded and $X = \{x_1, \ldots, x_N\} \subseteq \Omega$ are given. An open and bounded covering $\{\Omega_j\}_{j=1}^{M}$ is called* regular for (Ω, X) *if the following properties are satisfied.*

1) *For every $x \in \Omega$, the number of cells Ω_j with $x \in \Omega_j$ is bounded by a global constant K.*
2) *There exists a constant $C_r > 0$ and an angle $\theta \in (0, \pi/2)$ such that every patch Ω_j satisfies an interior cone condition with angle θ and radius $r = C_r h_{X,\Omega}$.*
3) *There exists a constant $0 < C_{ae} < C_r$ such that $h_{X_j,\Omega_j} \leq C_{ae} h_{X,\Omega}$, where $X_j = X \cap \Omega_j$.*

This looks technical at first sight. But a closer look at each property shows that these requirements are more or less natural. For example, the first property is necessary for making sure that the outer sum in (3) is actually a sum over at most K summands. Since K is independent of N, while M is usually proportional to N, this is essential for not losing convergence orders. Moreover, it is crucial for an efficient evaluation of the global approximant that only a constant number of local approximants have to be evaluated. To this end, it also has to be possible to locate those K indices in constant time. The second and third property are important for employing our estimates on radial basis function interpolants, as we will see very soon. We will also need that C_r has to be reasonably larger than C_{ae}. This is in particular necessary to ensure that Ω_j contains enough points to allow a unique interpolant. Another consequence of these last two properties is that the union of all patches $\tilde{\Omega} = \cup_{j=1}^{M} \Omega_j$ cannot overlap Ω too much, i.e. there exists a constant $C_{ov} > 0$, such that

$$\mathrm{dist}(\tilde{\Omega}, \Omega) := \sup_{\tilde{x} \in \tilde{\Omega}} \inf_{x \in \Omega} \|\tilde{x} - x\|_2 \leq C_{ov} h_{X,\Omega}.$$

We are now able to state the main result concerning the approximation properties of a partition of unity method built on radial basis

functions. The bottom line is that it inherits the same approximation
order as the global method.

Theorem 5. *Suppose* $\Omega \subseteq \mathbb{R}^d$ *is open and bounded, and that* $X = \{x_1, \ldots, x_N\} \subseteq \Omega$. *Let* $\Phi \in C_\nu^k(\mathbb{R}^d)$ *be conditionally positive definite of order* m. *Let* $\{\Omega_j\}$ *be a regular covering for* (Ω, X) *and let* $\{w_j\}$ *be* k-*stable for* $\{\Omega_j\}$. *Let* C_0 *denote the constant from Theorem 3. If* $C_r \geq C_0 C_{ae}$, *then the error between* $f \in \mathcal{N}_\Phi(\Omega)$ *and its partition of unity interpolant* $s_f(x) = \sum_j s_{f,X_j}(x) w_j(x)$ *can be bounded by*

$$|D^\alpha f(x) - D^\alpha s_f(x)| \leq C h_{X,\Omega}^{\frac{k+\nu}{2} - |\alpha|} |f|_{\mathcal{N}_\Phi(\Omega)},$$

for all $x \in \Omega$ *and all* $|\alpha| \leq k/2$.

Proof: The function f has a norm preserving extension $Ef \in \mathcal{N}_\Phi(\mathbb{R}^d)$. Moreover, it is known that the restriction $f_j = Ef|\Omega_j$ satisfies $|f_j|_{\mathcal{N}_\Phi(\Omega_j)} \leq |Ef|_{\mathcal{N}_\Phi(\mathbb{R}^d)} = |f|_{\mathcal{N}_\Phi(\Omega)}$. Hence, if we denote all these functions by f again, we have $|f|_{\mathcal{N}_\Phi(\Omega_j)} \leq |f|_{\mathcal{N}_\Phi(\Omega)}$.

Furthermore, we find estimates for the local interpolants by Theorem 3. Setting $h = C_{ae} h_{X,\Omega}$, $\tilde{\theta} = \theta$, $\tilde{\Omega} = \Omega_j$, and $\tilde{X} = X_j$, we see that $h_{X_j,\Omega_j} \leq h$ and that Ω_j satisfies an interior cone condition with angle θ and radius $r = \frac{C_r}{C_{ae}} h$. Since $C_r > C_0 C_{ae}$, Theorem 3 yields

$$|D^\alpha f(x) - D^\alpha s_{f,X_j}(x)| \leq C h_{X_j,\Omega_j}^{\frac{k+\nu}{2} - |\alpha|} |f|_{\mathcal{N}_\Phi(\Omega_j)} \leq C h_{X,\Omega}^{\frac{k+\nu}{2} - |\alpha|} |f|_{\mathcal{N}_\Phi(\Omega)}$$

for $x \in \Omega_j$ and $|\alpha| \leq k/2$ with C being independent of j. To apply (3) we need two more ingredients. Since every patch Ω_j satisfies an interior cone condition with radius $C_r h_{X,\Omega}$, we have $\delta_j = \mathrm{diam}(\Omega_j) \geq C_r h_{X,\Omega}$. Moreover, every $x \in \Omega$ is contained in at most K patches Ω_j. Hence, the error bound (3) leads to

$$|D^\alpha f(x) - D^\alpha s_f(x)| \leq \sum_{j \in I(x)} \sum_{\beta \leq \alpha} \binom{\alpha}{\beta} C_{\alpha-\beta} \delta_j^{|\beta|-|\alpha|} \varepsilon_j(\beta)$$

$$\leq K \sum_{\beta \leq \alpha} \binom{\alpha}{\beta} C_{\alpha-\beta} C_r^{|\beta|-|\alpha|} h_{X,\Omega}^{|\beta|-|\alpha|} C h_{X,\Omega}^{\frac{k+\nu}{2} - |\beta|} |f|_{\mathcal{N}_\Phi(\Omega)}$$

$$= C h_{X,\Omega}^{\frac{k+\nu}{2} - |\alpha|} |f|_{\mathcal{N}_\Phi(\Omega)}$$

for all $x \in \Omega$ and all $|\alpha| \leq k/2$. $\square$

§4. Data Structures for the Centers

To take advantage of the locality of our method, we have to show that this diminishes the computational complexity substantially. To this end we have to investigate data structures for both the covering Ω_j and the data sites X. Since this is a dual problem, we concentrate here on data structures for the data sites.

To be useful for our method, the data structure for X has to be able to answer range query problems efficiently. A range query problem is the following. Suppose we are given a set of points X in $\mathbb{R}^d$ and a range $R \subseteq \mathbb{R}^d$. Then, the task is to report all points $x_j \in X$ with $x_j \in R$. Fortunately, there exist several good data structures to handle this problem, in particular if R is an axis-parallel box. Examples are kd-trees, bd-trees, and range trees. We will give some information about the latter and about a very simple data structure, called fixed grid, which is very favourable in case of quasi-uniform data. Details on kd- and bd-trees can be found in [9] and [1], respectively.

For the fixed grid method, one first searches the bounding box BB of the given data $X = \{x_1, \ldots, x_N\}$. Then, a fine grid is defined on BB. This grid consists of $\lfloor N^{1/d} \rfloor^d$ axis-parallel boxes, also called cells. For each cell we keep a list that contains the indices of the points in that specific cell. If the cells have side length k, then the indices of the cell containing a specific point can efficiently be found by multiplying the components of the point by $1/k$ and taking integer parts, which is often faster than dividing by k.

After the data structure has been built, a range query can then simply be answered in two steps. In the first step we find all cells that have common points with the query region R. Then, we test the data points in these cells explicitly. To execute both steps, we will assume that it is possible in constant time to decide whether R and a cell are disjoint or not, and whether a specific point belongs to R or not. This is obviously true if R is a box or a ball or something similar.

As said before, the fixed grid method is favourable in case of quasi-uniform data. Quasi-uniformity means that the fill distance is asymptotically proportional to the separation distance $q_X = \min_{j \neq k} \|x_j - x_k\|_2$. If this is true, it is easy to see that both are also asymptotically behaving like $N^{-1/d}$. Moreover, it follows almost immediately that the number of points in each cell of the fixed grid is bounded by a fixed constant.

Proposition 6. *If the data sites X are quasi-uniform, the fixed grid data structure can be built in $\mathcal{O}(N)$ time and space. If R is a range of diameter proportional to $N^{-1/d}$, the range query problem for R can be answered in constant time.*

To apply this to our partition of unity method, we need an additional assumption on the locality of the covering. We will say that $\{\Omega_j\}$ is local if there is a constant C_{ba} such that $\text{diam}(\Omega_j) \leq C_{ba} h_{X,\Omega}$. This means that the cells are not too big. Hence, if $\{\Omega_j\}$ is local and regular, the size of each cell is proportional to $h_{X,\Omega}$. In case of a quasi-uniform data set this means that the number of cells M is proportional to the number of data sites N.

Theorem 7. *Let $X = \{x_1, \ldots, x_N\} \subseteq \Omega$ be quasi-uniform and $\{\Omega_j\}$ a regular and local covering for (Ω, X). Suppose $\{\Omega_j\}$ can be built in $\mathcal{O}(N)$ time and $\mathcal{O}(N)$ space and for every $x \in \Omega$ all K or fewer patches Ω_j with $x \in \Omega_j$ can be reported in constant time. Then the partition of unity method based on radial basis functions can be implemented in $\mathcal{O}(N)$ space with $\mathcal{O}(N)$ time needed for the prepocessing step. Furthermore, each evaluation of the global interpolant needs $\mathcal{O}(1)$ time.*

Proof: If we use the fixed grid to build our data structure for X, we know that this can be done in $\mathcal{O}(N)$ time and space. Since the number of centers in each patch is bounded by a constant, we need constant space and constant time for each patch to solve the local interpolation problem. Furthermore, the points in each patch can be reported in constant time. Since the number M of patches Ω_j is bounded by $\mathcal{O}(N)$, this adds up to $\mathcal{O}(N)$ space and time for solving all local problems in all patches. By assumption we can determine $I(x) = \{j : x \in \Omega_j\}$ in constant time, and the cardinality of $I(x)$ is also bounded by a constant. Thus, we have to add up a constant number of local interpolants to get the value of the global interpolant. This can be done in constant time. $\square$

In case of data that is not quasi-uniform, the number of points in the cells may vary dramatically. Hence, the data structure might need more than linear space. In the worst case, all N points are concentrated in one of the cells. In this situation, the data structure needs $\mathcal{O}(N^2)$ space. Nonetheless, it still can be built in $\mathcal{O}(N)$ time. The time for a range query problem can also vary.

If R is always an axis-parallel box, another data structure is better suited for highly irregular data sites: the range tree. In one dimension a range tree is simply a balanced binary search tree, where all data sites are stored in the leafs and the inner nodes contain splitting information.

A d-dimensional range tree can be built in the following way. Firstly, a balanced binary search tree is built on the first coordinate of the points. This is the first-level or main tree. To each node ν of this tree we associate a canonical set of points $P(\nu)$ containing all the points stored in the leaves of the subtree rooted at ν. For each node ν we construct an associated data structure $\mathcal{T}_{\text{assoc}}(\nu)$. This second-level tree is a $(d-1)$-dimensional range tree for the points in $P(\nu)$ restricted to their last $d-1$ coordinates. This $(d-1)$-dimensional range tree is constructed recursively. The recursion stops when we have reached the last coordinate. Here, the points are stored in a one-dimensional range tree.

The query algorithm now works on each coordinate. Given a range $R = [\alpha_1, \beta_1] \times \ldots \times [\alpha_d, \beta_d]$, it selects $\mathcal{O}(\log N)$ canonical subsets whose union contains the points whose first coordinates lie in the range $[\alpha_1, \beta_1]$. On these canonical subsets it performs a range query on the second level tree. This gives subsets of points whose first and second coordinates lie in the range. Continuing recursively, this search procedure results in the points in the range.

A more thorough description of range trees and the following result can be found in [6].

Proposition 8. *Given N points in $\mathbb{R}^d$, a range tree can be built using $\mathcal{O}(N \log^{d-1} N)$ space and time. It is possible to report the points that lie in a rectangular query range in $\mathcal{O}(\log^d N + k)$ time, where k is the number of reported points.*

The query time can be reduced to $\mathcal{O}(\log^{d-1} N + k)$ if a technique called fractional cascading is employed.

§5. Stability

So far we know that our method inherits the convergence properties of the global radial basis function interpolation. We also know that in case of regular data it can be implemented efficiently leading to an interpolant that can be evaluated in constant time.

The next question that comes up naturally is stability. Unfortunately, the condition numbers of the interpolation matrices depend more on the separation distance than on the number of points. For most of the cells, the local separation distance q_{X_j} will be of the same size (or smallness) as the global separation distance q_X. Hence, our method seems to be as stable as the global one. Fortunately, this is not a problem in case of basis functions of moderate smoothness. For example, the compactly supported basis functions used as global functions of smoothness up to order 4 work nicely, and a direct solver for the local problems can be applied.

But for certain classes of basis functions including thin plate splines, we have a different possibility. Instead of using the basis function itself to form the interpolation matrix, a modified kernel can be used. If the basis function $\Phi(x, y)$ satisfies a homogeneous condition, one can use the kernel

$$\kappa(x, y) = \Phi(x, y) - \sum_{j=1}^{Q} p_j(x)\Phi(\xi_j, y) - \sum_{k=1}^{Q} p_k(y)\Phi(x, \xi_k)$$

$$+ \sum_{j,k=1}^{Q} p_j(x)p_k(y)\Phi(\xi_j, \xi_k).$$

Here Q is the dimension of the linear space of d-variate polynomials of degree less than m, if Φ is conditionally positive definite of order m, the points $\xi_1, \ldots, \xi_Q$ are different from those in X, and $\{p_j\}$ is a polynomial Lagrangian basis with respect to these points. For this kernel it can be shown that the condition number of the involved interpolation matrices is rather independent of the separation distance, but depends only on the number of points. Details can be found in [5].

§6. A Numerical Example

To demonstrate the numerical behaviour of our method, we give one "best case" example. The data sites are a regular $(2^j + 1)^2$ grid on $\Omega = [0, 1]^2$ for different values of j. The function values come from Franke's test function, which is a sum of four exponential terms. For the covering we have chosen overlapping axis-parallel boxes on a fine grid with a certain offset such that these boxes overlap each other and also the boundary of the domain. The partition of unity was formed by a Shepard function. The radial basis function was given by $\phi(r) = (1 - r)^6_+ (35r^2 + 18r + 3)$. The theory for this basis function allows us to expect a convergence order of at least 2.5. The error was measured on a fine 2000×2000 grid. The results are

j	$N = (2^j + 1)^2$	L_∞	Order
3	81	4.51084×10^{-2}	
4	289	3.32996×10^{-3}	3.76
5	1089	2.81404×10^{-4}	3.56
6	4225	3.56702×10^{-5}	2.98
7	16641	4.49974×10^{-6}	2.99
8	66049	5.56833×10^{-7}	3.01
9	263169	6.83149×10^{-8}	3.03

References

1. Arya, S., D. Mount, N. Netanyahu, R. Silverman, and A. Wu, An optimal algorithm for approximate nearest neighbor searching, 5th Ann. ACM-SIAM Symposium on Discrete Algorithms, 1994, 573–582.

2. Beatson, R. K., and G. N. Newsam, Fast evaluation of radial basis functions: I, Comput. Math. Appl. **24** (1992), 7–19.

3. Beatson, R. K., G. Goodsell, and M. J. D. Powell, On multigrid techniques for thin plate spline interpolation in two dimensions, Lectures in Applied Mathematics **32** (1996), 77–97.

4. Beatson, R. K., and G. N. Newsam, Fast evaluation of radial basis functions: moment-based methods, SIAM J. Sci. Comp. **19** (1998), 1428–1449.

5. Beatson, R. K., S. Billings, and W. A. Light, Fast Solution of the radial basis function interpolation equations: domain decomposition methods, SIAM J. Sci. Comp. **22** (2000), 1717–1740.

6. Berg, M. de, M. van Kreveld, M. Overmars, and O. Schwarzkopf, *Computational Geometry*, Springer, New York, 1991.

7. Buhmann, M. D., Radial basis functions, in *Acta Numerica 2000*, Cambridge University Press, Cambridge, 2000, 1–38.

8. Faul, A. and M. J. D. Powell, Proof of convergence of an iterative technique for thin plate spline interpolation in two dimensions, Advances in Comp. Math. **11** (1999), 183–192.

9. Friedman, J., J. L. Bentley, and R. A. Finkel, An Algorithm for finding best matches in logarithmic expected time, ACM Transactions on Mathematical Software **3** (1977), 209–226.

10. Schaback, R., Native Hilbert spaces for radial basis functions I, in *New Developments in Approximation Theory*, Müller, M. W. et al. (eds.), Birkhäuser Verlag, Basel, 1999, 255–282.

11. Wu, Z., and R. Schaback, Local error estimates for radial basis function interpolation of scattered data, IMA J. of Num. Analysis **13** (1993), 13–27.

Holger Wendland
Institut für Numerische und Angewandte Mathematik
Universität Göttingen
Lotzestr. 16-18
37083 Göttingen
wendland@math.uni-goettingen.de